T0291092

Transport Properties and Potential Energy Models for Monatomic Gases

International Series of Monographs on Physics

INTERNATIONAL SERIES OF MONOGRAPHS ON PHYSICS

174. H. Li, F. R. W. McCourt: *Transport Properties and Potential Energy Models for Monatomic Gases*
173. G. H. Fredrickson, K. T. Delaney: *Field-Theoretic Simulations in Soft Matter and Quantum Fluids*
172. J. Kübler: *Theory of itinerant electron magnetism, Second edition*
171. J. Zinn-Justin: *Quantum field theory and critical phenomena, Fifth edition*
170. V. Z. Kresin, S. G. Ovchinnikov, S. A. Wolf: *Superconducting state - mechanisms and materials*
169. P. T. Chruściel: *Geometry of black holes*
168. R. Wigmans: *Calorimetry – Energy measurement in particle physics, Second edition*
167. B. Mashhoon: *Nonlocal gravity*
166. N. Horing: *Quantum statistical field theory*
165. T. C. Choy: *Effective medium theory, Second edition*
164. L. Pitaevskii, S. Stringari: *Bose-Einstein condensation and superfluidity*
163. B. J. Dalton, J. Jeffers, S. M. Barnett: *Phase space methods for degenerate quantum gases*
162. W. D. McComb: *Homogeneous, isotropic turbulence - phenomenology, renormalization, and statistical closures*
160. C. Barrabès, P. A. Hogan: *Advanced general relativity - gravity waves, spinning particles, and black holes*
159. W. Barford: *Electronic and optical properties of conjugated polymers, Second edition*
158. F. Strocchi: *An introduction to non-perturbative foundations of quantum field theory*
157. K. H. Bennemann, J. B. Ketterson: *Novel superfluids, Volume 2*
156. K. H. Bennemann, J. B. Ketterson: *Novel superfluids, Volume 1*
155. C. Kiefer: *Quantum gravity, Third edition*
154. L. Mestel: *Stellar magnetism, Second edition*
153. R. A. Klemm: *Layered superconductors, Volume 1*
152. E. L. Wolf: *Principles of electron tunneling spectroscopy, Second edition*
151. R. Blinc: *Advanced ferroelectricity*
150. L. Berthier, G. Biroli, J.-P. Bouchaud, W. van Saarloos, L. Cipelletti: *Dynamical heterogeneities in glasses, colloids, and granular media*
149. J. Wesson: *Tokamaks, Fourth edition*
148. H. Asada, T. Futamase, P. Hogan: *Equations of motion in general relativity*
147. A. Yaouanc, P. Dalmas de Réotier: *Muon spin rotation, relaxation, and resonance*
146. B. McCoy: *Advanced statistical mechanics*
145. M. Bordag, G. L. Klimchitskaya, U. Mohideen, V. M. Mostepanenko: *Advances in the Casimir effect*
144. T. R. Field: *Electromagnetic scattering from random media*
143. W. Götze: *Complex dynamics of glass-forming liquids - a mode-coupling theory*
142. V. M. Agranovich: *Excitations in organic solids*
141. W. T. Grandy: *Entropy and the time evolution of macroscopic systems*
140. M. Alcubierre: *Introduction to 3+1 numerical relativity*
139. A. L. Ivanov, S. G. Tikhodeev: *Problems of condensed matter physics - quantum coherence phenomena in electron-hole and coupled matter-light systems*
138. I. M. Vardavas, F. W. Taylor: *Radiation and climate*
137. A. F. Borghesani: *Ions and electrons in liquid helium*
135. V. Fortov, I. Iakubov, A. Khrapak: *Physics of strongly coupled plasma*
134. G. Fredrickson: *The equilibrium theory of inhomogeneous polymers*
133. H. Suhl: *Relaxation processes in micromagnetics*
132. J. Terning: *Modern supersymmetry*
131. M. Mariño: *Chern–Simons theory, matrix models, and topological strings*
130. V. Gantmakher: *Electrons and disorder in solids*
129. W. Barford: *Electronic and optical properties of conjugated polymers*

128. R. E. Raab, O. L. de Lange: *Multipole theory in electromagnetism*
127. A. Larkin, A. Varlamov: *Theory of fluctuations in superconductors*
126. P. Goldbart, N. Goldenfeld, D. Sherrington: *Stealing the gold*
125. S. Atzeni, J. Meyer-ter-Vehn: *The physics of inertial fusion*
123. T. Fujimoto: *Plasma spectroscopy*
122. K. Fujikawa, H. Suzuki: *Path integrals and quantum anomalies*
121. T. Giamarchi: *Quantum physics in one dimension*
120. M. Warner, E. Terentjev: *Liquid crystal elastomers*
119. L. Jacak, P. Sitko, K. Wieczorek, A. Wojs: *Quantum Hall systems*
117. G. Volovik: *The Universe in a helium droplet*
116. L. Pitaevskii, S. Stringari: *Bose–Einstein condensation*
115. G. Dissertori, I. G. Knowles, M. Schmelling: *Quantum chromodynamics*
114. B. DeWitt: *The global approach to quantum field theory*
112. R. M. Mazo: *Brownian motion - fluctuations, dynamics, and applications*
111. H. Nishimori: *Statistical physics of spin glasses and information processing - an introduction*
110. N. B. Kopnin: *Theory of nonequilibrium superconductivity*
109. A. Aharoni: *Introduction to the theory of ferromagnetism, Second edition*
108. R. Dobbs: *Helium three*
105. Y. Kuramoto, Y. Kitaoka: *Dynamics of heavy electrons*
104. D. Bardin, G. Passarino: *The Standard Model in the making*
103. G. C. Branco, L. Lavoura, J. P. Silva: *CP Violation*
101. H. Araki: *Mathematical theory of quantum fields*
100. L. M. Pismen: *Vortices in nonlinear fields*
99. L. Mestel: *Stellar magnetism*
98. K.H. Bennemann: *Nonlinear optics in metals*
94. S. Chikazumi: *Physics of ferromagnetism*
91. R. A. Bertlmann: *Anomalies in quantum field theory*
90. P. K. Gosh: *Ion traps*
87. P. S. Joshi: *Global aspects in gravitation and cosmology*
86. E. R. Pike, S. Sarkar: *The quantum theory of radiation*
83. P. G. de Gennes, J. Prost: *The physics of liquid crystals*
73. M. Doi, S. F. Edwards: *The theory of polymer dynamics*
69. S. Chandrasekhar: *The mathematical theory of black holes*
51. C. Møller: *The theory of relativity*
46. H. E. Stanley: *Introduction to phase transitions and critical phenomena*
32. A. Abragam: *Principles of nuclear magnetism*
27. P. A. M. Dirac: *Principles of quantum mechanics*
23. R. E. Peierls: *Quantum theory of solids*

Transport Properties and Potential Energy Models for Monatomic Gases

Hui Li
Jilin University, China

Frederick R.W. McCourt
University of Waterloo, Canada

OXFORD
UNIVERSITY PRESS

Great Clarendon Street, Oxford, OX2 6DP,
United Kingdom

Oxford University Press is a department of the University of Oxford.
It furthers the University's objective of excellence in research, scholarship,
and education by publishing worldwide. Oxford is a registered trade mark of
Oxford University Press in the UK and in certain other countries

Published in the United States of America by Oxford University Press
198 Madison Avenue, New York, NY 10016, United States of America

British Library Cataloguing in Publication Data
Data available

Library of Congress Control Number: 2023943566

ISBN 9780198888253

DOI: 10.1093/oso/9780198888253.001.0001

Printed and bound by
CPI Group (UK) Ltd, Croydon, CR0 4YY

To the days of inspiring discussion and coffee time, where we shared our thoughts and ideas, and learned from each other.

Preface

The present project represents an update and outgrowth of the first two chapters of the monograph *Nonequilibrium Phenomena in Polyatomic Gases* by one of the present authors together with J. J. M. Beenakker, W. E. Köhler, and I. Kuščer, published in 1990 by OUP. In particular, Chapter 1, which dealt exclusively with dilute monatomic gases, has been completely rewritten and updated, while retaining the notation of the original chapter. A number of errors and misprints have also been eliminated.

Similarly, while Chapter 2 of the present work retains the notation of the original chapter, the present treatment extends the description beyond the first Chapman–Cowling approximation for the various transport coefficients, giving explicit expressions at the level of the second approximation. It also gives expressions for all effective cross-sections that appear at the level of the second approximation. It includes both classical and quantum mechanical evaluations of effective cross-sections, as well as a discussion of dynamical models for binary atomic collisions, the moment method, and its extension to binary mixtures.

The update to Chapters 1 and 2 of the *Nonequilibrium Phenomena in Polyatomic Gases* monograph is a response to considerable improvements in the experimental determination of the transport properties of dilute gases that have taken place during the past 30 years. Whereas in 1990 it was sufficient to calculate these properties in the first Chapman–Cowling approximation, the experimental determination has improved sufficiently that it has become necessary to carry out calculations at the level of the second approximation in order to give computed results that lie within the current experimental uncertainties now being reported. Third and higher approximations contribute an order of magnitude smaller corrections than does the second approximation. The present volume is restricted to atomic systems, largely because second Chapman–Cowling approximation expressions for the transport of molecular gases have proven to be extremely challenging due to the as-yet unknown role that will be played by the rotational angular momentum vector associated with the internal (rotational) states of polyatomic molecules.

Chapter 3 is devoted to realistic interatomic potential energy functions, and begins with a discussion of the need for more accurate representations of these functions. Following an historical discussion of the Mie/Lennard-Jones potential energy model for interatomic interactions and its improvement, the basic Hartree–Fock plus damped-dispersion (HFD) and exchange-Coulomb semi-empirical models, and the employment of empirical multi-property fitting procedures, the Morse/Long-Range (MLR) representation of the potential energy function (PEF) is developed. It discusses direct inversion of both microscopic (spectroscopic transition frequencies and atomic beam scattering) and bulk property (pressure and acoustic second virial coefficients, transport properties) data in detail. It covers the quantum chemical *ab initio* determination of binary atomic interaction energies and their analytical representation, followed by detailed considerations of the interaction energies between pairs of noble gas atoms.

Chapter 4 is concerned with connections between theory and experiment. It first considers the pure noble gases (helium, neon, argon, krypton, xenon) and presents extensive comparisons between calculated and experimental bulk properties. It goes on to describe the correlation concept and examines the universal viscosity correlation using both experimental and computed (*ab initio*) values. Finally, it provides a detailed discussion of binary mixture properties.

Chapter 5 focuses step by step on how to obtain the spectroscopic and thermophysical properties of a specific molecular system theoretically and provides a reference for the specific theoretical calculation work. It uses a consistent set of MLR functions to describe interactions between pairs of noble gas atoms, presents a full set of fitting parameters for MLR fits to noble gas atomic interactions, which may be employed to make sensible comparisons between experimental and computed results. Jia Liu was the first author of Chapter 5, and some computer program in this chapter came from Yu Zhai, who are both members of the research group of Hui Li.

The Appendices present related mathematical material.

This monograph will be of interest to researchers and graduate students in physics, chemistry, and chemical engineering in particular. It can provide more profound knowledge and some practical guidance for them with their work on nonequilibrium phenomena. At the same time, this book can also serve as a supplementary reference material for the course of nonequilibrium statistical mechanics.

Changchun, Jilin, China — H. L.
Waterloo, Ontario, Canada — F. R. W. M.

Acknowledgments

When Hui Li visited Waterloo in 2018, Frederick R. W. McCourt invited him to work together on this book. Hui Li takes this opportunity to express his sincere thanks to Frederick R. W. McCourt, not only for bringing him into this field and giving him the opportunity to collaborate, but also for the amount of time and wisdom that he invested in the creation of this book. The book was mainly written and created at Waterloo, Canada and Changchun, China. We have all put much effort into it.

Our deepest gratitude and admiration would be to Professor McCourt, who led the way in finishing the book despite his poor health. We thank our dearest friend, the late Professor Robert J. Le Roy, whose attitude to science and life impacted us greatly, for his support and help in both academia and life. We also thank our colleagues, Zhong-Yuan Lu, Wing-Ki Liu, Marcel Nooijen, and Pierre-Nicholas Roy, who provided important advice and wonderful inspiration for our work. Chapters 1 and 2 of this book are an update and outgrowth of the monograph *Nonequilibrium Phenomena in Polyatomic Gases* published in 1990, the figures are from this monograph, with thanks for permission from Oxford University Press. Part of this manuscript was edited and organized by Jia Liu and Yu Zhai.

The book could not have been written without the support from our families. In the past decades, our wives, Janet L. McCourt and Yuan-Yuan Zhao, have given us irreplaceable support in work, life, and spirit. Professor McCourt and his wife celebrated their fiftieth wedding anniversary while we were writing this book. We were moved and blessed by their happy marriage and long company. Mrs. McCourt and Jennifer McCourt, their daughter, helped to write emails to Hui Li when Professor McCourt was in poor health, thereby building a bridge of communication on this work between the two authors. The good news is that Professor McCourt is much recovered and we witness the publication of this book together.

Contents

1
The Monatomic Boltzmann Equation

1.1 The Boltzmann Equation for Dilute Monatomic Gases

The state of a fluid composed of N spherically symmetric atoms of mass m, positions $\mathbf{r}_i$, and velocities $\mathbf{c}_i$, $i = 1, \ldots N$, may be characterized by a set of n-particle classical distribution functions $f_n(\mathbf{c}_1, \ldots, \mathbf{c}_n, \mathbf{r}_1, \ldots, \mathbf{r}_n, t)$, $n = 1, \ldots, N$. The lower the density, the smaller is the maximum number n required in order to describe the state of the gas with sufficient accuracy. The notational convention that the single-particle distribution function carries no subscripts is employed, so that the second particle in two-particle distribution functions is labeled by the subscript "1." It has been well established that a description of the state of the gas in terms of the single-particle distribution $f(\mathbf{c}, \mathbf{r}, t)$ suffices for densities that lie in what is referred to as the dilute gas regime: in practice, this regime typically extends over the pressure range 10^2–10^6 Pa. In this description, $f(\mathbf{c}, \mathbf{r}, t)\mathrm{d}\mathbf{r}\mathrm{d}\mathbf{c}$ represents the average number of atoms at time t in a volume element $\mathrm{d}\mathbf{r}\mathrm{d}\mathbf{c}$ of the single-particle phase space. Equivalently, $f\mathrm{d}\mathbf{c}\mathrm{d}\mathbf{r}$ may be considered to represent the probability of finding a particle within the phase space volume element $\mathrm{d}\mathbf{c}\mathrm{d}\mathbf{r}$. It is consistent to assume that all higher distribution functions may then be expressed as products of the single-particle distribution functions, so that $f_2(\mathbf{c}, \mathbf{c}_1, \mathbf{r}, \mathbf{r}_1, t) = f(\mathbf{c}, \mathbf{r}, t)f(\mathbf{c}_1, \mathbf{r}_1, t)$, for example.

The local particle number density at time t and position $\mathbf{r}$ is obtained upon integrating f over $\mathbf{c}$ to obtain

$$n(\mathbf{r}, t) = \int \mathrm{d}\mathbf{c} f(\mathbf{c}, \mathbf{r}, t)\,, \tag{1.1}$$

while the total number of atoms, N, in the volume V at time t is obtained upon integrating $n(\mathbf{r}, t)$ over $\mathbf{r}$, thereby giving

$$N = \int_V \mathrm{d}\mathbf{r}\; n(\mathbf{r}, t) = \int_V \mathrm{d}\mathbf{r} \int \mathrm{d}\mathbf{c}\; f(\mathbf{c}, \mathbf{r}, t)\,. \tag{1.2}$$

The ratio $f(\mathbf{c}, \mathbf{r}, t)/N$ thus represents a probability density in the single-particle phase space. The local nonequilibrium average, $\langle \psi \rangle_{\mathrm{ne}}$, of an arbitrary function of the molecular velocity $\mathbf{c}$ (such as the energy $\psi = \frac{1}{2}mc^2$) may then be obtained from $f(\mathbf{c}, \mathbf{r}, t)$ as

$$\langle \psi \rangle_{\mathrm{ne}}(\mathbf{r}, t) = n^{-1}(\mathbf{r}, t) \int \mathrm{d}\mathbf{c}\; f(\mathbf{c}, \mathbf{r}, t)\psi(\mathbf{c})\,. \tag{1.3}$$

Transport Properties and Potential Energy Models for Monatomic Gases. Hui Li and Frederick R.W. McCourt, Oxford University Press.
 DOI: 10.1093/oso/9780198888253.003.0001

The study of dynamical processes in a gas, such as energy and momentum flow, requires an equation that describes the time evolution of f. A preliminary step is to examine the behavior of a collisionless gas in which all particles move independently of one another under the influence of an external force $\mathbf{F}$. With $\mathbf{r}$ and $\mathbf{c}$ denoting, respectively, position and velocity at time t, while $\mathbf{r}'$ and $\mathbf{c}'$ denote position and velocity at a later time t', conservation of the number of particles is given by

$$f(\mathbf{c}', \mathbf{r}', t')\, \mathrm{d}\mathbf{c}' \mathrm{d}\mathbf{r}' = f(\mathbf{c}, \mathbf{r}, t)\, \mathrm{d}\mathbf{c} \mathrm{d}\mathbf{r}\,, \tag{1.4}$$

with $\mathbf{c}'$ and $\mathbf{r}'$ evolving from $\mathbf{c}, \mathbf{r}$ via

$$\mathbf{r}' = \mathbf{r} + \int \mathrm{d}t''\; \mathbf{c}(t'') \tag{1.5a}$$

and

$$\mathbf{c}' = \mathbf{c} + \int \mathrm{d}t''\; \frac{1}{m} \mathbf{F}(t'')\,. \tag{1.5b}$$

For systems that obey Hamiltonian mechanics, the Liouville theorem expressing the conservation of phase space becomes

$$\mathrm{d}\mathbf{c}' \mathrm{d}\mathbf{r}' = \mathrm{d}\mathbf{c} \mathrm{d}\mathbf{r}\,, \tag{1.6}$$

a result that, taken together with eqn (1.4), gives

$$f(\mathbf{c}', \mathbf{r}', t) = f(\mathbf{c}, \mathbf{r}, t)\,. \tag{1.7}$$

For infinitesimal time steps, eqn (1.7) implies that the substantial time derivative of f in phase space vanishes, giving rise to the single-particle Liouville equation, which is

$$\frac{\mathrm{D}f}{\mathrm{D}t} \equiv \frac{\partial f}{\partial t} + \mathbf{c} \cdot \nabla f + \frac{1}{m} \mathbf{F} \cdot \frac{\partial f}{\partial \mathbf{c}} = 0\,. \tag{1.8}$$

As this equation is based strictly upon reversible mechanics, it is itself reversible. Thus, as the form of eqn (1.8) remains unchanged under time-reversal, that is, $t \to -t$, $\mathbf{c} \to -\mathbf{c}$, and $\mathbf{F} \to \mathbf{F}$, the time-reversed distribution function $f = f(-\mathbf{c}, \mathbf{r}, -t)$ is also a solution of eqn (1.8). For this reason, the single-particle Liouville equation cannot describe irreversible processes and the approach to thermal equilibrium unless a direction of time is introduced via boundary conditions. The Liouville equation is often employed in this manner, together with appropriate boundary conditions (such as thermalization at the walls), to study the behavior of highly rarefied, typically Knudsen (collisionless) gases.

1.1.1 The Boltzmann "Stosszahlansatz"

Atoms of a dilute gas do not move freely under normal conditions, but undergo collisions with one another. The terminology "dilute gas" is normally understood to imply that a pair of colliding atoms may be considered to be isolated from all other atoms, so that only binary collisions are important for the evolution of the gas. The effects of these binary collisions are accounted for by the addition on the right-hand side of eqn (1.8) of a term, denoted $(\delta f/\delta t)_{\text{coll}}$, that accounts for the collisional rate of change of the distribution function. It was Boltzmann's ingenious idea in 1872 to obtain an expression for $(\delta f/\delta t)_{\text{coll}}$ that, on the one hand, depends upon f, thus giving rise to a closed equation for f while, on the other hand, brings the microscopic molecular dynamics into play, thereby establishing a connection with the interatomic potential energy dependence (Cercignani, 1975).

To follow the Boltzmann arguments (Boltzmann, 1872), preliminary kinematical and dynamical aspects of atomic collisions must first be considered. In first instance, this involves consideration of a pair of point-like atoms that interact via a spherically symmetric potential energy. Velocities prior to the collision are denoted by $\mathbf{c}', \mathbf{c}_1'$ and those following completion of the collision are denoted by $\mathbf{c}, \mathbf{c}_1$. Conservation of linear momentum in the collision (see Fig. 1.1) is expressed by

$$m\mathbf{c}' + m\mathbf{c}_1' = m\mathbf{c} + m\mathbf{c}_1 \,, \tag{1.9a}$$

while the corresponding conservation of (kinetic) energy, due to elasticity of the collision process, is given by

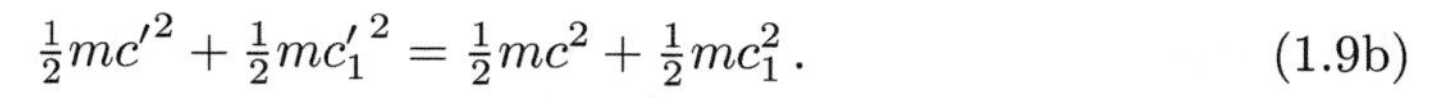

$$\tfrac{1}{2}mc'^2 + \tfrac{1}{2}mc_1'^2 = \tfrac{1}{2}mc^2 + \tfrac{1}{2}mc_1^2 \,. \tag{1.9b}$$

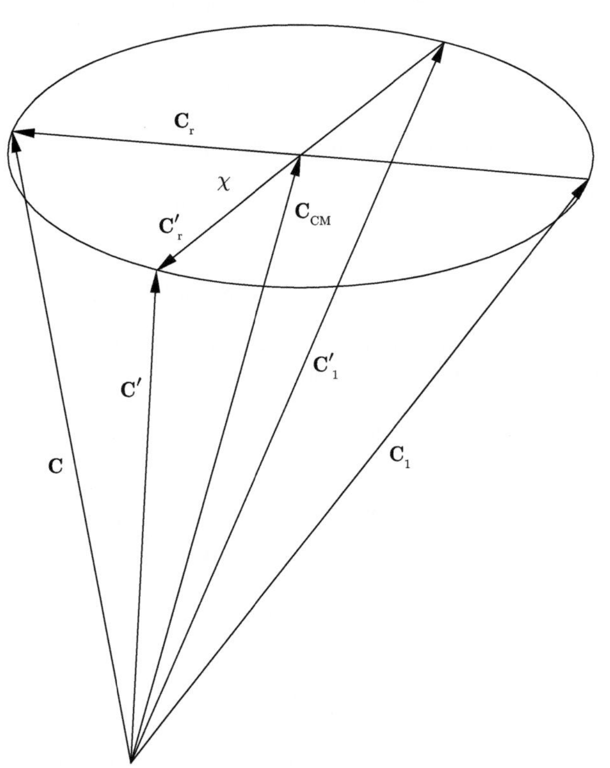

Fig. 1.1 Velocity diagram for a binary collision.

Upon introducing center-of-mass velocities $\mathbf{c}'_{\rm CM}$ and $\mathbf{c}_{\rm CM}$ via

$$\mathbf{c}'_{\rm CM} = \tfrac{1}{2}(\mathbf{c}'_1 + \mathbf{c}')\,, \qquad \mathbf{c}_{\rm CM} = \tfrac{1}{2}(\mathbf{c}_1 + \mathbf{c})\,, \tag{1.10a}$$

and relative velocities $\mathbf{c}'_{\rm r}$ and $\mathbf{c}_{\rm r}$ (with associated directions $\mathbf{e}'$ and $\mathbf{e}$) via

$$\mathbf{c}'_{\rm r} = \mathbf{c}' - \mathbf{c}'_1 = c'_{\rm r}\mathbf{e}'\,, \qquad \mathbf{c}_{\rm r} = \mathbf{c} - \mathbf{c}_1 = c_{\rm r}\mathbf{e}\,, \tag{1.10b}$$

eqns (1.9a, 1.9b) imply that

$$\mathbf{c}'_{\rm CM} = \mathbf{c}_{\rm CM}\,, \qquad c'_{\rm r} = c_{\rm r}\,. \tag{1.10c}$$

By utilizing eqns (1.10b) and (1.10c), $\mathbf{c}'$, $\mathbf{c}'_1$, $\mathbf{c}$, $\mathbf{c}_1$ may be expressed in terms of $\mathbf{c}_{\rm CM}$, $c_{\rm r}$, $\mathbf{e}'$, and $\mathbf{e}$ as

$$\mathbf{c}' = \mathbf{c}_{\rm CM} + \tfrac{1}{2}c_{\rm r}\mathbf{e}'\,, \qquad \mathbf{c}'_1 = \mathbf{c}_{\rm CM} - \tfrac{1}{2}c_{\rm r}\mathbf{e}'\,, \tag{1.11a}$$

and

$$\mathbf{c} = \mathbf{c}_{\rm CM} + \tfrac{1}{2}c_{\rm r}\mathbf{e}\,, \qquad \mathbf{c}_1 = \mathbf{c}_{\rm CM} - \tfrac{1}{2}c_{\rm r}\mathbf{e}\,. \tag{1.11b}$$

These equations are often referred to as the kinematical collision equations.

It is a straightforward task to evaluate the respective 6×6 Jacobian determinants

$$\frac{\partial(\mathbf{c}_{\rm r}, \mathbf{c}_{\rm CM})}{\partial(\mathbf{c}, \mathbf{c}_1)} = 1\,, \qquad \frac{\partial(\mathbf{c}'_{\rm r}, \mathbf{c}_{\rm CM})}{\partial(\mathbf{c}', \mathbf{c}'_1)} = 1 \tag{1.12a}$$

from eqns (1.10a, 1.10b), thereby leading to the conclusions

$$\mathrm{d}\mathbf{c}\mathrm{d}\mathbf{c}_1 = \mathrm{d}\mathbf{c}_{\rm r}\mathrm{d}\mathbf{c}_{\rm CM} = c_{\rm r}^2\mathrm{d}c_{\rm r}\mathrm{d}\mathbf{e}\mathrm{d}\mathbf{c}_{\rm CM} \tag{1.12b}$$

and

$$\mathrm{d}\mathbf{c}'\mathrm{d}\mathbf{c}'_1 = \mathrm{d}\mathbf{c}'_{\rm r}\mathrm{d}\mathbf{c}_{\rm CM} = c_{\rm r}^2\mathrm{d}c_{\rm r}\mathrm{d}\mathbf{e}'\mathrm{d}\mathbf{c}_{\rm CM}. \tag{1.12c}$$

When the first of these two identities is multiplied by the solid angle $\mathrm{d}\mathbf{e}'$ and the second by $\mathrm{d}\mathbf{e}$, their right-hand sides become equal. This gives the equality

$$\mathrm{d}\mathbf{c}'\mathrm{d}\mathbf{c}'_1\mathrm{d}\mathbf{e} = \mathrm{d}\mathbf{c}\mathrm{d}\mathbf{c}_1\mathrm{d}\mathbf{e}'\,, \tag{1.13}$$

which is sometimes referred to as the Liouville theorem for elastic collisions .

A heuristic derivation of the collisional rate of change, $(\delta f/\delta t)_{\rm coll}$, of the distribution function may be obtained from consideration of two crossed monoenergetic atomic beams with respective particle densities n and n_1. The number $\mathrm{d}\dot{N}_{\rm e}$ of scattering events per unit time per unit volume leading into the solid angle element $\mathrm{d}\mathbf{e}$ in the center-of-mass (CM) system is

$$\mathrm{d}\dot{N}_{\rm e} = \sigma\mathrm{d}\mathbf{e}\; c_{\rm r}nn_1\,. \tag{1.14}$$

The differential scattering cross-section, σ, defined in this manner, depends only upon $c_{\rm r}$ and the scattering angle χ in the CM system (see Fig. 1.1). The intent is ultimately to determine σ from a knowledge of the interatomic potential (see, for example, Section

2.5). The appearance of the number densities n and n_1 in the form of a product implies that there is no correlation between atoms in the separate beams.

The collisional rate of change of f may often be written as the difference between "gain" and "loss" terms as

$$\left(\frac{\delta f}{\delta t}\right)_{\mathrm{coll}} = \left(\frac{\delta f}{\delta t}\right)_{\mathrm{gain}} - \left(\frac{\delta f}{\delta t}\right)_{\mathrm{loss}}, \tag{1.15}$$

with $(\delta f/\delta t)_{\mathrm{gain}}\mathrm{d}\mathbf{c}$ representing the number of atoms scattered per unit time and unit volume into the velocity range $\mathrm{d}\mathbf{c}$ while $(\delta f/\delta t)_{\mathrm{loss}}\mathrm{d}\mathbf{c}$ counts the number of atoms scattered away from the velocity range $\mathrm{d}\mathbf{c}$. Expression (1.14) enables the gain and loss terms to be obtained in terms of the distribution functions for scatterers and scattered atoms.

Upon employing the abbreviations $f' \equiv f(\mathbf{c}', \mathbf{r}, t)$ and $f_1' \equiv f(\mathbf{c}_1', \mathbf{r}, t)$, and replacing n, n_1 by $f'\mathrm{d}\mathbf{c}'$, $f_1'\mathrm{d}\mathbf{c}_1'$, the result

$$\left(\frac{\delta f}{\delta t}\right)_{\mathrm{gain}} \mathrm{d}\mathbf{c} = \left[\int\!\!\int \mathrm{d}\mathbf{c}'\mathrm{d}\mathbf{c}'_1\, \sigma c_{\mathrm{r}} f' f_1'\right] \mathrm{d}\mathbf{e} \tag{1.16}$$

is obtained for the gain term following integration over the pre-collisional velocities. This implies that there is no correlation between the two groups of atoms. Finally, by using eqn (1.13), the gain term may be rewritten in the form

$$\left(\frac{\delta f}{\delta t}\right)_{\mathrm{gain}} = \int\!\!\int \mathrm{d}\mathbf{e}'\mathrm{d}\mathbf{c}_1\, \sigma(\mathbf{e}' \to \mathbf{e}, c_{\mathrm{r}}) c_{\mathrm{r}} f' f_1', \tag{1.17}$$

in which $\sigma(\mathbf{e}' \to \mathbf{e}, c_{\mathrm{r}})$ is the differential cross-section for the "direct" collision process $\mathbf{e}' \to \mathbf{e}$.

The number of atoms, $f\mathrm{d}\mathbf{c}$, lost from the beam per unit time due to scattering by atoms with arbitrary velocities $\mathbf{c}_1$ into arbitrary directions $\mathbf{e}'$ of relative velocity is given by

$$\left(\frac{\delta f}{\delta t}\right)_{\mathrm{loss}} \mathrm{d}\mathbf{c} = \left[\int\!\!\int \mathrm{d}\mathbf{e}'\mathrm{d}\mathbf{c}_1 \sigma(\mathbf{e} \to \mathbf{e}', c_{\mathrm{r}}) c_{\mathrm{r}} f f_1\right] \mathrm{d}\mathbf{c}, \tag{1.18}$$

in which $\sigma(\mathbf{e} \to \mathbf{e}', c_{\mathrm{r}})$ is the differential cross-section for the "inverse" collision process $\mathbf{e} \to \mathbf{e}'$. However, as the $\mathbf{e}$ and $\mathbf{e}'$ dependence enters only via

$$\mathbf{e} \cdot \mathbf{e}' = \cos\chi \tag{1.19}$$

for spherically symmetric interaction potentials, the differential cross-sections for direct and inverse collisions are equal, so that

$$\sigma(\mathbf{e}' \to \mathbf{e}, c_{\mathrm{r}}) = \sigma(\mathbf{e} \to \mathbf{e}', c_{\mathrm{r}}) \equiv \sigma(\mathbf{e} \cdot \mathbf{e}', c_{\mathrm{r}}). \tag{1.20}$$

Upon combining the gain-term (1.17) with the loss-term (1.18), the famous Boltzmann Stosszahlansatz, namely,

$$\left(\frac{\delta f}{\delta t}\right)_{\mathrm{coll}} = \int\!\!\int \mathrm{d}\mathbf{e}'\mathrm{d}\mathbf{c}_1\, (f' f_1' - f f_1) \sigma c_{\mathrm{r}} \equiv \mathcal{C}(f f_1), \tag{1.21}$$

is obtained for the collision term. The resulting Boltzmann equation,

$$\frac{\partial f}{\partial t} + \mathbf{c} \cdot \nabla f + \frac{1}{m}\mathbf{F} \cdot \frac{\partial f}{\partial \mathbf{c}} = \mathcal{C}(f f_1)\,, \tag{1.22}$$

is a quadratic integro-differential equation for the single-particle distribution function f. It is worth noting, however, that the variables $\mathbf{c}'$, $\mathbf{c}_1'$ in f', f_1' must be expressed in terms of $\mathbf{c}$, $\mathbf{c}_1$, and $\mathbf{e}'$ according to the kinematical collision equations (1.11a).

The Boltzmann eqn (1.22) has two shortcomings. Firstly, because collisions have been treated as point-like events, the collision term is instantaneous and local. The assumption of point-like interparticle interaction also dictates that the equilibrium equation of state be that of an ideal gas. For hard spheres, a generalization of the Boltzmann equation that considers finite particle size has been constructed by Enskog (1922) (see Section 2.6.1). Secondly, collisions produce correlations among particles. As the particles in a pair drift apart, however, the correlation assumes an increasingly long-range character, so that it cannot conceivably affect the next collision. Similar arguments are typically employed to justify the neglect of correlations for all times, as has been done in obtaining eqn (1.21). It is this "persistent chaos" assumption that is responsible for the irreversibility of the Boltzmann equation, manifested by a change in sign under time reversal of the left-hand side of eqn (1.22) and no change in sign of the right-hand side.

Such reasoning clearly demonstrates that an assumption of persistent chaos cannot be precise. Although it is believed to hold well for dilute gases, in which the mean free path greatly exceeds the particle dimensions, it becomes doubtful for higher densities, in which there is a finite probability for a colliding pair of atoms to re-collide following a few collisions with other partners, with the consequence that correlations can no longer be neglected. Any generalization to higher densities must therefore consider such correlated events as well as multiple collisions. Although some such generalizations have been formulated, they have had only limited success thus far (Cohen, 1972; Cohen, 1971; Dorfman and van Beijeren, 1977; Resibois and de Leener, 1977), and are not considered further in the case of low-density gases (Cercignani, 1975; Hirschfelder *et al.*, 1964; Ferziger and Kaper, 1972).

More sophisticated derivations of the Boltzmann equation, which usually start with the Bogoliubov–Born–Green–Kirkwoood–Yvon (BBGKY) hierarchy of equations for multi-particle distribution functions, also rely in one place or another upon the assumption of persistent chaos. For this, however, the reader is referred to the literature, particularly Cercignani (1975), Hirschfelder *et al.* (1964), and Ferziger and Kaper (1972).

1.1.2 Some quantum mechanical considerations

It is perhaps of some concern that quantum mechanics has thus far been ignored. It might have been considered to be more appropriate to begin from a quantum mechanical description, from which the classical result could be obtained as an appropriate limit. There are three modifications that should be considered. Firstly, the evaluation of cross-sections from interatomic potential energy functions should either be carried out using quantum mechanical scattering or, minimally, classical cross-section calculations should be validated. However, as cross-sections serve as input data for the

Boltzmann equation, it becomes irrelevant whether or not they are obtained from experiment, from quantum mechanical calculations, or even from classical mechanical calculations (when such calculations can be shown to provide a satisfactory approximation), such a discussion may be postponed. Secondly, because the concept of a distribution function (or probability density) that depends simultaneously upon velocity (strictly, momentum) and position violates the spirit of the Heisenberg uncertainty principle, a reconciliation is required. To address this question requires a discussion of the general nonrelativistic description of a state of N structureless particles in terms of a statistical operator (or density matrix) ρ_N (ter Haar, 1960; Tolman, 1979; Fano, 1957; McWeeney, 1960). Thirdly, and perhaps the most serious concern, indistinguishability of identical particles strictly requires the application of Bose–Einstein statistics for bosons (particles that possess intrinsic integral spins) or Fermi–Dirac statistics for fermions (particles that possess intrinsic half odd-integral spins).

In order to determine whether or not the concept of a distribution function $f(\mathbf{c}, \mathbf{r}, t)$ in phase space is legitimate, it suffices to consider the density matrix, or statistical operator, which is a self-adjoint positive semi-definite operator that allows for arbitrary linear superpositions and for incoherent mixing of states. In the Schrödinger representation for structureless particles, the density matrix ρ_N is traditionally represented by a function of a double set of coordinates and time, that is, $\rho_N = \rho_N(\mathbf{r}_1, \mathbf{r}_2, \cdots, \mathbf{r}_N; \bar{\mathbf{r}}_1\bar{\mathbf{r}}_2, \cdots, \bar{\mathbf{r}}_N; t)$. For identical particles, ρ_N is invariant with respect to a simultaneous permutation of the two sets of coordinate vectors. For $\mathbf{r}_1 = \bar{\mathbf{r}}_1, \cdots, \mathbf{r}_N = \bar{\mathbf{r}}_N$, ρ_N represents a probability density in $3N$-dimensional configuration space.

By projecting out unwanted variables, reduced statistical operators may be constructed for single particles, for pairs, for triples, and so on. Only the single-particle statistical operator, defined as

$$\rho_1(\mathbf{r}; \bar{\mathbf{r}}; t) \equiv \int \cdots \int \mathrm{d}\mathbf{r}_2 \cdots \mathrm{d}\mathbf{r}_N \rho_N(\mathbf{r}, \mathbf{r}_2, \cdots, \mathbf{r}_N; \bar{\mathbf{r}}, \mathbf{r}_2, \cdots, \mathbf{r}_N; t), \tag{1.23}$$

is required for a dilute gas. The number density of the particles, denoted by $n(\mathbf{r})$, is defined in terms of the single-particle statistical operator by $n(\mathbf{r}) \equiv N\rho_1(\mathbf{r}; \mathbf{r}; t)$. However, it is important to note that the statistical operator is a distribution in configuration space, rather than in phase space, and that kinetic theory typically utilizes distributions in phase space rather than in configuration space. A procedure introduced by Wigner (1932) provides a satisfactory resolution of this quandary, at least from a practical standpoint. A Fourier transformation "across the diagonal" of the statistical operator, a process now referred to as "Wignerization," produces the so-called Wigner distribution function that resembles a distribution in phase space.

If $f(\mathbf{c}, \mathbf{r}, t)$ is defined as

$$f(\mathbf{c}, \mathbf{r}, t) = N\left(\frac{m}{2\pi\hbar}\right)^3 \int_V \mathrm{d}\mathbf{s}\, \mathrm{e}^{im\mathbf{c}\cdot\mathbf{s}/\hbar} \rho_1(\mathbf{r} - \tfrac{1}{2}\mathbf{s}; \mathbf{r} + \tfrac{1}{2}\mathbf{s}; t), \tag{1.24}$$

it may readily be verified that f is both real and normalized by the total number of particles, that is,

$$\int_V \mathrm{d}\mathbf{r} \int \mathrm{d}\mathbf{c}\; f(\mathbf{c}, \mathbf{r}, t) = N,$$

as is customary in kinetic theory. In addition, any single-particle dynamical quantity represented by an operator $\mathcal{A}$ involving momentum and position operators may be transformed in a manner similar to that employed in (1.24) to produce its Wignerized counterpart $\mathcal{A}(\mathbf{c}, \mathbf{r})$. It can then be shown that the average of $\mathcal{A}$ is given by

$$\langle \mathcal{A} \rangle_{\text{ne}} = \frac{1}{N} \iint \mathrm{d}\mathbf{c}\mathrm{d}\mathbf{r}\, f(\mathbf{c}, \mathbf{r}, t) \mathcal{A}(\mathbf{c}, \mathbf{r})\,. \tag{1.25}$$

Although this result is very encouraging, it does not provide sufficient proof that $N^{-1} f(\mathbf{c}, \mathbf{r}, t)$ is a genuine probability density: examples for which f becomes negative in certain regions of phase space have been found, thereby ruling out a strict interpretation in this sense.

In particular, negative values of f may occur when a particle is localized within a region having dimensions of the order of its de Broglie wavelength, such as may occur for electrons bound within an atom. The de Broglie wavelength for an atom itself is of the order of or smaller than an atomic diameter. However, such narrow localization of an atom does not arise in the kinetic theory of dilute gases, for which particle densities are allowed to vary significantly only over distances that are hundreds or more atomic diameters. It is, therefore, plausible that under such conditions, the Wigner function $f(\mathbf{c}, \mathbf{r}, t)$ does not become negative. Moreover, a reassuring example that is certainly pertinent to kinetic theory is that of a Gaussian wave packet in free space, for which a straightforward calculation leads to an $f(\mathbf{c}, \mathbf{r}, t)$ that has a Gaussian shape and is manifestly non-negative. It thus seems reasonable to assume that the Boltzmann $f(\mathbf{c}, \mathbf{r}, t)$ does indeed correspond to a Wignerized single-particle statistical operator.

It remains to be argued through quantum mechanical reasoning that the function $f(\mathbf{c}, \mathbf{r}, t)$ defined via eqn (1.24) approximately obeys the Boltzmann equation. However, rather than attempting to proceed in a rigorous manner, it seems sufficient to employ plausibility arguments based upon consideration of a pure state of a dilute gas in an ideal box. The wavefunction $\psi(\mathbf{r}_1, \cdots, \mathbf{r}_N, t)$ may initially be assumed to be composed of wave packets in $3N$ dimensions, more or less separated except when they are scattered from one another: an assumption of largely non-overlapping wave packets has been made in order to avoid difficulties that could arise due to the bosonic or fermionic natures of quantum particles. Wave packets in the six-dimensional $(\mathbf{r}, \overline{\mathbf{r}})$ and $(\mathbf{c}, \mathbf{r})$ spaces obtained by carrying out a reduction (1.23) followed by Wignerization via (1.24) will behave almost like particles, in that the centers-of-mass will move according to classical laws, as determined by the Ehrenfest theorem, except during the scattering process. Such a picture is far too detailed for the purposes of kinetic theory, as it is not necessary to account for individual particles and individual scattering events. It is therefore possible to imagine an incoherent mixture of many one-particle statistical operators of the relevant kind (or of their Wigner transforms) such that the final $f(\mathbf{c}, \mathbf{r}, t)$ is a smooth and slowly varying function. Apart from some changes in wording, the Boltzmann Stosszahlansatz and all ensuing steps carry over to give rise to the Boltzmann equation. It may, however, still be necessary to employ quantum mechanics for the description of scattering and for the evaluation of cross-sections.

It will be useful to discuss the question of proper quantum statistics by starting from the known Bose–Einstein and Fermi–Dirac equilibrium distributions, which have the forms

$$f_0(\mathbf{c}) = \frac{1}{\theta}\,\frac{1}{\mathrm{e}^{\beta(\frac{1}{2}mc^2-\mu)} \mp 1}\,, \tag{1.26}$$

with $\beta \equiv (k_\mathrm{B}T)^{-1}$ and k_B the Boltzmann constant. The upper sign in expression (1.26) applies to bosons, the lower sign to fermions, while θ is defined as $\theta \equiv (2\pi\hbar/m)^3$, and $\mu(n,T)$ is the chemical potential, which may be determined from the normalization requirement

$$\int \mathrm{d}\mathbf{c}\; f_0(\mathbf{c}) = n\,.$$

For simplicity, the gas may be considered to be at rest. In order for $f_0(\mathbf{c})$ to be obtained from a vanishing at equilibrium of the collision integral (1.21), the term $f'f_1' - ff_1$ must be replaced by

$$f'f_1'(1 \pm \theta f)(1 \pm \theta f_1) - ff_1(1 \pm \theta f')(1 \pm \theta f_1')\,, \tag{1.27}$$

with the upper signs applying to bosons, the lower signs to fermions. The collision term is now of third order in the distribution. The Boltzmann equation with the fermion collision term (1.27) is known as the Uehling–Uhlenbeck equation (Uehling and Uhlenbeck, 1933). The collision term containing (1.27) ensures that in accordance with the Pauli exclusion principle no fermion can be scattered into a state that is already occupied. In addition, the product $f[1 - \theta f]$ is more or less sharply peaked about μ (which becomes the Fermi energy for low temperatures) such that only energies in a shell around μ contribute significantly to the collision integral. For bosons, the product $f[1 + \theta f]$ ensures a preferential scattering of particles into states that are already occupied.

The identity

$$\ln(1 \pm \theta f_0) = \beta(\tfrac{1}{2}mc^2 - \mu) + \ln(\theta f_0) \tag{1.28}$$

obtained from (1.26) may be employed to establish that (1.27) is consistent with the equilibrium distribution (1.26). Comparison of the logarithms the two terms in (1.27), shows that μ drops out. Moreover, due to the conservation of energy, namely,

$$\tfrac{1}{2}mc^2 + \tfrac{1}{2}mc_1^2 = \tfrac{1}{2}mc'^2 + \tfrac{1}{2}mc_1'^2\,, \tag{1.29}$$

the two logarithms are equal, and expression (1.27) does indeed vanish.

To see under which conditions quantum statistics should be considered, it suffices to consider an ideal gas of spinless atoms, with the chemical potential given approximately as (ter Haar, 1960)

$$\mu(n,T) = k_\mathrm{B}T\ln(n\lambda_\mathrm{th}^3)\,, \tag{1.30}$$

in terms of the thermal de Broglie wavelength

$$\lambda_\mathrm{th} \equiv \left(\frac{2\pi}{mk_\mathrm{B}T}\right)^{\frac{1}{2}}\hbar\,. \tag{1.31}$$

For $n\lambda_{\rm th}^3 \ll 1$, which corresponds to low density and/or high temperature, $\beta|\mu|$ becomes very large, and the classical Maxwell–Boltzmann distribution (see Section 1.4.1), that is,

$$f_0(\mathbf{c}) = \left(\frac{m}{2\pi\hbar}\right)^3 \exp\left\{\frac{2\mu - mc^2}{2k_{\rm B}T}\right\} = n\left(\frac{m}{2\pi k_{\rm B}T}\right)^{\frac{3}{2}} \exp\left\{\frac{-mc^2}{2k_{\rm B}T}\right\}, \tag{1.32}$$

is again obtained at equilibrium. Note also that the terms $\pm\theta f$ in (1.27) may be neglected under such conditions.

Deviations from classical statistics become important for high densities (at which the Boltzmann equation is not applicable), for light atoms such as helium and neon at low temperatures, and for particles of very small mass, such as conduction electrons in metals. Although the Uehling–Uhlenbeck equation is frequently employed for such systems, some serious objections have been raised by Waldmann (1958*a*) with respect to its use in this context. More specifically, because fermions or bosons possessing non-zero spin s have $(2s+1)$-fold degenerate energy levels, a Boltzmann-like equation for particles possessing degenerate internal degrees of freedom, rather than the Uehling–Uhlenbeck equation, should be employed, particularly if spin effects are to be considered (Waldmann, 1957; Waldmann and Kupatt, 1963; McCourt *et al.*, 1990). Further, as the Uehling–Uhlenbeck collision term is of third order in f, a question arises as to whether a consistent treatment should also consider contributions stemming from triple collisions.

1.2 Equations of Change and Collisional Invariants

The temporal and spatial behavior of the average (1.3) of a dynamical quantity $\psi(\mathbf{c})$, which is referred to as an equation of change may be derived from the Boltzmann equation upon multiplying both sides of the Boltzmann eqn (1.22) by $\psi(\mathbf{c})$ and integrating over $\mathbf{c}$. In the absence of an external force $\mathbf{F}$, the resulting equation is

$$\frac{\partial}{\partial t}(n\langle\psi\rangle_{\rm ne}) + \nabla\cdot(n\langle\mathbf{c}\psi\rangle_{\rm ne}) = \left(\frac{\delta n\langle\psi\rangle_{\rm ne}}{\delta t}\right)_{\rm coll}, \tag{1.33}$$

with the first term on the left-hand side of eqn (1.33) representing the rate of change of the density of $\langle\psi\rangle_{\rm ne}$ and the second term representing the divergence of the flux $n\langle\mathbf{c}\psi\rangle_{\rm ne}$. The right-hand side of eqn (1.33) represents the collisional rate of change of $n\langle\psi\rangle_{\rm ne}$ as

$$\left(\frac{\delta n\langle\psi\rangle_{\rm ne}}{\delta t}\right)_{\rm coll} = \iiint \mathrm{d}\mathbf{e}'\mathrm{d}\mathbf{c}_1\mathrm{d}\mathbf{c}\,(f'f_1' - ff_1)\sigma c_{\rm r}\psi(\mathbf{c}) = \int \mathrm{d}\mathbf{c}\psi(\mathbf{c})\mathcal{C}(ff_1)\,. \tag{1.34}$$

The collisional rate of change of $n\langle\psi\rangle_{\rm ne}$ of an $\mathbf{r}$-independent quantity $\psi(\mathbf{c})$ may be cast into a more convenient form by employing invariances of the collision operator $\mathcal{C}(ff_1)$. As the collisional change is invariant under an interchange of collision partners, eqn (1.34) may be symmetrized to give

$$\left(\frac{\delta n\langle\psi\rangle_{\rm ne}}{\delta t}\right)_{\rm coll} = \tfrac{1}{2}\iiint \mathrm{d}\mathbf{e}'\mathrm{d}\mathbf{c}_1\mathrm{d}\mathbf{c}\,(f'f_1' - ff_1)\sigma c_{\rm r}[\psi(\mathbf{c}) + \psi(\mathbf{c}_1)]\,. \tag{1.35}$$

Because σ in eqn (1.35) depends only upon $c_\mathrm{r} = c_\mathrm{r}'$ and $\mathbf{e}' \cdot \mathbf{e}$, it remains unchanged upon an interchange of primed and unprimed variables. Further, because the integration element $\mathrm{d}\mathbf{e}'\mathrm{d}\mathbf{c}_1\mathrm{d}\mathbf{c}$ remains unchanged [as can be seen from the Liouville theorem (1.18)], the collisional rate of change of $n\langle\psi\rangle_\mathrm{ne}$ may also be expressed in the fully symmetrized form

$$\left(\frac{\delta n\langle\psi\rangle_\mathrm{ne}}{\delta t}\right)_\mathrm{coll} = \tfrac{1}{4}\iiint \mathrm{d}\mathbf{e}'\mathrm{d}\mathbf{c}_1\mathrm{d}\mathbf{c}(f'f_1' - ff_1)\sigma c_\mathrm{r}\Delta\psi\,, \tag{1.36}$$

in which $\Delta\psi$, defined by

$$\Delta\psi \equiv \psi(\mathbf{c}) + \psi(\mathbf{c}_1) - \psi(\mathbf{c}') - \psi(\mathbf{c}_1')\,, \tag{1.37}$$

represents the microscopic collisional change of ψ.

For dynamical quantities χ for which $\Delta\chi = 0$, referred to as collisional invariants, the collision term vanishes, giving rise to a local conservation law,

$$\frac{\partial}{\partial t}(n\langle\chi\rangle_\mathrm{ne}) + \nabla \cdot (n\langle\mathbf{c}\chi\rangle_\mathrm{ne}) = 0\,. \tag{1.38}$$

Atoms have precisely five linearly independent collisional invariants, consisting of two scalars and one vector, specifically the particle number, the kinetic energy, and the linear momentum vector (Cercignani, 1975; Kennard, 1938; Grad, 1949). This set of five collisional invariants may be referred to in terms of the five-dimensional array

$$(1, m\mathbf{c}, \tfrac{1}{2}mc^2) \equiv (\chi, \boldsymbol{\chi}, \chi_E)\,. \tag{1.39}$$

For $\chi = 1$, eqn (1.38) gives the equation of continuity for the particle number density n, namely,

$$\frac{\partial n}{\partial t} + \nabla \cdot (n\mathbf{v}) = 0\,, \tag{1.40}$$

a result that is well known from hydrodynamics. The quantity

$$\mathbf{v} \equiv \langle\mathbf{c}\rangle_\mathrm{ne} \tag{1.41}$$

is called the flow velocity of the gas. Introduction of the mass density ρ, defined by

$$\rho \equiv nm\,, \tag{1.42}$$

allows eqn (1.40) to be rewritten as

$$\frac{\partial\rho}{\partial t} + \nabla \cdot (\rho\mathbf{v}) = \frac{\mathrm{d}\rho}{\mathrm{d}t} + \rho\nabla \cdot \mathbf{v} = 0 \tag{1.43}$$

in terms of an operator $\mathrm{d}/\mathrm{d}t$ that denotes the time rate-of-change for an observer moving with the fluid. In hydrodynamics, the operator

$$\frac{\mathrm{d}}{\mathrm{d}t} = \frac{\partial}{\partial t} + \mathbf{v} \cdot \nabla \tag{1.44}$$

is known as the Stokes, or substantial time, derivative. It will be clear from eqn (1.43) that the incompressibility condition $\mathrm{d}\rho/\mathrm{d}t = 0$ is equivalent to the condition $\nabla \cdot \mathbf{v} = 0$.

Substitution of the linear momentum $\boldsymbol{\chi} = m\mathbf{v}$ into eqn (1.38) gives the local momentum conservation law,

$$\frac{\partial}{\partial t}(\rho\mathbf{v}) + \nabla\cdot(\rho\langle\mathbf{cc}\rangle_{\rm ne}) = 0\,, \tag{1.45}$$

also known from hydrodynamics as the equation of motion of the fluid (sometimes called Newton's law). If the peculiar velocity

$$\mathbf{C}(\mathbf{r},t) \equiv \mathbf{c} - \mathbf{v}(\mathbf{r},t) \tag{1.46}$$

representing the deviation $\mathbf{C}$ of the microscopic velocity $\mathbf{c}$ from the flow velocity $\mathbf{v}$ is introduced, then eqn (1.45) takes the form

$$\frac{\partial}{\partial t}(\rho\mathbf{v}) = -\nabla\cdot(\rho\mathbf{vv} + \mathsf{P})\,. \tag{1.47}$$

In eqn (1.47) the quantity $\rho\mathbf{vv}$ represents the convective flux of momentum through the gas, while P, defined by

$$\mathsf{P} \equiv \rho\langle\mathbf{CC}\rangle_{\rm ne}\,, \tag{1.48}$$

is called the pressure tensor. Expression (1.47) may also be written in terms of the substantial time derivative as

$$\rho\frac{\mathrm{d}\mathbf{v}}{\mathrm{d}t} = -\nabla\cdot\mathsf{P}\,. \tag{1.49}$$

Formation of the scalar product between $\mathbf{v}$ and eqn (1.49) gives an equivalent expression

$$\rho\frac{\mathrm{d}}{\mathrm{d}t}(\tfrac{1}{2}\mathrm{v}^2) = -\mathbf{v}\cdot(\nabla\cdot\mathsf{P})\,. \tag{1.50}$$

For later purposes, note that the pressure tensor P may be split into an isotropic component, $p\,\boldsymbol{\delta}$, in which p is the equilibrium pressure of the gas and $\boldsymbol{\delta}$ is the isotropic second-rank tensor (with components given by the Kronecker delta $\delta_{ij}; i,j = \{x,y,z\}$), plus a traceless component $\boldsymbol{\Pi}$ (referred to as the viscous pressure tensor) as

$$\mathsf{P} = p\boldsymbol{\delta} + \boldsymbol{\Pi}\,. \tag{1.51}$$

The local energy conservation equation is obtained from eqn (1.38) when χ_E is considered. It has the form

$$\frac{\partial}{\partial t}(\tfrac{1}{2}\rho\langle c^2\rangle_{\rm ne}) + \nabla\cdot(\tfrac{1}{2}\rho\langle c^2\mathbf{c}\rangle_{\rm ne}) = 0 \tag{1.52}$$

which, upon employing $\mathbf{v}+\mathbf{C}$, rather than $\mathbf{c}$, becomes

$$\frac{\partial}{\partial t}[\rho(\tfrac{1}{2}\mathrm{v}^2 + u)] + \nabla\cdot[\rho\mathbf{v}(\tfrac{1}{2}\mathrm{v}^2 + u)] + \nabla\cdot(\mathsf{P}\cdot\mathbf{v}) + \nabla\cdot\mathbf{q} = 0\,, \tag{1.53}$$

with u, defined as

$$u \equiv \tfrac{1}{2}\langle C^2\rangle_{\rm ne}\,, \tag{1.54}$$

the specific internal energy (i.e. in the thermodynamic sense as energy per unit mass rather than in the microscopic sense associated with internal states), while $\mathbf{q}$ is the heat flux vector, defined by

$$\mathbf{q} \equiv \tfrac{1}{2}\rho\langle C^2\mathbf{C}\rangle_{\mathrm{ne}} . \tag{1.55}$$

Upon employing eqns (1.43, 1.50), the local energy conservation equation may be expressed in the form

$$\rho\frac{\mathrm{d}u}{\mathrm{d}t} = -\mathsf{P} : \nabla\mathbf{v} - \nabla\cdot\mathbf{q} . \tag{1.56}$$

To obtain this expression requires a double application of the identity

$$\frac{\partial(\rho\phi)}{\partial t} + \nabla\cdot(\rho\mathbf{v}\phi) = \rho\frac{\mathrm{d}\phi}{\mathrm{d}t} , \tag{1.57}$$

that can be obtained from eqn (1.43), and is valid for any (not necessarily conserved) quantity ϕ. The local energy conservation equation expressed as eqn (1.56) is known in hydrodynamics as the energy balance equation, with the first term on the right, that is, $\mathsf{P} : \nabla\mathbf{v}$, accounting for internal friction effects.

The local energy conservation equation is often expressed in terms of the divergence of the total energy flux $\mathbf{J}^E$ as

$$\frac{\partial}{\partial t}[\rho(\tfrac{1}{2}\mathrm{v}^2 + u)] + \nabla\cdot\mathbf{J}^E = 0 ,$$

with $\mathbf{J}^E$ defined by

$$\mathbf{J}^E \equiv \tfrac{1}{2}\rho\langle c^2\mathbf{c}\rangle_{\mathrm{ne}} = \tfrac{1}{2}\rho\mathrm{v}^2\mathbf{v} + \rho u\mathbf{v} + \mathsf{P}\cdot\mathbf{v} + \mathbf{q} . \tag{1.58}$$

The final three terms of expression (1.58) constitute what may be called the flux of internal energy, $\mathbf{J}^u$, namely,

$$\mathbf{J}^u = \rho u\mathbf{v} + \mathsf{P}\cdot\mathbf{v} + \mathbf{q} . \tag{1.59}$$

If, following the substitution of eqn (1.51) for P into eqn (1.59) for $\mathbf{J}^u$, the term $\boldsymbol{\Pi}\cdot\mathbf{v}$ is recognized to represent a second-order correction for deviations from equilibrium (small in the sense that $\boldsymbol{\Pi}_{ij} \ll p$), and noting that $v \ll c_0 \equiv (2k_{\mathrm{B}}T/m)^{\frac{1}{2}}$, then it is possible to approximate the heat-flux vector $\mathbf{q}$ as

$$\mathbf{q} \simeq \mathbf{J}^u - \rho h\mathbf{v} , \tag{1.60}$$

in which $h = u + p/\rho$ is the specific enthalpy (i.e. the enthalpy per unit mass). It will also later prove convenient to rewrite definition (1.55) as

$$\mathbf{q} = n\langle(\tfrac{1}{2}mC^2 - \tfrac{5}{2}k_{\mathrm{B}}T)\mathbf{C}\rangle_{\mathrm{ne}} , \tag{1.61}$$

in which the temperature is defined via $u = \frac{3}{2}k_{\mathrm{B}}T/m$: it should be noted that the subtraction of $\frac{5}{2}k_{\mathrm{B}}T$ makes no contribution to $\mathbf{q}$, as $\langle\mathbf{C}\rangle_{\mathrm{ne}} = 0$ by definition. The situation will, however, be different for mixtures, (see Section 1.6).

1.3 Entropy Production

In 1872, Boltzmann succeeded in explaining the validity of the second law of thermodynamics for gases by deriving his H-theorem. The name refers to the function $H = -S/k_{\mathrm{B}}$, so that Boltzmann's statement that $\mathrm{d}H/\mathrm{d}t < 0$ for an isolated system is equivalent to $\mathrm{d}S/\mathrm{d}t > 0$. It will suffice for the time being to consider the local H-theorem,[1] which states that the density of entropy source within the gas is always non-negative, that is, that the entropy production ς is given by

$$\varsigma \equiv \frac{\partial}{\partial t}(\rho s) + \nabla \cdot \mathbf{J}^s \geq 0\,; \tag{1.62}$$

the equality sign holds only should the molecular distribution in velocity space at the point of interest be Maxwellian. By proceeding along much the same lines as did Boltzmann, the entropy density $s(\mathbf{r}, t)$ and the (total) entropy flux $\mathbf{J}^s$ are given by

$$\rho s(\mathbf{r}, t) = -k_{\mathrm{B}} \int \mathrm{d}\mathbf{c}\; f \ln f + nK \tag{1.63}$$

and

$$\mathbf{J}^s = -k_{\mathrm{B}} \int \mathrm{d}\mathbf{c}\, \mathbf{c}\; f \ln f + nK\mathbf{v}\,, \tag{1.64}$$

with K a constant. Equivalently, ρs and $\mathbf{J}^s$ may be expressed as $\rho s = -nk_{\mathrm{B}} \langle \ln f \rangle_{\mathrm{ne}} + nK$ and $\mathbf{J}^s = -nk_{\mathrm{B}} \langle \mathbf{c} \ln f \rangle_{\mathrm{ne}} + nK\mathbf{v}$.

The "entropy-balance" equation is obtained upon substituting $\ln f$ for ψ in eqn (1.33) as

$$\varsigma = \left(\frac{\delta(\rho s)}{\delta t} \right)_{\mathrm{coll}}\,. \tag{1.65}$$

This expression represents the entropy produced per unit time in unit volume of the gas by collisions between pairs of atoms. According to (1.34), ς is thus obtained via

$$\varsigma = -k_{\mathrm{B}} \int \mathrm{d}\mathbf{c}\; (\ln f) \mathcal{C}(ff_1)\,, \tag{1.66}$$

which may be expressed more conveniently by utilizing eqns (1.36, 1.37). Indeed, substitution of $\Delta\psi$ by

$$\Delta\psi = \Delta(\ln f) = \ln \left\{ \frac{f(\mathbf{c}) f(\mathbf{c}_1)}{f(\mathbf{c}') f(\mathbf{c}_1')} \right\} \tag{1.67}$$

gives the entropy production as

$$\varsigma = \tfrac{1}{4} k_{\mathrm{B}} \iiint \mathrm{d}\mathbf{e}' \mathrm{d}\mathbf{c}_1 \mathrm{d}\mathbf{c}\; (f' f_1' - f f_1) \ln \left(\frac{f' f_1'}{f f_1} \right) \sigma c_{\mathrm{r}}\,. \tag{1.68}$$

The positive semi-definiteness of the entropy production now becomes obvious: as $\sigma c_{\mathrm{r}} \mathrm{d}\mathbf{e}'$, $\mathrm{d}\mathbf{c}$, and $\mathrm{d}\mathbf{c}_1$ are all clearly positive, expression (1.68) is positive should the

[1] The local H-theorem, rather than a more general H-theorem, suffices, as neither is expansion of a gas cloud into a void nor are the effects of boundaries being considered.

product $(f'f_1' - ff_1)\ln(f'f_1'/ff_1)$ be positive. Upon defining x as $x \equiv f'f_1'/(ff_1)$, the integrand in eqn (1.68) can be seen to take the form $ff_1(x-1)\ln x$, a function that is readily seen to be positive for $x \neq 1$, so that the condition $(\delta(\rho s)/\delta t)_{\text{coll}} > 0$ is always fulfilled. As the case $x = 1$ corresponds to thermal equilibrium, it is examined in Section 1.4.

Another form for the entropy production may be obtained by carrying out only the first of the two symmetrizations, as in eqn (1.35), to obtain

$$\varsigma = -\tfrac{1}{2}k_{\text{B}} \iiint \mathrm{d}\mathbf{c}\mathrm{d}\mathbf{c}_1\mathrm{d}\mathbf{e}' c_{\text{r}}\sigma (f'f_1' - ff_1)\ln(ff_1),$$

rather than eqn (1.66). Upon exchanging primed and unprimed variables in the loss term only, then employing the identity (1.18), ς becomes

$$\varsigma = -\tfrac{1}{2}k_{\text{B}} \iiint \mathrm{d}\mathbf{c}\mathrm{d}\mathbf{c}_1\mathrm{d}\mathbf{e}' c_{\text{r}}\sigma [f'f_1'\ln(ff_1) - f'f_1'\ln(f'f_1')] \,.$$

Addition of the identity

$$0 = \tfrac{1}{2}k_{\text{B}} \iiint \mathrm{d}\mathbf{c}\mathrm{d}\mathbf{c}_1\mathrm{d}\mathbf{e}' c_{\text{r}}\sigma (ff_1 - f'f_1') \tag{1.69}$$

(obtained, for example, from eqn (1.34) for the collisional invariant 1) to expression (1.69) gives a second form, namely,

$$\varsigma = \tfrac{1}{2}k_{\text{B}} \iiint \mathrm{d}\mathbf{c}\mathrm{d}\mathbf{c}_1\mathrm{d}\mathbf{e}' c_{\text{r}}\sigma f'f_1' \left[\frac{ff_1}{f'f_1'} - 1 - \ln\left(\frac{ff_1}{f'f_1'}\right)\right] \tag{1.70}$$

for the entropy production. The factor within square brackets behaves like the function $x - 1 - \ln x$, which is positive for positive x, and vanishes for $x = 1$. Hence, the same conclusions obtained from eqn (1.68) regarding ς may be drawn.

It is reasonable to ask why so much effort has been expended to obtain two equivalent formulae. The reason for doing this is that the second form is useful should a polyatomic gas be considered, as in that case, the traditional method for obtaining (1.68) fails, while the alternative approach proves to be successful in obtaining an equation that is analogous to (1.70).

1.4 The Equilibrium State

In general, the full spatial and temporal evolution of the distribution function $f(\mathbf{c}, \mathbf{r}, t)$ for an atomic gas is determined by the Boltzmann eqn (1.22) together with an appropriate set of boundary conditions. Such a non-linear differential equation is in general rather difficult to solve, and thus we must seek some means of approximation. As powerful solution procedures have been developed for linear equations, it will be much easier to effect a solution to the non-linear integro-differential eqn (1.22) should it be possible to replace it by an appropriate (i.e. nearly equivalent) linear integro-differential equation. Linearization of the Boltzmann equation becomes a viable option when the state of the gas is not far removed from an equilibrium state that may be characterized by the requirement that its collision term vanish.

1.4.1 Thermal equilibrium

Thermal equilibrium is characterized by zero entropy production which, in turn, implies via eqn (1.68) that the corresponding (absolute) equilibrium distribution functions f_0 and f_{10} must be such that

$$f_0' f_{10}' = f_0 f_{10} \,. \tag{1.71a}$$

An alternative version of the equilibrium condition may be obtained by taking the natural logarithm of both sides of eqn (1.71a) to obtain

$$\ln f_0' + \ln f_{10}' = \ln f_0 + \ln f_{10} \,, \tag{1.71b}$$

from which it follows that the natural logarithm of the equilibrium distribution function must be a collisional invariant, and will thus consist of a linear combination of the five fundamental collisional invariants (1.39). This observation allows $\ln f_0$ to be expressed as

$$\ln f_0 = \alpha - \beta(\tfrac{1}{2}mc^2) + \boldsymbol{\gamma} \cdot m\mathbf{c} \,, \tag{1.72}$$

in which the scalars α, β, and the vector $\boldsymbol{\gamma}$ are determined by fixing the equilibrium particle number density n, the equilibrium temperature T, and the equilibrium velocity $\mathbf{v}$ of the gas. As the equilibrium velocity $\mathbf{v}$ may always be set to zero using a Galilean transformation from a frame at rest to a frame moving with the gas, this is typically done in practice, so that the flow velocity of the gas is simply given by the non-equilibrium average of $\mathbf{c}$. With n and T given, respectively, by

$$n = \int \mathrm{d}\mathbf{c} f_0 \,, \tag{1.73a}$$

and

$$\tfrac{3}{2} n k_{\mathrm{B}} T = \int \mathrm{d}\mathbf{c} \, \tfrac{1}{2} m C^2 f_0 \,, \tag{1.73b}$$

the equilibrium distribution function for a gas at rest is explicitly given by

$$f_0(\mathbf{c}) = n \left(\frac{m}{2\pi k_{\mathrm{B}} T} \right)^{\frac{1}{2}} \exp\left\{ -\frac{mc^2}{2k_{\mathrm{B}}T} \right\} \,. \tag{1.74}$$

As the heat flux vector (1.55) is an odd function of $\mathbf{c}$, it vanishes for absolute equilibrium. The pressure tensor (1.48) is isotropic, and is hence equivalent to the scalar pressure, that is,

$$\mathsf{P} = n k_{\mathrm{B}} T \boldsymbol{\delta} = p \boldsymbol{\delta} \,. \tag{1.75}$$

Note that even though an interatomic interaction is necessary in order to bring the gas to thermal equilibrium via binary collisions between atom pairs, it does not appear in the equation of state for the equilibrium gas.

Averages of functions of velocity over a Maxwellian distribution will often be employed. They will be denoted in general by

$$\langle \psi \rangle = \frac{1}{n} \int d\mathbf{c}\, f_0(\mathbf{c})\psi(\mathbf{c}) \,, \tag{1.76}$$

which is to be distinguished from the nonequilibrium value $\langle \psi \rangle_{\text{ne}}$ introduced earlier. Important examples are the average molecular speed $\langle c \rangle$, namely,

$$\langle c \rangle \equiv \bar{c} = \left(\frac{8k_{\text{B}}T}{\pi m} \right)^{\frac{1}{2}} = \frac{2}{\sqrt{\pi}}\, c_0 \,, \tag{1.77a}$$

with $c_0 \equiv (2k_{\text{B}}T/m)^{\frac{1}{2}}$, and averages related to the kinetic energy, such as

$$\langle \tfrac{1}{2}mc_x^2 \rangle = \langle \tfrac{1}{2}mc_y^2 \rangle = \langle \tfrac{1}{2}mc_z^2 \rangle = \tfrac{1}{2}k_{\text{B}}T \,, \tag{1.77b}$$

and

$$\langle \tfrac{1}{2}mc^2 \rangle = \tfrac{3}{2}k_{\text{B}}T \,. \tag{1.77c}$$

1.4.2 Local Maxwellians

The Boltzmann collision operator has been shown to vanish for the (global) equilibrium distribution. It is also mathematically possible to impose a condition of the form of eqn (1.71) with the number density, temperature, and flow velocity still functions of $\mathbf{r}$ and t. Because the distribution function defined in this manner has the structure of a Maxwellian distribution, but with local values of n, T, and $\mathbf{v}$, it is referred to as a local Maxwellian. Upon introducing a dimensionless reduced peculiar velocity $\mathbf{W}(\mathbf{r}, t)$, defined by

$$\mathbf{W}(\mathbf{r}, t) \equiv \left(\frac{m}{2k_{\text{B}}T(\mathbf{r}, t)} \right)^{\frac{1}{2}} [\mathbf{c} - \mathbf{v}(\mathbf{r}, t)] \,, \tag{1.78}$$

the local Maxwellian takes the form

$$f^{(0)}(\mathbf{c}, \mathbf{r}, t) = n(\mathbf{r}, t) \left(\frac{m}{2\pi k_{\text{B}}T(\mathbf{r}, t)} \right)^{\frac{1}{2}} \exp\{-W^2\} \,. \tag{1.79}$$

This local Maxwellian must satisfy the condition

$$\frac{\partial f^{(0)}}{\partial t} + \mathbf{c} \cdot \nabla f^{(0)} = 0 \,, \tag{1.80}$$

which contrasts to the condition that both $\partial f_0/\partial t$ and ∇f_0 must vanish for the global thermal equilibrium distribution. While it is, of course, irrelevant in principle whether a global or local Maxwellian is involved in defining the equilibrium average $\langle \psi \rangle$, it is important to remember that, should linearization be carried out around a local equilibrium distribution, averages $\langle \psi \rangle$ defined via eqn (1.76) using $f^{(0)}(\mathbf{c}; \mathbf{r}, t)$ rather than $f_0(\mathbf{c})$, do assume dependences on $\mathbf{r}$ and t through the (local) Maxwellian (1.79).

Note that the results (1.77b, 1.77c) are equivalent, respectively, to the averages $\langle W_x^2 \rangle = \langle W_y^2 \rangle = \langle W_z^2 \rangle = \frac{1}{2}$, $\langle W^2 \rangle = \frac{3}{2}$ involving the reduced peculiar velocity $\mathbf{W}$ of eqn (1.78). Of course, for global equilibrium, for which $\mathbf{v} = 0$, the reduced peculiar velocity defined for local equilibrium via eqn (1.78) becomes simply the dimensionless velocity $\mathbf{W} = \mathbf{c}/c_0$. Additional Maxwellian averages relevant to transport phenomena in atomic gases may be found in Appendix A.1.

The Chapman–Enskog solution scheme to be discussed in Chapter 2 is based upon an assumption that after a small number of collisions have occurred, a so-called local equilibrium (which results in a distribution of atoms that approximates a local Maxwellian distribution) is established, after which the approach to global equilibrium occurs on a macroscopic (i.e. hydrodynamic) timescale. In the hydrodynamical stage, the state of the gas always remains close to a local equilibrium state, with the deviation from global equilibrium governed by irreversible first-order hydrodynamic equations, specifically, the Fourier law of heat conduction and the Navier–Stokes equations.

1.5 Linearization of the Boltzmann Equation

Linearization of the full Boltzmann equation is carried out with respect to a local Maxwellian distribution if a Chapman–Enskog procedure(Chapman and Cowling, 1970) is employed (as in Section 2.2), or with respect to a global equilibrium distribution if the Grad moment method (Grad, 1952), is used (cf. Section 2.4). In the Chapman–Enskog procedure, an Ansatz

$$f(\mathbf{c},\mathbf{r},t) = f^{(0)}(\mathbf{c},\mathbf{r},t)[1+\phi(\mathbf{c},\mathbf{r},t)] \tag{1.81a}$$

is made, in which $f^{(0)}(\mathbf{c},\mathbf{r},t)$ is the local Maxwellian given by eqn (1.79) and ϕ represents a small nonequilibrium perturbation. In a first approximation the $\mathbf{r}, t$ dependence of ϕ may be ignored, so that it can be treated as a function of velocity alone, to give

$$f(\mathbf{c},\mathbf{r},t) = f^{(0)}(\mathbf{c},\mathbf{r},t)[1+\phi(\mathbf{c})]\,. \tag{1.81b}$$

If it is assumed that $|\phi| \ll 1$ within the thermal range (i.e. for energies up to a few multiples of $k_\mathrm{B}T$), so that terms quadratic in ϕ may be disregarded in the collision integral and its derivatives may be neglected in the flow terms of eqn (1.22), then the linearized Boltzmann equation,

$$\frac{\mathrm{D}}{\mathrm{D}t}\ln f^{(0)} + \mathcal{R}\phi = 0\,, \tag{1.82}$$

is obtained. The streaming operator $\mathrm{D}/\mathrm{D}t$ is as in eqn (1.8), and the linearized collision operator $\mathcal{R}$ is defined by

$$\mathcal{R}\phi = -\frac{1}{f^{(0)}}\left[\mathcal{C}(f^{(0)}\phi f_1^{(0)}) + \mathcal{C}(f^{(0)}f_1^{(0)}\phi_1)\right] = \iint \mathrm{d}\mathbf{e}'\mathrm{d}\mathbf{c}_1\, f_1^{(0)}\Delta\phi\,\sigma c_\mathrm{r}\,, \tag{1.83}$$

with $\Delta\phi$ given by (1.37). The symbol $\mathcal{R}$ has been chosen (Snider, 1964) by analogy with the generalized resistance in the linear response theory of irreversible processes (de Groot and Mazur, 1962). Note that $\mathcal{R}$ as defined here is proportional to the gas density.

It shall be assumed henceforth that external forces $\mathbf{F}$ are absent, so that the streaming operator may also be written in terms of the substantial derivative $\mathrm{d}/\mathrm{d}t$ of eqn (1.44) as

$$\frac{\mathrm{D}}{\mathrm{D}t} = \frac{\partial}{\partial t} + \mathbf{c} \cdot \nabla = \frac{\mathrm{d}}{\mathrm{d}t} + \mathbf{C} \cdot \nabla \,. \tag{1.84}$$

It will be convenient to introduce a Hilbert space[2] of functions generated by the inner product

$$\langle \phi | \psi \rangle = n^{-1} \int \mathrm{d}\mathbf{c} \; \phi^* f^{(0)} \psi \tag{1.85}$$

between two elements ϕ and ψ, with ϕ^* denoting the complex conjugate of ϕ. Averages over the Maxwellian may now be interpreted as inner products, namely,

$$\langle \psi \rangle = \langle \psi^* | 1 \rangle = \langle 1 | \psi \rangle \,. \tag{1.86}$$

As ψ is normally a real quantity, the asterisk will often be omitted unless it is specifically required or is contained in general formulae. Should ψ be such that $\langle \psi \rangle = 0$, then the nonequilibrium average may be represented by an inner product

$$\langle \psi \rangle_{\mathrm{ne}} = \langle \psi^* | \phi \rangle = n^{-1} \int \mathrm{d}\mathbf{c} \, \psi f^{(0)} \phi \,. \tag{1.87}$$

Matrix elements of the linearized collision operator are defined by

$$\langle \phi | \mathcal{R} \psi \rangle = n^{-1} \int \mathrm{d}\mathbf{c} \, \phi^* f^{(0)} \mathcal{R} \psi \tag{1.88a}$$

$$= n^{-1} \iiint \mathrm{d}\mathbf{e}' \mathrm{d}\mathbf{c}_1 \mathrm{d}\mathbf{c} \; f^{(0)} f_1^{(0)} \phi^* \Delta \psi \sigma c_{\mathrm{r}} \,. \tag{1.88b}$$

By applying the same permutation procedure that led to eqn (1.36) to ϕ^*, the matrix element (1.88b) may be cast into a more convenient form, namely,

$$\langle \phi | \mathcal{R} \psi \rangle = (4n)^{-1} \iiint \mathrm{d}\mathbf{e}' \mathrm{d}\mathbf{c}_1 \mathrm{d}\mathbf{c} \, f^{(0)} f_1^{(0)} \Delta \phi^* \Delta \psi \sigma c_{\mathrm{r}} \,. \tag{1.89}$$

The positive semi-definiteness of the linearized collision operator

$$\langle \phi | \mathcal{R} \phi \rangle \geq 0 \tag{1.90}$$

becomes obvious from eqn (1.89), with the equality sign holding only should ϕ be one of the five collisional invariants (1.39) or some linear combination of them. The important relation

$$\langle \psi | \mathcal{R} \phi \rangle = \langle \mathcal{R} \psi | \phi \rangle \tag{1.91}$$

may also be inferred from eqn (1.89). This symmetry is consistent with the linearized collision operator $\mathcal{R}$ being a self-adjoint operator, a property that arises from the

[2]In fact, a set of Hilbert spaces is involved because the weight $f^{(0)}$ varies with $\mathbf{r}$ and t. However, this is of little consequence as in the end, it is the behavior of the gas in the vicinity of a specific, but arbitrarily chosen space-time point that will be examined.

equality of the differential cross-sections for direct and inverse collisions, eqn (1.20). The matrix elements are Hermitian symmetric, giving rise to the Onsager symmetry relations (de Groot and Mazur, 1962). It is worth noting, however, that these relations remain valid provided that the weaker micro-reversibility relation,

$$\sigma(\mathbf{e}' \to \mathbf{e}, c_{\mathrm{r}}) = \sigma(-\mathbf{e} \to -\mathbf{e}', c_{\mathrm{r}}), \tag{1.92}$$

expressing the equality of the differential cross-sections for direct and time-reversed (or reverse) collision processes holds. For point-like particles with spherical interactions, however, eqns (1.20) and (1.92) are identical, because $\sigma = \sigma(\mathbf{e} \cdot \mathbf{e}', c_{\mathrm{r}})$. For molecules, the situation is different (McCourt *et al.*, 1990).

Because $\mathcal{R}$ is an unbounded operator, its symmetry does not automatically ensure that it is also self-adjoint, as the domains in which $\mathcal{R}$ and $\mathcal{R}^\dagger$ operate could conceivably differ. However, positive semi-definiteness of $\mathcal{R}$ allows the conclusion (Riesz and Nagy, 1955) that this operator may suitably be extended (upon extending the definition of $\mathcal{R}$ to a wider domain) so as to become self-adjoint. It shall always tacitly be assumed that such an extension has been carried out, so that

$$\mathcal{R} = \mathcal{R}^\dagger. \tag{1.93}$$

An important property of the linearized collision operator is its rotational invariance, which is a consequence of the undisturbed gas being isotropic, as may also be inferred from the form of definition (1.83). Thus, if $\widehat{D}$ is any rotation operator in three dimensions, then

$$\widehat{D}^{-1}\mathcal{R}\widehat{D} = \mathcal{R}. \tag{1.94}$$

A consequence of the rotational invariance of $\mathcal{R}$ is that when it acts upon an irreducible tensor, another such tensor of the same rank and weight and with the same orientation is produced. This result provides the basis for the Curie principle (de Groot and Mazur, 1962). Consequently, if a matrix element $\langle\boldsymbol{\phi}|\mathcal{R}\boldsymbol{\phi}\rangle$ is formed from two such tensors, it vanishes should the tensor ranks differ. If the tensor ranks are equal, and if the orientations coincide, then by the Wigner–Eckart theorem (Messiah, 1966; Edmonds, 1974; Zare, 1988; Thompson, 1994), the result is an isotropic tensor proportional to $\boldsymbol{\Delta}^{(p)}$, a tensor of rank $2p$ that projects out of any tensor of rank p its traceless symmetric part. The scalar coefficient is referred to as the reduced matrix element, and is denoted by placing double bars on either side of the operator. Thus, for example, for ϕ a tensor of rank p, the matrix element $\langle\phi|\mathcal{R}\phi\rangle$ takes the form[3]

$$\langle\boldsymbol{\phi}|\mathcal{R}\boldsymbol{\phi}\rangle = \langle\boldsymbol{\phi}\|\mathcal{R}\|\boldsymbol{\phi}\rangle\boldsymbol{\Delta}^{(p)}. \tag{1.95a}$$

As the full trace $\boldsymbol{\Delta}^{(p)} \odot \boldsymbol{\Delta}^{(p)}$ of $\boldsymbol{\Delta}^{(p)}$ equals $2p + 1$, the reduced matrix element is thus given by

$$\langle\phi\|\mathcal{R}\|\phi\rangle = \frac{1}{2p+1}\langle\boldsymbol{\phi}\odot|\mathcal{R}\boldsymbol{\phi}\rangle. \tag{1.95b}$$

[3]Only two values of p occur for pure noble gases and mixtures of noble gases: $p = 1$ for transport phenomena associated with temperature or concentration gradients, and $p = 2$ for transport phenomena associated with a gradient of the flow velocity.

Note that the entropy production ς from (1.66) is given in the linearized theory by

$$\varsigma = nk_{\mathrm{B}} \langle \ln f | \mathcal{R}\phi \rangle \,. \tag{1.96}$$

In the spirit of linearization of f, $\ln f$ may be approximated as

$$\ln f = \ln f^{(0)} + \ln(1+\phi) \simeq \ln f^{(0)} + \phi \,, \tag{1.97}$$

and substituted into the expression for ς. Because $\ln f^{(0)}$ is a collisional invariant, it makes no contribution, and the result will then be

$$\varsigma = nk_{\mathrm{B}} \langle \phi | \mathcal{R}\phi \rangle \,. \tag{1.98}$$

Thus, the positive semi-definiteness of $\mathcal{R}$ is intimately linked to non-negativity of the entropy production.

1.5.1 Spectrum of the linearized collision operator

It is useful to consider the spectrum of the linearized collision operator, which consists of all values of λ for which the operator $(\mathcal{R} - \lambda)$ does not have a bounded inverse. In particular, it includes the eigenvalues λ associated with those members ϕ of the Hilbert space that satisfy the (eigen)equation

$$\mathcal{R}\phi = \lambda\phi \,. \tag{1.99}$$

Because $\mathcal{R}$ is a self-adjoint operator, its spectrum lies on the real axis, and because it is a positive semi-definite operator, its spectrum is non-negative. Moreover, $\mathcal{R}$ has a precisely fivefold degenerate (Grad, 1949) eigenvalue $\lambda = 0$ that corresponds to a set of five linearly independent eigenfunctions, namely, the five collisional invariants $1, m\mathbf{c}$, and $\frac{1}{2}mc^2$ (equivalently, $1, m\mathbf{C}, \frac{1}{2}mC^2$). This quintuple degeneracy of $\mathcal{R}$ may be deduced directly from eqn (1.83). The behavior for constant ϕ may also be verified in a more physical sense by examining eqns (1.81): constant ϕ implies a modification in density of the original Maxwellian, with the modified Maxwellian still obeying the equation

$$\mathcal{C}(f^{(0)} f_1^{(0)}) = 0 \,. \tag{1.100}$$

This result corresponds to $\mathcal{R}$ operating upon a constant to give zero. Similarly, $\phi \propto m\mathbf{c}$ and $\phi \propto \frac{1}{2}mc^2$ (with small coefficients) correspond, in linear approximation, to changes in flow velocity and temperature, respectively, without altering the shape of the Maxwellian. In operator notation, these statements correspond to

$$\mathcal{R}1 = 0 \,, \quad \mathcal{R}m\mathbf{c} = \mathbf{0} \,, \quad \mathcal{R}\tfrac{1}{2}mc^2 = 0 \,. \tag{1.101}$$

The zero eigenvalue of $\mathcal{R}$ is often referred to as the hydrodynamic eigenvalue, and the corresponding eigenfunctions are said to span the hydrodynamic subspace, $\mathcal{H}_{\mathrm{h}}$, of the full Hilbert space. In mathematics, $\mathcal{H}_{\mathrm{h}}$ would be termed the null space of the operator $\mathcal{R}$. The orthogonal complement of the hydrodynamic subspace, termed the nonhydrodynamic subspace $\mathcal{H}_{\mathrm{nh}}$, consists of all functions that are orthogonal to the collisional invariants. Every element in $\mathcal{H}$ can be written as the sum of an element from $\mathcal{H}_{\rm h}$ and an element from $\mathcal{H}_{\rm nh}$. This decomposition of the Hilbert space will be found to be relevant in Chapter 2.

The nonzero values of λ for $\mathcal{R}$ depend upon the individual properties of the collision operator, and may differ from gas to gas or from model to model. The spectrum is entirely discrete for some classical models with infinite-range interactions. In particular, for a Maxwell gas (which has a repulsive R^{-4} interatomic interaction potential), the spectrum and all eigenfunctions are completely known (Waldmann, 1958*b*). It is a characteristic of such models that the gain and loss terms in the Boltzmann equation cannot be treated separately, so that the integral in eqn (1.21) must be treated as a single entity: this is no longer a requirement should the interaction range have a cut-off, or should quantum mechanics be employed for the evaluation of the differential cross-section. An interatomic potential energy function that falls off faster than R^{-3} has a finite total scattering cross-section (Landau and Lifshitz, 1977), which enables a separation of the gain and loss terms, with the consequence that $\mathcal{R}\phi(\mathbf{c})$ may be expressed in the form

$$\mathcal{R}\phi(\mathbf{c}) = \nu(\mathbf{c})\phi(\mathbf{c}) - \int \mathrm{d}\mathbf{c}'\sigma(\mathbf{c} \to \mathbf{c}')\phi(\mathbf{c}') \equiv \nu\phi - \mathcal{K}\phi\,, \tag{1.102}$$

with $\nu(\mathbf{c})$ known as the collision rate, and $\mathcal{K}$ termed the scattering operator. Comparison of this defining relation with eqn (1.83) for the linearized Boltzmann equation shows that $\nu(\mathbf{c})$ and $\mathcal{K}\phi(\mathbf{c})$ correspond respectively to

$$\nu(\mathbf{c}) = \iint \mathrm{d}\mathbf{c}_1 \mathrm{d}\mathbf{e}'\sigma c_{\rm r} f_1^{(0)} \tag{1.103}$$

and

$$\mathcal{K}\phi(\mathbf{c}) = \iint \mathrm{d}\mathbf{c}_1 \mathrm{d}\mathbf{e}'\sigma c_{\rm r} f_1^{(0)} (\phi' + \phi_1' - \phi_1)\,. \tag{1.104}$$

Examination of the three terms making up $\mathcal{K}\phi(\mathbf{c})$ shows that the first term corresponds to the scattering of atoms associated with the deviation $f^{(0)}(\mathbf{c}')\phi(\mathbf{c}')$ into the velocity group near $\mathbf{c}$ by atoms associated with $f^{(0)}(\mathbf{c}_1')$, the second term corresponds to atoms associated with the Maxwellian $f^{(0)}(\mathbf{c}')$ colliding with "disturbed" atoms associated with $f^{(0)}(\mathbf{c}_1')\phi(\mathbf{c}_1')$ and being scattered into the velocity group near $\mathbf{c}$. An interchange of particle labels for the collision partners shows that these two contributions to $\mathcal{K}\phi(\mathbf{c})$ are equal. Finally, the third term corresponds to atoms associated with the Maxwellian $f^{(0)}(\mathbf{c})$ colliding with atoms associated with $f^{(0)}(\mathbf{c}_1)\phi(\mathbf{c}_1)$ and being scattered away from $\mathbf{c}$, thereby creating a negative disturbance. This third contribution thus constitutes part of the loss term in the original Boltzmann equation.

Neutral atoms are always "harder" than Maxwell molecules, as the interaction energy between a pair of such atoms decreases more rapidly with separation R than it does for the Maxwell model. The multiplicative operator ν has a continuous spectrum that covers an interval $[\nu(0), \infty)$, with $\nu(0) > 0$, rather than a spectrum composed of discrete eigenvalues. An answer to the question "How does inclusion of the scattering operator $\mathcal{K}$ affect the spectrum of the full collision operator?" is straightforward should $\mathcal{K}$ be a compact operator (which can have only a discrete set of eigenvalues that accumulate at 0), in which case the Weyl–von Neumann theorem applies (Riesz and

Nagy, 1955), so that the inclusion of $\mathcal{K}$ can only inject eigenvalues, and does not affect the continuous spectrum produced by ν. It has been shown, for example, that the hard-sphere model has this property (Dorfman, 1963; Grad, 1963). Of course, in any case, the fivefold zero eigenvalue must appear. For hard spheres, a set of additional eigenvalues that accumulate at $\nu(0)$, the edge of the continuous spectrum, have been located in the gap $(0, \nu(0))$ (Kuščer and Williams, 1967). The first nonhydrodynamic eigenvalue, λ_1 given by $0.666\nu(0)$, is not much smaller than $\nu(0)$. Ultimately, it may be anticipated that such qualitative aspects obtained for the hard-sphere gas will also carry over for more general forms of the pair interaction.

1.5.2 Linearization about global equilibrium

Rather than expanding the distribution function about a local Maxwellian distribution, it is also possible to expand it about the global equilibrium distribution $f_0(\mathbf{c})$ of eqn (1.75) for a gas at rest. Thus, by writing $f(\mathbf{c}, \mathbf{r}, t)$ in terms of $f_0(\mathbf{c})$ as

$$f(\mathbf{c}, \mathbf{r}, t) = f_0(\mathbf{c})[1 + \Phi(\mathbf{c}, \mathbf{r}, t)]\,, \tag{1.105}$$

a corresponding linearized Boltzmann equation

$$\frac{\partial \Phi}{\partial t} + \mathbf{c} \cdot \nabla \Phi + \mathcal{R}\Phi = 0 \tag{1.106}$$

is obtained for the perturbation Φ. The functional form of the linearized collision operator $\mathcal{R}$ is, apart from $f^{(0)}$ being replaced by f_0, the same as that given in eqn (1.83). As the matrix elements (1.88) are the same in both cases, the integrations over $\mathbf{c}, \mathbf{c}_1$ can as well be performed over $\mathbf{C}, \mathbf{C}_1$. There is, however, one major difference between the two approaches: while eqn (1.82) is inhomogeneous, eqn (1.106) is homogeneous. Methods developed for solving linearized versions of the Boltzmann equation are discussed in greater detail in Chapter 2.

1.6 The Boltzmann Equations for Mixtures

A multicomponent gas mixture of N chemically nonreactive atoms can be described in terms of a set of distribution functions $f_\mathrm{k}(\mathbf{c}, \mathbf{r}, t)$, $\mathrm{k} = 1, \ldots, N$. Collisional changes of of a particular f_k will be due partly to collisions of an atom of chemical species k with atoms of the same chemical species and partly to collisions between atoms of chemical species k with atoms of chemical species $\mathrm{k}' \neq \mathrm{k}$. This form of the collision operator $\mathcal{C}$ represents these contributions through the differential collision cross-sections $\sigma_{\mathrm{kk}'}$ that govern binary collisions between atoms of species k and k′, and leads to a system of coupled Boltzmann equations

$$\frac{\mathrm{D}f_\mathrm{k}}{\mathrm{D}t} = \sum_{\mathrm{k}'=1}^{N} \mathcal{C}_{\mathrm{kk}'}(f_\mathrm{k} f_{\mathrm{k}'1})\,, \tag{1.107}$$

with the streaming operator $\mathrm{D}/\mathrm{D}t$ given in terms of the substantial derivative $\mathrm{d}/\mathrm{d}t$ by eqn (1.84).

To obtain the form of the collision terms $\mathcal{C}_{\mathrm{kk}'}$, some of the collision kinematics of Section 1.1 must be slightly modified in order to include differences in the species masses in the conservation eqns (1.9a, 1.9b). In particular, the center-of-mass velocities before and after the collision of an atom of species k with an atom of species k′ become

$$\mathbf{c}'_{\mathrm{CM}} = \frac{m_{\mathrm{k}}\mathbf{c}' + m_{\mathrm{k}'}\mathbf{c}'_1}{m_{\mathrm{k}} + m_{\mathrm{k}'}}, \qquad \mathbf{c}_{\mathrm{CM}} = \frac{m_{\mathrm{k}}\mathbf{c} + m_{\mathrm{k}'}\mathbf{c}_1}{m_{\mathrm{k}} + m_{\mathrm{k}'}}. \tag{1.108}$$

However, as the relative velocities $\mathbf{c}'_{\mathrm{r}} = c_{\mathrm{r}}\mathbf{e}'$ and $\mathbf{c}_{\mathrm{r}} = c_{\mathrm{r}}\mathbf{e}$ are defined in the same manner as in Section 1.1, the Jacobians $\partial(\mathbf{c}_{\mathrm{r}}, \mathbf{c}_{\mathrm{CM}})/\partial(\mathbf{c}, \mathbf{c}_1)$ prior to and following a binary collision remain equal to unity. Moreover, as the magnitude of $\mathbf{c}_{\mathrm{r}}$ remains unchanged, relation (1.13),

$$\mathrm{d}\mathbf{c}\mathrm{d}\mathbf{c}_1\mathrm{d}\mathbf{e}' = \mathrm{d}\mathbf{c}'\mathrm{d}\mathbf{c}'_1\mathrm{d}\mathbf{e}$$

remains valid and, as all further reasoning remains precisely as in Section 1.1, the collision term

$$\mathcal{C}_{\mathrm{kk}'}(f_{\mathrm{k}}f_{\mathrm{k}'1}) = \iint \mathrm{d}\mathbf{c}_1\mathrm{d}\mathbf{e}'\,[f_{\mathrm{k}}(\mathbf{c}')f_{\mathrm{k}'}(\mathbf{c}'_1) - f_{\mathrm{k}}(\mathbf{c})f_{\mathrm{k}'}(\mathbf{c}_1)]c_{\mathrm{r}}\sigma_{\mathrm{kk}'}(c_{\mathrm{r}}, \mathbf{e}\cdot\mathbf{e}') \tag{1.109}$$

between an atom of species k and an atom of species k′ assumes essentially the same form as that of a collision operator for a pair of like atoms, but in terms of the relative speed c_{r} of the two unlike (colliding) atoms and the differential cross-section $\sigma_{\mathrm{kk}'}$ for binary collisions between an atom of species k and one of species k′.

1.6.1 The conservation laws for mixtures

Prior to formulating the conservation laws, it is useful to obtain the general equation of change for a quantity ψ that may depend upon both velocity and species. This dependence can be formulated in terms of an array of functions $\psi(\mathbf{c}) = (\psi_1(\mathbf{c}), \psi_2(\mathbf{c}), \ldots, \psi_N(\mathbf{c}))$ for which the nonequilibrium average density and average flux of ψ may then be defined as

$$n\langle\psi\rangle_{\mathrm{ne}} \equiv \sum_{\mathrm{k}} \int \mathrm{d}\mathbf{c}\, f_{\mathrm{k}}\psi_{\mathrm{k}} \tag{1.110a}$$

and

$$n\langle\mathbf{c}\psi\rangle_{\mathrm{ne}} \equiv \sum_{\mathrm{k}} \int \mathrm{d}\mathbf{c}\, f_{\mathrm{k}}\mathbf{c}\psi_{\mathrm{k}}\,. \tag{1.110b}$$

The equation of change may now be obtained by multiplying each of equations (1.107) by the appropriate ψ_{k}, summing over k, and integrating over $\mathbf{c}$. The final result is analogous to eqn (1.33) for a pure gas, namely,

$$\frac{\partial}{\partial t}(n\langle\psi\rangle_{\mathrm{ne}}) + \nabla\cdot(n\langle\mathbf{c}\psi\rangle_{\mathrm{ne}}) = \left(\frac{\delta n\langle\psi\rangle_{\mathrm{ne}}}{\delta t}\right)_{\mathrm{coll}}, \tag{1.111}$$

with the collisional contribution now given, however, by

$$\left(\frac{\delta n\langle\psi\rangle_{\mathrm{ne}}}{\delta t}\right)_{\mathrm{coll}} = \sum_{\mathrm{k,k}'} \iiint \mathrm{d}\mathbf{e}'\mathrm{d}\mathbf{c}_1\mathrm{d}\mathbf{c}\,[f_{\mathrm{k}}(\mathbf{c}')f_{\mathrm{k}'}(\mathbf{c}'_1) - f_{\mathrm{k}}(\mathbf{c})f_{\mathrm{k}'}(\mathbf{c}_1)]\sigma_{\mathrm{kk}'}c_{\mathrm{r}}\psi_{\mathrm{k}}(\mathbf{c})\,. \tag{1.112}$$

An array for which the collisional term vanishes is termed a collisional invariant. There are now $N + 4$ such invariants, labelled $\chi_1, \ldots, \chi_N, \boldsymbol{\chi}, \chi_E$, corresponding to particle conservation for each of the N species, to conservation of the three linear momentum components, and to conservation of the kinetic energy. Stated explicitly, the collisional invariants for N-component atomic mixtures are $\chi_1 = (1,\ 0,\ \ldots\ ,\ 0)$, $\chi_2 = (0,\ 1,\ \ldots\ ,\ 0)$, $\ldots$, $\chi_N = (0,\ 0,\ \ldots\ ,\ 1)$, $\boldsymbol{\chi} = (m_1\mathbf{c}, m_2\mathbf{c}, \ldots, m_N\mathbf{c})$, $\chi_E = (\frac{1}{2}m_1c^2, \frac{1}{2}m_2c^2, \ldots, \frac{1}{2}m_Nc^2)$. When the right-hand side of eqn (1.111) vanishes, it becomes an equation expressing a local conservation law. One such equation will be obtained for each collisional invariant.

The conservation of particle numbers gives

$$\frac{\partial n_\mathrm{k}}{\partial t} + \nabla \cdot (n_\mathrm{k}\mathbf{v}_\mathrm{k}) = 0, \qquad \mathrm{k} = 1, \cdots, \mathrm{N}. \tag{1.113a}$$

For mass densities ρ_k defined by

$$\rho_\mathrm{k} = m_\mathrm{k}n_\mathrm{k} = nx_\mathrm{k}m_\mathrm{k}, \qquad n \equiv \sum_\mathrm{k} n_\mathrm{k}\,, \tag{1.114}$$

with $x_\mathrm{k} = n_\mathrm{k}/n$ the mole fraction of species k in the mixture, the equations of continuity may also be rewritten as

$$\frac{\partial \rho_\mathrm{k}}{\partial t} + \nabla \cdot (\rho_\mathrm{k}\mathbf{v}_\mathrm{k}) = 0\,.$$

By analogy with a pure gas, the number density n_k and the mean velocity $\mathbf{v}_\mathrm{k}$ (individual flow velocity) for each species k are given in terms of f_k by

$$n_\mathrm{k} = \int \mathrm{d}\mathbf{c}\, f_\mathrm{k} \tag{1.115a}$$

and

$$\mathbf{v}_\mathrm{k} = n_\mathrm{k}^{-1} \int \mathrm{d}\mathbf{c}\, f_\mathrm{k}\mathbf{c}\,. \tag{1.115b}$$

Conservation laws are often formulated in terms of the peculiar velocity

$$\mathbf{C} = \mathbf{c} - \mathbf{v}\,, \tag{1.116}$$

with the barycentric flow velocity $\mathbf{v}$ and the mass density ρ of the N-component mixture given by

$$\rho\mathbf{v} = \sum_\mathrm{k} \rho_\mathrm{k}\mathbf{v}_\mathrm{k}\,, \qquad \sum_\mathrm{k} \rho_\mathrm{k} = \rho\,. \tag{1.117}$$

The average velocity[4] with which atoms of species k diffuse through a multicomponent mixture relative to the barycentric velocity $\mathbf{v}$ is given by

$$\mathbf{v}_\mathrm{k} - \mathbf{v} = n_\mathrm{k}^{-1} \int \mathrm{d}\mathbf{c}\ \mathbf{C} f_\mathrm{k}\,. \tag{1.118a}$$

Although diffusive fluxes

[4]The average velocity $\mathbf{v}_\mathrm{k} - \mathbf{v}$ is also referred to (Monchick *et al.*, 1963) as the diffusion velocity of species k in the gas mixture.

$$\widehat{\mathbf{J}}_{\mathrm{k}}^{d} = \rho_{\mathrm{k}}(\mathbf{v}_{\mathrm{k}} - \mathbf{v}) \tag{1.118b}$$

may be introduced as an alternative, only N − 1 of these fluxes will be independent, as their sum vanishes by definition. Thus, for example, in a binary mixture (N = 2, species labels A, B), only one such flux is required, namely,

$$\widehat{\mathbf{J}}^{d} = \rho_{\mathrm{A}}(\mathbf{v}_{\mathrm{A}} - \mathbf{v}) = -\rho_{\mathrm{B}}(\mathbf{v}_{\mathrm{B}} - \mathbf{v}) = \frac{\rho_{\mathrm{A}}\rho_{\mathrm{B}}}{\rho}(\mathbf{v}_{\mathrm{A}} - \mathbf{v}_{\mathrm{B}}) \,. \tag{1.118c}$$

For a binary mixture in which the pressure is uniform, it is often more convenient to employ an alternative diffusive flux defined by

$$\mathbf{J}^{d} \equiv \mathbf{v}_{\mathrm{A}} - \mathbf{v}_{\mathrm{B}} \,, \tag{1.118d}$$

(see, for example, Section 2.2). This diffusive flux is simply related to $\widehat{\mathbf{J}}^{d}$ by

$$\mathbf{J}^{d} = \frac{\rho}{\rho_{\mathrm{A}}\rho_{\mathrm{B}}}\widehat{\mathbf{J}}^{d} \,. \tag{1.118e}$$

The equations of continuity may be rewritten in terms of the individual flow and diffusion velocities as

$$\frac{\mathrm{d}\rho_{\mathrm{k}}}{\mathrm{d}t} = -\rho_{\mathrm{k}}\nabla\cdot\mathbf{v}_{\mathrm{k}} - (\mathbf{v}_{\mathrm{k}} - \mathbf{v})\cdot\nabla\rho_{\mathrm{k}} \,, \tag{1.119a}$$

or equivalently, as

$$\frac{\mathrm{d}\rho_{\mathrm{k}}}{\mathrm{d}t} = -\rho_{\mathrm{k}}\nabla\cdot\mathbf{v} - \nabla\cdot\widehat{\mathbf{J}}_{\mathrm{k}}^{d} \,. \tag{1.119b}$$

Upon summing over the species in the mixture, the equation of continuity for the mixture as a whole becomes simply

$$\frac{\mathrm{d}\rho}{\mathrm{d}t} = -\rho\nabla\cdot\mathbf{v} \,. \tag{1.120}$$

The equation of motion, which arises from the conservation of momentum, is given by

$$\frac{\partial}{\partial t}(\rho\mathbf{v}) + \nabla\cdot\sum_{\mathrm{k}}\int \mathrm{d}\mathbf{c}\, m_{\mathrm{k}}\mathbf{c}\mathbf{c}f_{\mathrm{k}} = 0 \,. \tag{1.121}$$

while the energy balance equation obtained from the conservation of energy is given by

$$\frac{\partial}{\partial t}\sum_{\mathrm{k}}\int \mathrm{d}\mathbf{c}\,\tfrac{1}{2}m_{\mathrm{k}}c^{2}f_{\mathrm{k}} + \nabla\cdot\sum_{\mathrm{k}}\int \mathrm{d}\mathbf{c}\,\tfrac{1}{2}m_{\mathrm{k}}c^{2}\mathbf{c}f_{\mathrm{k}} = 0 \,. \tag{1.122}$$

If the pressure tensor P, the specific internal energy u, and the heat flux vector $\widehat{\mathbf{q}}$ of the gas mixture are introduced via the defining relations

$$\mathsf{P} \equiv \sum_{\mathrm{k}}\int \mathrm{d}\mathbf{c}\; m_{\mathrm{k}}\mathbf{C}\mathbf{C}f_{\mathrm{k}} \,, \tag{1.123}$$

$$u \equiv \rho^{-1} \sum_{\rm k} \int \mathrm{d}\mathbf{c}\ \tfrac{1}{2} m_{\rm k} C^2 f_{\rm k}\,, \tag{1.124}$$

and

$$\widehat{\mathbf{q}} \equiv \sum_{\rm k} \int \mathrm{d}\mathbf{c}\ \tfrac{1}{2} m_{\rm k} C^2 \mathbf{C} f_{\rm k}\,, \tag{1.125}$$

then the equations of motion and energy balance for the mixture take the forms

$$\rho \frac{\mathrm{d}\mathbf{v}}{\mathrm{d}t} = -\nabla \cdot \mathsf{P}\,, \tag{1.126}$$

and

$$\rho \frac{\mathrm{d}u}{\mathrm{d}t} = -\mathsf{P} : \nabla \mathbf{v} - \nabla \cdot \widehat{\mathbf{q}}\,, \tag{1.127}$$

with precisely the same forms as eqn (1.49) and eqn (1.56) obtained for a pure gas.

An alternative definition for the heat flux, namely,

$$\mathbf{q} = \sum_{\rm k} \int \mathrm{d}\mathbf{c}\ \left(\tfrac{1}{2} m_{\rm k} C^2 - \tfrac{5}{2} k_{\rm B} T\right) \mathbf{C} f_{\rm k}\,, \tag{1.128}$$

with T determined by $\rho u = \frac{3}{2} n k_{\rm B} T$, is suggested by eqn (1.61). Note that for a mixture the heat flux thereby defined is not the same as the heat flux $\widehat{\mathbf{q}}$ defined in eqn (1.125). For a binary mixture, in particular, the difference between these two heat fluxes can be related to the diffusion flux $\widehat{\mathbf{J}}^d$ via (de Groot and Mazur, 1962; Monchick *et al.*, 1963)

$$\begin{aligned} \widehat{\mathbf{q}} - \mathbf{q} &= \tfrac{5}{2} k_{\rm B} T [n_{\rm A}(\mathbf{v}_{\rm A} - \mathbf{v}) + n_{\rm B}(\mathbf{v}_{\rm B} - \mathbf{v})] \\ &= \rho_{\rm A} h_{\rm A}(\mathbf{v}_{\rm A} - \mathbf{v}) + \rho_{\rm B} h_{\rm B}(\mathbf{v}_{\rm B} - \mathbf{v}) \\ &= \widehat{\mathbf{J}}^d (h_{\rm A} - h_{\rm B})\,, \end{aligned} \tag{1.129}$$

in which $h_{\rm k} = \frac{5}{2} k_{\rm B} T / m_{\rm k}$ is the partial specific enthalpy of species k corresponding to temperature T. As it turns out that definition (1.128) is more suitable for gaseous mixtures than is definition (1.125), $\mathbf{q}$ will normally be employed hereafter, except for the energy balance eqn (1.52), which is more conveniently expressed in terms of $\widehat{\mathbf{q}}$.

As the heat flux vector $\widehat{\mathbf{q}}$ plays the same role for a gas mixture that $\mathbf{q}$ plays for a pure gas [see eqn (1.60)], it may also be expressed approximately as

$$\widehat{\mathbf{q}} = \mathbf{J}^u - \rho h \mathbf{v}\,, \tag{1.130}$$

while the corresponding approximation for $\mathbf{q}$ is

$$\mathbf{q} = \mathbf{J}^u - \rho_{\rm A} h_{\rm A} \mathbf{v}_{\rm A} - \rho_{\rm B} h_{\rm B} \mathbf{v}_{\rm B}\,. \tag{1.131}$$

The difference between these two heat fluxes can be reduced to the interpretation of the convective enthalpy flux, which must be subtracted from the flux of internal energy. The relevant convective enthalpy flux for $\widehat{\mathbf{q}}$ is given in terms of the specific enthalpy h of the mixture and the mixture flow velocity $\mathbf{v}$. In order to compare these two heat fluxes, it will suffice to consider a binary mixture, in which $\mathbf{q}$ consists of the sum of heat fluxes $\mathbf{q}$ obtained by treating the two constituents as flowing independently, each having its own diffusion velocity $\mathbf{v}_\mathrm{k}$. These alternative concepts thus correspond to two possible definitions of heat received by the mixture, namely,

$$\mathrm{d}\widehat{Q} = T\,\mathrm{d}S - Ts\,\mathrm{d}m = \mathrm{d}U + p\,\mathrm{d}V - h\,\mathrm{d}m\,, \tag{1.132}$$

and

$$\begin{aligned}\mathrm{d}Q &= T\,\mathrm{d}S - T(s_\mathrm{A}\,\mathrm{d}m_\mathrm{A} + s_\mathrm{B}\,\mathrm{d}m_\mathrm{B}) \\ &= \mathrm{d}U + p\,\mathrm{d}V - h_\mathrm{A}\,\mathrm{d}m_\mathrm{A} - h_\mathrm{B}\,\mathrm{d}m_\mathrm{B}\,,\end{aligned} \tag{1.133}$$

in which s, s_A, s_B are the specific entropies of the mixture and its components: $\widehat{\mathbf{q}}$ and $\mathbf{q}$, could be referred to, respectively, as conventional and reduced heat fluxes.

1.6.2 Entropy production

As the entropy density ρs for a gas mixture is an additive property, it is given by

$$\rho s = \sum_\mathrm{k} \rho_\mathrm{k} s_\mathrm{k} = -k_\mathrm{B} \sum_\mathrm{k} \int \mathrm{d}\mathbf{c}\ f_\mathrm{k} \ln f_\mathrm{k} + nK\,, \tag{1.134}$$

in which the entropy densities for the individual components are defined by eqn (1.63). In a similar manner, the entropy flux $\mathbf{J}_s$ is given as

$$\mathbf{J}_s = -k_\mathrm{B} \sum_\mathrm{k} \int \mathrm{d}\mathbf{c}\ f_\mathrm{k} \ln f_\mathrm{k} \mathbf{c} + nK\mathbf{v}\,, \tag{1.135}$$

[cf. eqn (1.64)]. The local entropy balance equation is thus given by

$$\frac{\partial}{\partial t}(\rho s) + \nabla \cdot \mathbf{J}_s = \varsigma\,, \tag{1.136}$$

in which the entropy production ς is given by

$$\varsigma = \left(\frac{\delta(\rho s)}{\delta t}\right)_\mathrm{coll} = \tfrac{1}{4} k_\mathrm{B} \sum_{\mathrm{kk}'} \iiint \mathrm{d}\mathbf{e}'\mathrm{d}\mathbf{c}_1\mathrm{d}\mathbf{c}\ [f'_\mathrm{k} f'_{\mathrm{k}'1} - f_\mathrm{k} f_{\mathrm{k}'1}] \ln \left\{\frac{f'_\mathrm{k} f'_{\mathrm{k}'1}}{f_\mathrm{k} f_{\mathrm{k}'1}}\right\} \sigma_{\mathrm{kk}'} c_\mathrm{r}\,. \tag{1.137}$$

As for a pure gas (see Section 1.3), an H-theorem holds, so that the entropy production is non-negative.

1.6.3 Thermal equilibrium and linearization

Global equilibrium is obtained when the condition

$$f'_{\mathrm{k}0} f'_{\mathrm{k}'10} = f_{\mathrm{k}0} f_{\mathrm{k}'10} \tag{1.138}$$

holds for the distribution functions associated with the mixture components. It follows in complete analogy with the pure gas case that this condition gives rise to a set of Maxwellian functions

$$f_{k0}(\mathbf{c}) = n_k \left(\frac{m_k}{2\pi k_B T} \right)^{3/2} \exp\left\{ - \frac{m_k(\mathbf{c} - \mathbf{v})^2}{2k_B T} \right\}, \tag{1.139}$$

in which n, $\mathbf{v}$, and T are all independent of $\mathbf{r}$, t. A condition equivalent to eqn (1.138) holds for the local Maxwellians $f_k^{(0)}$, except that n, T, and $\mathbf{v}$ become functions of $\mathbf{r}$ and t. Either global Maxwellians f_{k0} or local Maxwellians $f_k^{(0)}$ may be utilized to define equilibrium averages for the mixture constituents that are akin to those introduced for pure gases in Section 1.4. One additional average that is needed, because it plays an important role in the analysis of binary collisions, is the average relative speed of the colliding pair which, for species k and k′, is defined by

$$\bar{c}_{kk'} = \frac{1}{n_k n_{k'}} \iint d\mathbf{c}_1 d\mathbf{c} \, c_r f_k^{(0)}(\mathbf{c}) f_{k'}^{(0)}(\mathbf{c}_1), \tag{1.140}$$

in which $c_r = |\mathbf{c} - \mathbf{c}_1|$. Each Maxwellian is centered around $\mathbf{v} = 0$ and refers to the common temperature T. As the Jacobian is unity, the differential $d\mathbf{c}_1 d\mathbf{c}$ may be replaced by $d\mathbf{c}_r d\mathbf{c}_{CM}$. Multiplication of the two Maxwellians gives an exponential of the total kinetic energy of the atom pair in the exponential. By converting the total kinetic energy expression into a sum of the kinetic energies associated with the center-of-mass and relative motions via

$$\tfrac{1}{2} m_k c^2 + \tfrac{1}{2} m_{k'} c_1^2 = \tfrac{1}{2}(m_k + m_{k'}) c_{CM}^2 + \tfrac{1}{2} m_{kk'} c_r^2 ,$$

in which $m_{kk'} = m_k m_{k'}/(m_k + m_{k'})$ is the reduced mass of the colliding pair, then carrying out the requisite integrations, the mass factors cancel to give the average thermal speed corresponding to a Maxwellian for the reduced mass, namely,

$$\bar{c}_{kk'} = \left(\frac{8 k_B T}{\pi m_{kk'}} \right)^{\frac{1}{2}} . \tag{1.141}$$

Two special cases come to mind. If one species is much heavier than the other, then $\bar{c}_{kk'}$ approximates the average speed $\bar{c}$ of the light partner. For a pure gas (equal partners), $m_{kk'} \equiv m_r = m/2$, the average relative speed becomes $\bar{c}_r = [16 k_B T/(\pi m)]^{\frac{1}{2}}$.

Linearization of the system of Boltzmann equations for the mixture may be carried out either with respect to a local or a global Maxwellian. By starting from a set of local Maxwellians $\{f_k^{(0)} |\ k = 1, \cdots, N\}$, expanding each f_k as

$$f_k = f_k^{(0)}(1 + \phi_k), \tag{1.142}$$

then proceeding precisely as in Section 1.5, a system of linearized Boltzmann equations

$$\frac{D}{Dt} \ln f_k^{(0)} + \sum_{k'=1}^{N} \mathcal{R}_{kk'} \phi_{k'} = 0, \qquad k = 1, \cdots, N, \tag{1.143}$$

is obtained, with the linearized collision operators $\mathcal{R}_{kk'}$ given by

$$\mathcal{R}_{\mathrm{kk'}}\phi_{\mathrm{k'}} = -(1/f_{\mathrm{k}}^{(0)})\left\{\delta_{\mathrm{kk'}}\sum_{\mathrm{k''}}\mathcal{C}_{\mathrm{kk''}}(f_{\mathrm{k}}^{(0)}\phi_{\mathrm{k}}f_{\mathrm{k''}1}^{(0)}) + \mathcal{C}_{\mathrm{kk'}}(f_{\mathrm{k}}^{(0)}f_{\mathrm{k'}1}^{(0)}\phi_{\mathrm{k'}1})\right\} \tag{1.144a}$$

$$= \delta_{\mathrm{kk'}}\sum_{\mathrm{k''}}\iint \mathrm{d}\mathbf{c}_1\mathrm{d}\mathbf{e}'\, f_{\mathrm{k''}}^{(0)}(\mathbf{c}_1)c_{\mathrm{r}}\sigma_{\mathrm{kk''}}[\phi_{\mathrm{k}}(\mathbf{c}) - \phi_{\mathrm{k}}(\mathbf{c}')]$$

$$+ \iint \mathrm{d}\mathbf{c}_1\mathrm{d}\mathbf{e}'\, f_{\mathrm{k'}}^{(0)}(\mathbf{c}_1)c_{\mathrm{r}}\sigma_{\mathrm{kk'}}[\phi_{\mathrm{k'}}(\mathbf{c}_1) - \phi_{\mathrm{k'}}(\mathbf{c}_1')]\,. \tag{1.144b}$$

The state of the mixture may now be described in terms of an array of functions, $\phi = (\phi_1, \phi_2, \cdots, \phi_{\mathrm{N}})$. As in the preceding section, physical quantities defined for each species may also be represented in terms of arrays, such as $A = (A_1, A_2, \cdots, A_{\mathrm{N}})$ for the physical quantity A. It is then convenient to introduce a Hilbert space of such arrays, with the inner product

$$\langle A|\phi\rangle = n_{-1}\sum_{\mathrm{k}}\int \mathrm{d}\mathbf{c}\, A_{\mathrm{k}}^{*}f_{\mathrm{k}}^{(0)}\phi_{\mathrm{k}}\,. \tag{1.145}$$

Note that an asterisk has been retained for complex conjugation should there be a need to employ complex functions. Note also that should the distribution functions be given by eqn (1.142), and should the real quantity A be such that its equilibrium average vanishes, that is, $\langle A\rangle = \langle A|1\rangle = 0$, then expression (1.145) represents the nonequilibrium average $\langle A\rangle_{\mathrm{ne}}$.

By employing such arrays, the system of linearized Boltzmann equations may be written as the single equation

$$\frac{\mathrm{D}}{\mathrm{D}t}\ln f^{(0)} + \boldsymbol{\mathcal{R}}\phi = 0\,, \tag{1.146}$$

in which the operator $\boldsymbol{\mathcal{R}}$ is represented by the square matrix of operators $\mathcal{R}_{\mathrm{kk'}}$ (k, $\mathrm{k'} = 1, \cdots, \mathrm{N}$). Equation (1.146) brings the analogy with the pure gas into clearer focus.

For a gaseous mixture, the additivity of the entropy density allows the entropy production to be written as

$$\varsigma = \sum_{\mathrm{k}}\varsigma_{\mathrm{k}} = -k_{\mathrm{B}}\sum_{\mathrm{kk'}}\int \mathrm{d}\mathbf{c}\, ln f_{\mathrm{k}}\mathcal{C}_{\mathrm{kk'}}(f_{\mathrm{k}}f_{\mathrm{k'}1})\,, \tag{1.147}$$

which becomes, upon utilizing the results of linearization,

$$\varsigma = k_{\mathrm{B}}\sum_{\mathrm{k}}\int \mathrm{d}\mathbf{c}\, f_{\mathrm{k}}^{(0)}\ln f_{\mathrm{k}}\sum_{\mathrm{k'}}\mathcal{R}_{\mathrm{kk'}}\phi_{\mathrm{k'}}\,. \tag{1.148a}$$

In terms of arrays, the positive-definite nature of ς may be expressed in the same form as that for a pure gas, that is,

$$\varsigma = nk_{\mathrm{B}}\langle\boldsymbol{\phi}|\boldsymbol{\mathcal{R}}\boldsymbol{\phi}\rangle \geq 0\,. \tag{1.148b}$$

Linearization can equally well be carried out around thermal equilibrium, which leads to a homogeneous system of integro-differential equations,

$$\left(\frac{\partial}{\partial t} + \mathbf{c} \cdot \nabla + \boldsymbol{\mathcal{R}}\right) \boldsymbol{\Phi}(\mathbf{c}, \mathbf{r}, t) = 0\,, \tag{1.149}$$

in which $\boldsymbol{\mathcal{R}}$ is a matrix of collision operators, as before. A binary mixture of gases A and B will give rise to a pair of equations for Φ_{A} and Φ_{B}. They further simplify if one component (B), referred to as the carrier gas, is in equilibrium, and the other component (A), referred to as the test-particle gas, is present only in very low concentration. The test particles are then incapable of noticeably disturbing the carrier gas and, moreover, collisions between test particles are so rare that they may be ignored. Only the equation governing the test particles is needed and only collisions between the test particles and the carrier gas need to be considered. Upon dropping the index A to give $\Phi_{\mathrm{A}} \equiv \Phi$ and $\mathcal{R}_{\mathrm{AA}} \equiv \mathcal{R}_1$, the resulting equation takes the same form as that for a pure gas, namely,

$$\left(\frac{\partial}{\partial t} + \mathbf{c} \cdot \nabla + \mathcal{R}_1\right) \Phi(\mathbf{c}, \mathbf{r}, t) = 0\,. \tag{1.150}$$

In this case, however, the collision operator, given by

$$\mathcal{R}_1 \Phi(\mathbf{c}) = \iint \mathrm{d}\mathbf{c}_1 \mathrm{d}\mathbf{e}'\, c_{\mathrm{r}} \sigma_{\mathrm{AB}} f_{\mathrm{B}0}(\mathbf{c}_1)[\Phi(\mathbf{c}) - \Phi(\mathbf{c}')]\,, \tag{1.151}$$

is simpler. As for a pure gas, cf. eqn (1.102), $\mathcal{R}_1$ can in most cases be expressed as the difference between a multiplicative and an integral operator, for example, as $\mathcal{R}_1 = \nu - \mathcal{K}_1$, with

$$\nu(\mathbf{c}) = \iint \mathrm{d}\mathbf{c}_1 \mathrm{d}\mathbf{e}'\, c_{\mathrm{r}} \sigma_{\mathrm{AB}} f_{\mathrm{B}0}(\mathbf{c}_1)\,, \tag{1.152}$$

and

$$\mathcal{K}_1 \Phi(\mathbf{c}) = \iint \mathrm{d}\mathbf{c}_1 \mathrm{d}\mathbf{e}'\, c_{\mathrm{r}} \sigma_{\mathrm{AB}} f_{\mathrm{B}0}(\mathbf{c}_1) \Phi(\mathbf{c}')\,. \tag{1.153}$$

Should the A and B particles each have the same mass and the same interaction potential, the collision rate $\nu(\mathbf{c})$ of the test particles has the same form as that of the particles in a pure gas, cf. eqn (1.103). However, the scattering operators differ even in this case. Indeed, a comparison of eqn (1.153) with eqn (1.104) shows that, in such a case, $\mathcal{K}$ can be written as

$$\mathcal{K} = 2\mathcal{K}_1 - \mathcal{K}'\,, \tag{1.154}$$

with $\mathcal{K}'$ accounting for the term with Φ_1 in an equation like (1.104). The kernel $\mathcal{K}_1(\mathbf{c} \to \mathbf{c}')$, being proportional to a probability density in $\mathbf{c}'$-space, is non-negative. The single-gas kernel $\mathcal{K}(\mathbf{c} \to \mathbf{c}')$ does not in general have this property, nor does it allow such a simple interpretation (Case and Zweifel, 1967; Williams, 1972; Duderstadt and Martin, 1979; Chandrasekhar, 1950).

Such a simplified description of the test-particle problem does not account for any gain or loss of momentum or energy in the carrier gas, as the focus is exclusively on the test particles, for which the only conserved quantity is their number. Consequently, 0 becomes a simple eigenvalue, as $\mathcal{R}_1$ has only one hydrodynamic eigenfunction, namely unity, so that

$$\mathcal{R}_1 1 = 0\,, \tag{1.155a}$$

and $\nu(\mathbf{c})$ is given by

$$\nu(\mathbf{c}) = \int \mathrm{d}\mathbf{c}'\, \mathcal{K}_1(\mathbf{c} \to \mathbf{c}')\,. \tag{1.155b}$$

The test-particle formalism has been used extensively in describing neutron diffusion and the diffusion of light in scattering media (Case and Zweifel, 1967; Williams, 1972; Duderstadt and Martin, 1979; Chandrasekhar, 1950).

2
Solutions of the Boltzmann Equation

It is not uncommon to find that the equations that govern a physical phenomenon are much easier to derive than they are to solve. This has certainly proven to be true of the nonlinear Boltzmann equation, for which the only generally valid exact solution is the equilibrium solution. Although the development of fast and efficient numerical methods for the solution of the nonlinear Boltzmann equation, starting from a given set of initial conditions and proceeding via step-by-step time integration is currently an area of active research, it has not yet attained a stage at which general physical insight can be obtained. It may, however, be drawn from solutions to the Boltzmann equation obtained by considering a gas that is never far removed from an equilibrium state, so that solutions of the linearized version of the Boltzmann collision integral described in Section 1.5 can be obtained and studied. The present chapter thus examines procedures for solving the linearized Boltzmann equation, thereby relying upon well-developed techniques for solving linear equations, as relatively little is yet known about obtaining general solutions to equations that involve nonlinear integral operators.

2.1 Chapman–Enskog Solution for Pure Monatomic Gases

A successful approximation procedure was proposed initially by Chapman (1916) and slightly later, but independently, by Enskog (1917). Their approximation procedure begins with the linearized Boltzmann eqn (1.22), to which well-established mathematical solution procedures for linearized integro-differential equations may be applied. In particular, both Chapman and Enskog assumed that the desired solution can be approximated as

$$f(\mathbf{c}, \mathbf{r}, t) = f^{(0)}[1 + \phi]\,, \tag{2.1}$$

in which $f^{(0)}$ is a local Maxwellian corresponding to given fields of pressure $p(\mathbf{r}, t)$, temperature $T(\mathbf{r}, t)$, and flow velocity $\mathbf{v}(\mathbf{r}, t)$, subject to specified constraints. They also assumed that deviations from the local Maxwellian are small, in the sense that $|\phi|$ is much smaller than unity within the range of thermal energies, thereby allowing ϕ^2 to be neglected in first approximation. Finally, they assumed that both the Maxwellian and the unknown perturbation ϕ varied slowly in space and in time, in the sense that variations over distances of the order of a mean free path and over times of the order of mean free flight are small. This final assumption was utilized to justify the neglect of derivatives of ϕ (in addition to products of ϕ with derivatives of $f^{(0)}$) in first approximation.

Transport Properties and Potential Energy Models for Monatomic Gases. Hui Li and Frederick R.W. McCourt, Oxford University Press.
 DOI: 10.1093/oso/9780198888253.003.0002

Substitution of expression (2.1) into the nonlinear Boltzmann equation

$$\frac{\mathrm{D}}{\mathrm{D}t} f(\mathbf{c}, \mathbf{r}, t) = \mathcal{C}(f f_1)\,, \tag{2.2}$$

followed by division by $f^{(0)}$ and the neglect of all quantities of second order gives rise to the linearized Boltzmann eqn (1.82) for ϕ, that is,

$$\mathcal{R}\phi = -\frac{1}{f^{(0)}} \frac{\mathrm{D}f^{(0)}}{\mathrm{D}t}\,. \tag{2.3}$$

The right-hand side of eqn (2.3) may be converted, upon employing the logarithm of $f^{(0)}$, given by

$$\ln f^{(0)} = \ln p - \tfrac{5}{2} \ln T - \frac{mC^2}{2k_{\mathrm{B}}T} + \text{const}\,, \tag{2.4}$$

into

$$\frac{1}{f^{(0)}} \frac{\mathrm{D}f^{(0)}}{\mathrm{D}t} = \frac{1}{p} \frac{\mathrm{D}p}{\mathrm{D}t} + \frac{1}{T} \left(\frac{mC^2}{2k_{\mathrm{B}}T} - \frac{5}{2} \right) \frac{\mathrm{D}T}{\mathrm{D}t} + \frac{m}{k_{\mathrm{B}}T} \mathbf{C} \cdot \frac{\mathrm{D}\mathbf{v}}{\mathrm{D}t}\,. \tag{2.5}$$

This expression is often referred to as the streaming term, and serves as the known inhomogeneous term in eqn (2.3). Because the linearized collision operator $\mathcal{R}$ may also be expressed as a combination of a multiplicative and an integral operator, the end result is thus an inhomogeneous linear integral equation.

Reliable methods, such as matrix inversion following a suitable expansion of ϕ, a variational (Hirschfelder *et al.*, 1964) (Rayleigh–Ritz) method, or equivalently, the Galerkin technique (Kantorovich and Krylov, 1958) (also used by Chapman and Cowling 1970), for example, are available for solving inhomogeneous linear integral equations. The general formulation of transport equations and transport coefficients depends neither upon the choice of a particular solution procedure nor upon its method of execution, but it does depend upon whether or not solutions can be found. It will thus be assumed for present purposes that solutions to the linearized Boltzmann equation can be obtained, and that the result may be represented symbolically in terms of the inverse, $\mathcal{R}^{-1}$, of the collision operator $\mathcal{R}$.

As no operator that possesses a zero eigenvalue can be inverted, the existence of nontrivial solutions to the homogeneous integral equation has the consequence that the nonhomogeneous integral equation is solvable only under special conditions, and never uniquely. Indeed, as 0 is a fivefold degenerate eigenvalue of $\mathcal{R}$, corresponding to the hydrodynamic eigenfunctions 1, $m\mathbf{C}$ and $\frac{1}{2}mC^2$, inversion of $\mathcal{R}$ is only possible in what is referred to as the nonhydrodynamic subspace of the full Hilbert space: this condition requires that $\mathcal{R}$ be restricted to operate only upon functions that are orthogonal to the hydrodynamic eigenfunctions. Moreover, the solutions themselves must also be orthogonal to these functions. Of course, these two requirements do not differ from the traditional constraints and subsidiary conditions that arise in the traditional Chapman–Enskog procedure or, even more generally, in the theory of integral equations. Specifically, these subsidiary conditions require ϕ to be chosen so that it leaves the particle density, flow velocity, and average kinetic energy all unaffected.

In terms of eqn (2.3), the first set of conditions requires the right-hand side of eqn (2.5) to be orthogonal to 1 and to the (center-of-mass) momentum $m\mathbf{C}$ and energy $\frac{1}{2}mC^2$. Equation (2.3) is thus solvable only for pressure fields $p(\mathbf{r},t)$, flow velocity fields $\mathbf{v}(\mathbf{r},t)$ and temperature fields $T(\mathbf{r},t)$ that satisfy these conditions. Moreover, these three fields cannot be constructed arbitrarily, but must be adjusted to meet the required orthogonality conditions, as otherwise the linearized Boltzmann eqn (2.3) does not have a valid solution.

The inner products with the streaming term, namely $\langle\cdots|1\rangle$, $\langle\cdots|m\mathbf{C}\rangle$ and $\langle\cdots|\frac{1}{2}mC^2\rangle$, with the ellipsis representing the streaming term, must now be evaluated. These evaluations are made easier upon replacing $\mathbf{c}$ by $\mathbf{C}+\mathbf{v}$ in $\mathbf{c}\cdot\nabla$ and by separating terms that are even in $\mathbf{C}$ from those that are odd in $\mathbf{C}$. Such a calculation is straightforward, and gives rise to the three conditions

$$\frac{1}{p}\frac{\mathrm{d}p}{\mathrm{d}t}-\frac{1}{T}\frac{\mathrm{d}T}{\mathrm{d}t}+\nabla\cdot\mathbf{v}=0\,, \tag{2.6}$$

$$\nabla p+nm\frac{\mathrm{d}\mathbf{v}}{\mathrm{d}t}=0\,, \tag{2.7}$$

and

$$\frac{1}{p}\frac{\mathrm{d}p}{\mathrm{d}t}+\tfrac{5}{3}\nabla\cdot\mathbf{v}=0\,, \tag{2.8}$$

involving the substantial derivative (1.44). It is worth remarking that for stationary subsonic flow, the second term in eqn (2.7) is small, so that $\nabla p\approx 0$.

Substitution of the equilibrium pressure, $p=nk_{\mathrm{B}}T$, into eqn (2.6) gives the equation of continuity,

$$\frac{\mathrm{d}n}{\mathrm{d}t}+n\nabla\cdot\mathbf{v}=0\,,$$

or equivalently,

$$\frac{\partial n}{\partial t}+\nabla\cdot(n\mathbf{v})=0\,.$$

Condition (2.7) coincides with the equation of motion for an inviscid fluid, otherwise known as the Euler equation, while the final condition, eqn (2.8), is simply a version of the equation for adiabatic change, namely, $p/n^{5/3}=$ const (recall that $c_p/c_V=\frac{5}{3}$ for a monatomic gas). These three equations, taken together, may be called the Euler system of equations. Thus, in the first Chapman–Enskog approximation, the Boltzmann equation is solvable only for macroscopic pressure, temperature, and velocity fields $p(\mathbf{r},t)$, $T(\mathbf{r},t)$, $\mathbf{v}(\mathbf{r},t)$ that satisfy the Euler equations.

Conditions (2.6–2.8) may be utilized to eliminate the substantial time derivatives of the field variables p, T, $\mathbf{v}$ from eqn (2.5). Upon recalling that

$$\frac{D}{Dt}\equiv\frac{\partial}{\partial t}+\mathbf{c}\cdot\nabla=\frac{\mathrm{d}}{\mathrm{d}t}+\mathbf{C}\cdot\nabla\,,$$

and upon taking advantage of a number of cancellations, the linearized Boltzmann equation assumes the form

$$\mathcal{R}\phi = -\left(\frac{mC^2}{2k_\mathrm{B}T} - \frac{5}{2}\right)\mathbf{C}\cdot\frac{\nabla T}{T} - \frac{m}{k_\mathrm{B}T}\mathbf{C}\cdot(\mathbf{C}\cdot\nabla)\mathbf{v} + \frac{mC^2}{3k_\mathrm{B}T}\nabla\cdot\mathbf{v}\,.$$

The final two terms in this expression may be combined to give $-(k_\mathrm{B}T)^{-1}[m\mathbf{CC} - \frac{1}{3}mC^2\boldsymbol{\delta}] : \nabla\mathbf{v}$. Because the factor $m\mathbf{CC} - \frac{1}{3}mC^2\boldsymbol{\delta}$ is symmetric and traceless, the asymmetric and trace parts of the second factor can be deleted. In terms of the usual notation for symmetric traceless tensors (see, for example, Appendix A.3.4), the final two terms in the linearized Boltzmann equation can be rewritten as $(k_\mathrm{B}T)^{-1}m\overline{\mathbf{CC}} : \overline{\nabla\mathbf{v}}$. In terms of the microscopic heat flux $\boldsymbol{\Psi}^E$ (a vector) and the microscopic stress $\boldsymbol{\Psi}^\pi$ (a second rank tensor) given by

$$\boldsymbol{\Psi}^E = \mathbf{C}\left(\tfrac{1}{2}mC^2 - \tfrac{5}{2}k_\mathrm{B}T\right)\,, \qquad \boldsymbol{\Psi}^\pi = m\overline{\mathbf{CC}}\,, \tag{2.9}$$

eqn (2.3) can be rewritten as

$$\mathcal{R}\phi = -\frac{1}{k_\mathrm{B}T}\left(\boldsymbol{\Psi}^E\cdot\frac{\nabla T}{T} + \boldsymbol{\Psi}^\pi : \overline{\nabla\mathbf{v}}\right) \equiv \frac{1}{k_\mathrm{B}T}\sum_\alpha \boldsymbol{\Psi}^\alpha \odot \mathsf{X}^\alpha\,. \tag{2.10}$$

Even though the solution ϕ may contain the spatial coordinate $\mathbf{r}$ and the time t as parameters via the gradients ∇T and $\overline{\nabla\mathbf{v}}$, the only true variable appearing in eqn (2.10) is in fact $\mathbf{C}$. For later convenience, a general symbol X^α has been introduced into eqn (2.10) for the thermodynamic forces, namely, $\mathsf{X}^E \equiv \mathbf{X}^E = -\nabla T/T$ and $\mathsf{X}^\pi = -\overline{\nabla\mathbf{v}}$. A general tensor notation, in which the symbol $\odot$ denotes a full scalar product (equivalently, a full tensor contraction), has also been employed in eqn (2.10) as one of the thermodynamic forces, $\mathbf{X}^E$, is a vector while the other, X^π, is a second-rank tensor.

With the knowledge that the right-hand side of eqn (2.10) lies in the nonhydrodynamic subspace, the collision operator can be inverted to obtain a formal solution of eqn (2.3) as

$$\phi = \frac{1}{k_\mathrm{B}T}\mathcal{R}^{-1}\sum_\alpha \boldsymbol{\Psi}^\alpha \odot \mathsf{X}^\alpha\,. \tag{2.11}$$

This formal solution enables the derivation of important relations for the heat flux $\mathbf{q}$ and the traceless part $\boldsymbol{\Pi}$ of the pressure tensor, given by the nonequilibrium averages of the microscopic fluxes (2.9) multiplied by the number density of the gas, that is,

$$\mathbf{q} = n\langle\boldsymbol{\Psi}^E\rangle_\mathrm{ne} = n\langle\boldsymbol{\Psi}^E|\phi\rangle\,, \qquad \boldsymbol{\Pi} = n\langle\boldsymbol{\Psi}^\pi\rangle_\mathrm{ne} = n\langle\boldsymbol{\Psi}^\pi|\phi\rangle\,. \tag{2.12a}$$

It is again convenient to introduce a general notation for the thermodynamic fluxes $\mathbf{q} = \mathbf{J}^E \equiv \mathsf{J}^E$ of energy and $\boldsymbol{\Pi} = \mathsf{J}^\pi$ of momentum, so that eqns (2.12a) may be restated as

$$\mathsf{J}^\alpha = n\langle\boldsymbol{\Psi}^\alpha|\phi\rangle\,, \qquad \alpha = E, \pi\,. \tag{2.12b}$$

Note that had $\boldsymbol{\Psi}^\alpha$ been a complex quantity, the equivalent expression for the generalized flux J^α would have required the substitution of $\langle\boldsymbol{\Psi}^{\alpha*}|\phi\rangle$. However, if all quantities remain real, it will be unnecessary to bother with complex conjugation.

Substitution of the formal solution (2.11) into the definition (2.12b) gives the phenomenological equations

$$\mathsf{J}^\alpha = \sum_\beta \mathsf{L}^{\alpha\beta} \odot \mathsf{X}^\beta \tag{2.13}$$

connecting thermodynamic fluxes and forces. As these results have been obtained from the linearized Boltzmann equation, they represent linear approximations to the more general (nonlinear) relations. The transport coefficients, given by

$$\mathsf{L}^{\alpha\beta} = \frac{n}{k_\mathrm{B}T}\langle \mathbf{\Psi}^\alpha | \mathcal{R}^{-1} \mathbf{\Psi}^\beta \rangle , \tag{2.14}$$

are, in general, tensorial in nature. However, due to rotational invariance of the collision operator (see Section 1.5), they are isotropic, and hence, equivalent to scalars. In fact, only two such coefficients exist for a pure gas, specifically,

$$\mathsf{L}^{EE} = \frac{n}{k_\mathrm{B}T}\langle \mathbf{\Psi}^E | \mathcal{R}^{-1} \mathbf{\Psi}^E \rangle = L^{EE} \boldsymbol{\delta} \tag{2.15a}$$

and

$$\mathsf{L}^{\pi\pi} = \frac{n}{k_\mathrm{B}T}\langle \mathbf{\Psi}^\pi | \mathcal{R}^{-1} \mathbf{\Psi}^\pi \rangle = L^{\pi\pi} \mathbf{\Delta} . \tag{2.15b}$$

The projection tensor $\mathbf{\Delta} \equiv \mathbf{\Delta}^{(2)}$ has already been encountered in Section 1.5; see also Appendix A.2. No cross-effects arise because $\mathsf{L}^{E\pi} = \mathsf{L}^{\pi E}$ vanishes by symmetry: equivalently, it may be said that the microscopic fluxes $\mathbf{\Psi}^E$ and $\mathbf{\Psi}^\pi$ cannot couple.

Taking traces on both sides of eqns (2.15) or, equivalently, substituting reduced matrix elements leads to the final expressions

$$L^{EE} = \frac{n}{k_\mathrm{B}T}\langle \mathbf{\Psi}^E \| \mathcal{R}^{-1} \| \mathbf{\Psi}^E \rangle = \frac{1}{3}\frac{n}{k_\mathrm{B}T}\langle \mathbf{\Psi}^E \cdot | \mathcal{R}^{-1} \mathbf{\Psi}^E \rangle = \lambda T \tag{2.16a}$$

and

$$L^{\pi\pi} = \frac{n}{k_\mathrm{B}T}\langle \mathbf{\Psi}^\pi \| \mathcal{R}^{-1} \| \mathbf{\Psi}^\pi \rangle = \frac{1}{5}\frac{n}{k_\mathrm{B}T}\langle \mathbf{\Psi}^\pi : | \mathcal{R}^{-1} \mathbf{\Psi}^\pi \rangle = 2\eta , \tag{2.16b}$$

for the nonzero diagonal transport coefficients. Identification of the phenomenological coefficients L^{EE} and $L^{\pi\pi}$ with the conventional thermal conductivity and (shear) viscosity coefficients λ and η requires a matching of the definitions of the individual fluxes and forces with the standard forms for the phenomenological equations, namely,

$$\mathbf{q} = -\lambda \nabla T , \qquad \mathbf{\Pi} = -2\eta \overline{\nabla \mathbf{v}} . \tag{2.17}$$

Note that as $\mathcal{R}$ is proportional to n, both λ and η from eqns (2.16) are independent of the gas number density.

The connection with the Onsager theory of irreversible thermodynamics is established by examining expression (1.98) for the entropy production rate, and substituting the right-hand side of the Boltzmann eqn (2.10) for $\mathcal{R}\phi$. Upon recalling definition

(2.12b) of the fluxes J^α, a bilinear sum of products of conjugate pairs of thermodynamic fluxes and forces is obtained as (de Groot and Mazur, 1962)

$$T\varsigma = \sum_\alpha \mathsf{J}^\alpha \odot \mathsf{X}^\alpha = -\mathbf{q} \cdot \frac{\nabla T}{T} - \mathbf{\Pi} : \overline{\nabla \mathbf{v}} . \tag{2.18}$$

The formulation (2.13) of the phenomenological equations is quite natural in this context.[1]

The heat flux and viscous pressure tensor obtained in eqns (2.17) lead to correction terms for the hydrodynamic equations. A substitution of $\nabla \cdot \mathsf{P} = \nabla p + \nabla \cdot \mathbf{\Pi}$ into eqns (1.49) and (1.46) into (1.56), gives the Navier–Stokes system, namely the Navier–Stokes equation proper,

$$nm \frac{\mathrm{d}\mathbf{v}}{\mathrm{d}t} = -\nabla p + \eta \left[\nabla^2 \mathbf{v} + \tfrac{1}{3} \nabla (\nabla \cdot \mathbf{v}) \right] , \tag{2.19}$$

plus the heat conduction equation

$$\tfrac{3}{2} n k_\mathrm{B} \frac{\mathrm{d}T}{\mathrm{d}t} = -p \nabla \cdot \mathbf{v} - \mathbf{\Pi} : \overline{\nabla \mathbf{v}} + \nabla \cdot (\lambda \nabla T) . \tag{2.20}$$

The specific internal energy has been expressed as $u = 3k_\mathrm{B}T/(2m)$ while the equation of continuity remains unchanged.

It may seem paradoxical that the initial system of field eqns (the Euler equations) gave rise to a solution that leads naturally to a modified (in this case, the Navier–Stokes) system of field equations. Such situations are, however, not uncommon when mathematical techniques that involve successive approximations are utilized. Moreover, consider that the corrections are small whenever the fields are such that the solution procedure is justified.

As the original Boltzmann equation is a nonlinear equation, while that for ϕ is a linearized equation, the question of further approximations cannot be avoided. The second Chapman–Enskog approximation is obtained upon making the Ansatz

$$f = f^{(0)} (1 + \phi^{[1]} + \phi^{[2]}) , \tag{2.21}$$

in which $\phi^{[1]}$ is the previously obtained ϕ. Suppose that the calculation is repeated for different $\mathbf{r}$ and t, so that $\phi^{[1]}$ is available as a function of these variables. Upon accounting for the terms deleted in the solution of $\phi^{[1]}$ and retaining only the first power of the second-order term $\phi^{[2]}$, but none of its derivatives, nor its products with $\phi^{[1]}$ or with the derivatives of $f^{(0)}$, it is possible to see, without explicitly writing it all down, that the outcome is an inhomogeneous equation for $\phi^{[2]}$, in which $\phi^{[1]}$ has become part of the known term. The constraints (i.e. the conditions of solvability) in this case are now identical to the Navier–Stokes system of equations. Utilization of operator inversion again in order to obtain a more precise solution then leads to the so-called Burnett corrections to the hydrodynamic equations.

[1]Because of the factor T on the left of eqn (2.18), the present definitions of the thermodynamic forces differ by the same factor from the traditional ones found, for example, in de Groot and Mazur (1962).

It would naturally be imprudent to press on blindly, hoping to obtain ever higher Chapman–Enskog approximations, as neither is convergence assured nor is it known if each successive step actually improves the accuracy of the solution. Indeed, such a procedure is legitimate only for slowly varying fields, in which case, the first approximation, with the Navier–Stokes equations derived from it, is sufficient for most practical purposes. Only in exceptional cases does it make sense to proceed to the next (i.e. Burnett) approximation.

The aim of Chapman and Enskog was almost exclusively the development of a theory of transport phenomena in gases and the derivation of expressions for and values of the transport coefficients themselves. For such a purpose the choice of the fields p, $\mathbf{v}$, T is irrelevant, and does not even need to be stated explicitly. A radically different situation occurs, however, should a solution of the Boltzmann equation be sought for a given set of initial and boundary conditions, in which case the fields are then not known in advance, thereby making it much more involved to find a solution.

2.2 Chapman–Enskog Solution for Binary Mixtures

The Chapman–Enskog procedure can be carried out for mixtures in much the same way as was done for the pure gas. Little of interest is lost by restricting the present analysis to binary mixtures (k = A,B). Upon expanding the mixture distribution functions f_{k} around local Maxwellians $f_{\mathrm{k}}^{(0)}$ (see Section 1.6) as

$$f_{\mathrm{k}} = f_{\mathrm{k}}^{(0)}(1+\phi_{\mathrm{k}})\,, \tag{2.22}$$

with the partial pressures $p_{\mathrm{k}} = p_{\mathrm{k}}(\mathbf{r},t)$, in addition to $T = T(\mathbf{r},t)$ and $\mathbf{v} = \mathbf{v}(\mathbf{r},t)$ regarded as known, linearization gives eqn (1.140). Thus, the linearized Boltzmann equation for a binary mixture can be put into the form

$$\mathcal{R}\phi = -\,\frac{\mathrm{D}\ln f^{(0)}}{\mathrm{D}t}\,, \tag{2.23}$$

in which $\mathcal{R}$ is a 2×2 matrix of operators and ϕ and $f^{(0)}$ are binary arrays of functions. The right-hand side of eqn (2.23) has components

$$\frac{\mathrm{D}\ln f_{\mathrm{k}}^{(0)}}{\mathrm{D}t} = \frac{\mathrm{D}\ln p_{\mathrm{k}}}{\mathrm{D}t} + \left(\frac{m_{\mathrm{k}}C^2}{2k_{\mathrm{B}}T} - \frac{5}{2}\right)\frac{\mathrm{D}\ln T}{\mathrm{D}t} + \frac{m_{\mathrm{k}}\mathbf{C}}{k_{\mathrm{B}}T}\cdot\frac{\mathrm{D}\mathbf{v}}{\mathrm{D}t}\,, \tag{2.24}$$

k = A,B, analogous to the pure gas eqn (2.5).

Inversion of the matrix $\mathcal{R}$ of operators will allow a formal expression to be obtained for the array $\phi = (\phi_{\mathrm{A}}, \phi_{\mathrm{B}})$ representing the formal solution of eqn (2.23). However, this may only be achieved for an inhomogeneous term that is orthogonal to the hydrodynamic subspace, now six-dimensional, since it is spanned by the arrays (1,0), (0,1), $(m_{\mathrm{A}}\mathbf{C},\, m_{\mathrm{B}}\mathbf{C})$, and $(\frac{1}{2}m_{\mathrm{A}}C^2,\, \frac{1}{2}m_{\mathrm{B}}C^2)$. Proceeding along the same lines followed for a pure gas gives the Euler eqns (2.7), (2.8) as two constraints. Similarly, two continuity equations analogous to eqn (2.6) for a pure gas, given by

$$\frac{\partial n_{\mathrm{k}}}{\partial t} + \nabla\cdot(n_{\mathrm{k}}\mathbf{v}) = 0\,, \qquad \mathrm{k} = \mathrm{A},\,\mathrm{B}\,, \tag{2.25}$$

are obtained as additional constraints. Moreover, the mixture mass density ρ is given by

$$\rho = n_{\mathrm{A}} m_{\mathrm{A}} + n_{\mathrm{B}} m_{\mathrm{B}} \equiv \rho_{\mathrm{A}} + \rho_{\mathrm{B}} ,$$

in terms of the mass densities ρ_{A} and ρ_{B} of the mixture constituents. As for a pure gas, the constraints may be employed to eliminate the substantial time derivatives from expression (2.24). By introducing microscopic fluxes via the arrays

$$\boldsymbol{\Psi}^E = (\tfrac{1}{2} m_{\mathrm{A}} C^2 - \tfrac{5}{2} k_{\mathrm{B}} T,\ \tfrac{1}{2} m_{\mathrm{B}} C^2 - \tfrac{5}{2} k_{\mathrm{B}} T)\mathbf{C} , \tag{2.26a}$$

$$\boldsymbol{\Psi}^\pi = (m_{\mathrm{A}},\ m_{\mathrm{B}})\overline{\mathbf{CC}} , \tag{2.26b}$$

$$\widehat{\boldsymbol{\Psi}}^d = \frac{m_{\mathrm{A}} m_{\mathrm{B}}}{\rho}(n_{\mathrm{B}},\ -n_{\mathrm{A}})\mathbf{C} , \tag{2.26c}$$

the linearized Boltzmann eqn (2.23) may be rewritten as

$$\begin{aligned} \mathcal{R}\phi &= -\frac{1}{k_{\mathrm{B}} T}\left[\boldsymbol{\Psi}^E \cdot \frac{\nabla T}{T} + \boldsymbol{\Psi}^\pi : \overline{\nabla \mathbf{v}} + \widehat{\boldsymbol{\Psi}}^d \cdot \left(\frac{\nabla p_{\mathrm{A}}}{\rho_{\mathrm{A}}} - \frac{\nabla p_{\mathrm{B}}}{\rho_{\mathrm{B}}}\right)\right] \\ &\equiv \frac{1}{k_{\mathrm{B}} T} \sum_\alpha \boldsymbol{\Psi}^\alpha \odot \mathsf{X}^\alpha . \end{aligned} \tag{2.27}$$

Equation (2.27) has the formal solution

$$\phi = \frac{1}{k_{\mathrm{B}} T} \mathcal{R}^{-1} \sum_\alpha \boldsymbol{\Psi}^\alpha \odot \mathsf{X}^\alpha . \tag{2.28}$$

With a slight extension of eqns (2.12a) macroscopic fluxes may be introduced individually via

$$\mathbf{q} \equiv n\langle \boldsymbol{\Psi}^E | \phi \rangle , \quad \boldsymbol{\Pi} \equiv n\langle \boldsymbol{\Psi}^\pi | \phi \rangle , \quad \widehat{\mathbf{J}}^d \equiv n\langle \widehat{\boldsymbol{\Psi}}^d | \phi \rangle , \tag{2.29a}$$

or collectively via

$$\mathsf{J}^\alpha = n\langle \boldsymbol{\Psi}^\alpha | \phi \rangle . \tag{2.29b}$$

These macroscopic fluxes have been encountered previously, including the diffusion flux $\widehat{\mathbf{J}}^d$ given in eqn (1.118c). Notice, however, that $\mathbf{q}$ is the reduced heat flux from Section 1.6.

Substitution of the formal solution (2.28) into expressions (2.29) leads to the phenomenological equations relating the fluxes $\mathbf{J}^E = \mathbf{q}$, $\widehat{\mathsf{J}}^d \equiv \widehat{\mathbf{J}}^d$, and $\mathsf{J}^\pi = \boldsymbol{\Pi}$ to the driving gradients $\mathbf{X}^E = -\nabla T/T$, $\widehat{\mathbf{X}}^d = -(\nabla p_{\mathrm{A}}/\rho_{\mathrm{A}} - \nabla p_{\mathrm{B}}/\rho_{\mathrm{B}})$, and $\mathsf{X}^\pi = -\overline{\nabla \mathbf{v}}$. In symbolic form, these equations resemble those obtained for a pure gas, that is,

$$\mathsf{J}^\alpha = \sum_\beta \mathsf{L}^{\alpha\beta} \odot \mathsf{X}^\beta = \sum_\beta L^{\alpha\beta} \mathsf{X}^\beta . \tag{2.30}$$

The somewhat lengthier form for $\widehat{\mathsf{X}}^d$ should not pose a problem, as it is directly related to the gradient of the chemical potential, which should be regarded as the true driving force for diffusion (de Groot and Mazur, 1962).

Written out in detail using the traditional notation for the transport coefficients, the equations become more cumbersome, namely,

$$\mathbf{q} = -\lambda T \frac{\nabla T}{T} - \frac{n^2 m_\mathrm{A} m_\mathrm{B}}{\rho} \widetilde{D}_\mathrm{T} \left(\frac{\nabla p_\mathrm{A}}{\rho_\mathrm{A}} - \frac{\nabla p_\mathrm{B}}{\rho_\mathrm{B}} \right) , \tag{2.31a}$$

$$\widehat{\mathbf{J}}^d = -\frac{n^2 m_\mathrm{A} m_\mathrm{B}}{\rho} D_\mathrm{T} \frac{\nabla T}{T} - \frac{n m_\mathrm{A} m_\mathrm{B} \rho_\mathrm{A} \rho_\mathrm{B}}{\rho^2 k_\mathrm{B} T} D \left(\frac{\nabla p_\mathrm{A}}{\rho_\mathrm{A}} - \frac{\nabla p_\mathrm{B}}{\rho_\mathrm{B}} \right) , \tag{2.31b}$$

and

$$\mathbf{\Pi} = -2\eta \overline{\nabla \mathbf{v}} . \tag{2.31c}$$

The first two equations contain the coefficients for diffusion (D), thermal diffusion (D_T), and the Dufour coefficient ($\widetilde{D}_\mathrm{T}$).

Should the pressure of the mixture be uniform, the formalism may be simplified considerably by expressing $\widehat{\mathbf{X}}^d$ in terms of $\mathbf{X}^d = -\nabla p_\mathrm{A} = \nabla p_\mathrm{B}$, and $\widehat{\mathbf{J}}^d$ by

$$\mathbf{J}^d = \langle n \mathbf{\Psi}^d | \phi \rangle = \mathbf{v}_\mathrm{A} - \mathbf{v}_\mathrm{B} = \frac{\rho}{\rho_\mathrm{A} \rho_\mathrm{B}} \widehat{\mathbf{J}}^d \tag{2.32}$$

(see Section 1.6), derived from the modified microscopic diffusive flux

$$\mathbf{\Psi}^d = (n_\mathrm{A}^{-1}, \, -n_\mathrm{B}^{-1}) \mathbf{C} = \frac{\rho}{\rho_\mathrm{A} \rho_\mathrm{B}} \widehat{\mathbf{\Psi}}^d . \tag{2.33}$$

Indeed, for this case, $\widehat{\mathbf{X}}^d = -[\rho/(\rho_\mathrm{A} \rho_\mathrm{B})] \nabla p_\mathrm{A}$. By making this substitution, then multiplying eqn (2.31b) by $\rho/(\rho_\mathrm{A} \rho_\mathrm{B})$, eqns (2.31a, 2.31b) can be reformulated as

$$\mathbf{q} = -\lambda T \frac{\nabla T}{T} - \frac{1}{x_\mathrm{A} x_\mathrm{B}} \widetilde{D}_\mathrm{T} \nabla p_\mathrm{A} , \tag{2.34}$$

and

$$\mathbf{J}^d \equiv \mathbf{v}_\mathrm{A} - \mathbf{v}_\mathrm{B} = -\frac{1}{x_\mathrm{A} x_\mathrm{B}} \left[D_\mathrm{T} \frac{\nabla T}{T} + \frac{D}{p} \nabla p_\mathrm{A} \right] , \tag{2.35}$$

with $x_\mathrm{i} = n_\mathrm{i}/n = p_\mathrm{i}/p$ for i = A,B. It is also worth noting that the identities

$$\widehat{\mathbf{\Psi}}^d \cdot \widehat{\mathbf{X}}^d = \mathbf{\Psi}^d \cdot \mathbf{X}^d , \qquad \widehat{\mathbf{J}}^d \cdot \widehat{\mathbf{X}}^d = \mathbf{J}^d \cdot \mathbf{X}^d , \tag{2.36}$$

are obtained.

Since the first Chapman–Enskog approximation presupposes the validity of condition (2.7), which for strongly subsonic stationary flows implies that $\nabla p = 0$, there is good reason to adopt the simplified formalism and to abandon $\widehat{\mathbf{\Psi}}^d$, $\widehat{\mathbf{J}}^d$, and $\widehat{\mathbf{X}}^d$ henceforth. The general transport coefficients $\mathsf{L}^{\alpha\beta}$ and eqns (2.29b, 2.30) shall from now on be understood as complying with this simplification. The new scalar coefficients are thus

$$L^{EE} = \lambda T = \frac{n}{3k_\mathrm{B}T}\langle \mathbf{\Psi}^E \cdot |\mathcal{R}^{-1}\mathbf{\Psi}^E\rangle \,, \tag{2.37a}$$

$$L^{Ed} = \frac{1}{x_\mathrm{A}x_\mathrm{B}}\widetilde{D}_\mathrm{T} = \frac{n}{3k_\mathrm{B}T}\langle \mathbf{\Psi}^E \cdot |\mathcal{R}^{-1}\mathbf{\Psi}^d\rangle \,, \tag{2.37b}$$

$$L^{dE} = \frac{1}{x_\mathrm{A}x_\mathrm{B}}D_\mathrm{T} = \frac{n}{3k_\mathrm{B}T}\langle \mathbf{\Psi}^d \cdot |\mathcal{R}^{-1}\mathbf{\Psi}^E\rangle \,, \tag{2.37c}$$

$$L^{dd} = \frac{1}{x_\mathrm{A}x_\mathrm{B}p}D = \frac{n}{3k_\mathrm{B}T}\langle \mathbf{\Psi}^d \cdot |\mathcal{R}^{-1}\mathbf{\Psi}^d\rangle \,, \tag{2.37d}$$

and

$$L^{\pi\pi} = 2\eta = \frac{n}{5k_\mathrm{B}T}\langle \mathbf{\Psi}^\pi : |\mathcal{R}^{-1}\mathbf{\Psi}^\pi\rangle \,. \tag{2.37e}$$

All coefficients are scalars due to the isotropy of the undisturbed mixture.

Strictly speaking, the thermal conductivity coefficient λ defined by eqn (2.37a) represents the thermal conductivity that would be determined in a hypothetical experiment in which no unmixing occurs from the action of thermal diffusion. Both thermal diffusion and an energy flux caused by diffusion forces occur immediately upon the application of a temperature gradient to a gas mixture. However, as thermal conduction is normally measured under steady-state conditions, diffusion is stopped by the walls of the measuring vessel, so that $\mathbf{J}^d$ vanishes. The vanishing of $\mathbf{J}^d$ has the consequence that ∇p_A in eqn (2.35) can be replaced in terms of ∇T, thereby allowing the heat flux vector $\mathbf{q}$ to be expressed as

$$\mathbf{q} = -\lambda_\mathrm{s}\nabla T \,, \tag{2.38a}$$

in which λ_s is now the steady-state thermal conductivity that is determined experimentally. It is related to λ and the other mixture phenomenological coefficients via

$$\lambda_\mathrm{s} = \lambda - \frac{nk_\mathrm{B}}{x_\mathrm{A}x_\mathrm{B}}\frac{\widetilde{D}_\mathrm{T}D_\mathrm{T}}{D} \,. \tag{2.38b}$$

Because the steady-state mixture thermal conductivity coefficient λ_s typically differs from the true thermal conductivity coefficient λ by less than 1%, the distinction between λ_s and λ is often ignored.

As $\mathcal{R}$ is self-adjoint, expressions (2.37b, 2.37c) are equal, and $\widetilde{D}_\mathrm{T} = D_\mathrm{T}$. This is a typical case of Onsager symmetry, which relates the phenomenological coefficients for conjugate pairs of transport phenomena (de Groot and Mazur, 1962), which are, in the present case, the Dufour coefficient $\widetilde{D}_\mathrm{T}$ and the coefficient of thermal diffusion D_T. The first coefficient is associated with the heat flux driven by a gradient of partial pressure, the second with a diffusive flux driven by a temperature gradient.

To complete the analysis and to verify that J^α and X^α are indeed properly conjugate to one another, it will also be useful to obtain an expression for the entropy source density ς for a near-equilibrium mixture. Following substitution of the right-hand side of the linearized Boltzmann eqn (2.27) into expression (1.148b), to obtain

$$T\varsigma = -n \sum_\alpha \langle \mathbf{\Psi}^\alpha | \phi \rangle \odot \mathsf{X}^\alpha = \sum_\alpha \mathsf{J}^\alpha \odot \mathsf{X}^\alpha ,$$

eqns (2.13, 2.15) may be utilized to obtain the familiar result

$$T\varsigma = \sum_\alpha \mathsf{J}^\alpha \odot \mathsf{X}^\alpha = \sum_{\alpha\beta} L^{\alpha\beta} \mathsf{X}^\alpha \odot \mathsf{X}^\beta \tag{2.39}$$

for the entropy production. This sum specifically includes the term $\widehat{\mathbf{J}}^d \cdot \widehat{\mathbf{X}}^d$; however, when the pressure, p, of the mixture is constant, it becomes irrelevant whether it is substituted by the product $\mathbf{J}^d \cdot \mathbf{X}^d = -(\mathbf{v}_\mathrm{A} - \mathbf{v}_\mathrm{B}) \cdot \nabla p_\mathrm{A}$ or not, as the two products are then equal.

2.3 Matrix Approximations for the Inverse Collision Operator

Once a general formalism giving expressions for the transport coefficients has been established, it only remains to find a means for their actual evaluation. The traditional method that is often employed for this purpose may be found in the standard texts by Chapman and Cowling (1970), Hirschfelder, Curtiss, and Bird (1964), and Ferziger and Kaper (1972).

The basic idea is to approximate the solution ϕ of the linearized Boltzmann eqn (2.10), namely,

$$\mathcal{R}\phi = \frac{1}{k_\mathrm{B}T} \sum_\alpha \mathbf{\Psi}^\alpha \odot \mathsf{X}^\alpha , \tag{2.40}$$

by a truncated expansion in terms of a specified set of basis functions contained in the nonhydrodynamic subspace $\mathcal{H}_\mathrm{nh}$. In the traditional version of this method, the coefficients are optimized using the Galerkin principle (Kantorovich and Krylov, 1958) by demanding that the difference between the approximated $\mathcal{R}\phi$ and the right-hand side of eqn (2.40) be orthogonal to each of the basis functions employed. Because the collision operator for a monatomic gas is self-adjoint, the Galerkin principle is equivalent to the Rayleigh–Ritz variational method. The same result is obtained if both the unknown ϕ and the right-hand side of the equation are expanded, and the resulting system of algebraic equations for the coefficients is solved via matrix inversion. By following this modified approach, it will be possible to make full use of the operator formalism that has been developed here. Thus, rather than attempting an inversion of the true collision operator, the matrix by which the operator is approximated is inverted in the spirit of the method that is often utilized in quantum mechanics.

2.3.1 Pure gases

It is useful to group the basis functions into traceless symmetric tensors built up from tensor (direct) products of the vector $\mathbf{W}$ [the reduced peculiar velocity from eqn (1.78)] with itself. The choice of basis functions

$$\mathbf{\Phi}^{ps}(\mathbf{W}) = (-1)^s \left[\frac{2^{p+s} s!(2p+1)!!}{p!(2p+2s+1)!!}\right]^{\frac{1}{2}} L_s^{(p+\frac{1}{2})}(W^2)\overline{\mathbf{W}^p}\,, \tag{2.41}$$

in which the $L_s^{(p+\frac{1}{2})}(W^2)$ are associated Laguerre polynomials (Szegö, 1959), otherwise referred to as Sonine polynomials $S_{p+\frac{1}{2}}^{(s)}(W^2)$ in much of the traditional literature (Hirschfelder *et al.*, 1964; Chapman and Cowling, 1970), has been widely accepted. The double factorial $(2n+1)!!$ employed in eqn (2.41) is defined as

$$(2n+1)!! \equiv 1 \cdot 3 \cdot 5 \cdots (2n+1) = \frac{(2n+1)!}{2^n n!}\,.$$

Moreover, as $\overline{\mathbf{W}^p}$ represents the completely traceless p-fold tensor product of $\mathbf{W}$, it is an irreducible tensor of rank p and highest weight, with spherical components given by

$$(\overline{\mathbf{W}^p})_\mu = \left[\frac{4\pi p!}{(2p+1)!!}\right]^{\frac{1}{2}} W^p Y_{p\mu}(\widehat{\mathbf{W}})\,,$$

in which $\widehat{\mathbf{W}} = \mathbf{W}/W$ is a unit vector. The spherical harmonics $Y_{p\mu}(\widehat{\mathbf{W}})$ employed here are defined in accordance with the phase convention of Edmonds (1957) (see also Appendix A.4), and the factor $(-1)^s$ appears in order that the coefficient accompanying the highest power of W^2 be positive. The first few such basis functions, sufficient to carry out calculations up to the level of what will be referred to as second Chapman–Cowling approximations, are given explicitly in Table 2.1. The tensors (2.41) are normalized according to

$$\langle \mathbf{\Phi}^{ps} | \mathbf{\Phi}^{p'q'} \rangle = \frac{1}{n} \int \mathrm{d}\mathbf{W}\; \mathbf{\Phi}^{ps}(\mathbf{W}) f^{(0)} \mathbf{\Phi}^{p's'}(\mathbf{W}) = \delta_{pp'}\delta_{ss'}\mathbf{\Delta}^{(p)}\,. \tag{2.42a}$$

which can be stated in terms of spherical components (see Appendix A.4) as

$$\langle \Phi_\mu^{ps} | \Phi_{\mu'}^{p's'} \rangle = \delta_{pp'}\delta_{ss'}\delta_{\mu\mu'}\,. \tag{2.42b}$$

It often proves to be convenient to employ a collective label, such as α, to represent an index pair, such as $(p,\, s)$. The first three $\mathbf{\Phi}^\alpha$, namely $\Phi^{00} = 1$, $\mathbf{\Phi}^{10} = \sqrt{2}\,\mathbf{W}$ and $\Phi^{01} = (2/3)^{\frac{1}{2}}(W^2 - 3/2)$, whose elements correspond to the summational invariants, span the hydrodynamic subspace. The remaining $\mathbf{\Phi}^{ps}$ belong to the nonhydrodynamic subspace. A particular advantage of this basis set is that it contains the microscopic fluxes from Section 2.1 as members, namely

$$\mathbf{\Psi}^E = C^E \mathbf{\Phi}^{11}\,, \qquad \mathbf{\Psi}^\pi = C^\pi \mathbf{\Phi}^{20}\,, \tag{2.43a}$$

in which the C^α are defined as

$$C^E = k_\mathrm{B}T\left(\frac{5k_\mathrm{B}T}{2m}\right)^{\frac{1}{2}}\,, \qquad C^\pi = \sqrt{2}\,k_\mathrm{B}T \tag{2.43b}$$

Hence, the orthogonality relations

$$\langle \mathbf{\Phi}^\alpha | \mathbf{\Psi}^\beta \rangle = \delta_{\alpha\beta} C^\beta \mathbf{\Delta}^{(\beta)} \tag{2.44}$$

hold, with $\mathbf{\Delta}^{(\beta)} = \mathbf{\Delta}^{(p)}$; the latter expression identifies explicitly the value of the tensor rank p implied in the index β.

Table 2.1 The first nine basis functions for a pure monatomic gas*†

$p = 0$	$p = 1$
$\Phi^{00} = 1$	$\mathbf{\Phi}^{10} = \sqrt{2}\,\mathbf{W}$
$\Phi^{01} = \left(\frac{2}{3}\right)^{\frac{1}{2}} (W^2 - \frac{3}{2})$	$\mathbf{\Phi}^{11} = \left(\frac{4}{5}\right)^{\frac{1}{2}} (W^2 - \frac{5}{2})\mathbf{W}$
$\Phi^{02} = \left(\frac{2}{15}\right)^{\frac{1}{2}} (W^4 - 5W^2 + \frac{15}{4})$	$\mathbf{\Phi}^{12} = \left(\frac{4}{35}\right)^{\frac{1}{2}} (W^4 - 7W^2 + \frac{35}{4})\,\mathbf{W}$
$p = 2$	
$\mathbf{\Phi}^{20} = \sqrt{2}\,\overline{\mathbf{W}\mathbf{W}}$	
$\mathbf{\Phi}^{21} = \left(\frac{4}{7}\right)^{\frac{1}{2}} (W^2 - \frac{7}{2})\,\overline{\mathbf{W}\mathbf{W}}$	
$\mathbf{\Phi}^{22} = \left(\frac{4}{63}\right)^{\frac{1}{2}} (W^4 - 9W^2 + \frac{63}{4})\,\overline{\mathbf{W}\mathbf{W}}$	

* The scalars Φ^{00}, Φ^{01}, and the vector $\mathbf{\Phi}^{10}$ span the hydrodynamic subspace $\mathcal{H}_{\mathrm{h}}$.
† For mixtures, $\mathbf{\Phi}^{ps}$ is replaced by an array $\mathbf{\Phi}^{ps|\mathrm{k}}$ with a tensor of the same form as $\mathbf{\Phi}^{ps}$ in the kth position, and 0 elsewhere. Moreover, the substitution $\mathbf{W} \to \mathbf{W}_{\mathrm{k}}$ must also be carried out, cf. eqn (1.78).

With this choice of basis set, the transport coefficients (2.14) may be rewritten as

$$L^{\alpha\beta} = \frac{n}{k_{\mathrm{B}}T}\, C^{\alpha} \langle \mathbf{\Phi}^{\alpha} \| \mathcal{R}^{-1} \| \mathbf{\Phi}^{\beta} \rangle C^{\beta} \,. \tag{2.45}$$

Even though $L^{\alpha\beta} = 0$ for $\alpha \neq \beta$ at present, a general formalism shall be retained for the sake of subsequent extensions.

It can already be seen that the reduced matrix elements of the collision operator will play a central role in the approximation method to be applied. They have often been substituted by equivalent or related concepts, in particular, by "square bracket integrals" (Hirschfelder *et al.*, 1964) or by "omega integrals" (Chapman and Cowling, 1970). Another option, to be followed here, is to introduce so-called effective cross-sections, which are related to the differential collision cross-section of Section 1.1, to help visualize the processes involved. They are defined in terms of reduced matrix elements of $\mathcal{R}$ via

$$\langle \mathbf{\Phi}^{ps} \| \mathcal{R} \| \mathbf{\Phi}^{ps'} \rangle = n\bar{c}_{\mathrm{r}} \mathfrak{S}\binom{ps}{ps'} \,, \tag{2.46a}$$

in which $\bar{c}_{\mathrm{r}}$ is the average relative speed of a pair of atoms (see eqn (1.141) of Section 1.6.3) which, for a pure gas is $\bar{c}_{\mathrm{r}} = \sqrt{2}\,\bar{c} = [16k_{\mathrm{B}}T/(\pi m)]^{\frac{1}{2}}$. Because $\mathcal{R}$ is a real symmetric operator, it will be represented by a symmetric matrix, and hence,

$$\mathfrak{S}\binom{ps}{ps'} = \mathfrak{S}\binom{ps'}{ps} \,. \tag{2.46b}$$

It may also be useful to introduce a shorthand notation for the diagonal matrix elements, namely $\mathfrak{S}\binom{ps}{ps} \equiv \mathfrak{S}(ps)$, and the relation between $\mathfrak{S}\binom{ps}{ps'}$ and the standard square-bracket notation, which is

$$n\bar{c}_{\rm r}\mathfrak{S}\left({}^{ps}_{ps'}\right) = \left[{}^{ps}_{ps'}\right]. \tag{2.46c}$$

Because n has been extracted from the matrix elements of $\mathcal{R}$, effective cross-sections depend only upon temperature and molecular properties, as may be demonstrated by deriving a cross-section expression from eqn (1.83), specifically, the expression for the collisional coupling between the expansion tensors $\mathbf{\Phi}^{ps}$ and $\mathbf{\Phi}^{ps'}$, namely,

$$\begin{aligned}\mathfrak{S}\left({}^{ps}_{ps'}\right)\mathbf{\Delta}^{(p)} &= \frac{1}{\bar{c}_{\rm r}}\int \mathrm{d}\mathbf{c}\, n^{-1}f^{(0)}(\mathbf{c})\mathbf{\Phi}^{ps}(\mathbf{c})\iint \mathrm{d}\mathbf{c}_1\mathrm{d}\mathbf{e}'\, c_{\rm r}\sigma n^{-1}f^{(0)}(\mathbf{c}_1)\Delta\mathbf{\Phi}^{ps'} \\ &= \frac{1}{4\bar{c}_{\rm r}}\iiint \mathrm{d}\mathbf{c}\mathrm{d}\mathbf{c}_1\mathrm{d}\mathbf{e}'\, c_{\rm r}\sigma n^{-2}f^{(0)}(\mathbf{c})f^{(0)}(\mathbf{c}_1)\Delta\mathbf{\Phi}^{ps}\Delta\mathbf{\Phi}^{ps'}.\end{aligned} \tag{2.47}$$

The effective cross-section $\mathfrak{S}\left({}^{ps}_{ps'}\right)$ may be interpreted as describing the negative rate (per unit colliding flux) by which the $\mathbf{\Phi}^{ps}$-type deviation from equilibrium is produced from $\mathbf{\Phi}^{ps'}$. A diagonal cross-section $\mathfrak{S}(ps)$ then simply represents the rate of loss of $\mathbf{\Phi}^{ps}$ and may therefore be referred to as a decay or relaxation cross-section.

Should N basis-tensors be employed in the expansion of ϕ, then an $N \times N$ matrix $n\bar{c}_{\rm r}\boldsymbol{\mathfrak{S}}(ps)$ of reduced matrix elements, given by eqn (2.46a), is obtained. This matrix of cross-sections thus approximates the operator $\mathcal{R}$, so that the transport coefficient $L^{\alpha\beta}$ is then given (approximately) by an appropriate matrix element of the inverse matrix $(n\bar{c}_{\rm r}\boldsymbol{\mathfrak{S}})^{-1}$ as

$$L^{\alpha\beta} = \frac{1}{k_{\rm B}T\,\bar{c}_{\rm r}}C^{\alpha}\mathfrak{S}^{-1}\left({}^{\alpha}_{\beta}\right)C^{\beta}. \tag{2.48}$$

Note that only one inverse matrix element is required for a given transport coefficient, with the consequence that even the simplest approximations of this kind give accuracies typically exceeding 98% (see, for example, Chapter 4).

A more detailed derivation can be presented for the benefit of those who may distrust the simplicity of eqn (2.48). Consider an N-term approximation

$$\phi = \sum_{\alpha}\mathbf{\Phi}^{\alpha}\odot\mathsf{A}^{\alpha}, \tag{2.49}$$

in which each coefficient A^{α} is a symmetric traceless tensor of the same rank as the corresponding basis element $\mathbf{\Phi}^{\alpha}$. Substitution of approximation (2.49) into eqn (2.40) gives

$$\mathcal{R}\sum_{\alpha}\mathbf{\Phi}^{\alpha}\odot\mathsf{A}^{\alpha} = \frac{1}{k_{\rm B}T}\sum_{\alpha'}\mathbf{\Psi}^{\alpha'}\odot\mathsf{X}^{\alpha'}, \tag{2.50}$$

which may be expected to hold approximately. Scalar multiplication of both sides of this equation with $\mathbf{\Phi}^{\beta}$ then projects out a single term $C^{\beta}\mathsf{X}^{\beta}$ on the right-hand side via the orthogonality relation (2.44), while on the left-hand side it gives

$$\sum_{\alpha}\langle\mathbf{\Phi}^{\beta}|\mathcal{R}\mathbf{\Phi}^{\alpha}\rangle\odot\mathsf{A}^{\alpha} = \sum_{\alpha}\langle\mathbf{\Phi}^{\beta}\|\mathcal{R}\|\mathbf{\Phi}^{\alpha}\rangle\mathbf{\Delta}^{(\alpha)}\odot\mathsf{A}^{\alpha} = n\bar{c}_{\rm r}\sum_{\alpha}\mathfrak{S}\left({}^{\beta}_{\alpha}\right)\mathsf{A}^{\alpha}. \tag{2.51}$$

The defining relation (2.46a) for effective cross-sections in terms of reduced matrix elements of the linearized Boltzmann collision operator has been utilized in the final step of this result. The approximated Boltzmann equation thus becomes

$$n\bar{c}_r \sum_\alpha \mathfrak{S}\binom{\beta}{\alpha}\mathsf{A}^\alpha = \frac{1}{k_B T} C^\beta \mathsf{X}^\beta . \tag{2.52}$$

Only those terms on the left-hand side having the same tensorial rank as X^β differ from zero. Inversion of the array $\mathfrak{S}$ yields

$$\mathsf{A}^\alpha = \frac{1}{nk_B T\,\bar{c}_r} \sum_\beta \mathfrak{S}^{-1}\binom{\alpha}{\beta} C^\beta \mathsf{X}^\beta \tag{2.53}$$

for the coefficients A^α. Substitution of this result into the approximate solution (2.49) and its use in expression (2.12b) for the thermodynamic flux yields J^α as

$$\mathsf{J}^\alpha = n\langle \mathbf{\Psi}^\alpha | \phi \rangle = nC^\alpha \langle \mathbf{\Phi}^\alpha | \mathbf{\Phi}^\alpha \rangle \odot \mathsf{A}^\alpha = nC^\alpha \mathsf{A}^\alpha$$

or, equivalently, as

$$\mathsf{J}^\alpha = \frac{1}{k_B T\,\bar{c}_r} \sum_\beta C^\alpha \mathfrak{S}^{-1}\binom{\alpha}{\beta} C^\beta \odot \mathsf{X}^\beta . \tag{2.54}$$

Now, since by eqn (2.13) this result must equal $\sum_\beta \mathsf{L}^{\alpha\beta} \odot \mathsf{X}^\beta = \sum_\beta L^{\alpha\beta}\mathsf{X}^\beta$, the result (2.48) is established.

2.3.2 Gas mixtures

Section 1.6.3 established that each of the coupled Boltzmann equations obtained for a multicomponent gas mixture can be linearized in the same manner as the Boltzmann equation for a pure gas. Indeed, upon introducing an appropriate Hilbert space, the set of linearized Boltzmann equations could be written in array format as a single linearized equation of the same form as that for a pure gas. Clearly, the Chapman–Enskog procedure can be applied in carrying out expansions about the local Maxwellians.

Much of the following material could readily be extended to N-component monatomic gas mixtures. Eventually, however, a number of complications arise in working with multicomponent diffusion and heat fluxes because of the difficulty in choosing consistent sets of linearly independent driving forces for diffusion. As binary mixtures were the only ones considered for polyatomic gases and because the initial three chapters deal with pure monatomic gases primarily in order to familiarize readers with important concepts and definitions, uncluttered by additional complications (which certainly arise in polyatomic systems), multicomponent mixtures are not the main focus of this section. At an appropriate point, the focus turns to binary mixtures. Moreover, multicomponent monatomic gas mixtures have been adequately treated elsewhere in the traditional fashion (Hirschfelder *et al.*, 1964; Chapman and Cowling, 1970; Ferziger and Kaper, 1972; Ern and Giovangigli, 1994; Maitland *et al.*, 1980).

A basis of normalized tensors may be defined for each species in a multicomponent mixture in the same manner as that employed for a pure gas. Thus, an appropriate set of normalized tensors $\mathbf{\Phi}^{ps|k}$ in velocity space is defined by

$$\mathbf{\Phi}^{ps|\mathrm{k}} = (-1)^s \left[\frac{2^{p+s} s!(2p+1)!!}{p!(2p+2s+1)!!}\right]^{\frac{1}{2}} L_s^{(p+\frac{1}{2})}(W_\mathrm{k}^2)\,\overline{\mathbf{W}_\mathrm{k}^p}\,. \tag{2.55}$$

The reduced peculiar velocity $\mathbf{W}_\mathrm{k}$ for species k is given by

$$\mathbf{W}_\mathrm{k} = \left(\frac{m_\mathrm{k}}{2k_\mathrm{B}T}\right)^{\frac{1}{2}} (\mathbf{c} - \mathbf{v})\,, \tag{2.56}$$

with $\mathbf{v}$ the barycentric flow velocity of the mixture [eqn (1.117)]. The first few such basis tensors are as in Table 2.1. Without changing notation, $\mathbf{\Phi}^{ps|\mathrm{k}}$ shall henceforth be regarded as an array that has the tensor (2.55) in the kth position and zeroes elsewhere. When understood in this sense, the basis tensors are mutually orthogonal in the Hilbert space introduced in Section 1.6.3, that is,

$$\langle \mathbf{\Phi}^{ps|\mathrm{k}} | \mathbf{\Phi}^{p's'|\mathrm{k}'} \rangle = x_\mathrm{k} \delta_{pp'} \delta_{ss'} \delta_{\mathrm{kk}'} \mathbf{\Delta}^{(p)}\,. \tag{2.57}$$

Note, however, that these basis tensors have normalized to x_k rather than to unity in order to obtain a concentration-independent basis set. The microscopic flux arrays can be expressed as linear combinations of basis tensor arrays as

$$\mathbf{\Psi}^{\alpha} = \sum_{\mathrm{k}=1}^{\mathrm{N}} C^{\alpha|\mathrm{k}} \mathbf{\Phi}^{\alpha|\mathrm{k}}\,, \tag{2.58a}$$

using the scalar coefficients $C^{\alpha|\mathrm{k}}$ listed in Table 2.2, so that

$$\langle \mathbf{\Psi}^{\alpha} | \mathbf{\Phi}^{\alpha|\mathrm{k}} \rangle = C^{\alpha|\mathrm{k}} x_\mathrm{k} \mathbf{\Delta}^{(p)}\,. \tag{2.58b}$$

As for the pure gas case, matrix elements of $\mathcal{R}^{-1}$ may be obtained approximately in terms of matrix elements of the linearized collision operator $\mathcal{R}$. Note, however, that the matrix elements of $\mathcal{R}$ will have a structure that differs from those for a pure gas, as they depend upon the mole fraction $x_\mathrm{k} = n_\mathrm{k}/n$. The matrix elements of $\mathcal{R}$ may be expressed as

$$\langle \mathbf{\Phi}^{ps|\mathrm{k}} | \mathcal{R} \mathbf{\Phi}^{p's'|\mathrm{k}'} \rangle = \langle \mathbf{\Phi}^{ps|\mathrm{k}} | \mathcal{R}_{\mathrm{kk}'} \mathbf{\Phi}^{p's'|\mathrm{k}'} \rangle = \delta_{pp'} \mathbf{\Delta}^{(p)} n S\left(\begin{smallmatrix} ps \\ ps' \end{smallmatrix} \middle| \begin{smallmatrix} \mathrm{k} \\ \mathrm{k}' \end{smallmatrix}\right), \tag{2.59}$$

in which the Kronecker delta, $\delta_{pp'}$, arises from the isotropy of the collision operator $\mathcal{R}$. Inspection of expression (1.144b) for $\mathcal{R}_{\mathrm{kk}'}$ shows how $S\left(\begin{smallmatrix} ps \\ ps' \end{smallmatrix} \middle| \begin{smallmatrix} \mathrm{k} \\ \mathrm{k}' \end{smallmatrix}\right)$ can be obtained in terms of effective cross-sections that are independent of mole fractions by employing the defining relations

$$S\left(\begin{smallmatrix} ps \\ ps' \end{smallmatrix} \middle| \begin{smallmatrix} \mathrm{k} \\ \mathrm{k}' \end{smallmatrix}\right) \equiv x_\mathrm{k} x_{\mathrm{k}'} \bar{c}_{\mathrm{kk}'} \mathfrak{S}\left(\begin{smallmatrix} ps \\ ps' \end{smallmatrix} \middle| \begin{smallmatrix} \mathrm{k} \\ \mathrm{k}' \end{smallmatrix}\right)_{\mathrm{kk}'}\,, \qquad \mathrm{k} \neq \mathrm{k}' \tag{2.60a}$$

$$S\left(\begin{smallmatrix} ps \\ ps' \end{smallmatrix} \middle| \begin{smallmatrix} \mathrm{k} \\ \mathrm{k} \end{smallmatrix}\right) \equiv \sum_{\ell} x_\mathrm{k} x_\ell \bar{c}_{\mathrm{k}\ell} \mathfrak{S}\left(\begin{smallmatrix} ps \\ ps' \end{smallmatrix} \middle| \begin{smallmatrix} \mathrm{k} \\ \mathrm{k} \end{smallmatrix}\right)_{\mathrm{k}\ell}\,, \tag{2.60b}$$

for basis tensors belonging to different, respectively, the same species. The average relative thermal speed $\bar{c}_{\mathrm{kk}'} = [8k_\mathrm{B}T/(\pi m_{\mathrm{kk}'})]^{\frac{1}{2}}$ for the collision pair has been defined in eqn (1.141).

Table 2.2 Thermodynamic fluxes J^α and forces X^α, plus related data corresponding to the microscopic fluxes† $\boldsymbol{\Psi}^\alpha = \sum_k C^{\alpha|k}\boldsymbol{\Phi}^{\alpha|k}$ for binary noble gas mixtures.

α	$\boldsymbol{\Psi}^\alpha$	J^α	X^α
$11 \equiv E$	$\frac{1}{2}(m_A C^2 - 5k_B T,\ m_B C^2 - 5k_B T)\mathbf{C}$	$\mathbf{q}$	$-\nabla \ln T$
$10 \equiv d$	$\left(\frac{1}{n_A}, -\frac{1}{n_B}\right)\mathbf{C}$	$\mathbf{v}_A - \mathbf{v}_B$	$-\nabla p_A$
$20 \equiv \pi$	$(m_A, m_B)\overline{\mathbf{CC}}$	$\mathbf{\Pi}$	$-\overline{\nabla \mathbf{v}}$
α	$C^{\alpha\|A}$	$C^{\alpha\|B}$	$N^{\alpha\|A}$ *
$11 \equiv E$	$\left(\frac{5k_B^3T^3}{2m_A}\right)^{\frac{1}{2}}$	$\left(\frac{5k_B^3T^3}{2m_B}\right)^{\frac{1}{2}}$	$\frac{5}{2}(k_B T)^3\left(\frac{x_A}{m_A} + \frac{x_B}{m_B}\right)$
$10 \equiv d$	$\frac{1}{nx_A}\left(\frac{k_B T}{m_A}\right)^{\frac{1}{2}}$	$\frac{-1}{nx_B}\left(\frac{k_B T}{m_B}\right)^{\frac{1}{2}}$	$\frac{k_B T \rho}{n\rho_A\rho_B}$
$20 \equiv \pi$	$k_B T\sqrt{2}$	$k_B T\sqrt{2}$	

† For a pure gas, the arrays $\boldsymbol{\Psi}^{11}$ and $\boldsymbol{\Psi}^{20}$ reduce to single components, $x_A = 1$, $x_B = 0$, and the rows identified by $10 \equiv d$ in this table no longer apply.

* For vectorial fluxes, $\langle \boldsymbol{\Psi}^\alpha | \boldsymbol{\Psi}^\beta \rangle = \delta_{\alpha\beta} N^\alpha \boldsymbol{\delta}$, with $\boldsymbol{\delta}$ the second rank isotropic tensor.

Expression (2.60ba) explicitly accounts for the production of bulk polarization $\boldsymbol{\Phi}^{ps}$ in a binary mixture by the conversion of a microscopic polarization $\boldsymbol{\Phi}^{ps'|B}$ into a microscopic polarization $\boldsymbol{\Phi}^{ps|A}$ via a collision between a B atom and an A atom. This type of effective cross-section may also be designated, following Curtiss (1981), by $\mathfrak{S}''\left({ps \atop ps'}\middle|{A \atop B}\right)_{AB}$, that is,

$$\mathfrak{S}\left({ps \atop ps'}\middle|{A \atop B}\right)_{AB} \equiv \mathfrak{S}''\left({ps \atop ps'}\middle|{A \atop B}\right)_{AB}. \tag{2.60c}$$

A similar examination of a specific example of eqn (2.60b), namely,

$$S\left({ps \atop ps'}\middle|{A \atop A}\right) = x_A^2 \bar{c}_{AB} \mathfrak{S}\left({ps \atop ps'}\middle|{A \atop A}\right)_{AA} + x_A x_B \bar{c}_{AB} \mathfrak{S}\left({ps \atop ps'}\middle|{A \atop A}\right)_{AB},$$

which contributes to the production of a bulk polarization $\boldsymbol{\Phi}^{ps}$ by conversion of the bulk polarization $\boldsymbol{\Phi}^{ps'}$ via binary collisions, may also be expressed in the form

$$S\left({ps \atop ps'}\middle|{A \atop A}\right) = x_A^2 \bar{c}_{AA} \mathfrak{S}\left({ps \atop ps'}\middle|{A \atop A}\right)_{AA} + x_A x_B \bar{c}_{AB} \mathfrak{S}'\left({ps \atop ps'}\middle|{A \atop A}\right)_{AB}, \tag{2.60d}$$

in which $\mathfrak{S}'\left({ps \atop ps'}\middle|{A \atop A}\right)_{AB}$ designates the production of the microscopic polarization $\boldsymbol{\Phi}^{ps|A}$ from the microscopic polarization $\boldsymbol{\Phi}^{ps'|A}$ via collision between an A atom and a B atom. Of course, although it is possible to express the pure gas effective cross-section $\mathfrak{S}\left(\begin{smallmatrix} ps \\ ps' \end{smallmatrix}\middle|\begin{smallmatrix} \mathrm{A} \\ \mathrm{A} \end{smallmatrix}\right)_{\mathrm{AA}}$ as

$$\mathfrak{S}\left(\begin{smallmatrix} ps \\ ps' \end{smallmatrix}\middle|\begin{smallmatrix} \mathrm{A} \\ \mathrm{A} \end{smallmatrix}\right)_{\mathrm{AA}} = \mathfrak{S}'\left(\begin{smallmatrix} ps \\ ps' \end{smallmatrix}\middle|\begin{smallmatrix} \mathrm{A} \\ \mathrm{A} \end{smallmatrix}\right)_{\mathrm{AA}} + \mathfrak{S}''\left(\begin{smallmatrix} ps \\ ps' \end{smallmatrix}\middle|\begin{smallmatrix} \mathrm{A} \\ \mathrm{A} \end{smallmatrix}\right)_{\mathrm{AA}},$$

in which $\mathfrak{S}'\left(\begin{smallmatrix} ps \\ ps' \end{smallmatrix}\middle|\begin{smallmatrix} \mathrm{A} \\ \mathrm{A} \end{smallmatrix}\right)_{\mathrm{AA}}$ represents a contribution to $\mathbf{\Phi}^{ps|\mathrm{A}}$ obtained from $\mathbf{\Phi}^{ps'|\mathrm{A}}$ associated with the atom of interest and $\mathfrak{S}''\left(\begin{smallmatrix} ps \\ ps' \end{smallmatrix}\middle|\begin{smallmatrix} \mathrm{A} \\ \mathrm{A} \end{smallmatrix}\right)_{\mathrm{AA}}$ represents $\mathbf{\Phi}^{ps|\mathrm{A}}$ produced from microscopic polarization $\mathbf{\Phi}^{ps'|\mathrm{A}}$ associated with the collision partner (another A atom), there seems to be no compelling reason to do so.

Expressions for the effective cross-sections obtained using eqn (1.144), namely,

$$\mathfrak{S}\left(\begin{smallmatrix} ps \\ ps' \end{smallmatrix}\middle|\begin{smallmatrix} \mathrm{k} \\ \mathrm{k} \end{smallmatrix}\right)_{\mathrm{kk}'}\mathbf{\Delta}^{(p)} = (\bar{c}_{\mathrm{kk}'})^{-1}\int \mathrm{d}\mathbf{c}\; n_{\mathrm{k}}^{-1} f_{\mathrm{k}}^{(0)}(\mathbf{c})\mathbf{\Phi}^{ps|\mathrm{k}}(\mathbf{c}) \iint \mathrm{d}_1\mathrm{d}\mathbf{e}'\; c_{\mathrm{r}}\sigma_{\mathrm{kk}'} n_{\mathrm{k}'}^{-1} f_{\mathrm{k}'}^{(0)}(\mathbf{c}_1)$$
$$\times\; \{[\mathbf{\Phi}^{ps'|\mathrm{k}}(\mathbf{c}) - \mathbf{\Phi}^{ps'|\mathrm{k}}(\mathbf{c}')] + \delta_{\mathrm{kk}'}[\mathbf{\Phi}^{ps'|\mathrm{k}}(\mathbf{c}_1) - \mathbf{\Phi}^{ps'|\mathrm{k}}(\mathbf{c}_1')]\}\,. \tag{2.61a}$$

$$\mathfrak{S}\left(\begin{smallmatrix} ps \\ ps' \end{smallmatrix}\middle|\begin{smallmatrix} \mathrm{k} \\ \mathrm{k}' \end{smallmatrix}\right)_{\mathrm{kk}'}\mathbf{\Delta}^{(p)} = (\bar{c}_{\mathrm{kk}'})^{-1}\int \mathrm{d}\mathbf{c}\; n_{\mathrm{k}}^{-1} f_{\mathrm{k}}^{(0)}(\mathbf{c})\mathbf{\Phi}^{ps|\mathrm{k}}(\mathbf{c}) \iint \mathrm{d}\mathbf{c}_1\mathrm{d}\mathbf{e}'\; c_{\mathrm{r}}\sigma_{\mathrm{kk}'}\, n_{\mathrm{k}'}^{-1}$$
$$\times\; f_{\mathrm{k}'}^{(0)}(\mathbf{c}_1)\,[\mathbf{\Phi}^{ps'|\mathrm{k}'}(\mathbf{c}_1) - \mathbf{\Phi}^{ps'|\mathrm{k}'}(\mathbf{c}_1')]\,, \qquad \mathrm{k} \neq \mathrm{k}'\,, \tag{2.61b}$$

may be examined in order to verify that they do indeed depend only upon temperature and properties of the colliding atoms. As for the pure gas expressions, these effective cross-sections describe the negative rates (per unit colliding flux) by which $\mathbf{\Phi}^{ps}$-type deviations from equilibrium are produced from one another. The effective cross-section $\mathfrak{S}\left(\begin{smallmatrix} ps \\ ps' \end{smallmatrix}\middle|\begin{smallmatrix} \mathrm{k} \\ \mathrm{k}' \end{smallmatrix}\right)_{\mathrm{kk}'}$ accounts for the production of $\mathbf{\Phi}^{ps}$ in species k from $\mathbf{\Phi}^{ps'}$ in species k′ by collisions between the two species. Similarly, $\mathfrak{S}\left(\begin{smallmatrix} ps \\ ps' \end{smallmatrix}\middle|\begin{smallmatrix} \mathrm{k} \\ \mathrm{k} \end{smallmatrix}\right)_{\mathrm{kk}'}$ accounts for the production of $\mathbf{\Phi}^{ps}$ from $\mathbf{\Phi}^{ps'}$ within the same species, either by collisions between atoms of just this species (k′ = k) or by collisions with another species (k′ ≠ k). Note that although diagonal mixture effective cross-sections $\mathfrak{S}\left(\begin{smallmatrix} ps \\ ps \end{smallmatrix}\middle|\begin{smallmatrix} \mathrm{k} \\ \mathrm{k} \end{smallmatrix}\right)_{\mathrm{kk}'} \equiv \mathfrak{S}(ps|\mathrm{k})_{\mathrm{kk}'}$ are positive, nondiagonal mixture effective cross-sections may have either sign. Further, note that the effective cross-sections $\mathfrak{S}\left(\begin{smallmatrix} ps \\ ps' \end{smallmatrix}\middle|\begin{smallmatrix} \mathrm{k} \\ \mathrm{k} \end{smallmatrix}\right)_{\mathrm{kk}}$ are simply those of the pure species encountered in the previous section.

It is useful to set up the general formalism prior to embarking upon the calculation of specific transport coefficients. There are two differences to observe relative to the procedure employed for a pure gas. Firstly, arrays of basis tensors $\mathbf{\Phi}^{ps|\mathrm{k}}$ replace the single basis tensors employed for pure gases. Moreover, they are normalized differently [see eqn (2.57)]. Secondly, the microscopic fluxes $\mathbf{\Psi}^{\alpha}$ are not equal to particular basis arrays, but rather to linear combinations of them, as shown by eqn (2.58a). An adjustment will be made to the procedure developed in the previous subsection in order to accommodate the fact that it will be necessary to employ the reduced matrix elements as intermediaries to the effective cross-sections due to the complexity of eqn (2.57).

By analogy with eqn (2.49), the array ϕ may be expanded into a finite double sum in terms of the chemical-species-dependent basis elements $\mathbf{\Phi}^{\alpha|\mathrm{k}}$, namely,

$$\phi = \sum_{\alpha,\mathrm{k}} \boldsymbol{\Phi}^{\alpha|\mathrm{k}} \odot \mathsf{A}^{\alpha|\mathrm{k}}\,, \tag{2.62}$$

and this expression then substituted into the linearized Boltzmann equation (2.27). Formation of the inner product with $\boldsymbol{\Phi}^{\beta|\mathrm{k}'}$ on both sides of the resultant equation gives, upon utilizing relations (2.58b) and (2.59), a result

$$n \sum_{\alpha,\mathrm{k}} S\big({}^{\beta}_{\alpha}\big|{}^{\mathrm{k}'}_{\mathrm{k}}\big)\mathsf{A}^{\alpha|\mathrm{k}} = \frac{1}{k_\mathrm{B}T} C^{\beta|\mathrm{k}'} x_{\mathrm{k}'} \mathsf{X}^{\beta} \tag{2.63}$$

that is analogous to eqn (2.52) for a pure gas. The $S\big({}^{\beta}_{\alpha}\big|{}^{\mathrm{k}'}_{\mathrm{k}}\big)$ may be regarded as multi-labeled elements of a matrix S that is intended to approximate the operator $\mathcal{R}/n$. It may thus be inverted to obtain $\mathsf{A}^{\alpha|\mathrm{k}}$ as

$$\mathsf{A}^{\alpha|\mathrm{k}} = \frac{1}{p} \sum_{\beta,\mathrm{k}'} S^{-1}\big({}^{\alpha}_{\beta}\big|{}^{\mathrm{k}}_{\mathrm{k}'}\big) C^{\beta|\mathrm{k}'} x_{\mathrm{k}'} \mathsf{X}^{\beta}\,, \tag{2.64}$$

with $S^{-1}\big({}^{\alpha}_{\beta}\big|{}^{\mathrm{k}}_{\mathrm{k}'}\big)$ understood to be a matrix element of the inverse S^{-1} of the matrix S. This result is first substituted into (2.62), thence into

$$\mathsf{J}^{\alpha} = n\langle \boldsymbol{\Psi}^{\alpha} | \phi \rangle = n \sum_{\mathrm{k}} C^{\alpha|\mathrm{k}} \langle \boldsymbol{\Phi}^{\alpha|\mathrm{k}} | \phi \rangle\,. \tag{2.65}$$

Finally, a comparison between eqn (2.30) and the result obtained from this substitution, following its simplification using eqn (2.57), shows that the transport coefficients may be expressed as

$$L^{\alpha\beta} = \frac{1}{k_\mathrm{B}T} \sum_{\mathrm{k},\mathrm{k}'} C^{\alpha|\mathrm{k}} x_{\mathrm{k}} S^{-1}\big({}^{\alpha}_{\beta}\big|{}^{\mathrm{k}}_{\mathrm{k}'}\big) C^{\beta|\mathrm{k}'} x_{\mathrm{k}'}\,. \tag{2.66}$$

This generalization relative to eqn (2.48) for pure gases appears natural. Once all elements of S^{-1} have been expressed in terms of the $S\big({}^{\alpha}_{\beta}\big|{}^{\mathrm{k}}_{\mathrm{k}'}\big)$, the only step remaining in the derivation of an expression for the transport coefficients $L^{\alpha\beta}$ in terms of effective cross-sections is to utilize eqns (2.60b).

For a binary mixture, formulae (2.60b) may be rewritten in a form that allows $\bar{c}_{\mathrm{AB}}$ to be extracted explicitly, namely,

$$S\big({}^{ps}_{ps'}\big|{}^{\mathrm{A}}_{\mathrm{A}}\big) = \bar{c}_{\mathrm{AB}} x_{\mathrm{A}} \widetilde{\mathfrak{S}}\big({}^{ps}_{ps'}\big|{}^{\mathrm{A}}_{\mathrm{A}}\big)\,, \tag{2.67a}$$

$$S\big({}^{ps}_{ps'}\big|{}^{\mathrm{B}}_{\mathrm{A}}\big) = \bar{c}_{\mathrm{AB}} x_{\mathrm{A}} x_{\mathrm{B}} \mathfrak{S}\big({}^{ps}_{ps'}\big|{}^{\mathrm{B}}_{\mathrm{A}}\big)_{\mathrm{AB}}\,, \tag{2.67b}$$

$$S\big({}^{ps}_{ps'}\big|{}^{\mathrm{B}}_{\mathrm{B}}\big) = \bar{c}_{\mathrm{AB}} x_{\mathrm{B}} \widetilde{\mathfrak{S}}\big({}^{ps}_{ps'}\big|{}^{\mathrm{B}}_{\mathrm{B}}\big)\,, \tag{2.67c}$$

in which mass ratios y_{A} and y_{B} defined via

$$y_{\mathrm{A}}^2 = m_{\mathrm{A}}/(m_{\mathrm{A}} + m_{\mathrm{B}})\,, \qquad y_{\mathrm{B}}^2 = m_{\mathrm{B}}/(m_{\mathrm{A}} + m_{\mathrm{B}})\,, \tag{2.68}$$

and a pair of mixture cross-sections $\widetilde{\mathfrak{S}}$, defined as

$$\widetilde{\mathfrak{S}}\left(\begin{smallmatrix} ps \\ ps' \end{smallmatrix}\middle|\begin{smallmatrix} \mathrm{A} \\ \mathrm{A} \end{smallmatrix}\right) \equiv \sqrt{2}\, x_{\mathrm{A}} y_{\mathrm{B}} \mathfrak{S}\left(\begin{smallmatrix} ps \\ ps' \end{smallmatrix}\middle|\begin{smallmatrix} \mathrm{A} \\ \mathrm{A} \end{smallmatrix}\right)_{\mathrm{AA}} + x_{\mathrm{B}} \mathfrak{S}\left(\begin{smallmatrix} ps \\ ps' \end{smallmatrix}\middle|\begin{smallmatrix} \mathrm{A} \\ \mathrm{A} \end{smallmatrix}\right)_{\mathrm{AB}} \tag{2.69a}$$

and

$$\widetilde{\mathfrak{S}}\left(\begin{smallmatrix} ps \\ ps' \end{smallmatrix}\middle|\begin{smallmatrix} \mathrm{B} \\ \mathrm{B} \end{smallmatrix}\right) \equiv x_{\mathrm{A}} \mathfrak{S}\left(\begin{smallmatrix} ps \\ ps' \end{smallmatrix}\middle|\begin{smallmatrix} \mathrm{B} \\ \mathrm{B} \end{smallmatrix}\right)_{\mathrm{AB}} + \sqrt{2}\, x_{\mathrm{B}} y_{\mathrm{A}} \mathfrak{S}\left(\begin{smallmatrix} ps \\ ps' \end{smallmatrix}\middle|\begin{smallmatrix} \mathrm{B} \\ \mathrm{B} \end{smallmatrix}\right)_{\mathrm{BB}}, \tag{2.69b}$$

have been introduced.

2.4 The Transport Coefficients

In Section 2.1 two transport coefficients were seen to occur in pure gases: one, denoted by η and called the (shear) viscosity coefficient, is associated with the transport of linear momentum through the gas via binary collisions, while the other, denoted by λ and called the coefficient of thermal conductivity, is associated with the collisional transport of energy through the gas. A richer set of transport coefficients occurs in gas mixtures. As for pure gases, the transport of linear momentum and energy through the gas mixture via binary collisions of like and unlike species gives rise to a (shear) viscosity coefficient, denoted η_{mix}, and a thermal conductivity coefficient, denoted λ_{mix}. However, because there is now a set of new vectorial thermodynamic forces present, namely, the concentration gradients associated with the various species making up the mixture, there will necessarily also be $N(N-1)/2$ linearly independent diffusion coefficients for an N-component mixture associated with mass transports through the gas mixture via binary collisions between the various unlike species. Moreover, because these new thermodynamic driving forces are, like the temperature gradient, (polar) vectors, it is possible for a temperature gradient to give rise to mass fluxes and, vice versa, for concentration gradients to give rise to an energy (heat) flux. For a binary mixture, the transport coefficients associated with these cross-effects are called the thermal diffusion coefficient, denoted D_{T}, and the Dufour (or diffusion-thermal) coefficient, denoted $\widetilde{D}_{\mathrm{T}}$. However, as already pointed out in Section 2.2, these two new transport coefficients are equal due to the self-adjoint symmetry of the linearized collision operator $\mathcal{R}$. The orbiting collisions also effect the calculation of transport coefficients; more details can be found in the references [Hirschfelder *et al.* 1964; Taylor 1979; O'Hara and Smith 1971; Barker *et al.* 1964; Zhai *et al.* 2023].

Because the thermodynamic force $-\overline{\nabla \mathbf{v}}$ couples tensorially neither to $-\nabla T$ nor to $-\nabla p$, viscous flow may be treated separately, following which the tensorially coupled system associated with diffusion, heat conduction, and their cross-effects can be considered.

2.4.1 Shear viscosity for pure gases

Upon describing the viscosity in terms of a single expansion tensor $\mathbf{\Phi}^{20}$, the first Chapman–Cowling approximation $[\eta]_1$ to the viscosity coefficient η is obtained as

$$[\eta]_1 = \frac{1}{2k_{\mathrm{B}}T} \frac{(C^{\pi})^2}{\overline{c}_{\mathrm{r}} \mathfrak{S}(20)} = \frac{k_{\mathrm{B}}T}{\overline{c}_{\mathrm{r}} \mathfrak{S}(20)} \tag{2.70a}$$

or simply as

$$[\eta]_1 \equiv \frac{k_\mathrm{B}T}{\overline{c}_\mathrm{r}\mathfrak{S}_\eta} \,. \tag{2.70b}$$

In terms of a shorthand notation $\mathfrak{S}_\eta \equiv \mathfrak{S}(20)$ that has been introduced via eqn (2.70b) in order to indicate that this effective cross-section may be considered to characterize the viscosity. It is unfortunate that this effective cross-section is sometimes (mistakenly) referred to simply as *the* kinetic cross-section. It is worth noting also that this result would be exact were $\boldsymbol{\Phi}^{20}$ an eigenfunction of the linearized collision operator: however, as it is not an eigenfunction of $\mathcal{R}$ except for certain models, it is commonly referred to as the first Chapman–Cowling approximation,[2] $[\eta]_1$, to η. Section 2.9 shows how this approximation is, in general, surprisingly good.

The second Chapman–Cowling approximation to the viscosity coefficient includes a second orthonormal basis element of the nonhydrodynamic subspace. Of course, in order to be coupled to $\boldsymbol{\Phi}^{20}$ by $\mathcal{R}$, this second basis element must be a tensor of the same rank and parity as that of the microscopic momentum flux $\boldsymbol{\Psi}^\pi$. The simplest such appropriate basis element that satisfies this condition is $\boldsymbol{\Phi}^{21}$. Thus, if the two-member basis set

$$\boldsymbol{\Phi}^{20} = \sqrt{2}\,\overline{\mathbf{W}\mathbf{W}}\,, \qquad \boldsymbol{\Phi}^{21} = (\tfrac{4}{7})^{\frac{1}{2}}(W^2 - \tfrac{7}{2})\,\overline{\mathbf{W}\mathbf{W}}\,, \tag{2.71}$$

from Table 2.1 is employed, the linearized collision operator $\mathcal{R}$ may be represented approximately in terms of a 2×2 effective-cross-section matrix as

$$\mathcal{R} \to n\overline{c}_\mathrm{r}\boldsymbol{\mathfrak{S}}_\eta = n\overline{c}_\mathrm{r}\begin{pmatrix} \mathfrak{S}(20) & \mathfrak{S}\binom{20}{21} \\ \mathfrak{S}\binom{21}{20} & \mathfrak{S}(21) \end{pmatrix} . \tag{2.72}$$

The inverse operator $\mathcal{R}^{-1}$ may similarly be represented approximately in terms of the inverse of the effective-cross-section matrix $\boldsymbol{\mathfrak{S}}_\eta$. In doing so, note that, according to eqn (2.48), only a single element of the inverse $\mathfrak{S}_\eta$-matrix is required, namely,

$$\mathfrak{S}_\eta^{-1}(20) = \frac{\mathfrak{S}(21)}{\mathfrak{S}(20)\mathfrak{S}(21) - \mathfrak{S}^2\binom{20}{21}} \,, \tag{2.73a}$$

which may also be expressed as

$$\mathfrak{S}_\eta^{-1}(20) = \frac{1}{\mathfrak{S}(20)}\left(1 - \frac{\mathfrak{S}^2\binom{20}{21}}{\mathfrak{S}(20)\mathfrak{S}(21)}\right)^{-1} . \tag{2.73b}$$

The symmetry relation (2.46b) has also been employed in order to obtain this expression. Equation (2.73b) displays more clearly the manner in which higher-order approximations develop as corrections to the first-order result.

[2] Higher-order Chapman–Enskog approximations, which involve gradients other than the first gradient or powers and/or products of first gradients, are distinguished from Chapman–Cowling approximations, which apply to transport coefficients determined within a given Chapman–Enskog approximation.

The various orders of Chapman–Cowling approximation to η are denoted by $[\eta]_n$. In particular, the second Chapman–Cowling approximation $[\eta]_2$ may be expressed as the product

$$[\eta]_2 = [\eta]_1 f_\eta^{[2]} \tag{2.74a}$$

of the first Chapman–Cowling approximation $[\eta]_1$ of eqn (2.70b) and a second approximation Chapman–Cowling correction factor $f_\eta^{[2]}$ defined as

$$f_\eta^{[2]} \equiv (1 - \delta_\eta^{[2]})^{-1} , \tag{2.74b}$$

with $\delta_\eta^{[2]}$ expressed in terms of effective cross-sections as

$$\delta_\eta^{[2]} \equiv \frac{\mathfrak{S}^2\binom{20}{21}}{\mathfrak{S}(20)\mathfrak{S}(21)} . \tag{2.74c}$$

Higher Chapman–Cowling approximations $[\eta]_n$ to η have the same structure, namely,

$$[\eta]_n = [\eta]_1 f_\eta^{[n]} , \tag{2.74d}$$

essentially because the inverse matrix cross-section $\mathfrak{S}^{-1}\binom{\pi}{\pi}$ is given by the ratio of a determinant of rank $n-1$ to a determinant of rank n. The traditional expression (Hirschfelder *et al.*, 1964; Chapman and Cowling, 1970) for the second approximation Chapman–Cowling correction $f_\eta^{[2]}$ is typically expressed in terms of quantities referred to as reduced omega integrals $\Omega^{(l,s)\bigstar}$, which in turn are ratios of the values for omega integrals (see Section 2.5.1) obtained from a realistic binary interaction potential $V(R)$ to the corresponding values for a hard-sphere gas. It becomes a fairly straightforward task to establish that $f_\eta^{[2]}$ and $f_\eta^{(2)}$ of Hirschfelder, Curtiss and Bird (Hirschfelder, Curtiss and Bird, 1964) are equivalent by expressing them both in terms of omega integrals. This means that should corrections of order greater than two be needed, the higher-order expressions provided by Viehland and co-workers (1995) may be utilized. Because the collisional couplings between higher members, $\mathbf{\Phi}^{2\ell}$ $(\ell > 0)$, of the basis set and $\mathbf{\Phi}^{20}$ turn out to be very weak, this successive approximation procedure proves to be quite efficient, and the Laguerre/Sonine expansion actually converges quite rapidly (see, for example, Chapter 4). Note, however, that while the second approximation expression may represent a reasonable approximation for pure neutral gases and their binary mixtures, it may prove to be a relatively poor approximation for ion-neutral mixtures, such as weak plasmas (Viehland *et al.*, 1995).

2.4.2 Thermal conductivity for pure gases

As has already been established for the viscosity coefficient, a simplest matrix approximation to $\mathcal{R}$ may be made in terms of a single reduced matrix element. The thermal conductivity coefficient may also in first instance be described in terms of the single basis element $\mathbf{\Phi}^{11}$, which is contained in $\mathbf{\Psi}^E$. In the language of the variational method, the use of a single trial function $\mathbf{\Phi}^{11}$ is based upon the hope that it is close to being an eigenfunction of $\mathcal{R}$, so that in such an approximation, only $n\bar{c}_r\mathfrak{S}(11)$ and its inverse (here, reciprocal) will be required in order to obtain the relevant first

Chapman–Cowling approximation $[\lambda]_1$ to λ. In this lowest-level approximation, $[\lambda]_1$ is obtained as

$$[\lambda]_1 T = \frac{1}{k_B T}(C^E)^2[\bar{c}_r \mathfrak{S}(11)]^{-1} = \frac{5(k_B T)^2}{2m\bar{c}_r \mathfrak{S}(11)}, \tag{2.75a}$$

or simply as

$$[\lambda]_1 \equiv \frac{5k_B^2 T}{2m\bar{c}_r \mathfrak{S}_\lambda} \tag{2.75b}$$

where, as in eqn (2.70b) for the first Chapman–Cowling approximation to η, the shorthand notation $\mathfrak{S}_\lambda \equiv \mathfrak{S}(11)$ has been introduced for the effective cross-section that may be considered to characterize the thermal conductivity.

The second Chapman–Cowling approximation to the thermal conductivity coefficient may be obtained upon inclusion of a second basis element having the same tensor rank and parity as $\mathbf{\Psi}^E$. In this case, the appropriate additional basis element is $\mathbf{\Phi}^{12}$. Hence, upon employing the two-member basis set

$$\mathbf{\Phi}^{11} = \left(\frac{4}{5}\right)^{\frac{1}{2}} (W^2 - \tfrac{5}{2})\,\mathbf{W}, \qquad \mathbf{\Phi}^{12} = \left(\frac{4}{35}\right)^{\frac{1}{2}} (W^4 - 7W^2 + \tfrac{35}{4})\,\mathbf{W}, \tag{2.76}$$

from Table 2.1, the corresponding 2×2 matrix representation of $\mathcal{R}$ is obtained as

$$\mathcal{R} \to n\bar{c}_r \boldsymbol{\mathfrak{S}}_\lambda = n\bar{c}_r \begin{pmatrix} \mathfrak{S}(11) & \mathfrak{S}\binom{11}{12} \\ \mathfrak{S}\binom{12}{11} & \mathfrak{S}(12) \end{pmatrix}. \tag{2.77}$$

The appropriate inverse matrix element for λ is thus given by

$$\mathfrak{S}_\lambda^{-1}(11) = \frac{1}{\mathfrak{S}(11)}\left(1 - \frac{\mathfrak{S}^2\binom{11}{12}}{\mathfrak{S}(11)\mathfrak{S}(12)}\right)^{-1}, \tag{2.78}$$

so that the second Chapman–Cowling approximation, $[\lambda]_2$, to λ becomes

$$[\lambda]_2 = [\lambda]_1 f_\lambda^{[2]}, \tag{2.79a}$$

in which $[\lambda]_1$ is given by eqn (2.75a), and $f_\lambda^{[2]}$ is given by

$$f_\lambda^{[2]} = (1 - \delta_\lambda^{[2]})^{-1}, \tag{2.79b}$$

with $\delta_\lambda^{[2]}$ defined in terms of effective cross-sections as

$$\delta_\lambda^{[2]} \equiv \frac{\mathfrak{S}^2\binom{11}{12}}{\mathfrak{S}(11)\mathfrak{S}(12)}. \tag{2.79c}$$

As for the shear viscosity, higher Chapman–Cowling approximations $[\lambda]_n$ to λ may also be expressed as

$$[\lambda]_n = [\lambda]_1 f_\lambda^{[n]}. \tag{2.79d}$$

The second Chapman–Cowling approximation for λ is less reliable than that for the viscosity, as the correction term $f_\lambda^{[2]}$ may be as much as twice the size of the viscosity correction term $f_\eta^{[2]}$. However, as $f_\lambda^{[2]}$ of eqn (2.79) is equivalent to $f_\lambda^{(2)}$ of Hirschfelder *et al.* (1964), the higher-order approximation expressions obtained by Viehland and co-workers (1995) may also be utilized for thermal conductivity calculations.

2.4.3 Shear viscosity for binary mixtures

The shear viscosity coefficient η_{mix} is given in eqn (2.37e) in terms of the reduced matrix element of the inverse of the linearized collision operator $\mathcal{R}$ with the microscopic momentum-flux array $\boldsymbol{\Psi}^\pi$. The lowest-order Chapman–Cowling approximation for a binary mixture necessarily involves the two basis arrays $\boldsymbol{\Phi}^{20|\text{A}}$ and $\boldsymbol{\Phi}^{20|\text{B}}$, from which the matrix representation

$$\mathsf{S}^{[1]}(\eta_{\text{mix}}) = \begin{pmatrix} S(20|\text{A}) & S(\begin{smallmatrix}20\\20\end{smallmatrix}|\begin{smallmatrix}\text{A}\\\text{B}\end{smallmatrix}) \\ S(\begin{smallmatrix}20\\20\end{smallmatrix}|\begin{smallmatrix}\text{B}\\\text{A}\end{smallmatrix}) & S(20|\text{B}) \end{pmatrix},$$

is obtained for $\mathcal{R}$. Matrix inversion of $\mathsf{S}^{[1]}(\eta_{\text{mix}})$ can readily be carried out to provide a 2×2 matrix representation of the inverse collision operator $\mathcal{R}^{-1}$. By employing this matrix representation, together with $C^{\pi|\text{A}} = C^{\pi|\text{B}} = \sqrt{2}\,k_\text{B}T$ from Table 2.2, expression (2.66) may be evaluated, thereby giving the result

$$L^{\pi\pi} = 2[\eta_{\text{mix}}]_1 = 2k_\text{B}T\frac{x_\text{A}^2 S(20|\text{B}) - 2x_\text{A}x_\text{B}S(\begin{smallmatrix}20\\20\end{smallmatrix}|\begin{smallmatrix}\text{B}\\\text{A}\end{smallmatrix}) + x_\text{B}^2 S(20|\text{A})}{S(20|\text{A})S(20|\text{B}) - S^2(\begin{smallmatrix}20\\20\end{smallmatrix}|\begin{smallmatrix}\text{B}\\\text{A}\end{smallmatrix})} \tag{2.80}$$

for $L^{\pi\pi}$. To convert this first Chapman–Cowling approximation $[\eta_{\text{mix}}]_1$ into an equivalent expression in terms of effective cross-sections, relations (2.67, 2.69) (with $p=2$, $s=s'=0$) may be utilized to obtain $[\eta_{\text{mix}}]_1$ as

$$[\eta_{\text{mix}}]_1 = \frac{k_\text{B}T}{\bar{c}_\text{AB}}\,\frac{x_\text{A}\widetilde{\mathfrak{S}}(20|\text{B}) - 2x_\text{A}x_\text{B}\mathfrak{S}(\begin{smallmatrix}20\\20\end{smallmatrix}|\begin{smallmatrix}\text{B}\\\text{A}\end{smallmatrix})_\text{AB} + x_\text{B}\widetilde{\mathfrak{S}}(20|\text{A})}{\widetilde{\mathfrak{S}}(20|\text{A})\widetilde{\mathfrak{S}}(20|\text{B}) - x_\text{A}x_\text{B}\mathfrak{S}^2(\begin{smallmatrix}20\\20\end{smallmatrix}|\begin{smallmatrix}\text{B}\\\text{A}\end{smallmatrix})_\text{AB}}, \tag{2.81a}$$

or as

$$[\eta_{\text{mix}}]_1 = \frac{k_\text{B}T}{\bar{c}_\text{AB}\Delta_{\text{mix}}(20)}\left[x_\text{A}\widetilde{\mathfrak{S}}(20|\text{B}) - 2x_\text{A}x_\text{B}\mathfrak{S}(\begin{smallmatrix}20\\20\end{smallmatrix}|\begin{smallmatrix}\text{B}\\\text{A}\end{smallmatrix})_\text{AB} + x_\text{B}\widetilde{\mathfrak{S}}(20|\text{A})\right], \tag{2.81b}$$

with $\Delta_{\text{mix}}(20) \equiv \Delta^{[1]}(\eta_{\text{mix}})$ given by

$$\Delta_{\text{mix}}(20) = \widetilde{\mathfrak{S}}(20|\text{A})\widetilde{\mathfrak{S}}(20|\text{B}) - x_\text{A}x_\text{B}\mathfrak{S}^2(\begin{smallmatrix}20\\20\end{smallmatrix}|\begin{smallmatrix}\text{B}\\\text{A}\end{smallmatrix})_\text{AB}\,. \tag{2.81c}$$

As may readily be anticipated from the definitions of the mixture cross-sections $\widetilde{\mathfrak{S}}$ given in eqns (2.69), the shear viscosity of a dilute binary gas mixture of molecules A and B depends both upon the unlike interatomic interaction (energy) $V_\text{AB}(R)$ and upon the like interaction energies $V_\text{AA}(R)$ and $V_\text{BB}(R)$. Hirschfelder and colleagues (1964) showed that $[\eta_{\text{mix}}]_1$ could be expressed in terms of the pure-gas component viscosities $[\eta_\text{A}]_1$ and $[\eta_\text{B}]_1$ plus the binary diffusion coefficient $D_\text{AB} \equiv [D_\text{AB}]_1$ (see Section 2.4.4) and an "interaction viscosity" η_AB, each of which depends solely upon the unlike interaction $V_{\mathrm{AB}}(R)$. The interaction viscosity η_{AB} can be expressed explicitly in terms of effective cross-sections as

$$\eta_{\mathrm{AB}} = \frac{2y_{\mathrm{B}}^2 k_{\mathrm{B}}T}{\bar{c}_{\mathrm{AB}}[\mathfrak{S}(20|\mathrm{A})_{\mathrm{AB}} + \mathfrak{S}(\begin{smallmatrix}20\\20\end{smallmatrix}|\begin{smallmatrix}\mathrm{A}\\\mathrm{B}\end{smallmatrix})_{\mathrm{AB}}]} \tag{2.82a}$$

or, equivalently, as

$$\eta_{\mathrm{AB}} = \frac{2y_{\mathrm{A}}^2 k_{\mathrm{B}}T}{\bar{c}_{\mathrm{AB}}[\mathfrak{S}(20|\mathrm{B})_{\mathrm{AB}} + \mathfrak{S}(\begin{smallmatrix}20\\20\end{smallmatrix}|\begin{smallmatrix}\mathrm{A}\\\mathrm{B}\end{smallmatrix})_{\mathrm{AB}}]} \,. \tag{2.82b}$$

However, even though the interaction viscosity η_{AB} contains all the information on the unlike intermolecular interaction that may be obtained from mixture viscosity measurements, it cannot be measured directly, and must be extracted from the measured mixture viscosity data: the extraction process necessarily lowers the final accuracy with which η_{AB} can be determined from the experimental data. Moreover, it is important to note (Maitland *et al.*, 1980) that although η_{AB} has all the attributes of a shear viscosity coefficient, it cannot actually be measured experimentally, because it represents the shear viscosity of a hypothetical gas of atoms with masses twice the reduced mass of an A–B pair that interact via an interatomic potential energy function that is equal to $V_{\mathrm{AB}}(R)$.

That the first Chapman–Cowling approximation $[\eta_{\mathrm{mix}}]_1$ to the mixture viscosity is given by a complicated combination of effective cross-sections involving both like-pair interactions $V_{\mathrm{AA}}(R)$ and $V_{\mathrm{BB}}(R)$ as well as the unlike-pair interaction $V_{\mathrm{AB}}(R)$ becomes evident upon utilizing definitions (2.59) to express $\widetilde{\mathfrak{S}}(20|\mathrm{A})$ and $\widetilde{\mathfrak{S}}(20|\mathrm{B})$ as

$$\widetilde{\mathfrak{S}}(20|\mathrm{A}) = x_{\mathrm{A}}\sqrt{2}y_{\mathrm{B}}\mathfrak{S}(20|\mathrm{A})_{\mathrm{AA}} + x_{\mathrm{B}}\mathfrak{S}(20|\mathrm{A})_{\mathrm{AB}}$$

and

$$\widetilde{\mathfrak{S}}(20|\mathrm{B}) = x_{\mathrm{A}}\mathfrak{S}(20|\mathrm{B})_{\mathrm{AB}} + x_{\mathrm{B}}\sqrt{2}y_{\mathrm{A}}\mathfrak{S}(20|\mathrm{B})_{\mathrm{BB}} \,.$$

Of these four effective cross-sections, $\mathfrak{S}(20|\mathrm{A})_{\mathrm{AA}}$ and $\mathfrak{S}(20|\mathrm{B})_{\mathrm{BB}}$ are the pure gas effective cross-sections that determine the first Chapman–Cowling approximations to the pure gas viscosities η_{A} and η_{B}, while $\mathfrak{S}(20|\mathrm{A})_{\mathrm{AB}}$ and $\mathfrak{S}(20|\mathrm{B})_{\mathrm{AB}}$, together with $\mathfrak{S}(\begin{smallmatrix}20\\20\end{smallmatrix}|\begin{smallmatrix}\mathrm{A}\\\mathrm{B}\end{smallmatrix})_{\mathrm{AB}}] = \mathfrak{S}(\begin{smallmatrix}20\\20\end{smallmatrix}|\begin{smallmatrix}\mathrm{B}\\\mathrm{A}\end{smallmatrix})_{\mathrm{AB}}$, depend upon the unlike-atom interaction. Moreover, as $\mathfrak{S}(20|\mathrm{A})_{\mathrm{AB}}$ is related to the diffusion effective cross-sections $\mathfrak{S}(10|\mathrm{A})_{\mathrm{AB}}$ and $\mathfrak{S}(\begin{smallmatrix}20\\20\end{smallmatrix}|\begin{smallmatrix}\mathrm{A}\\\mathrm{B}\end{smallmatrix})_{\mathrm{AB}}]$ by an exact relation, the first Chapman–Cowling approximation, $[D_{\mathrm{AB}}]_1$, to the binary diffusion coefficient D is implicitly involved. It appears in the form of $A_{12}^{\star}$, traditionally defined as (Hirschfelder *et al.*, 1964; Ferziger and Kaper, 1972; Maitland *et al.*, 1980)

$$A_{12}^{\star} \equiv \frac{5nm_{\mathrm{r}}(\mathrm{A},\mathrm{B})[D_{\mathrm{AB}}]_1}{6\eta_{\mathrm{AB}}} \,, \tag{2.83a}$$

in which $m_{\mathrm{r}}(\mathrm{A},\mathrm{B})$ is the reduced mass of the A–B atom pair. It is worth noting that, as $A_{12}^{\star}$ is proportional to the ratio of two (mixture) transport coefficients, it turns out to be relatively insensitive to the actual form of the interaction $V_{\mathrm{AB}}(R)$. In terms of effective cross-sections, $A_{12}^{\star}$ is given by

$$A_{12}^{\star} = \frac{5}{6}\,\frac{\mathfrak{S}(20|\mathrm{A})_{\mathrm{AB}} + \mathfrak{S}\big({}^{20}_{20}\big|{}^{\mathrm{A}}_{\mathrm{B}}\big)_{\mathrm{AB}}}{\mathfrak{S}(10|\mathrm{A})_{\mathrm{AB}}}\,. \tag{2.83b}$$

Although the first-approximation expression [given explicitly in eqns (8.2–8.22) of Hirschfelder *et al.*, 1964] may be employed in principle to obtain estimates of $[D_{\mathrm{AB}}]_1$ from sets of viscosity measurements obtained at a given temperature for the pure components and a number of their binary mixtures, it is essentially impossible in practice (as also mentioned in Hirschfelder *et al.*, 1964) to obtain precision values of η_{AB} or D_{AB} from such experimental data: indeed, this is especially the case for high-precision measurements of the shear viscosity (in which experimental uncertainties are of the order of or smaller than $\pm 0.2\%$).

The second Chapman–Cowling approximation to the shear viscosity of a diatomic mixture requires the addition of a second set of two orthogonal basis functions for the nonhydrodynamic subspace, specifically $\mathbf{\Phi}^{21|\mathrm{A}}$ and $\mathbf{\Phi}^{21|\mathrm{B}}$, in which case the matrix approximation to $\mathcal{R}^{-1}$ involves the 4×4 matrix $\mathsf{S}^{[2]}(\eta_{\mathrm{mix}})$ of reduced matrix elements of $\mathcal{R}$ given by

$$\mathsf{S}^{[2]}(\eta_{\mathrm{mix}}) = \begin{pmatrix} S(20|\mathrm{A}) & S\big({}^{20}_{20}\big|{}^{\mathrm{A}}_{\mathrm{B}}\big) & S\big({}^{20}_{21}\big|{}^{\mathrm{A}}_{\mathrm{A}}\big) & S\big({}^{20}_{21}\big|{}^{\mathrm{A}}_{\mathrm{B}}\big) \\ S\big({}^{20}_{20}\big|{}^{\mathrm{B}}_{\mathrm{A}}\big) & S(20|\mathrm{B}) & S\big({}^{20}_{21}\big|{}^{\mathrm{B}}_{\mathrm{A}}\big) & S\big({}^{20}_{21}\big|{}^{\mathrm{B}}_{\mathrm{B}}\big) \\ S\big({}^{21}_{20}\big|{}^{\mathrm{A}}_{\mathrm{A}}\big) & S\big({}^{21}_{20}\big|{}^{\mathrm{A}}_{\mathrm{B}}\big) & S(21|\mathrm{A}) & S\big({}^{21}_{21}\big|{}^{\mathrm{A}}_{\mathrm{B}}\big) \\ S\big({}^{21}_{20}\big|{}^{\mathrm{B}}_{\mathrm{A}}\big) & S\big({}^{21}_{20}\big|{}^{\mathrm{B}}_{\mathrm{B}}\big) & S\big({}^{21}_{21}\big|{}^{\mathrm{B}}_{\mathrm{A}}\big) & S(21|\mathrm{B}) \end{pmatrix}.$$

Utilization of eqn (2.59,2.60b) leads to an expression for the second Chapman–Cowling approximation, $[\eta_{\mathrm{mix}}]_2$, to the (binary) mixture viscosity, namely,

$$[\eta_{\mathrm{mix}}]_2 = \frac{k_{\mathrm{B}}T}{\bar{c}_{\mathrm{AB}}\Delta^{[2]}(\eta_{\mathrm{mix}})}\left\{x_{\mathrm{A}}\Delta_{11}^{[2]} + x_{\mathrm{B}}\Delta_{12}^{[2]} + x_{\mathrm{A}}\Delta_{21}^{[2]} + x_{\mathrm{B}}\Delta_{22}^{[2]}\right\}, \tag{2.84}$$

in which the $\Delta_{ij}^{[2]} \equiv \Delta_{ij}^{[2]}(\eta_{\mathrm{mix}})$ are the ij^{th} cofactors of the 4×4 determinant, $\Delta^{[2]}(\eta_{\mathrm{mix}})$, of mixture effective cross-sections defined as

$$\Delta^{[2]}(\eta_{\mathrm{mix}}) \equiv \begin{vmatrix} \widetilde{\mathfrak{S}}(20|\mathrm{A}) & x_{\mathrm{A}}\mathfrak{S}\big({}^{20}_{20}\big|{}^{\mathrm{A}}_{\mathrm{B}}\big)_{\mathrm{AB}} & \widetilde{\mathfrak{S}}\big({}^{20}_{21}\big|{}^{\mathrm{A}}_{\mathrm{A}}\big) & x_{\mathrm{A}}\mathfrak{S}\big({}^{20}_{21}\big|{}^{\mathrm{A}}_{\mathrm{B}}\big)_{\mathrm{AB}} \\ x_{\mathrm{B}}\mathfrak{S}\big({}^{20}_{20}\big|{}^{\mathrm{B}}_{\mathrm{A}}\big)_{\mathrm{AB}} & \widetilde{\mathfrak{S}}(20|\mathrm{B}) & x_{\mathrm{B}}\mathfrak{S}\big({}^{20}_{21}\big|{}^{\mathrm{B}}_{\mathrm{A}}\big)_{\mathrm{AB}} & \widetilde{\mathfrak{S}}\big({}^{20}_{21}\big|{}^{\mathrm{B}}_{\mathrm{B}}\big) \\ \widetilde{\mathfrak{S}}\big({}^{21}_{20}\big|{}^{\mathrm{A}}_{\mathrm{A}}\big) & x_{\mathrm{A}}\mathfrak{S}\big({}^{21}_{20}\big|{}^{\mathrm{A}}_{\mathrm{B}}\big)_{\mathrm{AB}} & \widetilde{\mathfrak{S}}(21|\mathrm{A}) & x_{\mathrm{A}}\mathfrak{S}\big({}^{21}_{21}\big|{}^{\mathrm{A}}_{\mathrm{B}}\big)_{\mathrm{AB}} \\ x_{\mathrm{B}}\mathfrak{S}\big({}^{21}_{20}\big|{}^{\mathrm{B}}_{\mathrm{A}}\big)_{\mathrm{AB}} & \widetilde{\mathfrak{S}}\big({}^{21}_{20}\big|{}^{\mathrm{B}}_{\mathrm{B}}\big) & x_{\mathrm{B}}\mathfrak{S}\big({}^{21}_{21}\big|{}^{\mathrm{B}}_{\mathrm{A}}\big)_{\mathrm{AB}} & \widetilde{\mathfrak{S}}(21|\mathrm{B}) \end{vmatrix}. \tag{2.85}$$

2.4.4 Diffusion, thermal diffusion, and thermal conductivity for binary mixtures

More care is required in dealing with the vectorial phenomena, as the arrays $\mathbf{\Phi}^{10|\mathrm{A}}$ and $\mathbf{\Phi}^{10|\mathrm{B}}$, of which the microscopic diffusive flux is composed, do not belong to the nonhydrodynamic subspace. Operation with $\mathcal{R}$ annihilates their hydrodynamic components, with the consequence that $\mathcal{R}\boldsymbol{\Phi}^{10|\mathrm{A}}$ and $\mathcal{R}\boldsymbol{\Phi}^{10|\mathrm{B}}$ are not linearly independent, so that inversion of $\mathcal{R}$ cannot trivially be performed: as a consequence, it is not possible to include both arrays $\boldsymbol{\Phi}^{10|\mathrm{k}}$ (k = A, B) in the basis set. Either basis vector or, equivalently, any nonhydrodynamic linear combination of the two is, however, acceptable. Thus, so long as care is taken with respect to the normalization prescription, $\boldsymbol{\Psi}^d$ itself may be utilized. The normalization constant N^d, given by

$$N^d = \frac{1}{3}\langle \boldsymbol{\Psi}^d \cdot \boldsymbol{\Psi}^d \rangle = (C^{d|\mathrm{A}})^2 x_\mathrm{A} + (C^{d|\mathrm{B}})^2 x_\mathrm{B} = \frac{k_\mathrm{B} T \rho}{n \rho_\mathrm{A} \rho_\mathrm{B}}, \tag{2.86}$$

is then required in addition to the reduced matrix elements of $\mathcal{R}$ that also involve the unnormalized $\boldsymbol{\Psi}^d$: these matrix elements are given by

$$\langle \boldsymbol{\Phi}^{\alpha|\mathrm{k}} \| \mathcal{R} \| \boldsymbol{\Psi}^d \rangle = nS\left(\begin{smallmatrix}\alpha\\ d\end{smallmatrix}\middle|\,\mathrm{k}\right), \qquad \langle \boldsymbol{\Psi}^d \| \mathcal{R} \| \boldsymbol{\Psi}^d \rangle = nS\left(\begin{smallmatrix}d\\ d\end{smallmatrix}\right) \equiv nS(d). \tag{2.87}$$

The procedure that led to eqn (2.66) must be repeated, but with $\boldsymbol{\Psi}^d$ included in the basis set. Since it couples only to vectors, it may safely be assumed that only the vectorial forces X^d and X^E will be involved. The revised expansion (2.62) thus becomes

$$\phi = \sum_{\alpha,\mathrm{k}} \boldsymbol{\Phi}^{\alpha|\mathrm{k}} \odot \mathsf{A}^{\alpha|\mathrm{k}} + \boldsymbol{\Psi}^d \cdot \mathsf{A}^d .$$

When this expression is substituted into the Boltzmann eqn (2.27) and an inner product with the arrays $\boldsymbol{\Phi}^{\beta|\mathrm{k}}$ or $\boldsymbol{\Psi}^d$ is carried out, all terms involving tensors of rank other than 1 drop out. Further steps are carried out as previously, but now replacing the summation $\sum_\mathrm{k} C^{\alpha|\mathrm{k}} x_\mathrm{k}$ by the normalization constant N^d whenever $\boldsymbol{\Psi}^d$ is involved. As the matrix S then also contains the elements specified in eqn (2.87), it is reasonable to label the elements of the inverse matrix accordingly as $S^{-1}\left(\begin{smallmatrix}\alpha\\ \beta\end{smallmatrix}\middle|\begin{smallmatrix}\mathrm{k}\\ \mathrm{k}'\end{smallmatrix}\right)$, $S^{-1}\left(\begin{smallmatrix}\alpha\\ d\end{smallmatrix}\middle|\,\mathrm{k}\right)$, and $S^{-1}\left(\begin{smallmatrix}d\\ d\end{smallmatrix}\right)$. There is no need to provide every step of such a derivation explicitly, as the outcome will normally be clear. Whereas L^{EE} retains the form (2.66), two additional transport coefficients, namely,

$$L^{dE} = L^{Ed} = \frac{1}{k_\mathrm{B} T} \sum_\mathrm{k} C^\mathrm{k} x_\mathrm{k} S^{-1}\left(\begin{smallmatrix}E\\ d\end{smallmatrix}\middle|\,\mathrm{k}\right) N^d \tag{2.88a}$$

and

$$L^{dd} = \frac{1}{k_\mathrm{B} T} N^d S^{-1}\left(\begin{smallmatrix}d\\ d\end{smallmatrix}\right) N^d, \tag{2.88b}$$

are also obtained.

According to this scheme, the lowest matrix approximation for the coefficients $D \equiv L^{dd}$, $L^{dE} \equiv D_\mathrm{T} = \widetilde{D}_\mathrm{T}$, and λ would already require three basis arrays, namely $\boldsymbol{\Psi}^d$, $\boldsymbol{\Phi}^{11|\mathrm{A}}$, and $\boldsymbol{\Phi}^{11|\mathrm{B}}$, which would lead to a 3×3 matrix S. It is not difficult to imagine that the final expressions will become quite unwieldy. A cruder (and simpler) approximation is feasible for the diagonal coefficients D and λ if it is assumed that $\boldsymbol{\Phi}^{11|\mathrm{A}}$ and $\boldsymbol{\Phi}^{11|\mathrm{B}}$ do not couple to $\boldsymbol{\Psi}^d$: such an assumption implies that the cross-effects are sufficiently weak that they have negligible influence upon the outcome.

Should this assumption be made, then the 3×3 matrix S splits into a block diagonal matrix consisting of a 1×1 element $S(d)$ and a 2×2 matrix with elements $S(^{11}_{11}|^{\mathrm{k}}_{\mathrm{k}'})$, k, k$' =$ A, B.

For the evaluation of D using $\mathbf{\Psi}^d$ as the sole basis array, the result is much the same as those obtained for η and λ for a pure gas (cf. Section 4.1, Section 4.2), with L^{dd} given in terms of $S(d)$ only as

$$L^{dd} = \frac{D}{px_{\mathrm{A}}x_{\mathrm{B}}} = \frac{(N^d)^2}{k_{\mathrm{B}}TS(d)}\,. \tag{2.89}$$

By noting that $S(d)$ is given by

$$S(d) = (C^{d|\mathrm{A}})^2 S(10|\mathrm{A}) + 2C^{d|\mathrm{A}}C^{d|\mathrm{B}}S(^{10}_{10}|^{\mathrm{B}}_{\mathrm{A}}) + (C^{d|\mathrm{B}})^2 S(10|\mathrm{B})\,, \tag{2.90a}$$

and that the pure gas contributions $\mathfrak{S}(10|\mathrm{A})$ to $S(10|\mathrm{A})$ and $\mathfrak{S}(10|\mathrm{B})$ to $S(10|\mathrm{B})$ must vanish (because they are proportional to matrix elements involving a pure-gas linearized collision operator operating upon a hydrodynamic basis function $\mathbf{\Phi}^{10|\mathrm{k}}$), the result

$$S(d) = \frac{k_{\mathrm{B}}T\bar{c}_{\mathrm{AB}}}{n^2}\left[\frac{x_{\mathrm{B}}\mathfrak{S}(10|\mathrm{A})_{\mathrm{AB}}}{x_{\mathrm{A}}m_{\mathrm{A}}} - \frac{2\mathfrak{S}(^{10}_{10}|^{\mathrm{B}}_{\mathrm{A}})_{\mathrm{AB}}}{(m_{\mathrm{A}}m_{\mathrm{B}})^{\frac{1}{2}}} + \frac{x_{\mathrm{A}}\mathfrak{S}(10|\mathrm{B})_{\mathrm{AB}}}{x_{\mathrm{B}}m_{\mathrm{B}}}\right] \tag{2.90b}$$

is obtained upon utilizing eqns (2.67) and the values for $C^{d|\mathrm{k}}$ from Table 2.2. Further simplification arises from two relations among the three effective cross-sections involved in this expression: Section 2.6 shows how these relations may be summarized as

$$y_{\mathrm{A}}^2\mathfrak{S}(10|\mathrm{A})_{\mathrm{AB}} = -y_{\mathrm{A}}y_{\mathrm{B}}\mathfrak{S}(^{10}_{10}|^{\mathrm{A}}_{\mathrm{B}})_{\mathrm{AB}} = y_{\mathrm{B}}^2\mathfrak{S}(10|\mathrm{B})_{\mathrm{AB}}\,, \tag{2.91}$$

once relations (2.91) have been substituted into in expression (2.90b), $S(d)$ may be shown to reduce to

$$S(d) = \frac{k_{\mathrm{B}}T\rho^2\bar{c}_{\mathrm{AB}}\mathfrak{S}(10|\mathrm{A})_{\mathrm{AB}}}{n^2\rho_{\mathrm{A}}\rho_{\mathrm{B}}m_{\mathrm{B}}}\,, \tag{2.92}$$

with $\rho \equiv \rho_{\mathrm{A}} + \rho_{\mathrm{B}}$ given by $\rho = nx_{\mathrm{A}}m_{\mathrm{A}} + nx_{\mathrm{B}}m_{\mathrm{B}}$. The well-known Chapman–Cowling first approximation $[D_{\mathrm{AB}}]_1$ for the binary diffusion coefficient D_{AB} then follows from eqns (2.89, 2.92) as

$$[D_{\mathrm{AB}}]_1 = \frac{k_{\mathrm{B}}T}{n\bar{c}_{\mathrm{AB}}m_{\mathrm{A}}\mathfrak{S}(10|\mathrm{A})_{\mathrm{AB}}}\,. \tag{2.93}$$

Before discussing higher-order approximations to the binary diffusion coefficient, it will be useful to pause briefly to examine its dependence on the masses m_{A} and m_{B}. The next section shows that the effective cross-sections obtained for an isotropic interaction PEF can be expressed in general as linear combinations of quantities $\Omega_{\mathrm{kk}'}^{(m,n)}(T)$ referred to as "omega integrals." In particular, the effective cross-section $\mathfrak{S}(10|\mathrm{A})_{\mathrm{AB}}$

that determines the binary diffusion coefficient is given [see eqn (2.139)] in terms of a single omega integral as

$$\mathfrak{S}(10|\mathrm{A})_{\mathrm{AB}} = \frac{16y_{\mathrm{B}}^2}{3\bar{c}_{\mathrm{AB}}}\,\Omega_{\mathrm{AB}}^{(1,1)}(T)\,.$$

The temperature dependence of the omega integral $\Omega_{\mathrm{AB}}^{(1,1)}(T)$ is given [see eqns (2.161)] by

$$\Omega_{\mathrm{AB}}^{(1,1)}(T) = (k_{\mathrm{B}}T)^{-\frac{5}{2}}\left(\frac{2\pi}{m_{\mathrm{r}}}\right)^{\frac{1}{2}}\int_0^\infty \mathrm{e}^{-\beta E}E^2 Q_{\mathrm{AB}}^{(1)}(E)\,\mathrm{d}E\,,$$

with $\beta \equiv (k_{\mathrm{B}}T)^{-1}$, m_{r} the reduced mass of the colliding atoms, E the collision energy, and $Q_{\mathrm{AB}}^{(1)}(E)$ an energy-dependent collision cross-section obtained from the differential scattering cross-section $\sigma_{\mathrm{AB}}(E,b)$ as

$$Q_{\mathrm{AB}}^{(1)}(E) = 2\pi\int_{-1}^{1}(1-\cos\chi)\sigma_{\mathrm{AB}}(E,\chi)\,\mathrm{d}\cos\chi\,.$$

The temperature, number density, and reduced mass dependence of $[D_{\mathrm{AB}}]_1$ may therefore be represented as

$$[D_{\mathrm{AB}}]_1 = \frac{f(T)}{n\sqrt{m_{\mathrm{r}}}}\,, \tag{2.94a}$$

in which $f(T)$ is determined by the interaction energy $V_{\mathrm{AB}}(R)$.

Apart from the reference to the interaction energy, the above expression does not specifically differentiate between like and unlike atoms. It is thus tempting to consider the prospect of what may be termed "self-diffusion" in a pure gas. Note, however, that because atoms of a pure gas are identical there exists no mechanism by which their interdiffusion can be followed, and hence, true "self-diffusion" is unobservable. Much the same conclusion results from considering the expression for the pure gas effective cross-section $\mathfrak{S}(10)$ obtained from eqn (2.122): employment of expressions (2.120) and the basis vector $\mathbf{\Phi}^{10} = \sqrt{2}\mathbf{W}$ to show that, in accordance with the conservation of linear momentum in a binary collision, $\Delta\mathbf{\Phi}^{10}$, hence, $\mathfrak{S}(10)$, vanishes. The vanishing of $\mathfrak{S}(10|\mathrm{A})_{\mathrm{AB}}$ in eqn (2.93) implies an infinite diffusion coefficient or, equivalently, instantaneous mixing. This result also implies that a diffusion coefficient cannot properly be defined for a pure gas. However, should two different isotopes, A and A′, of the same atomic species be brought into contact, their interdiffusion can be followed using tracer techniques or nuclear spin diffusion, for example, to determine an isotopic diffusion coefficient $D_{\mathrm{AA'}}(T)$ for the gas. Moreover, as the reduced mass m_{r} that appears in eqn (2.94a) differs slightly from the value $\frac{1}{2}m_{\mathrm{A}}$, the isotopic diffusion coefficient will also differ slightly from the value that a self-diffusion coefficient would have, could it actually be measured.

The fact that thermal equilibrium is a dynamic, rather than a static, process means that the particles in an equilibrium gas always exhibit relative motion, which may be considered as a form of self-diffusion. Equation (2.94a) provides a means of determining the effective self-diffusion coefficient $D_{\mathrm{AA}}(T)$ for this process via observations made on the diffusive motion of tagged nuclei A′. It is thus possible to employ eqn (2.94a) to obtain values for the self-diffusion coefficient $D_{\mathrm{AA}}(T)$ from experimental values of the isotopic diffusion coefficient $D_{\mathrm{AA'}}(T)$ as[3]

$$D_{\mathrm{AA}}(T) = \left(\frac{2m_{\mathrm{A'}}}{m_{\mathrm{A}} + m_{\mathrm{A'}}}\right)^{\frac{1}{2}} D_{\mathrm{AA'}}(T)\,, \tag{2.94b}$$

in which m_{A} and $m_{\mathrm{A'}}$ are the masses of isotopes A and A'.

Higher-order Chapman–Cowling approximations to the diffusion coefficient may be generated as required, by adding additional basis vectors in a controlled fashion: the basis vectors involved in these approximations have the form $\boldsymbol{\Phi}^{1s|\mathrm{k}} \propto \mathbf{W}_{\mathrm{k}} L_s^{\frac{5}{2}}(W_{\mathrm{k}}^2)$, with $s = 1, 2, \cdots$. The second Chapman–Cowling approximation to D will thus be generated by including the collisional coupling between $\boldsymbol{\Psi}^d$ and the basis functions $\boldsymbol{\Phi}^{11|\mathrm{A}}$ and $\boldsymbol{\Phi}^{11|\mathrm{B}}$ required for the first Chapman–Cowling approximation to λ_{mix}. Addition of these two basis functions results in the relevant array of reduced matrix elements of $\mathcal{R}$ being augmented to give

$$\mathsf{S}_{D\lambda}^{[1]} = \begin{pmatrix} S(d) & S(\begin{smallmatrix} d \\ 11 \end{smallmatrix}|_{\mathrm{A}}) & S(\begin{smallmatrix} d \\ 11 \end{smallmatrix}|_{\mathrm{B}}) \\ S(\begin{smallmatrix} 11 \\ d \end{smallmatrix}|^{\mathrm{A}}) & S(11|\mathrm{A}) & S(\begin{smallmatrix} 11 \\ 11 \end{smallmatrix}|\begin{smallmatrix} \mathrm{A} \\ \mathrm{B} \end{smallmatrix}) \\ S(\begin{smallmatrix} 11 \\ d \end{smallmatrix}|^{\mathrm{B}}) & S(\begin{smallmatrix} 11 \\ 11 \end{smallmatrix}|\begin{smallmatrix} \mathrm{B} \\ \mathrm{A} \end{smallmatrix}) & S(11|\mathrm{B} \end{pmatrix}. \tag{2.95a}$$

The inverse matrix element $S^{-1}(\begin{smallmatrix} d \\ d \end{smallmatrix})$ required in eqn (2.88b) is now given by

$$S^{-1}(\begin{smallmatrix} d \\ d \end{smallmatrix}) \equiv S^{-1}(d) = \frac{x_{\mathrm{A}} x_{\mathrm{B}} \bar{c}_{\mathrm{AB}}^2 \Delta_{11}^{[1]}(D\lambda)}{\Delta^{[1]}(D\lambda)}\,, \tag{2.95b}$$

in which $\Delta_{11}^{[1]}(D\lambda)$ is the cofactor of $S(d)$ from the determinant,

$$\Delta^{[1]}(D\lambda) = \begin{vmatrix} S(d) & S(\begin{smallmatrix} d \\ 11 \end{smallmatrix}|_{\mathrm{A}}) & S(\begin{smallmatrix} d \\ 11 \end{smallmatrix}|_{\mathrm{B}}) \\ S(\begin{smallmatrix} 11 \\ d \end{smallmatrix}|^{\mathrm{A}}) & S(11|\mathrm{A}) & S(\begin{smallmatrix} 11 \\ 11 \end{smallmatrix}|\begin{smallmatrix} \mathrm{A} \\ \mathrm{B} \end{smallmatrix}) \\ S(\begin{smallmatrix} 11 \\ d \end{smallmatrix}|^{\mathrm{B}}) & S(\begin{smallmatrix} 11 \\ 11 \end{smallmatrix}|\begin{smallmatrix} \mathrm{B} \\ \mathrm{A} \end{smallmatrix}) & S(11|\mathrm{B} \end{vmatrix}\,, \tag{2.96}$$

of $\mathsf{S}_{D\lambda}^{[1]}$. Upon employing relations (2.67) and following a number of algebraic simplifications, the second Chapman–Cowling approximation $[D]_2$ may be obtained as

$$[D]_2 = \frac{[D]_1}{1 - \delta_D^{[2]}}\,, \tag{2.97}$$

in which $[D]_1$ is given by eqn (2.93), and $\delta_D^{[2]}$ is given by

[3]This relation appears to have been first obtained by Hutchinson (Meeks *et al.*, 1994).

$$\delta_D^{[2]} = \frac{1}{\Delta_{\text{mix}}(11)} \left\{ x_{\text{B}} \frac{\mathfrak{S}\left(\begin{smallmatrix}10\\11\end{smallmatrix}\middle|\begin{smallmatrix}\text{A}\\\text{A}\end{smallmatrix}\right)_{\text{AB}}}{\mathfrak{S}(10|\text{A})_{\text{AB}}} \begin{vmatrix} \mathfrak{S}\left(\begin{smallmatrix}11\\10\end{smallmatrix}\middle|\begin{smallmatrix}\text{A}\\\text{A}\end{smallmatrix}\right)_{\text{AB}} & x_{\text{A}}\mathfrak{S}\left(\begin{smallmatrix}11\\11\end{smallmatrix}\middle|\begin{smallmatrix}\text{A}\\\text{B}\end{smallmatrix}\right)_{\text{AB}} \\ \mathfrak{S}\left(\begin{smallmatrix}11\\10\end{smallmatrix}\middle|\begin{smallmatrix}\text{B}\\\text{A}\end{smallmatrix}\right)_{\text{AB}} & \widetilde{\mathfrak{S}}(11|\text{B}) \end{vmatrix} \right.$$

$$\left. - x_{\text{A}} \frac{\mathfrak{S}\left(\begin{smallmatrix}10\\11\end{smallmatrix}\middle|\begin{smallmatrix}\text{A}\\\text{B}\end{smallmatrix}\right)_{\text{AB}}}{\mathfrak{S}(10|\text{A})_{\text{AB}}} \begin{vmatrix} \mathfrak{S}\left(\begin{smallmatrix}11\\10\end{smallmatrix}\middle|\begin{smallmatrix}\text{A}\\\text{A}\end{smallmatrix}\right)_{\text{AB}} & \widetilde{\mathfrak{S}}(11|\text{A}) \\ \mathfrak{S}\left(\begin{smallmatrix}11\\10\end{smallmatrix}\middle|\begin{smallmatrix}\text{B}\\\text{A}\end{smallmatrix}\right)_{\text{AB}} & x_{\text{B}}\mathfrak{S}\left(\begin{smallmatrix}11\\11\end{smallmatrix}\middle|\begin{smallmatrix}\text{B}\\\text{A}\end{smallmatrix}\right)_{\text{AB}} \end{vmatrix} \right\}, \tag{2.98}$$

with $\Delta_{\text{mix}}(11)$ defined as

$$\Delta_{\text{mix}}(11) \equiv \widetilde{\mathfrak{S}}(11|\text{A})\widetilde{\mathfrak{S}}(11|\text{B}) - x_{\text{A}}x_{\text{B}}\mathfrak{S}^2\left(\begin{smallmatrix}11\\11\end{smallmatrix}\middle|\begin{smallmatrix}\text{A}\\\text{B}\end{smallmatrix}\right)_{\text{AB}}. \tag{2.99}$$

The defining relation

$$S\left(\begin{smallmatrix}d\\11\end{smallmatrix}\middle|\text{k}\right) = \frac{\sqrt{k_{\text{B}}T}\,\rho\bar{c}_{\text{AB}}}{n^2\sqrt{m_{\text{A}}\,m_{\text{B}}}}\,\mathfrak{S}\left(\begin{smallmatrix}10\\11\end{smallmatrix}\middle|\begin{smallmatrix}\text{A}\\\text{k}\end{smallmatrix}\right)_{\text{AB}}, \qquad \text{k} = \text{A, B}, \tag{2.100}$$

for effective cross-sections involving $\mathbf{\Psi}^d$ has been utilized, together with relations (2.67) to obtain $\delta_D^{[2]}$ in terms of effective cross-sections. This expression is similar to eqn (2.92) relating $S(d)$ to $\mathfrak{S}(10|\text{A})_{\text{AB}}$.

It is perhaps worth remarking also that even though the matrix representation of $\mathcal{R}$ has been denoted $\mathsf{S}^{[1]}_{D\lambda}$, the correction to $[D]_1$ is nevertheless referred to as $(1-\delta_D^{[2]})^{-1}$ in eqn (2.97), as it represents a true second approximation in the spirit of the Chapman–Cowling approximation scheme. This will not be the case, however, for the corresponding correction to the mixture thermal conductivity obtained using $\mathsf{S}^{[1]}(D\lambda)$. Expression (2.97) is equivalent to the second approximation expressions $[D]_2$ for D given in Hirschfelder *et al.* (1964), Chapman and Cowling (1970), Ferziger and Kaper (1972), and Maitland *et al.* (1980), except that it is couched in terms of effective cross-sections, rather than collision brackets.

Should the collisional coupling between $\mathbf{\Psi}^d$ and $\mathbf{\Phi}^{11|\text{A}}$, $\mathbf{\Phi}^{11|\text{B}}$ be ignored, as was done in arriving at the first Chapman–Cowling approximation to the diffusion coefficient, then following the same steps taken to arrive at eqn (2.80) for $[\eta_{\text{mix}}]_1$, gives L^{EE} as

$$L^{EE} = \tfrac{5}{2}(k_{\text{B}}T)^2 \frac{x_{\text{A}}^2 m_{\text{A}}^{-1} S(11|\text{B}) - 2x_{\text{A}}x_{\text{B}}(m_{\text{A}}m_{\text{B}})^{-\frac{1}{2}} S\left(\begin{smallmatrix}11\\11\end{smallmatrix}\middle|\begin{smallmatrix}\text{A}\\\text{B}\end{smallmatrix}\right) + x_{\text{B}}^2 m_{\text{B}}^{-1} S(11|\text{A})}{S(11|\text{A})S(11|\text{B}) - S^2\left(\begin{smallmatrix}11\\11\end{smallmatrix}\middle|\begin{smallmatrix}\text{A}\\\text{B}\end{smallmatrix}\right)}. \tag{2.101}$$

This expression may be converted into an expression giving λ_{mix} in terms of effective cross-sections upon recalling that $L^{EE} = \lambda_{\text{mix}}T$, then employing expressions (2.67) with $p = 1$, $s = s' = 1$, to obtain a first Chapman–Cowling approximation,

$$[\lambda_{\text{mix}}]_1 = \frac{5k_{\text{B}}^2T}{2m_{\text{A}}\bar{c}_{\text{AB}}} \frac{x_{\text{A}}\widetilde{\mathfrak{S}}(11|\text{B}) - 2x_{\text{A}}x_{\text{B}}\dfrac{y_{\text{A}}}{y_{\text{B}}}\mathfrak{S}\left(\begin{smallmatrix}11\\11\end{smallmatrix}\middle|\begin{smallmatrix}\text{A}\\\text{B}\end{smallmatrix}\right)_{\text{AB}} + \dfrac{y_{\text{A}}^2}{y_{\text{B}}^2}x_{\text{B}}\widetilde{\mathfrak{S}}(11|\text{A})}{\widetilde{\mathfrak{S}}(11|\text{A})\widetilde{\mathfrak{S}}(11|\text{B}) - x_{\text{A}}x_{\text{B}}\mathfrak{S}\left(\begin{smallmatrix}11\\11\end{smallmatrix}\middle|\begin{smallmatrix}\text{A}\\\text{B}\end{smallmatrix}\right)_{\text{AB}}}. \tag{2.102}$$

A more complete first Chapman–Cowling approximation for λ_{mix} will be obtained if $\mathbf{\Psi}^d$ is included in the basis set for the mixture thermal conductivity. The general defining relation eqn (2.66) for $L^{\alpha\beta}$ gives λ_{mix} as

$$\lambda_{\text{mix}} = \frac{1}{k_\text{B}T^2} \sum_{\text{k,k}'} C^{E|\text{k}} x_\text{k} S^{-1}\left({}^{E}_{E}\middle|{}^{\text{k}}_{\text{k}'}\right) C^{E|\text{k}'} x_{\text{k}'} \,, \tag{2.103a}$$

with $C^{E|\text{k}}$ obtained from Table 2.2. This expression becomes

$$\lambda_{\text{mix}} = \frac{5k_\text{B}^2 T}{2m_\text{A}} \left\{ x_\text{A}^2 S^{-1}(E|\text{A}) + 2x_\text{A}x_\text{B}\frac{y_\text{A}}{y_\text{B}} S^{-1}\left({}^{E}_{E}\middle|{}^{\text{A}}_{\text{B}}\right) + x_\text{B}^2 \frac{y_\text{A}^2}{y_\text{B}^2} S^{-1}(E|\text{B}) \right\} \tag{2.103b}$$

for a binary mixture of gases A and B, with $S^{-1}(E|\text{A})$, $S^{-1}(E|\text{B})$, $S^{-1}\left({}^{E}_{E}\middle|{}^{\text{A}}_{\text{B}}\right)$, and $S^{-1}\left({}^{E}_{E}\middle|{}^{\text{B}}_{\text{A}}\right)$ elements of the inverse matrix representation of $\mathcal{R}^{-1}$. The matrix inverse of concern for this first Chapman–Cowling approximation to λ_{mix}, is obtained from $\mathsf{S}^{[1]}(D\lambda)$, given in eqn (2.94). After some simplification, $[\lambda_{\text{mix}}]_1$ is obtained as

$$\begin{aligned}[\lambda_{\text{mix}}]_1 = \frac{5k_\text{B}^2 T S(d)}{2m_\text{A}|\mathsf{S}^{[1]}(D\lambda)|} &\left\{ x_\text{A}^2 S(11|\text{B}) - 2x_\text{A}x_\text{B}\frac{y_\text{A}}{y_\text{B}} S\left({}^{11}_{11}\middle|{}^{\text{A}}_{\text{B}}\right) + x_\text{B}^2 \frac{y_\text{A}^2}{y_\text{B}^2} S(11|\text{A}) \right. \\ &\left. -\frac{1}{S(d)}\left[x_\text{A}^2 S^2\left({}^{d}_{11}\middle|_{\text{B}}\right) - 2x_\text{A}x_\text{B}\frac{y_\text{A}}{y_\text{B}} S\left({}^{d}_{11}\middle|_{\text{A}}\right) S\left({}^{11}_{d}\middle|^{\text{B}}\right) + x_\text{B}^2\frac{y_\text{A}^2}{y_\text{B}^2} S^2\left({}^{d}_{11}\middle|_{\text{A}}\right)\right]\right\}.\end{aligned} \tag{2.104}$$

The determinant $|\mathsf{S}^{[1]}(D\lambda)|$ may also be written as

$$\begin{aligned}|\mathsf{S}^{[1]}(D\lambda)| &= S(d)x_\text{A}x_\text{B}\bar{c}_{\text{AB}}^3 \Delta_{\text{mix}}(11)\left(1 - \delta_\lambda^{[1]}\right) \\ &= \frac{k_\text{B}T\rho^2}{n^2\rho_\text{A}\rho_\text{B}m_\text{B}} x_\text{A}x_\text{B}\bar{c}_{\text{AB}}^3 \mathfrak{S}(10|\text{A})_{\text{AB}} \Delta_{\text{mix}}(11)\left(1 - \delta_\lambda^{[1]}\right) \,,\end{aligned} \tag{2.105a}$$

in which the correction term $\delta_\lambda^{[1]}$ is given by

$$\begin{aligned}\delta_\lambda^{[1]} = [&x_\text{B}\mathfrak{S}^2\left({}^{10}_{11}\middle|{}^{\text{A}}_{\text{A}}\right)_{\text{AB}}\widetilde{\mathfrak{S}}(11|\text{B}) - 2x_\text{A}x_\text{B}\mathfrak{S}\left({}^{10}_{11}\middle|{}^{\text{A}}_{\text{A}}\right)_{\text{AB}}\mathfrak{S}\left({}^{11}_{10}\middle|{}^{\text{B}}_{\text{A}}\right)_{\text{AB}}\mathfrak{S}\left({}^{11}_{11}\middle|{}^{\text{A}}_{\text{B}}\right)_{\text{AB}} \\ &+ x_\text{A}\mathfrak{S}^2\left({}^{10}_{11}\middle|{}^{\text{A}}_{\text{B}}\right)_{\text{AB}}\widetilde{\mathfrak{S}}(11|\text{A})]\,[\Delta_{\text{mix}}(11)\mathfrak{S}(10|\text{A})_{\text{AB}}]^{-1}\end{aligned} \tag{2.105b}$$

in terms of effective cross-sections. Further use of eqns (2.67) and eqns (2.92, 2.101) gives the final expression for $[\lambda_{\text{mix}}]_1$ as

$$\begin{aligned}[\lambda_{\text{mix}}]_1 = \frac{5k_\text{B}^2 T}{2m_\text{A}\bar{c}_{\text{AB}}\Delta_{\text{mix}}(11)(1-\delta_\lambda^{[1]})} &\left\{ x_\text{A}\widetilde{\mathfrak{S}}(11|\text{B}) - 2x_\text{A}x_\text{B}\frac{y_\text{A}}{y_\text{B}}\mathfrak{S}\left({}^{11}_{11}\middle|{}^{\text{A}}_{\text{B}}\right)_{\text{AB}} \right. \\ &+ x_\text{B}\frac{y_\text{A}^2}{y_\text{B}^2}\widetilde{\mathfrak{S}}(11|\text{A}) - \frac{1}{\mathfrak{S}(10|\text{A})_{\text{AB}}}\left[x_\text{A}^2\mathfrak{S}^2\left({}^{10}_{11}\middle|{}^{\text{A}}_{\text{B}}\right)_{\text{AB}}\right. \\ &\left.\left. -2x_\text{A}x_\text{B}\frac{y_\text{A}}{y_\text{B}}\mathfrak{S}\left({}^{10}_{11}\middle|{}^{\text{A}}_{\text{A}}\right)_{\text{AB}}\mathfrak{S}\left({}^{11}_{10}\middle|{}^{\text{B}}_{\text{A}}\right)_{\text{AB}} + x_\text{B}^2\frac{y_\text{A}^2}{y_\text{B}^2}\mathfrak{S}^2\left({}^{10}_{11}\middle|{}^{\text{A}}_{\text{A}}\right)_{\text{AB}}\right]\right\}.\end{aligned} \tag{2.106}$$

The first term of eqn (2.106) is the result (2.102) obtained upon assuming that $\mathbf{\Phi}^{11|\text{A}}$ and $\mathbf{\Phi}^{11|\text{B}}$ are not collisionally coupled to the microscopic diffusion flux $\mathbf{\Psi}^d$, and the second term gives the contribution to λ_{mix} associated with that collisional coupling. It is normally assumed to be rather small: Section 2.8 examines this assumption in greater detail.

The second Chapman–Cowling approximation to λ_{mix} for a binary gas mixture requires the addition of two basis vectors $\boldsymbol{\Phi}^{12|\mathrm{A}}$, $\boldsymbol{\Phi}^{12|\mathrm{B}}$ to the three-member basis $\{\boldsymbol{\Psi}^{d}, \boldsymbol{\Phi}^{11|\mathrm{A}}, \boldsymbol{\Phi}^{11|\mathrm{B}}\}$ used to determine $[\lambda_{\mathrm{mix}}]_1$. The most general expression for $[\lambda_{\mathrm{mix}}]_2$, obtained in a relatively straightforward process starting from eqn (2.103b), is given by

$$[\lambda_{\mathrm{mix}}]_2 = \frac{5k_{\mathrm{B}}^2 T}{2m_{\mathrm{A}}\bar{c}_{\mathrm{AB}}\Delta^{[2]}(D\lambda)}\left\{x_{\mathrm{A}}\Delta_{22}^{[2]}(D\lambda) + \frac{y_{\mathrm{A}}}{y_{\mathrm{B}}}\left[x_{\mathrm{B}}\Delta_{32}^{[2]}(D\lambda) + x_{\mathrm{A}}\Delta_{23}^{[2]}(D\lambda)\right]\right.$$
$$\left. + x_{\mathrm{B}}\frac{y_{\mathrm{A}}^2}{y_{\mathrm{B}}^2}\Delta_{33}^{[2]}(D\lambda)\right\}, \tag{2.107a}$$

in which the $\Delta_{ij}^{[2]}(D\lambda)$ are the ijth (signed) cofactors of the 5×5 determinant $\Delta^{[2]}(D\lambda)$ given by

$$\Delta^{[2]}(D\lambda) = \begin{vmatrix} \mathfrak{S}(10|\mathrm{A})_{\mathrm{AB}} & x_{\mathrm{B}}\mathfrak{S}\left({}^{10}_{11}\middle|{}^{\mathrm{A}}_{\mathrm{A}}\right)_{\mathrm{AB}} & x_{\mathrm{A}}\mathfrak{S}\left({}^{10}_{11}\middle|{}^{\mathrm{A}}_{\mathrm{B}}\right)_{\mathrm{AB}} & x_{\mathrm{B}}\mathfrak{S}\left({}^{10}_{12}\middle|{}^{\mathrm{A}}_{\mathrm{A}}\right)_{\mathrm{AB}} & x_{\mathrm{A}}\mathfrak{S}\left({}^{10}_{12}\middle|{}^{\mathrm{A}}_{\mathrm{B}}\right)_{\mathrm{AB}} \\ \mathfrak{S}\left({}^{11}_{10}\middle|{}^{\mathrm{A}}_{\mathrm{A}}\right)_{\mathrm{AB}} & \widetilde{\mathfrak{S}}(11|\mathrm{A}) & x_{\mathrm{A}}\mathfrak{S}\left({}^{11}_{11}\middle|{}^{\mathrm{A}}_{\mathrm{B}}\right)_{\mathrm{AB}} & \widetilde{\mathfrak{S}}\left({}^{12}_{12}\middle|{}^{\mathrm{A}}_{\mathrm{A}}\right) & x_{\mathrm{A}}\mathfrak{S}\left({}^{12}_{12}\middle|{}^{\mathrm{A}}_{\mathrm{B}}\right)_{\mathrm{AB}} \\ \mathfrak{S}\left({}^{11}_{10}\middle|{}^{\mathrm{B}}_{\mathrm{A}}\right)_{\mathrm{AB}} & \mathfrak{S}\left({}^{11}_{11}\middle|{}^{\mathrm{B}}_{\mathrm{A}}\right)_{\mathrm{AB}} & \widetilde{\mathfrak{S}}(11|\mathrm{B}) & x_{\mathrm{B}}\mathfrak{S}\left({}^{11}_{12}\middle|{}^{\mathrm{B}}_{\mathrm{A}}\right)_{\mathrm{AB}} & \widetilde{\mathfrak{S}}\left({}^{11}_{12}\middle|{}^{\mathrm{B}}_{\mathrm{B}}\right) \\ \mathfrak{S}\left({}^{12}_{10}\middle|{}^{\mathrm{A}}_{\mathrm{A}}\right)_{\mathrm{AB}} & \widetilde{\mathfrak{S}}\left({}^{12}_{11}\middle|{}^{\mathrm{A}}_{\mathrm{A}}\right) & x_{\mathrm{A}}\mathfrak{S}\left({}^{12}_{11}\middle|{}^{\mathrm{A}}_{\mathrm{B}}\right)_{\mathrm{AB}} & \widetilde{\mathfrak{S}}(12|\mathrm{A}) & x_{\mathrm{A}}\mathfrak{S}\left({}^{12}_{12}\middle|{}^{\mathrm{A}}_{\mathrm{B}}\right)_{\mathrm{AB}} \\ \mathfrak{S}\left({}^{12}_{10}\middle|{}^{\mathrm{B}}_{\mathrm{A}}\right)_{\mathrm{AB}} & x_{\mathrm{B}}\mathfrak{S}\left({}^{12}_{11}\middle|{}^{\mathrm{B}}_{\mathrm{A}}\right)_{\mathrm{AB}} & \widetilde{\mathfrak{S}}\left({}^{12}_{11}\middle|{}^{\mathrm{B}}_{\mathrm{B}}\right) & x_{\mathrm{B}}\mathfrak{S}\left({}^{12}_{12}\middle|{}^{\mathrm{B}}_{\mathrm{A}}\right)_{\mathrm{AB}} & \widetilde{\mathfrak{S}}(12|\mathrm{B}) \end{vmatrix}. \tag{2.107b}$$

For later use, it will be convenient to express $\Delta^{[2]}(D\lambda)$ in the form

$$\Delta^{[2]}(D\lambda) = \mathfrak{S}(10|\mathrm{A})_{\mathrm{AB}}\Delta_{\mathrm{mix}}^{[2]}(\lambda)(1 + \delta_{D\lambda}^{[2]}), \tag{2.107c}$$

with $\Delta_{\mathrm{mix}}^{[2]}(\lambda) \equiv \Delta_{11}^{[2]}(D\lambda)$, and $\delta_{D\lambda}^{[2]}$ expressed in terms of the $\Delta_{ij}^{[2]}(D\lambda)$ as

$$\delta_{D\lambda}^{[2]} = \frac{1}{\mathfrak{S}(10|\mathrm{A})_{\mathrm{AB}}\Delta_{\mathrm{mix}}^{[2]}(\lambda)}\sum_{i=2}^{5}\Delta_{1i}^{[2]}(D\lambda). \tag{2.107d}$$

Expression (2.107a) includes the collisional coupling between the microscopic diffusion flux $\boldsymbol{\Psi}^{d}$ and both the $\{\boldsymbol{\Phi}^{11|\mathrm{A}}, \boldsymbol{\Phi}^{11|\mathrm{B}}\}$ and $\{\boldsymbol{\Phi}^{12|\mathrm{A}}, \boldsymbol{\Phi}^{12|\mathrm{B}}\}$ sets of basis vectors from the nonhydrodynamic subspace, and required for the first and second Chapman–Cowling approximations to λ_{mix}.

If collisional coupling of $\boldsymbol{\Psi}^{d}$ to the first-order basis vectors $\{\boldsymbol{\Phi}^{11|\mathrm{A}}, \boldsymbol{\Phi}^{11|\mathrm{B}}\}$ can be assumed to be stronger than collisional coupling to the second-order basis vectors $\{\boldsymbol{\Phi}^{12|\mathrm{A}}, \boldsymbol{\Phi}^{12|\mathrm{B}}\}$, then a simpler version of $[\lambda_{\mathrm{mix}}]_2$, namely,

$$[\lambda_{\rm mix}]_2 = \frac{5k_{\rm B}^2T}{2m_{\rm A}\bar{c}_{\rm AB}\Delta_{\rm mix}^{[2]}(\lambda)}\Bigg\{x_{\rm A}\Delta_{11}^{[2]}(\lambda) - 2x_{\rm A}x_{\rm B}\frac{y_{\rm A}}{y_{\rm B}}\Delta_{\rm AB}^{[2]}(\lambda) + x_{\rm B}\Delta_{22}^{[2]}(\lambda)$$

$$-\frac{\Delta_{\rm mix}(12)}{\mathfrak{S}(10|{\rm A})_{\rm AB}}\Bigg[x_{\rm A}^2\mathfrak{S}^2\big({}^{11}_{10}|{}^{\rm B}_{\rm A}\big)_{\rm AB} - 2x_{\rm A}x_{\rm B}\frac{y_{\rm A}}{y_{\rm B}}\mathfrak{S}\big({}^{10}_{11}|{}^{\rm A}_{\rm B}\big)_{\rm AB} - 2x_{\rm A}x_{\rm B}\mathfrak{S}\big({}^{11}_{10}|{}^{\rm A}_{\rm A}\big)_{\rm AB}$$

$$+x_{\rm B}^2\frac{y_{\rm A}^2}{y_{\rm B}^2}\mathfrak{S}^2\big({}^{11}_{10}|{}^{\rm A}_{\rm A}\big)_{\rm AB}\Bigg]\Bigg\} \tag{2.108a}$$

may be obtained, in which $\Delta_{ij}^{[2]}(\lambda)$ is the ij$^{\rm th}$ (signed) cofactor of the 4×4 determinant $\Delta_{\rm mix}^{[2]}(\lambda)$, and $\Delta_{\rm AB}^{[2]}(\lambda)$ is given by

$$\Delta_{\rm AB}^{[2]}(\lambda) = \begin{vmatrix} \mathfrak{S}\big({}^{11}_{11}|{}^{\rm A}_{\rm B}\big)_{\rm AB} & \widetilde{\mathfrak{S}}\big({}^{11}_{12}|{}^{\rm A}_{\rm A}\big) & \mathfrak{S}\big({}^{11}_{12}|{}^{\rm A}_{\rm B}\big)_{\rm AB} \\ \mathfrak{S}\big({}^{12}_{11}|{}^{\rm A}_{\rm B}\big)_{\rm AB} & \widetilde{\mathfrak{S}}(12|{\rm A}) & \mathfrak{S}\big({}^{12}_{12}|{}^{\rm A}_{\rm B}\big)_{\rm AB} \\ \widetilde{\mathfrak{S}}\big({}^{12}_{11}|{}^{\rm A}_{\rm A}\big) & x_{\rm A}x_{\rm B}\mathfrak{S}\big({}^{12}_{12}|{}^{\rm A}_{\rm B}\big)_{\rm AB} & \widetilde{\mathfrak{S}}(12|{\rm B}) \end{vmatrix}. \tag{2.108b}$$

The first part of expression (2.108a), which is the binary mixture thermal conductivity analogon of the binary mixture viscosity second Chapman–Cowling approximation of eqn (2.85), results when all collisional couplings between $\boldsymbol{\Psi}^d$ and the basis elements for $\lambda_{\rm mix}$ are ignored, while the second part gives the additional correction associated with retention of the collisional coupling between $\boldsymbol{\Psi}^d$ and the basis elements $\boldsymbol{\Phi}^{11|{\rm A}}$, $\boldsymbol{\Phi}^{11|{\rm B}}$ for $[\lambda_{\rm mix}]_1$. The latter contribution to $[\lambda_{\rm mix}]_2$ differs slightly from the similar contribution to $[\lambda_{\rm mix}]_1$ generated by the collisional coupling to $\boldsymbol{\Psi}^d$, in that the contribution to $[\lambda_{\rm mix}]_1$ does not contain any contributions from the collisional coupling between the basis vectors $\{\boldsymbol{\Phi}^{11|{\rm A}}, \boldsymbol{\Phi}^{11|{\rm B}}\}$ and $\{\boldsymbol{\Phi}^{12|{\rm A}}, \boldsymbol{\Phi}^{12|{\rm B}}\}$.

There remains only the thermal diffusion coefficient to be evaluated. Upon starting from eqns (2.37c) and (2.66), with $\alpha\equiv d$ and $\beta\equiv E$, the thermal diffusion coefficient $D_{\rm T}$ is obtained as

$$\frac{D_{\rm T}}{x_{\rm A}x_{\rm B}} = \frac{1}{k_{\rm B}T}\sum_{\rm k=A,B} C^{\rm k}x_{\rm k}S^{-1}\big({}^{E}_{d}|{}^{\rm k}\big)N^d$$

$$= \frac{\rho}{n\rho_{\rm A}\rho_{\rm B}}\left(\frac{5k_{\rm B}^3T^3}{2}\right)^{\frac{1}{2}}\left[\frac{x_{\rm A}}{\sqrt{m_{\rm A}}}S^{-1}\big({}^{E}_{d}|{}^{\rm A}\big) + \frac{x_{\rm B}}{\sqrt{m_{\rm B}}}S^{-1}\big({}^{E}_{d}|{}^{\rm B}\big)\right]. \tag{2.109}$$

If, however, the basis set $\{\boldsymbol{\Psi}^d, \boldsymbol{\Phi}^{11|{\rm A}}, \boldsymbol{\Phi}^{11|{\rm B}}\}$ utilized to obtain the Chapman–Cowling first approximations $[D]_1$ and $[\lambda_{\rm mix}]_1$ is also employed to determine $D_{\rm T}$, then the relevant inverse matrix elements $S^{-1}\big({}^{E}_{d}|{}^{\rm A}\big)$ and $S^{-1}\big({}^{E}_{d}|{}^{\rm B}\big)$ are obtained from the determinant of $\mathsf{S}^{[1]}(D\lambda)$ as

$$S^{-1}\big({}^{E}_{d}|{}^{\rm A}\big) = -\frac{1}{|\mathsf{S}^{[1]}(D\lambda)|}\begin{vmatrix} S\big({}^{11}_{d}|{}^{\rm A}\big) & S\big({}^{11}_{11}|{}^{\rm A}_{\rm B}\big) \\ S\big({}^{11}_{d}|{}^{\rm B}\big) & S(11|{\rm B}) \end{vmatrix}$$

$$= \frac{\sqrt{k_{\rm B}T\rho}}{n^2\sqrt{m_{\rm A}m_{\rm B}}}\frac{x_{\rm B}\bar{c}_{\rm AB}^2}{|\mathsf{S}^{[1]}(D\lambda)|}\Delta_{12}^{[1]}(D\lambda),$$

that is,

$$S^{-1}\left(\begin{smallmatrix}E\\d\end{smallmatrix}\middle|\begin{smallmatrix}\mathrm{A}\\ \end{smallmatrix}\right) = -\frac{x_\mathrm{B}\sqrt{k_\mathrm{B}T}\rho\bar{c}_\mathrm{AB}^2}{n^2\sqrt{m_\mathrm{A}m_\mathrm{B}}|\mathsf{S}^{[1]}(D\lambda)|}\begin{vmatrix}\mathfrak{S}\left(\begin{smallmatrix}11\\10\end{smallmatrix}\middle|\begin{smallmatrix}\mathrm{A}\\\mathrm{A}\end{smallmatrix}\right)_\mathrm{AB} & x_\mathrm{A}\mathfrak{S}\left(\begin{smallmatrix}11\\11\end{smallmatrix}\middle|\begin{smallmatrix}\mathrm{A}\\\mathrm{B}\end{smallmatrix}\right)_\mathrm{AB}\\ \mathfrak{S}\left(\begin{smallmatrix}11\\10\end{smallmatrix}\middle|\begin{smallmatrix}\mathrm{B}\\\mathrm{A}\end{smallmatrix}\right)_\mathrm{AB} & \widetilde{\mathfrak{S}}(11\mathrm{B})\end{vmatrix}. \tag{2.110a}$$

Similarly,

$$S^{-1}\left(\begin{smallmatrix}E\\d\end{smallmatrix}\middle|\begin{smallmatrix}\mathrm{B}\\ \end{smallmatrix}\right) = \frac{1}{|\mathsf{S}^{[1]}(D\lambda)|}\begin{vmatrix}S\left(\begin{smallmatrix}11\\d\end{smallmatrix}\middle|\begin{smallmatrix}\mathrm{A}\\ \end{smallmatrix}\right) & S(11|\mathrm{A})\\ S\left(\begin{smallmatrix}11\\d\end{smallmatrix}\middle|\begin{smallmatrix}\mathrm{B}\\ \end{smallmatrix}\right) & S\left(\begin{smallmatrix}11\\11\end{smallmatrix}\middle|\begin{smallmatrix}\mathrm{B}\\\mathrm{A}\end{smallmatrix}\right)\end{vmatrix}$$

$$= \frac{x_\mathrm{A}\sqrt{k_\mathrm{B}T}\rho\bar{c}_\mathrm{AB}^2}{n^2\sqrt{m_\mathrm{A}m_\mathrm{B}}|\mathsf{S}^{[1]}(D\lambda)|}\begin{vmatrix}\mathfrak{S}\left(\begin{smallmatrix}11\\10\end{smallmatrix}\middle|\begin{smallmatrix}\mathrm{A}\\\mathrm{A}\end{smallmatrix}\right)_\mathrm{AB} & \widetilde{\mathfrak{S}}(11\mathrm{A})\\ \mathfrak{S}\left(\begin{smallmatrix}11\\10\end{smallmatrix}\middle|\begin{smallmatrix}\mathrm{B}\\\mathrm{A}\end{smallmatrix}\right)_\mathrm{AB} & x_\mathrm{B}\mathfrak{S}\left(\begin{smallmatrix}11\\11\end{smallmatrix}\middle|\begin{smallmatrix}\mathrm{B}\\\mathrm{A}\end{smallmatrix}\right)_\mathrm{AB}\end{vmatrix}. \tag{2.110b}$$

In these equations $|\mathsf{S}^{[1]}(D\lambda)|$ is given by eqn (2.105a), and $\Delta_{12}^{[1]}(D\lambda)$, $\Delta_{13}^{[1]}(D\lambda)$ are cofactors of the determinant $\Delta_\mathrm{mix}^{[1]}(D\lambda)$ defined by

$$\Delta_\mathrm{mix}^{[1]}(D\lambda) \equiv \begin{vmatrix}\mathfrak{S}(10|\mathrm{A}) & x_\mathrm{B}\mathfrak{S}\left(\begin{smallmatrix}10\\11\end{smallmatrix}\middle|\begin{smallmatrix}\mathrm{A}\\\mathrm{A}\end{smallmatrix}\right)_\mathrm{AB} & x_\mathrm{A}\mathfrak{S}\left(\begin{smallmatrix}10\\11\end{smallmatrix}\middle|\begin{smallmatrix}\mathrm{A}\\\mathrm{B}\end{smallmatrix}\right)_\mathrm{AB}\\ \mathfrak{S}\left(\begin{smallmatrix}11\\10\end{smallmatrix}\middle|\begin{smallmatrix}\mathrm{A}\\\mathrm{B}\end{smallmatrix}\right)_\mathrm{AB} & \widetilde{\mathfrak{S}}(11|\mathrm{A}) & x_\mathrm{A}\mathfrak{S}\left(\begin{smallmatrix}11\\11\end{smallmatrix}\middle|\begin{smallmatrix}\mathrm{A}\\\mathrm{B}\end{smallmatrix}\right)_\mathrm{AB}\\ \mathfrak{S}\left(\begin{smallmatrix}11\\10\end{smallmatrix}\middle|\begin{smallmatrix}\mathrm{B}\\\mathrm{A}\end{smallmatrix}\right)_\mathrm{AB} & x_\mathrm{B}\mathfrak{S}\left(\begin{smallmatrix}11\\11\end{smallmatrix}\middle|\begin{smallmatrix}\mathrm{B}\\\mathrm{A}\end{smallmatrix}\right)_\mathrm{AB} & \widetilde{\mathfrak{S}}(11|\mathrm{B})\end{vmatrix}. \tag{2.111}$$

The lowest Chapman–Cowling approximation, $[D_\mathrm{T}]_1$, to the thermal diffusion coefficient D_T is thus obtained as

$$[D_\mathrm{T}]_1 = \frac{x_\mathrm{A}x_\mathrm{B}k_\mathrm{B}T}{nm_\mathrm{A}\bar{c}_\mathrm{AB}\Delta_\mathrm{mix}^{[1]}(D\lambda)}\left(\frac{5}{2}\right)^{\frac{1}{2}}\left[\Delta_{12}^{[1]}(D\lambda) + \frac{y_\mathrm{A}}{y_\mathrm{B}}\Delta_{13}^{[1]}(D\lambda)\right]. \tag{2.112}$$

An expression for $[D_\mathrm{T}]_1$ in terms of effective cross-sections that more closely resembles the standard expression given in Hirschfelder *et al.* (1964), Chapman and Cowling (1970), Ferziger and Kaper (1972) and Maitland *et al.* (1980), may also be obtained as

$$[D_\mathrm{T}]_1 = -\frac{x_\mathrm{A}x_\mathrm{B}k_\mathrm{B}T}{nm_\mathrm{A}\bar{c}_\mathrm{AB}\mathfrak{S}(10|\mathrm{A})_\mathrm{AB}\Delta_\mathrm{mix}(11)(1-\delta_{D\lambda}^{[1]})}\left(\frac{5}{2}\right)^{\frac{1}{2}}\left\{\mathfrak{S}\left(\begin{smallmatrix}10\\11\end{smallmatrix}\middle|\begin{smallmatrix}\mathrm{A}\\\mathrm{A}\end{smallmatrix}\right)\left[\widetilde{\mathfrak{S}}(11|\mathrm{B}) - \frac{y_\mathrm{A}x_\mathrm{B}}{y_\mathrm{B}}\mathfrak{S}\left(\begin{smallmatrix}11\\11\end{smallmatrix}\middle|\begin{smallmatrix}\mathrm{A}\\\mathrm{B}\end{smallmatrix}\right)_\mathrm{AB}\right] + \mathfrak{S}\left(\begin{smallmatrix}10\\11\end{smallmatrix}\middle|\begin{smallmatrix}\mathrm{A}\\\mathrm{B}\end{smallmatrix}\right)_\mathrm{AB}\left[\frac{y_\mathrm{A}}{y_\mathrm{B}}\widetilde{\mathfrak{S}}(11|\mathrm{A}) - x_\mathrm{A}\mathfrak{S}\left(\begin{smallmatrix}11\\11\end{smallmatrix}\middle|\begin{smallmatrix}\mathrm{B}\\\mathrm{A}\end{smallmatrix}\right)_\mathrm{AB}\right]\right\}. \tag{2.113}$$

The main difference between equation eqn (2.113) and the expression for $[D_\mathrm{T}]_1$ appearing elsewhere (Hirschfelder *et al.*, 1964; Chapman and Cowling, 1970; Ferziger and Kaper, 1972; Maitland *et al.*, 1980) is that the present expression is presented in terms of effective cross-sections rather than collision brackets. Note also that eqn (2.113) contains an additional factor $(1-\delta_{D\lambda}^{[1]})$ in the denominator, arising from the factor $\Delta_\mathrm{mix}^{[1]}(D\lambda)$ in the denominator of eqn (2.112).

An examination of expression (2.113) for $[D_T]_1$ and eqn (2.93) for the binary diffusion coefficient $[D]_1$ leads to the conclusion that the thermal diffusion ratio $[k_T]_1$, defined by

$$[k_T]_1 = \frac{[D_T]_1}{[D]_1}, \tag{2.114}$$

has a slightly simpler structure than does $[D_T]_1$ itself. Moreover, upon dividing by the trivial $x_A x_B$ mole fraction dependence of $[D_T]_1$, a less-strongly varying function of concentration may be obtained. This quantity, termed the thermal-diffusion factor and denoted $[\alpha_T]_1$, is related to $[k_T]_1$ by

$$[\alpha_T]_1 = \frac{[k_T]_1}{x_A x_B}, \tag{2.115a}$$

and is given explicitly in terms of effective cross-sections as

$$[\alpha_T]_1 = \frac{1}{\Delta_{\mathrm{mix}}(11)(1-\delta_{D\lambda}^{[1]})}\left(\frac{5}{2}\right)^{\frac{1}{2}}\left\{\mathfrak{S}\left({}^{10}_{11}\middle|{}^{A}_{A}\right)_{AB}\left[\widetilde{\mathfrak{S}}(11|B) - \frac{y_A}{y_B}x_B\mathfrak{S}\left({}^{11}_{11}\middle|{}^{B}_{A}\right)_{AB}\right]\right.$$

$$\left. +\mathfrak{S}\left({}^{10}_{11}\middle|{}^{A}_{B}\right)_{AB}\left[\frac{y_A}{y_B}\widetilde{\mathfrak{S}}(11|A) - x_A\mathfrak{S}\left({}^{11}_{11}\middle|{}^{A}_{B}\right)_{AB}\right]\right\} \tag{2.115b}$$

for a binary mixture of monatomic gases A and B.

Mason (1957), Jones and Furry (1940), and Jones (1940) showed that the first Chapman–Cowling approximation expression for the thermal diffusion factor in an isotopic gas mixture can be expanded as a power series in the ratio $(m_{A'}-m_A)/(m_{A'}+m_A)$, $m_{A'} > m_A$, and that the leading term of the expansion provides a good approximation to $[\alpha_T]_1$. Thus, for isotopic thermal diffusion, $[\alpha_T]_1$ is given approximately by

$$[\alpha_T]_1 \simeq \alpha_0 \frac{m_{A'} - m_A}{m_{A'} + m_A}, \tag{2.116}$$

in which A$'$ represents the heavier isotope, and α_0 is given by (Hirschfelder *et al.*, 1964; Maitland *et al.*, 1980)

$$[\alpha_0]_1 = \frac{15(6C-5)(2A+5)}{2A(16A-12B+55)}. \tag{2.117a}$$

The quantities A, B, and C may be expressed in terms of effective cross-sections $\mathfrak{S}\left({}^{1s}_{1s'}\middle|{}^{A'}_{A}\right)_{AA'}$, $s, s' = \{0,1\}$ determined by binary collisions between atoms A and A$'$ as

$$A = \frac{5}{2y_{A'}^2}\left[\frac{1}{\mathfrak{S}(10|A)_{AA'}}\left(\mathfrak{S}(11|A)_{AA'} + \frac{y_{A'}^3}{y_A^3}\mathfrak{S}\left({}^{11}_{11}\middle|{}^{A}_{A'}\right)_{AA'}\right) + 3 - 4y_A^2\right], \tag{2.117b}$$

$$B = \frac{20}{9}\left\{\left[\frac{3}{16} + \frac{21}{8}y_A^2 - \frac{4y_A^4}{y_{A'}^4} + \frac{9y_A^2}{2y_{A'}^4}\right] + \left[\left(\frac{2}{5}\right)^{\frac{1}{2}}\frac{1}{y_{A'}^2}\left(\frac{y_A^4}{y_{A'}^4} - \frac{15}{8}\right)\right.\right. \tag{2.117c}$$

$$\left.\left. + \frac{3}{2y_A y_{A'}}\right]\frac{\mathfrak{S}\left({}^{11}_{11}\middle|{}^{A}_{A}\right)_{AA'}}{\mathfrak{S}(10|A)_{AA'}} + \frac{3}{2y_{A'}^4}\left(y_{A'}^2 - \frac{1}{4}\right)\frac{\mathfrak{S}(11|A)_{AA'}}{\mathfrak{S}(10|A)_{AA'}}\right\}, \tag{2.117d}$$

and

$$C = \frac{5}{6}\left[1 - \left(\frac{2}{5}\right)^{\frac{1}{2}} \frac{\mathfrak{S}\binom{10}{11}\big|\binom{\mathrm{A}}{\mathrm{A}}_{\mathrm{AA'}}}{y_{\mathrm{A'}}^2 \mathfrak{S}(10|\mathrm{A})_{\mathrm{AA'}}}\right]. \tag{2.117e}$$

The means for arriving at a second Chapman–Cowling approximation expression for the thermal diffusion coefficient has already been developed in obtaining the expression for $[\lambda_{\mathrm{mix}}]_2$, with the consequence that an expression for $[D_{\mathrm{T}}]_2$ is almost trivial to obtain. Starting from expression (2.109) for D_{T}, and considering the five-member basis $\{\mathbf{\Psi}^d, \mathbf{\Phi}^{11|\mathrm{A}}, \mathbf{\Phi}^{11|\mathrm{B}}, \mathbf{\Phi}^{12|\mathrm{A}}, \mathbf{\Phi}^{12|\mathrm{B}}\}$ from the nonhydrodynamic subspace leads, following evaluation of $S^{-1}\left(\begin{smallmatrix}E\\d\end{smallmatrix}\middle|\,\mathrm{A}\right)$ and $S^{-1}\left(\begin{smallmatrix}E\\d\end{smallmatrix}\middle|\,\mathrm{B}\right)$ in terms of the elements of the 5×5 determinant $\Delta_{\mathrm{mix}}^{[2]}(D\lambda)$ deriving from the matrix representation $\mathsf{S}^{[2]}(D\lambda)$ of the linearized collision operator $\mathcal{R}$, to the deceptively simple-looking expression

$$[D_{\mathrm{T}}]_2 = \frac{x_{\mathrm{A}} x_{\mathrm{B}} k_{\mathrm{B}} T}{n m_{\mathrm{A}} \bar{c}_{\mathrm{AB}} \Delta^{[2]}(D\lambda)} \left(\frac{5}{2}\right)^{\frac{1}{2}} \left[\Delta_{12}^{[2]}(D\lambda) + \frac{y_{\mathrm{A}}}{y_{\mathrm{B}}} \Delta_{13}^{[2]}(D\lambda)\right], \tag{2.118}$$

in which the $\Delta_{ij}^{[2]}(D\lambda)$ are cofactors of the determinant $\Delta^{[2]}(D\lambda)$ of eqns (2.107). It is likely needless to say that this expression is best left in determinantal form for processing once the set of effective cross-sections appearing in eqn (2.107b) have been evaluated.

2.5 Effective Cross-Sections

We now examine the nature of the reduced matrix elements of the linearized collision operator $\mathcal{R}$ (often referred to as (Hirschfelder *et al.*, 1964; Chapman and Cowling, 1970; Ferziger and Kaper, 1972) collision brackets). The relevant reduced matrix elements, denoted by $\langle \mathbf{\Phi}^{ps} \| \mathcal{R} \| \mathbf{\Phi}^{ps'} \rangle$, may be calculated from eqn (2.47) once the differential cross-section $\sigma(c_{\mathrm{r}}, \mathbf{e}{\cdot}\mathbf{e}')$ is known. Because the interaction potential depends only upon the relative separation of the colliding atoms, it is useful to carry out a transformation from the velocity variables $\mathbf{c}$, $\mathbf{c}_1$ to the center-of-mass velocity $\mathbf{c}_{\mathrm{CM}} = M^{-1}(m\mathbf{c}+m_1\mathbf{c}_1)$ and the relative velocity $\mathbf{c}_{\mathrm{r}} = \mathbf{c} - \mathbf{c}_1$, in which $M = m + m_1$ is the total mass of the colliding pair. Then, as the differential cross-section σ does is independent of $\mathbf{c}_{\mathrm{CM}}$, integration over this variable may be carried out. This goal is facilitated upon employing the reduced (dimensionless) velocity variables

$$\mathbf{W}_{\mathrm{CM}} = \left(\frac{M}{2k_{\mathrm{B}}T}\right)^{\frac{1}{2}} \mathbf{c}_{\mathrm{CM}}\,, \quad \mathbf{W}_{\mathrm{r}} = \left(\frac{m_{\mathrm{r}}}{2k_{\mathrm{B}}T}\right)^{\frac{1}{2}} c_{\mathrm{r}}\mathbf{e}, \quad \mathbf{W}_{\mathrm{r}}' = \left(\frac{m_{\mathrm{r}}}{2k_{\mathrm{B}}T}\right)^{\frac{1}{2}} c_{\mathrm{r}}'\mathbf{e}'\,, \tag{2.119}$$

expressed in terms of the reduced mass $m_{\mathrm{r}} = mm_1/M$ of the colliding pair.

2.5.1 Effective cross-sections for pure gases

When the masses of the colliding partners are equal, the reduced mass is given by $m_{\mathrm{r}} = m/2$, and eqns (1.11a, 1.11b) may then be rewritten as

$$\mathbf{W}' = \tfrac{1}{\sqrt{2}}(\mathbf{W}_{\mathrm{CM}} + W_{\mathrm{r}}'\mathbf{e}')\,, \qquad \mathbf{W}_1' = \tfrac{1}{\sqrt{2}}(\mathbf{W}_{\mathrm{CM}} - W_{\mathrm{r}}'\mathbf{e}')\,, \tag{2.120a}$$

$$\mathbf{W} = \tfrac{1}{\sqrt{2}}(\mathbf{W}_{\mathrm{CM}} + W_{\mathrm{r}}\mathbf{e}), \qquad \mathbf{W}_1' = \tfrac{1}{\sqrt{2}}(\mathbf{W}_{\mathrm{CM}} - W_{\mathrm{r}}\mathbf{e}). \tag{2.120b}$$

Even though conservation of energy in a gas of structureless particles requires that $W_{\mathrm{r}}' = W_{\mathrm{r}}$, the primed variables have nonetheless been retained in the following, as these same transformations may then also be applied to a polyatomic gas.

Formula (2.47) for the effective cross-sections calls for a six-dimensional integration. Through the substitutions (2.119, 2.120), the product of the Maxwellian weights and the differentials can be seen to transform as

$$\mathrm{d}\mathbf{c}\mathrm{d}\mathbf{c}_1\, f^{(0)}(\mathbf{c}_1)f^{(0)}(\mathbf{c}) = n^2\pi^{-3}\mathrm{d}\mathbf{W}_{\mathrm{CM}}\mathrm{d}W_{\mathrm{r}}\mathrm{d}\mathbf{e}\; W_{\mathrm{r}}^2\mathrm{e}^{-W_{\mathrm{CM}}^2-W_{\mathrm{r}}^2}. \tag{2.121}$$

With the abbreviation (1.37), and $\cos\chi = \mathbf{e}\cdot\mathbf{e}'$, the integral expressing the effective cross-section may be written as

$$\mathfrak{S}\left({ps \atop ps'}\right) = \frac{1}{(2p+1)\sqrt{\pi}}\int_0^\infty \mathrm{d}W_{\mathrm{r}}\; W_{\mathrm{r}}^3\mathrm{e}^{-W_{\mathrm{r}}^2}\int_{-1}^{1}\mathrm{d}(\cos\chi)\;\sigma(E,\chi)$$
$$\int \mathrm{d}\mathbf{W}_{\mathrm{CM}}\;\mathrm{e}^{-W_{\mathrm{CM}}^2}\Delta\mathbf{\Phi}^{ps}\odot\Delta\mathbf{\Phi}^{ps'}, \tag{2.122}$$

in which W_r is the total relative (collision) energy. From this expression, it is clear that the effective cross-sections $\mathfrak{S}\left({ps \atop ps'}\right)$ and $\mathfrak{S}\left({ps' \atop ps}\right)$ satisfy the symmetry relation

$$\mathfrak{S}\left({ps \atop ps'}\right) = \mathfrak{S}\left({ps' \atop ps}\right). \tag{2.123}$$

The single-term first Chapman–Cowling approximations (2.70, 2.75) to η and λ involve the quantities $\Delta\mathbf{\Phi}^{20}$ and $\Delta\mathbf{\Phi}^{11}$ given, respectively, by

$$\Delta\mathbf{\Phi}^{20} = \sqrt{2}\{\overline{\mathbf{W}_1\mathbf{W}_1} + \overline{\mathbf{W}\mathbf{W}} - \overline{\mathbf{W}_1'\mathbf{W}_1'} - \overline{\mathbf{W}'\mathbf{W}'}\} \tag{2.124a}$$

and

$$\Delta\mathbf{\Phi}^{11} = (\tfrac{4}{5})^{\frac{1}{2}}\{(W_1^2 - \tfrac{5}{2})\mathbf{W}_1 + (W^2 - \tfrac{5}{2})\mathbf{W} - (W_1'^2 - \tfrac{5}{2})\mathbf{W}_1' - (W'^2 - \tfrac{5}{2})\mathbf{W}'\}. \tag{2.124b}$$

Employment of relations (2.120) for the pure gas gives rise to considerable cancellation in these two expressions, so that they reduce to

$$\Delta\mathbf{\Phi}^{20} = \sqrt{2}\,W_{\mathrm{r}}^2(\overline{\mathbf{e}\mathbf{e}} - \overline{\mathbf{e}'\mathbf{e}'}) \tag{2.125a}$$

and

$$\Delta\mathbf{\Phi}^{11} = \left(\frac{8}{5}\right)^{\frac{1}{2}} W_{\mathrm{r}}^2\mathbf{W}_{\mathrm{CM}}\cdot(\mathbf{e}\mathbf{e} - \mathbf{e}'\mathbf{e}'). \tag{2.125b}$$

Note that as $\mathbf{e}$ and $\mathbf{e}'$ are unit vectors, the differences $\mathbf{e}\mathbf{e} - \mathbf{e}'\mathbf{e}'$ and $\overline{\mathbf{e}\mathbf{e}} - \overline{\mathbf{e}'\mathbf{e}'}$ are equal.

Although the product $\Delta\boldsymbol{\Phi}^{11}\cdot\Delta\boldsymbol{\Phi}^{11}$ appearing in the expression for $\mathfrak{S}(11)$ is more complicated than the product $\Delta\boldsymbol{\Phi}^{20}:\Delta\boldsymbol{\Phi}^{20}$ appearing in the expression for $\mathfrak{S}(20)$, it can be simplified into the expression

$$\Delta\boldsymbol{\Phi}^{11}\cdot\Delta\boldsymbol{\Phi}^{11}=\frac{8}{5}W_\mathrm{r}^4\{(\mathbf{W}_\mathrm{CM}\cdot\mathbf{e})^2-2\mathbf{e}'\cdot\mathbf{e}(\mathbf{W}_\mathrm{CM}\cdot\mathbf{e})(\mathbf{W}_\mathrm{CM}\cdot\mathbf{e}')+(\mathbf{W}_\mathrm{CM}\cdot\mathbf{e}')^2\}\,. \quad (2.126)$$

Integration over the angles of $\mathbf{W}_\mathrm{CM}$ is now straightforward in both cases, and gives

$$\int\mathrm{d}\widehat{\mathbf{W}}_\mathrm{CM}\,\Delta\boldsymbol{\Phi}^{20}:\Delta\boldsymbol{\Phi}^{20}=16\pi W_\mathrm{r}^4(1-\cos^2\chi) \quad (2.127\mathrm{a})$$

and

$$\int\mathrm{d}\widehat{\mathbf{W}}_\mathrm{CM}\,\Delta\boldsymbol{\Phi}^{11}\cdot\Delta\boldsymbol{\Phi}^{11}=\frac{64\pi}{15}\,W_\mathrm{r}^4W_\mathrm{CM}^2(1-\cos^2\chi)\,, \quad (2.127\mathrm{b})$$

in which $\widehat{\mathbf{W}}_\mathrm{CM}=\mathbf{W}_\mathrm{CM}/W_\mathrm{CM}$. Finally, integration over W_CM can be carried out according to

$$\int_0^\infty \mathrm{d}W_\mathrm{CM}\,\mathrm{e}^{-W_\mathrm{CM}^2}W_\mathrm{CM}^n=\tfrac{1}{2}\Gamma\left(\tfrac{n+1}{2}\right) \quad (2.128)$$

(see Appendix A.1). Complete integration over the center-of-mass velocity for $\mathfrak{S}(20)$ and $\mathfrak{S}(11)$ thus gives

$$\int\mathrm{d}\mathbf{W}_\mathrm{CM}\,\mathrm{e}^{-W_\mathrm{CM}^2}\,\Delta\boldsymbol{\Phi}^{20}:\Delta\boldsymbol{\Phi}^{20}=4\pi^{\frac{3}{2}}W_\mathrm{r}^4(1-\cos^2\chi) \quad (2.129\mathrm{a})$$

and

$$\int\mathrm{d}\mathbf{W}_\mathrm{CM}\,\mathrm{e}^{-W_\mathrm{CM}^2}\,\Delta\boldsymbol{\Phi}^{11}\cdot\Delta\boldsymbol{\Phi}^{11}=\frac{8}{5}\pi^{\frac{3}{2}}W_\mathrm{r}^4(1-\cos^2\chi)\,, \quad (2.129\mathrm{b})$$

respectively, so that the two effective cross-sections are obtained as

$$\mathfrak{S}(20)=\tfrac{3}{2}\mathfrak{S}(11)=\tfrac{2}{5}\int_0^\infty\mathrm{d}W_\mathrm{r}\,\mathrm{e}^{-W_\mathrm{r}^2}W_\mathrm{r}^7 2\pi\int_{-1}^1\mathrm{d}(\cos\chi)\,\sigma(W_r,\chi)(1-\cos^2\chi)\,. \quad (2.130)$$

It is traditional to identify the result of the integration over $\mathrm{d}\mathbf{e}$ as the energy-dependent cross-section $Q^{(m)}(W_r)$ according to (Hirschfelder *et al.*, 1964)

$$Q^{(m)}(W_r)=2\pi\int_{-1}^1\mathrm{d}(\cos\chi)\,(1-\cos^m\chi)\sigma(W_r,\chi)\,. \quad (2.131)$$

Once this definition has been introduced, it is also convenient to write expressions (2.130) in terms of the omega integral $\Omega^{(2,2)}$, which is a special case of the more general integrals (Hirschfelder *et al.*, 1964; Chapman and Cowling, 1970)

$$\Omega^{(m,n)}(T)=\tfrac{1}{4}\bar{c}_\mathrm{r}\int_0^\infty\mathrm{d}W_\mathrm{r}\,\mathrm{e}^{-W_\mathrm{r}^2}W_\mathrm{r}^{2n+3}Q^{(m)}(W_r)\,. \quad (2.132)$$

Thus, the final expressions for $\mathfrak{S}(20)$ and $\mathfrak{S}(11)$ are

$$\mathfrak{S}(20)=\tfrac{3}{2}\mathfrak{S}(11)=\frac{8}{5\bar{c}_\mathrm{r}}\Omega^{(2,2)}\,. \quad (2.133)$$

The effective cross-sections $\mathfrak{S}\binom{20}{21}=(n\bar{c}_\mathrm{r})^{-1}\langle 20\|\mathcal{R}\|21\rangle$, $\mathfrak{S}(21)=(n\bar{c}_\mathrm{r})^{-1}\langle 21\|\mathcal{R}\|21\rangle$, and $\mathfrak{S}\binom{11}{12}=(n\bar{c}_\mathrm{r})^{-1}\langle 11\|\mathcal{R}\|12\rangle$, $\mathfrak{S}(12)=(n\bar{c}_\mathrm{r})^{-1}\langle 12\|\mathcal{R}\|12\rangle$, additionally needed to evaluate $[\eta]_2$ and $[\lambda]_2$, may be determined along the lines outlined above for $\mathfrak{S}(20)$ and $\mathfrak{S}(11)$. Although the intermediate expressions are rather tedious, they may ultimately be distilled down to simple combinations of omega integrals, namely,

$$\mathfrak{S}(12) = \frac{16}{105\bar{c}_{\mathrm{r}}} \left(\frac{77}{4} \Omega^{(2,2)} - 7\Omega^{(2,3)} + \Omega^{(2,4)} \right) , \tag{2.134a}$$

$$\frac{3}{4}\sqrt{7}\,\mathfrak{S}\left(\begin{smallmatrix}11\\12\end{smallmatrix}\right) = \mathfrak{S}\left(\begin{smallmatrix}20\\21\end{smallmatrix}\right) = \frac{2}{5\bar{c}_{\mathrm{r}}} \left(7\Omega^{(2,2)} - 2\Omega^{(2,3)} \right) , \tag{2.134b}$$

$$\mathfrak{S}(21) = \frac{2}{5\bar{c}_{\mathrm{r}}} \left(\frac{301}{12} \Omega^{(2,2)} - 7\Omega^{(2,3)} + \Omega^{(2,4)} \right) . \tag{2.134c}$$

The effective cross-sections of eqns (2.133, 2.134) suffice to determine the second Chapman–Cowling approximations $[\eta]_2$ and $[\lambda]_2$ for the viscosity and thermal conductivity coefficients of pure monatomic gases. Should higher approximations be required (Viehland *et al.*, 1995), expressions for $f_\eta^{[n]}$ and $f_\lambda^{[n]}$ have been given in terms of omega integrals by Viehland *et al.* (1995) for $n \leq 5$.

2.5.2 Effective cross-sections for binary mixtures

Section 2.3 established the usefulness of expressing reduced matrix elements of the linearized collision operator in terms of effective cross-sections. Although the introduction of such cross-sections is not really much more than a formality for pure monatomic gases, it does provide some insight for gas mixtures. The reduced matrix elements $nS\left(\begin{smallmatrix}ps\\ps'\end{smallmatrix}\middle|\begin{smallmatrix}\mathrm{k}\\\mathrm{k}'\end{smallmatrix}\right)$ of the collision operator $\mathcal{R}$ have been expressed generally in terms of effective cross-sections in eqns (2.60).

That it is convenient to simplify the formal expressions for the effective cross-sections by integrating out the centre-of-mass velocity has already been established in Section 2.5.1. A similar simplification is accomplished for unlike-atom collisions by utilizing the generalization

$$\mathbf{W} = y_{\mathrm{A}}\mathbf{W}_{\mathrm{CM}} + y_{\mathrm{B}}W_{\mathrm{r}}\mathbf{e} , \qquad \mathbf{W}_1 = y_{\mathrm{B}}\mathbf{W}_{\mathrm{CM}} - y_{\mathrm{A}}W_{\mathrm{r}}\mathbf{e} , \tag{2.135}$$

and

$$\mathbf{W}' = y_{\mathrm{A}}\mathbf{W}_{\mathrm{CM}} + y_{\mathrm{B}}W'_{\mathrm{r}}\mathbf{e}' , \qquad \mathbf{W}'_1 = y_{\mathrm{B}}\mathbf{W}_{\mathrm{CM}} - y_{\mathrm{A}}W'_{\mathrm{r}}\mathbf{e}' , \tag{2.136}$$

of the transformation (2.120) in the expressions for $\mathfrak{S}\left(\begin{smallmatrix}ps\\ps'\end{smallmatrix}\middle|\begin{smallmatrix}\mathrm{k}\\\mathrm{k}'\end{smallmatrix}\right)_{\mathrm{kk}'}$. For collisions of like species, the mass factors $y_{\mathrm{A}} = (m_{\mathrm{A}}/M)^{\frac{1}{2}}$, $y_{\mathrm{B}} = (m_{\mathrm{B}}/M)^{\frac{1}{2}}$, each reduce to $\frac{1}{\sqrt{2}}$. Many more effective cross-sections appear in mixture expressions than have been encountered in pure gas expressions. There are also many more exact relations among effective cross-sections for binary mixtures than there are for pure gases. However, it is not necessary to dwell upon their derivations here, but only to take note of some of the relevant results (Hirschfelder *et al.*, 1964; Chapman and Cowling, 1970).

According to eqn (2.93), it is the three diffusion-type effective cross-sections, namely, $\mathfrak{S}(10|\mathrm{A})_{\mathrm{AB}}$, $\mathfrak{S}\left(\begin{smallmatrix}10\\10\end{smallmatrix}\middle|\begin{smallmatrix}\mathrm{A}\\\mathrm{B}\end{smallmatrix}\right)_{\mathrm{AB}}$, and $\mathfrak{S}(10|\mathrm{B})_{\mathrm{AB}}$, determine the first Chapman–Cowling approximation, $[D]_1$, to the binary diffusion coefficient in a mixture of gases A and B. The effective cross-section $\mathfrak{S}(10|\mathrm{A})_{\mathrm{AB}}$, traditionally referred to as *the* diffusion cross-section, is given explicitly by

$$\mathfrak{S}(10|\mathrm{A})_{\mathrm{AB}} = \tfrac{4}{3} y_{\mathrm{B}}^2 \int_0^\infty \mathrm{d}W_{\mathrm{r}}\, \mathrm{e}^{-W_{\mathrm{r}}^2} W_{\mathrm{r}}^5\, 2\pi \int_{-1}^{1} \mathrm{d}(\cos\chi)\, \sigma_{\mathrm{AB}}(E,\chi)(1-\cos\chi)\,. \tag{2.137}$$

This effective cross-section may also be expressed in terms of mixture omega-integrals $\Omega_{\mathrm{kk'}}^{(m,n)}(T) \equiv \Omega_{\mathrm{k'k}}^{(m,n)}(T)$, defined analogously to the pure gas omega-integrals of eqn (2.132) by

$$\Omega_{\mathrm{kk'}}^{(m,n)}(T) = \tfrac{1}{4}\bar{c}_{\mathrm{kk'}} \int_0^\infty \mathrm{d}W_{\mathrm{r}}\, \mathrm{e}^{-W_{\mathrm{r}}^2} W_{\mathrm{r}}^{2n+3}\, 2\pi \int_{-1}^{1} \mathrm{d}(\cos\chi)\, (1-\cos^m\chi)\sigma_{\mathrm{kk'}}(E,\chi)\,, \tag{2.138}$$

as

$$\mathfrak{S}(10|\mathrm{A})_{\mathrm{AB}} = \frac{16 y_{\mathrm{B}}^2}{3\bar{c}_{\mathrm{AB}}} \Omega_{\mathrm{AB}}^{(1,1)}(T)\,. \tag{2.139}$$

Hereafter, the explicit temperature dependence of $\Omega^{(m,n)}(T)$ will not be written, and it will be understood that the omega integrals are functions of T.

The kinematic conservation laws can be shown to establish relationships among some of the effective cross-sections. In particular, the conservation of momentum gives a set of relations (Köhler and 't Hooft, 1979) of the type

$$\mathfrak{S}\left(\begin{smallmatrix}10\\ps\end{smallmatrix}\middle|\begin{smallmatrix}\mathrm{k'}\\\mathrm{k}\end{smallmatrix}\right)_{\mathrm{kk'}} = -\frac{y_{\mathrm{k}}}{y_{\mathrm{k'}}} \mathfrak{S}\left(\begin{smallmatrix}10\\ps\end{smallmatrix}\middle|\begin{smallmatrix}\mathrm{k}\\\mathrm{k}\end{smallmatrix}\right)_{\mathrm{kk'}}, \quad \mathrm{k'} \neq \mathrm{k} = \mathrm{A},\mathrm{B}\,. \tag{2.140a}$$

For $ps = 10$, $\mathrm{k'} = \mathrm{B}$, $\mathrm{k} = \mathrm{A}$, this relation becomes

$$\mathfrak{S}\left(\begin{smallmatrix}10\\10\end{smallmatrix}\middle|\begin{smallmatrix}\mathrm{B}\\\mathrm{A}\end{smallmatrix}\right)_{\mathrm{AB}} = -\frac{y_{\mathrm{A}}}{y_{\mathrm{B}}} \mathfrak{S}(10|\mathrm{A})_{\mathrm{AB}}\,, \tag{2.140b}$$

while for $ps = 10$, $\mathrm{k} = \mathrm{B}$, $\mathrm{k'} = \mathrm{A}$, it gives

$$\mathfrak{S}(10|\mathrm{B})_{\mathrm{AB}} = -\frac{y_{\mathrm{A}}}{y_{\mathrm{B}}} \mathfrak{S}\left(\begin{smallmatrix}10\\10\end{smallmatrix}\middle|\begin{smallmatrix}\mathrm{A}\\\mathrm{B}\end{smallmatrix}\right)_{\mathrm{AB}}\,. \tag{2.140c}$$

Upon employing the symmetry relation $\mathfrak{S}\left(\begin{smallmatrix}10\\10\end{smallmatrix}\middle|\begin{smallmatrix}\mathrm{A}\\\mathrm{B}\end{smallmatrix}\right)_{\mathrm{AB}} = \mathfrak{S}\left(\begin{smallmatrix}10\\10\end{smallmatrix}\middle|\begin{smallmatrix}\mathrm{B}\\\mathrm{A}\end{smallmatrix}\right)_{\mathrm{AB}}$, together eqns (2.140b, 2.140c), $\mathfrak{S}(10|\mathrm{B})_{\mathrm{AB}}$ and $\mathfrak{S}(10|\mathrm{A})_{\mathrm{AB}}$ can be seen to be related by

$$y_{\mathrm{B}}^2\, \mathfrak{S}(10|\mathrm{B})_{\mathrm{AB}} = y_{\mathrm{A}}^2\, \mathfrak{S}(10|\mathrm{A})_{\mathrm{AB}}\,. \tag{2.140d}$$

Relation (2.140a) with $ps = 11$, coupled with the expression

$$\mathfrak{S}\left(\begin{smallmatrix}10\\11\end{smallmatrix}\middle|\begin{smallmatrix}\mathrm{A}\\\mathrm{A}\end{smallmatrix}\right)_{\mathrm{AB}} = \frac{16 y_{\mathrm{B}}^4}{3\bar{c}_{\mathrm{AB}}} \left(\frac{2}{5}\right)^{\frac{1}{2}} \left[\tfrac{5}{2}\Omega_{\mathrm{AB}}^{(1,1)} - \Omega_{\mathrm{AB}}^{(1,2)}\right] \tag{2.141}$$

giving $\mathfrak{S}\left(\begin{smallmatrix}10\\11\end{smallmatrix}\middle|\begin{smallmatrix}\mathrm{A}\\\mathrm{A}\end{smallmatrix}\right)_{\mathrm{AB}}$ in terms of omega-integrals, may be employed to obtain the other effective cross-sections representing the collisional coupling between $\boldsymbol{\Phi}^{10}$ and $\boldsymbol{\Phi}^{11}$ in terms of $\mathfrak{S}\left(\begin{smallmatrix}10\\11\end{smallmatrix}\middle|\begin{smallmatrix}\mathrm{A}\\\mathrm{A}\end{smallmatrix}\right)_{\mathrm{AB}}$. Thus, for $ps = 11$ with $\mathrm{k}' = \mathrm{A}$, $\mathrm{k} = \mathrm{B}$, and for $ps = 11$ with $\mathrm{k}' = \mathrm{B}$, $\mathrm{k} = \mathrm{A}$, eqn (2.140a) gives the relations

$$\mathfrak{S}\left(\begin{smallmatrix}10\\11\end{smallmatrix}\middle|\begin{smallmatrix}\mathrm{A}\\\mathrm{B}\end{smallmatrix}\right)_{\mathrm{AB}} = -\frac{y_{\mathrm{B}}}{y_{\mathrm{A}}}\mathfrak{S}\left(\begin{smallmatrix}10\\11\end{smallmatrix}\middle|\begin{smallmatrix}\mathrm{B}\\\mathrm{B}\end{smallmatrix}\right)_{\mathrm{AB}} \tag{2.142a}$$

and

$$\mathfrak{S}\left(\begin{smallmatrix}10\\11\end{smallmatrix}\middle|\begin{smallmatrix}\mathrm{B}\\\mathrm{A}\end{smallmatrix}\right)_{\mathrm{AB}} = -\frac{y_{\mathrm{A}}}{y_{\mathrm{B}}}\mathfrak{S}\left(\begin{smallmatrix}10\\11\end{smallmatrix}\middle|\begin{smallmatrix}\mathrm{A}\\\mathrm{A}\end{smallmatrix}\right)_{\mathrm{AB}}, \tag{2.142b}$$

respectively. An interchange of the roles of A and B in eqn (2.141) gives

$$\mathfrak{S}\left(\begin{smallmatrix}10\\11\end{smallmatrix}\middle|\begin{smallmatrix}\mathrm{B}\\\mathrm{B}\end{smallmatrix}\right)_{\mathrm{AB}} = \frac{y_{\mathrm{A}}^4}{y_{\mathrm{B}}^4}\,\mathfrak{S}\left(\begin{smallmatrix}10\\11\end{smallmatrix}\middle|\begin{smallmatrix}\mathrm{A}\\\mathrm{A}\end{smallmatrix}\right)_{\mathrm{AB}}, \tag{2.142c}$$

a result that can be combined with eqn (2.142a) for $\mathfrak{S}\left(\begin{smallmatrix}10\\11\end{smallmatrix}\middle|\begin{smallmatrix}\mathrm{A}\\\mathrm{B}\end{smallmatrix}\right)_{\mathrm{AB}}$ to give

$$\mathfrak{S}\left(\begin{smallmatrix}10\\11\end{smallmatrix}\middle|\begin{smallmatrix}\mathrm{A}\\\mathrm{B}\end{smallmatrix}\right)_{\mathrm{AB}} = -\frac{y_{\mathrm{A}}^3}{y_{\mathrm{B}}^3}\mathfrak{S}\left(\begin{smallmatrix}10\\11\end{smallmatrix}\middle|\begin{smallmatrix}\mathrm{A}\\\mathrm{A}\end{smallmatrix}\right)_{\mathrm{AB}}. \tag{2.142d}$$

The conservation of energy during a binary collision leads to the set of relations

$$\mathfrak{S}\left(\begin{smallmatrix}01\\ps\end{smallmatrix}\middle|\begin{smallmatrix}\mathrm{k}\\\mathrm{k}\end{smallmatrix}\right)_{\mathrm{kk}'} + \mathfrak{S}\left(\begin{smallmatrix}01\\ps\end{smallmatrix}\middle|\begin{smallmatrix}\mathrm{k}'\\\mathrm{k}\end{smallmatrix}\right)_{\mathrm{kk}'} = 0\,, \quad \mathrm{k} \neq \mathrm{k}' = \mathrm{A}, \mathrm{B}\,. \tag{2.143}$$

The choice $ps = 01$, $\mathrm{k} = \mathrm{A}$, $\mathrm{k}' = \mathrm{B}$, gives, for example, the relation

$$\mathfrak{S}(01|\mathrm{A})_{\mathrm{AB}} = -\mathfrak{S}\left(\begin{smallmatrix}01\\01\end{smallmatrix}\middle|\begin{smallmatrix}\mathrm{B}\\\mathrm{A}\end{smallmatrix}\right)_{\mathrm{AB}} \tag{2.144a}$$

and, upon examining the expression relating $\mathfrak{S}(01|\mathrm{A})_{\mathrm{AB}}$ to the integral scattering cross-section $\sigma_{\mathrm{AB}}(W_{\mathrm{r}}, \chi)$, namely,

$$\mathfrak{S}(01|\mathrm{A})_{\mathrm{AB}} = \frac{8}{3}\,y_{\mathrm{A}}^2 y_{\mathrm{B}}^2 \int_0^\infty \mathrm{d}W_{\mathrm{r}}\,\mathrm{e}^{-W_{\mathrm{r}}^2}\,W_{\mathrm{r}}^5\,2\pi \int_{-1}^{1} \mathrm{d}\cos\chi\,\sigma_{\mathrm{AB}}(E,\chi)(1-\cos\chi)\,, \tag{2.144b}$$

and comparing this result with eqn (2.135) for $\mathfrak{S}(10|\mathrm{A})_{\mathrm{AB}}$, it becomes clear that $\mathfrak{S}(01|\mathrm{A})_{\mathrm{AB}}$ may also be expressed as

$$\mathfrak{S}(01|\mathrm{A})_{\mathrm{AB}} = 2y_{\mathrm{A}}^2\mathfrak{S}(10|\mathrm{A})_{\mathrm{AB}}\,. \tag{2.145}$$

No additional relations may be obtained from the kinematic conservation laws.

Effective cross-sections governing the first Chapman–Cowling approximations to the thermal conductivity and shear viscosity of binary mixtures may be obtained by following the methods of calculation outlined above for pure gases. In particular, the effective cross-sections $\mathfrak{S}(20|\mathrm{A})_{\mathrm{AB}}$ and $\mathfrak{S}\left(\begin{smallmatrix}20\\20\end{smallmatrix}\middle|\begin{smallmatrix}\mathrm{A}\\\mathrm{B}\end{smallmatrix}\right)_{\mathrm{AB}}$, given by

$$\mathfrak{S}(20|\mathrm{A})_{\mathrm{AB}} = \frac{16y_{\mathrm{A}}^2y_{\mathrm{B}}^2}{15\bar{c}_{\mathrm{AB}}}\left[10\Omega_{\mathrm{AB}}^{(1,1)} + \frac{3y_{\mathrm{B}}^2}{y_{\mathrm{A}}^2}\Omega_{\mathrm{AB}}^{(2,2)}\right], \tag{2.146a}$$

and

$$\mathfrak{S}\left(\begin{smallmatrix}20\\20\end{smallmatrix}\middle|\begin{smallmatrix}\mathrm{A}\\\mathrm{B}\end{smallmatrix}\right)_{\mathrm{AB}} = \frac{16y_{\mathrm{A}}^2y_{\mathrm{B}}^2}{15\bar{c}_{\mathrm{AB}}}\left[-10\Omega_{\mathrm{AB}}^{(1,1)} + 3\Omega_{\mathrm{AB}}^{(2,2)}\right], \tag{2.146b}$$

together with expressions that can be deduced from them for the additional effective cross-sections occurring in eqn (2.81b), determine $[\eta_{\mathrm{mix}}]_1$. Similarly, the effective cross-sections $\mathfrak{S}(11|\mathrm{A})_{\mathrm{AB}}$ and $\mathfrak{S}\left(\begin{smallmatrix}11\\11\end{smallmatrix}\middle|\begin{smallmatrix}\mathrm{A}\\\mathrm{B}\end{smallmatrix}\right)_{\mathrm{AB}}$, given by

$$\mathfrak{S}(11|\mathrm{A})_{\mathrm{AB}} = \frac{16y_{\mathrm{B}}^2}{15\bar{c}_{\mathrm{AB}}}\left[\tfrac{5}{2}\left(6y_{\mathrm{A}}^4 + 5y_{\mathrm{B}}^4\right)\Omega_{\mathrm{AB}}^{(1,1)} - 10y_{\mathrm{B}}^4\Omega_{\mathrm{AB}}^{(1,2)} + 2y_{\mathrm{B}}^4\Omega_{\mathrm{AB}}^{(1,3)} + 4y_{\mathrm{A}}^2y_{\mathrm{B}}^2\,\Omega_{\mathrm{AB}}^{(2,2)}\right], \tag{2.147a}$$

$$\mathfrak{S}\left(\begin{smallmatrix}11\\11\end{smallmatrix}\middle|\begin{smallmatrix}\mathrm{A}\\\mathrm{B}\end{smallmatrix}\right)_{\mathrm{AB}} = \frac{8y_{\mathrm{A}}^3y_{\mathrm{B}}^3}{3\bar{c}_{\mathrm{AB}}}\left[-11\Omega_{\mathrm{AB}}^{(1,1)} + 4\Omega_{\mathrm{AB}}^{(1,2)} - \tfrac{4}{5}\Omega_{\mathrm{AB}}^{(1,3)} + \tfrac{8}{5}\Omega_{\mathrm{AB}}^{(2,2)}\right], \tag{2.147b}$$

together with expressions that can be deduced from them for the additional effective cross-sections occurring in eqn (2.106), determine $[\lambda_{\mathrm{mix}}]_1$.

Two interrelationships involving these effective cross-sections that have proven to be useful are (Köhler and 't Hooft, 1979)

$$\frac{y_{\mathrm{A}}^2}{y_{\mathrm{B}}^2}\mathfrak{S}(20|\mathrm{A})_{\mathrm{AB}} - \mathfrak{S}\left(\begin{smallmatrix}20\\20\end{smallmatrix}\middle|\begin{smallmatrix}\mathrm{A}\\\mathrm{B}\end{smallmatrix}\right)_{\mathrm{AB}} = 2\,\frac{y_{\mathrm{A}}^2}{y_{\mathrm{B}}^2}\mathfrak{S}(10|\mathrm{A})_{\mathrm{AB}} \tag{2.148a}$$

and

$$\begin{aligned}\mathfrak{S}(11|\mathrm{A})_{\mathrm{AB}} + \frac{y_{\mathrm{B}}^3}{y_{\mathrm{A}}^3}\mathfrak{S}\left(\begin{smallmatrix}11\\11\end{smallmatrix}\middle|\begin{smallmatrix}\mathrm{A}\\\mathrm{B}\end{smallmatrix}\right)_{\mathrm{AB}} &= \tfrac{4}{3}\mathfrak{S}(20|\mathrm{A})_{\mathrm{AB}} + \tfrac{5}{3}\mathfrak{S}(01|\mathrm{A})_{\mathrm{AB}} - 3\mathfrak{S}(10|\mathrm{A})_{\mathrm{AB}}\\ &= \frac{4y_{\mathrm{B}}^2}{3y_{\mathrm{A}}^2}\mathfrak{S}\left(\begin{smallmatrix}20\\20\end{smallmatrix}\middle|\begin{smallmatrix}\mathrm{A}\\\mathrm{B}\end{smallmatrix}\right)_{\mathrm{AB}} + \frac{5y_B^2}{3y_A^2}\mathfrak{S}\left(\begin{smallmatrix}01\\01\end{smallmatrix}\middle|\begin{smallmatrix}\mathrm{A}\\\mathrm{B}\end{smallmatrix}\right)_{\mathrm{AB}} + 3\mathfrak{S}(10|\mathrm{A})_{\mathrm{AB}}\,.\end{aligned} \tag{2.148b}$$

Examination of eqn (2.81b) shows that the specific effective cross-sections involved in the evaluation of $[\eta_{\mathrm{mix}}]_1$ are the mole-fraction dependent mixture effective cross-sections $\widetilde{\mathfrak{S}}(20|\mathrm{A})$ and $\widetilde{\mathfrak{S}}(20|\mathrm{B})$, plus the mixed effective cross-section $\mathfrak{S}\left(\begin{smallmatrix}20\\20\end{smallmatrix}\middle|\begin{smallmatrix}\mathrm{B}\\\mathrm{A}\end{smallmatrix}\right)_{\mathrm{AB}}$, none of which appear to be directly given by eqns (2.146). However, if it is noted that mixture effective cross-sections obey a symmetry relation analogous to eqn (2.123) for pure gas effective cross-sections, namely,

$$\mathfrak{S}\left(\begin{smallmatrix}ps\\ps'\end{smallmatrix}\middle|\begin{smallmatrix}\mathrm{A}\\\mathrm{B}\end{smallmatrix}\right)_{\mathrm{AB}} = \mathfrak{S}\left(\begin{smallmatrix}ps'\\ps\end{smallmatrix}\middle|\begin{smallmatrix}\mathrm{B}\\\mathrm{A}\end{smallmatrix}\right)_{\mathrm{AB}}, \tag{2.149}$$

then it can be seen that $\mathfrak{S}\left(\begin{smallmatrix}20\\20\end{smallmatrix}\middle|\begin{smallmatrix}\mathrm{B}\\\mathrm{A}\end{smallmatrix}\right)_{\mathrm{AB}} = \mathfrak{S}\left(\begin{smallmatrix}20\\20\end{smallmatrix}\middle|\begin{smallmatrix}\mathrm{A}\\\mathrm{B}\end{smallmatrix}\right)_{\mathrm{AB}}$. Moreover, the second term of the expression $\widetilde{\mathfrak{S}}(20|\mathrm{A}) \equiv x_{\mathrm{A}}\sqrt{2}y_{\mathrm{B}}\mathfrak{S}(20|\mathrm{A})_{\mathrm{AB}} + x_{\mathrm{B}}\mathfrak{S}(20|\mathrm{A})_{\mathrm{AB}}$ depends directly upon $\mathfrak{S}(20|\mathrm{A})_{\mathrm{AB}}$, while the first term is simply the pure gas effective cross-section $\mathfrak{S}(20)$ for chemical species A, but labeled explicitly as $\mathfrak{S}(20|\mathrm{A})_{\mathrm{AA}}$ in order to distinguish it from the same effective cross-section for species B, now labeled as $\mathfrak{S}(20|\mathrm{B})_{\mathrm{BB}}$. Finally, from the defining relation for $\widetilde{\mathfrak{S}}(20|\mathrm{B})$, namely, $\widetilde{\mathfrak{S}}(20|\mathrm{B}) \equiv x_{\mathrm{B}}\sqrt{2}y_{\mathrm{A}}\mathfrak{S}(20|\mathrm{B})_{\mathrm{BB}} + x_{\mathrm{A}}\mathfrak{S}(20|\mathrm{B})_{\mathrm{AB}}$, it can be seen that $\widetilde{\mathfrak{S}}(20|\mathrm{B})$ is obtained from $\widetilde{\mathfrak{S}}(20|\mathrm{A})$ upon interchanging the roles of species A and B. An appropriate expression for $\mathfrak{S}(20|\mathrm{B})_{\mathrm{AB}}$, analogous to eqn (2.146a) may also be obtained similarly by the interchange of the roles of species A and B. Precisely the same considerations may be applied to the effective cross-sections appearing in eqn (2.106) for $[\lambda]_1$. However, the symmetries just discussed make it unnecessary to provide explicit expressions for all mixture effective cross-sections required for evaluation of the mixture transport coefficients.

The set of effective cross-sections given in eqns (2.137)–(2.142), plus eqns (2.147) also suffice to calculate the second Chapman–Cowling approximation, $[D]_2$, to the diffusion coefficient via eqn (2.97), as well as the first Chapman–Cowling approximation, $[\alpha_\mathrm{T}]_1$, to the thermal diffusion factor via eqn (2.115b). The computation of second Chapman–Cowling approximations to the mixture viscosity, thermal conductivity, and thermal diffusion factor requires an additional set of effective cross-sections. For the reasons just discussed, however, only a minimal set of relevant expressions will be given explicitly.

In particular, for the computation of $[\eta_\mathrm{mix}]_2$, the relevant effective cross-sections needed in addition to the effective cross-sections $\mathfrak{S}(20|\mathrm{A})_\mathrm{AB}$ and $\mathfrak{S}(\begin{smallmatrix}20\\20\end{smallmatrix}|\begin{smallmatrix}\mathrm{A}\\\mathrm{B}\end{smallmatrix})_\mathrm{AB}$ given in eqns (2.146) are

$$\mathfrak{S}(\begin{smallmatrix}20\\21\end{smallmatrix}|\begin{smallmatrix}\mathrm{A}\\\mathrm{A}\end{smallmatrix})_\mathrm{AB} = \frac{32y_\mathrm{B}^4}{15\bar{c}_\mathrm{AB}}\left[70y_\mathrm{A}^2\Omega_\mathrm{AB}^{(1,1)} - 28y_\mathrm{A}^2\Omega_\mathrm{AB}^{(1,2)} + 21y_\mathrm{B}^2\Omega_\mathrm{AB}^{(2,2)} - 6y_\mathrm{B}^2\Omega_\mathrm{AB}^{(2,3)}\right], \tag{2.150a}$$

$$\mathfrak{S}(\begin{smallmatrix}20\\21\end{smallmatrix}|\begin{smallmatrix}\mathrm{A}\\\mathrm{B}\end{smallmatrix})_\mathrm{AB} = -\frac{32y_\mathrm{A}^4y_\mathrm{B}^2}{15\bar{c}_\mathrm{AB}}\left[70\Omega_\mathrm{AB}^{(1,1)} - 28\Omega_\mathrm{AB}^{(1,2)} - 21\Omega_\mathrm{AB}^{(2,2)} + 6\Omega_\mathrm{AB}^{(2,3)}\right], \tag{2.150b}$$

$$\begin{aligned}\mathfrak{S}(21|\mathrm{A})_\mathrm{AB} = \frac{16y_\mathrm{B}^2}{15\bar{c}_\mathrm{AB}}\Big[&2y_\mathrm{A}^2(140y_\mathrm{A}^4 + 245y_\mathrm{B}^2)\Omega_\mathrm{AB}^{(1,1)} - 392y_{\mathrm{A}^2}y_\mathrm{B}^2\Omega_\mathrm{AB}^{(1,2)} + 64y_\mathrm{A}^2y_\mathrm{B}^4\Omega_\mathrm{AB}^{(1,3)}\\ &+y_\mathrm{B}^2(154y_\mathrm{A}^4 + 147y_\mathrm{B}^4)\Omega_\mathrm{AB}^{(2,2)} - 84y_\mathrm{B}^6\Omega_\mathrm{AB}^{(2,3)} + 12y_\mathrm{B}^6\Omega_\mathrm{AB}^{(2,4)} + 24y_\mathrm{A}^2y_\mathrm{B}^4\Omega_\mathrm{AB}^{(3,3)}\Big],\end{aligned} \tag{2.150c}$$

and

$$\begin{aligned}\mathfrak{S}(\begin{smallmatrix}21\\21\end{smallmatrix}|\begin{smallmatrix}\mathrm{A}\\\mathrm{B}\end{smallmatrix})_\mathrm{AB} = -\frac{16y_\mathrm{A}^4y_\mathrm{B}^4}{15\bar{c}_\mathrm{AB}}\Big[&770\Omega_\mathrm{AB}^{(1,1)} - 392\Omega_\mathrm{AB}^{(1,2)} + 64\Omega_\mathrm{AB}^{(1,3)} - 301\Omega_\mathrm{AB}^{(2,2)}\\ &+84\Omega_\mathrm{AB}^{(2,3)} - 12\Omega_\mathrm{AB}^{(2,4)} + 24\Omega_\mathrm{AB}^{(3,3)}\Big],\end{aligned} \tag{2.150d}$$

plus effective cross-sections obtainable from them via symmetry arguments.

2.5.3 Classical evaluation of effective cross-sections

From the expressions for $\mathfrak{S}(11)$ and $\mathfrak{S}(20)$ for a pure gas, and similarly for other effective cross-sections $\mathfrak{S}(\begin{smallmatrix}ps\\ps'\end{smallmatrix})$, it becomes clear that their final evaluation hinges upon a knowledge of the differential cross-section $\sigma(c_\mathrm{r},\chi)$. For low-energy encounters, especially for light particles, the collision cross-section should only be expressed quantum-mechanically (see Section 2.5.4). However, useful approximations and, above all, simple understanding can be gained from classical arguments based upon trajectory calculations. The final goal is not to obtain a detailed description of the motion, but rather to obtain a relation between the parameters specifying the motion long before and long after the scattering event. These are the incoming and outgoing relative velocities $\mathbf{c}'_\mathrm{r} = c_\mathrm{r}\mathbf{e}'$ and $\mathbf{c}_\mathrm{r} = c_\mathrm{r}\mathbf{e}$, together with the impact parameter b, the hypothetical distance of closest approach were there no forces (see Fig. 2.1). The object is to calculate the deflection angle $\chi = \cos^{-1}(\mathbf{e}\cdot\mathbf{e}')$ as a function of c_r and b, after which the differential cross-section σ is evaluated as a function of c_r and χ.

Except for the different force field, the mathematics needed in a classical description of binary collisions of point particles is the same as in the Kepler problem of planetary motion. In a coordinate system attached to the center-of-mass, both particles move in a plane that is perpendicular to the orbital angular momentum. For coordinates, the separation R of the particles and the associated polar angle θ as illustrated in Fig. 2.1 suffice. The conservation of angular momentum is given, as in the Kepler problem, by

$$m_\mathrm{r} R^2 \frac{\mathrm{d}\theta}{\mathrm{d}t} = m_\mathrm{r} c_\mathrm{r} b\,. \tag{2.151}$$

Next, it is observed that the sum of kinetic and potential energies is constant, and hence, equal to the asymptotic value $E = \frac{1}{2} m_\mathrm{r} c_\mathrm{r}^2$ before and after scattering. Written in terms of R and θ, this statement becomes

$$\tfrac{1}{2} m_\mathrm{r} \left[\left(\frac{\mathrm{d}R}{\mathrm{d}t} \right)^2 + \left(R \frac{\mathrm{d}\theta}{\mathrm{d}t} \right)^2 \right] + V(R) = E\,. \tag{2.152}$$

Substitution of $\mathrm{d}\theta/\mathrm{d}t$ from (2.151) followed by some rearrangement gives

$$\frac{\mathrm{d}R}{\mathrm{d}t} = \pm \left[\frac{2}{m_\mathrm{r}} \left\{ E - V(R) - \frac{Eb^2}{R^2} \right\} \right]^{\frac{1}{2}}, \tag{2.153}$$

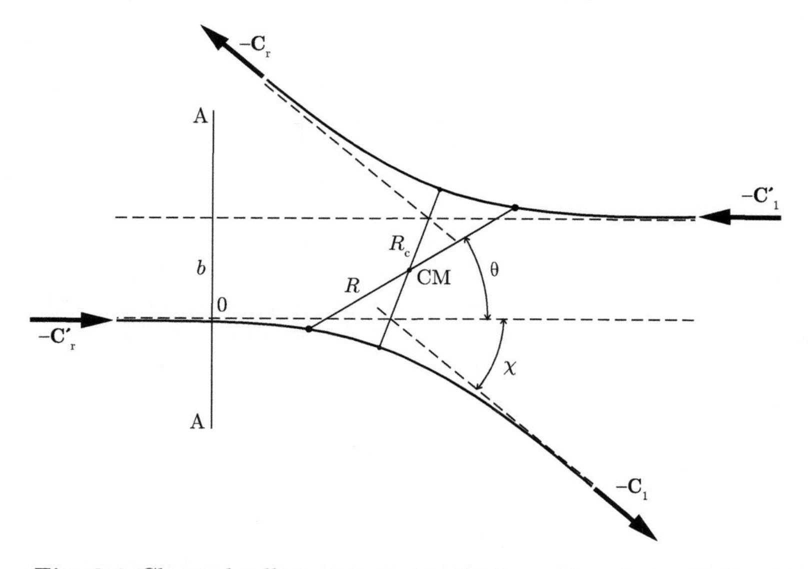

Fig. 2.1 Classical collision parameters in the center-of-mass system.

with the + and − signs corresponding, respectively, to incoming and outgoing parts of the trajectories. As the geometry of the encounter remains the primary interest, the time variable may be eliminated by again using $\mathrm{d}\theta/\mathrm{d}t$ from eqn (2.151) to obtain

$$\frac{\mathrm{d}\theta}{\mathrm{d}R} = \pm\frac{b}{R^2}\left[1-\left(\frac{b}{R}\right)^2-\frac{V(R)}{E}\right]^{-\frac{1}{2}}. \tag{2.154}$$

For any given interatomic potential energy $V(R)$ and for specified values of b and E, a simple integration gives the angle $\theta(b)$.

The derivative $\mathrm{d}\theta/\mathrm{d}R$ becomes infinite at the position of closest approach at which $R = R_\mathrm{c}$ (see Fig. 2.1), with R_c the positive root of the equation

$$1-\left(\frac{b}{R_\mathrm{c}}\right)^2-\frac{V(R_\mathrm{c})}{E} = 0. \tag{2.155}$$

Finally, note that the total change of θ from $t \to -\infty$ to $t \to \infty$ is $\pi - \chi$, with half the value due to the incoming portion, half to the outgoing portion of the trajectory. Thus, the deflection angle $\chi(b, E)$ is given by the expression

$$\chi(b,E) = \pi - 2\int_{R_\mathrm{c}}^{\infty} \mathrm{d}R\left(\frac{\mathrm{d}\theta}{\mathrm{d}R}\right) \tag{2.156a}$$

$$= \pi - 2b\int_{R_\mathrm{c}}^{\infty}\frac{1}{R^2}\left[1-\frac{b^2}{R^2}-\frac{V(R)}{E}\right]^{-\frac{1}{2}}\mathrm{d}R. \tag{2.156b}$$

The initial and final asymptotic states of motion differ only in that the relative velocity vector $\mathbf{c}'_\mathrm{r}$ has been rotated through an angle χ: this is the only potentially observable result of the collision, and the only result required for evaluation of the effective cross-sections.

To discuss the cross-sections, examine a plane (AA on Fig. 2.1) normal to the initial relative velocity $\mathbf{c}'_\mathrm{r}$ and choose the initial asymptote of one of the trajectories to mark the point O at which it pierces the plane. For a finite force range b_m, a circle of this radius may be drawn around O. The initial asymptote of the trajectory of the other particle must aim at the plane within this circle, say, at a distance $b < b_\mathrm{m}$ from O, if there is to be any deflection at all. In accord with the assumption that the particles are uncorrelated prior to collision, a uniform probability distribution of those second points within the circle may be assumed. Hence, should it be imagined that collisions with the same initial $\mathbf{c}'_\mathrm{r}$ are repeatedly recorded, a uniform flux J is obtained, thereby giving $\pi b_\mathrm{m}^2 J$ partner particles per unit time aimed at the circle. Division by the flux J gives, by definition, the integral cross-section

$$\sigma_\mathrm{int} = \pi b_\mathrm{m}^2. \tag{2.157}$$

Classical mechanics allows for such literal geometric interpretation also of the differential cross-section σ, which is the cross-section per unit solid angle $\mathrm{d}\mathbf{e}$ after scattering: note, however, that only partners aimed at the plane AA within a thin annulus of area $2\pi b\mathrm{d}b$ are to be counted. Post-scattering, they are found within a narrow solid angle $\mathrm{d}\mathbf{e} = 2\pi \mathrm{d}(\cos\chi)$ between two cones. By the same argument as that made earlier, the area of the annulus equals the differential $\sigma\mathrm{d}\mathbf{e}$ of the scattering cross-section. Following cancellation of the factor 2π, the result is

$$\sigma(E,b)\,\mathrm{d}(\cos\chi) = b\mathrm{d}b\,. \tag{2.158}$$

Having computed $\chi(E,b)$ from formula (2.156), the partial derivative $\partial(\cos\chi)/\partial b$ may then be employed to obtain the differential cross-section as

$$\sigma(E,b) = \left| b\left[\frac{\partial(\cos\chi)}{\partial b}\right]^{-1}\right| . \tag{2.159}$$

To convert to other variables, $b \equiv b(E,\chi)$ may be extracted from $\chi \equiv \chi(E,b)$ and the result substituted into (2.159) to obtain $\sigma \equiv \sigma(E,\chi)$, which may then be converted into $\sigma(c_\mathrm{r},\chi)$. The integral cross-section is then obtained as a function of the collision speed, c_r, as

$$\sigma_\mathrm{int}(c_\mathrm{r}) = \int \mathrm{d}\mathbf{e}\;\sigma(c_\mathrm{r},\chi) = 2\pi\int_{-1}^{1}\mathrm{d}(\cos\chi)\;\sigma(c_\mathrm{r},\chi)\,. \tag{2.160}$$

Upon returning to eqns (2.131, 2.132), eqn (2.158) can be utilized to replace $\sigma(E,b)\mathrm{d}(\cos\chi)$ by $b\mathrm{d}b$. This result allows the classical mechanical expression

$$Q^{(m)}(E) = \int_0^\infty [1-\cos^m\chi(E,b)]b\,\mathrm{d}b\,, \tag{2.161a}$$

to be obtained for the scattering cross-section $Q^{(m)}(E)$ and, in turn, this expression then enables the omega integral $\Omega^{(m,n)}$ given in eqn (2.132) to be expressed as

$$\begin{aligned}\Omega^{(m,n)}(T) &= \frac{\pi}{2}\bar{c}_\mathrm{r}\int_0^\infty \mathrm{e}^{-W_\mathrm{r}^2}W_\mathrm{r}^{2n+3}Q^{(m)}(E)\,\mathrm{d}W_\mathrm{r}\\ &= \frac{\pi\bar{c}_r}{4(k_\mathrm{B}T)^{n+2}}\int_0^\infty \mathrm{e}^{-\beta E}E^{n+1}Q^{(m)}(E)\,\mathrm{d}E\,,\end{aligned} \tag{2.161b}$$

with $\beta \equiv (k_\mathrm{B}T)^{-1}$.

2.5.4 Quantum evaluation of effective cross-sections

Classical calculations are subject to doubt, especially for light molecules and for small-angle scattering. A rigorous theory must take recourse to quantum mechanics, in which the scattering of two atoms is governed by the Schrödinger equation for the two-atom system. As in the classical treatment of the previous subsection, a transformation into center-of-mass and relative coordinates can be made in order to separate the two-particle Schrödinger equation. Stationary solutions of the form

$$\Psi(\mathbf{r},\mathbf{r}_1) = \exp\{iM\mathbf{c}_\mathrm{CM}\cdot\mathbf{R}_\mathrm{CM}/\hbar\}\psi(\mathbf{R}) \tag{2.162}$$

in which the factor $\psi(\mathbf{R})$ describes the relative motion and obeys the equation

$$\left[-\frac{\hbar^2}{2m_\mathrm{r}}\nabla_R^2 + V(R) - E_\mathrm{r}\right]\psi(\mathbf{R}) = 0 \tag{2.163}$$

are sought. As before, E is the total energy in the center-of-mass system. In the asymptotic region in which the interaction is no longer important, the scattering state may be expected to be described by the asymptotic form

$$\psi(\mathbf{R}) \sim \mathrm{e}^{i\mathbf{k}'\cdot\mathbf{R}} + a(E,\chi)\frac{\mathrm{e}^{ikR}}{R}\,, \tag{2.164}$$

in which $a(E,\chi)$ is the so-called scattering amplitude, and $\hbar\mathbf{k}' = m_\mathrm{r}\mathbf{c}_\mathrm{r}'$.

According to the general formula

$$\mathbf{J} = \frac{\hbar}{m_\mathrm{r}}\mathrm{Im}(\psi^*\nabla\psi) \tag{2.165}$$

for the quantum-mechanical probability flux, the first term on the right-hand side in eqn (2.164), which represents the incident wave function, corresponds to an incident flux per unit area, that is, $\mathbf{J}_\mathrm{inc} = \mathbf{c}_\mathrm{r}'$. The scattered wavefunction, described by the second term in (2.164), gives the magnitude of the scattered flux as $J_\mathrm{sc} = (\hbar k/m_\mathrm{r})|a|^2/R^2$. Hence, the scattered flux per unit solid angle equals $c_\mathrm{r}|a|^2$. Division by J_inc (with c_r' $= c_\mathrm{r}$ for atomic scattering) yields the differential cross-section as[4]

$$\sigma(E,\chi) = |a(E,\chi)|^2\,, \tag{2.166}$$

and thus the integral cross-section as

$$\sigma_\mathrm{int}(E) = 2\pi\int_{-1}^{1}\mathrm{d}(\cos\chi)|a(E,\chi)|^2\,. \tag{2.167}$$

An alternative relationship between the scattering amplitude $a(E,\chi)$ and the integral scattering cross-section $\sigma_\mathrm{int}(E)$ may be obtained from the conservation of the total quantum-mechanical probability (equivalently, number/mass conservation) for the scattering process (Roman, 1965). This relation, known as the optical theorem, is given by

$$\sigma_\mathrm{int}(E) = \frac{4\pi}{k}\mathrm{Im}\{a(E,0)\}\,, \tag{2.168}$$

and connects the integral scattering cross-section to the imaginary part of the scattering amplitude in the forward direction. One important conclusion to be drawn from this relation is that the scattering amplitude $a(E,\chi)$ is always complex, since $\sigma_\mathrm{int}(E)$ is positive unless $a(E,\chi)$ happens to be identically zero.

Because the operator in eqn (2.163) is rotationally invariant, a complete set of solutions of the form $\phi_{lm} = Z_l(R)Y_{lm}(\mathbf{R}/R)$ exists. Moreover, the asymptotic wave

[4]The differential cross-section $\sigma(E,\chi)$ is often written as a function of the physical scattering angle $\theta \equiv |\chi|$, rather than in terms of χ itself. So long as χ is restricted to the interval $[0,\pi]$, and σ is expressed in terms of $\cos\chi$, no further distinction need be drawn.

function (2.164) is axially symmetric, so that it can be expanded in terms of those ϕ_{lm} having $m = 0$, and the solution is given by

$$\psi(R,\chi) = \sum_{l=0}^{\infty} A_l Z_l(R) P_l(\cos\chi)\,. \tag{2.169}$$

The radial factor satisfies the equation

$$\left[-\hbar^2 2m_\mathrm{r}R\left(\frac{\mathrm{d}^2}{\mathrm{d}R^2}\right)R + \frac{l(l+1)\hbar^2}{2m_\mathrm{r}R^2} + V(R) - E\right] Z_l(R) = 0\,, \tag{2.170}$$

which normally requires numerical solution. The expansion coefficients A_l are determined by matching the asymptotic behavior (i.e. at $R \to \infty$) of the desired solution. This solution procedure is known as the method of partial waves (Roman, 1965; Newton, 1966).

Rather than dwell upon the computational methods to be employed for solving the radial eqn (2.170), the problem of asymptotic fitting is the focus of the present discussion. The proposed expansion can also be applied to the incident wave alone, via the well-known Rayleigh formula,

$$\begin{aligned}\exp\{ikR\cos\chi\} &= \sum_{l=0}^{\infty} i^l(2l+1)j_l(kR)P_l(\cos\chi)\\ &\sim \sum_{l=0}^{\infty} i^l(2l+1)\frac{1}{kR}\sin(kR - \tfrac{1}{2}l\pi)P_l(\cos\chi)\,,\end{aligned} \tag{2.171}$$

in which $j_l(kR)$ is a spherical Bessel function (Abramowitz and Stegun, 1972). The latter expression is an approximation for large kR. The sum can be pictured as a superposition of spherical standing waves (standing because the probability fluxes entering and leaving the scattering regions cancel). The same argument also applies to the total ψ, so that for large R, it can be expanded as

$$\mathrm{e}^{ikR\cos\chi} + a(E,\chi)\frac{\mathrm{e}^{ikR}}{R} = \sum_{l=0}^{\infty} A_l i^l(2l+1)\frac{\sin(kR - \frac{1}{2}l\pi + \delta_l)}{kR} P_l(\cos\chi)\,, \tag{2.172}$$

in which the amplitudes A_l and phase shifts δ_l are as yet unknown functions of E. The approximations used in the second line of eqn (2.171) and in eqn (2.172) are not uniform for all expansion terms, since $kR \gg l^2$ is required (Abramowitz and Stegun, 1972). When a large number of terms is considered, it is therefore better to refer back to the spherical Bessel functions $j_l(kR)$ for the matching. For approximation (2.172) to be valid, the separation R must be sufficiently large for the influence of the potential to become negligible (i.e. $|V(R)| \ll E$ is a necessary condition).

In contrast to the incident and total wave functions, the scattered part of ψ alone, as asymptotically described by the second term in eqn (2.164), contains only outgoing waves, behaving like $R^{-1}e^{ikR}P_l(\cos\chi)$ for large R. This condition results in a relation between the amplitudes and the partial-wave phase shifts, $\delta_l(E)$, namely

$$A_l(E) = e^{i\delta_l(E)}\,. \tag{2.173}$$

The phase shifts δ_l represent the differences between the asymptotic phases of the orbital angular momentum (partial wave) components of the scattered (or outgoing) waves in the presence of and in the absence of the interaction potential energy $V(R)$. These phase shifts are thus the crucial quantities to be computed using eqn (2.170).

The scattering amplitude is given in terms of partial scattering amplitudes $a_l(E)$ by

$$a(E,\chi) = \sum_{l=0}^{\infty} a_l(E)P_l(\cos\chi)\,, \tag{2.174}$$

in which the individual partial scattering amplitudes $a_l(E)$ are given by

$$a_l(E) = \frac{2l+1}{2ik}\left[e^{2i\delta_l} - 1\right] \tag{2.175a}$$

$$= \frac{2l+1}{2ik}[S_l(E) - 1]\,, \tag{2.175b}$$

with $S_l(E)$, defined as

$$S_l(E) \equiv e^{2i\delta_l(E)}\,, \tag{2.175c}$$

called the (scattering) S-matrix element for the lth partial wave scattering. Equivalently, $a_l(E)$ may be expressed as

$$a_l(E) = \frac{2l+1}{k}\,e^{i\delta_l}\sin\delta_l\,, \tag{2.176}$$

so that the scattering amplitude $a(E,\chi)$ takes the final form

$$a(E,\chi) = \frac{1}{k}\sum_{l=0}^{\infty}(2l+1)e^{i\delta_l}\sin\delta_l\,P_l(\cos\chi)\,. \tag{2.177}$$

The differential scattering cross-section of eqn (2.166) thus becomes

$$\sigma(E,\chi) = \frac{1}{k^2}\sum_{l,l'=0}^{\infty}(2l+1)(2l'+1)e^{i(\delta_l-\delta_{l'})}\sin\delta_l\sin\delta_{l'}P_l(\cos\chi)P_{l'}(\cos\chi)\,, \tag{2.178}$$

from which the structure of the differential scattering cross-section becomes evident. As the integral scattering cross-section, $\sigma_{\text{int}}(E)$, is related to the differential scattering cross-section $\sigma(E,\chi)$ by eqn (2.167), a final expression for the integral scattering cross-section $\sigma_{\text{int}}(E)$ is obtained in terms of the phase shifts $\delta_l(E)$ associated with the various orbital angular momentum components (partial waves) of the scattered wave by employing the orthogonality condition

$$\int_{-1}^{1} P_l(\cos\chi) P_{l'}(\cos\chi)\, \mathrm{d}\cos\chi = \frac{2}{2l+1}\, \delta_{l'l}$$

of the Legendre polynomials. The integral cross-section is thus given in terms of the partial-wave phase shifts as

$$\sigma_{\mathrm{int}}(E) = \frac{4\pi}{k^2} \sum_{l=0}^{\infty} (2l+1) \sin^2 \delta_l(E)\,. \tag{2.179}$$

The reason that the integral cross-section is generally found to be less sensitive to variations in the scattering energy E becomes evident upon examining expressions (2.178) and (2.179). Roman (1965), for example, pointed out that the difference in the structures of these two scattering cross-sections results from the fact that the partial waves interfere coherently to give the differential scattering cross-section, but add together incoherently to form the integral scattering cross-section.

Expression (2.178) for the differential cross-section may now be used in eqn (2.131) to obtain the quantum-mechanical cross-sections $Q^{(m)}(W_{\mathrm{r}})$ required in (2.132) for evaluation of the omega integrals. The most important omega integrals for the first and second Chapman–Cowling approximations to the transport coefficients involve (Massey and Mohr, 1933; Munn *et al.*, 1965; Fox, 1968*a*; Fox, 1968*b*; Meeks *et al.*, 1994)

$$Q^{(1)}(E) = \frac{4\pi}{k^2} \sum_l (l+1) \sin^2(\delta_{l+1} - \delta_l)\,, \tag{2.180a}$$

$$Q^{(2)}(E) = \frac{4\pi}{k^2} \sum_l \frac{(l+1)(l+2)}{2l+3} \sin^2(\delta_{l+2} - \delta_l) \equiv \sum_l Q_l^{(2)}(E)\,, \tag{2.180b}$$

$$Q^{(3)}(E) = \frac{4\pi}{k^2} \sum_l \left[\frac{(l+1)(l+2)(l+3)}{(2l+3)(2l+5)} \sin^2(\delta_{l+3} - \delta_l) + \frac{3(l^2+2l-1)(l+1)}{(2l-1)(2l+5)} \sin^2(\delta_{l+1} - \delta_l) \right], \tag{2.180c}$$

$$Q^{(4)}(E) = \frac{4\pi}{k^2} \sum_l \frac{(l+1)(l+2)}{(2l+3)(2l+7)} \left[\frac{(l+3)(l+4)}{2l+5} \sin^2(\delta_{l+4} - \delta_l) + \frac{4l^2+12l-6}{2l-1} \sin^2(\delta_{l+2} - \delta_l) \right] \equiv \sum_l Q_l^{(4)}(E)\,. \tag{2.180d}$$

These expressions are appropriate when spin-statistical effects are unimportant (i.e. when Maxwell–Boltzmann statistics apply). The individual cross-sections $Q_l^{(2)}$ and $Q_l^{(4)}$ that have been defined via eqns (2.180b, 2.180d) will still be required for identical particle scattering.

The quantum-mechanical cross-section expressions (2.180) must be modified when identical particle scattering is considered: in particular, an additional factor 2 arises from the replacement of $|a(\chi)|^2$ by $\frac{1}{2}|a(\chi) + a(\pi - \chi)|^2$ for spinless particles, and the sum is restricted to even values of l. Meeks and colleagues (1994) showed that for odd values of m and for either l and l' both even (BE statistics) or l and l' both odd (FD statistics) the relations

$$Q_{\mathrm{BE}}^{(m)}(E) = Q_{\mathrm{BE}}^{(1)}(E) = \sum_{l=\mathrm{even}} (2l+1)\sin^2\delta_l \,, m \text{ odd} \tag{2.181a}$$

and

$$Q_{\mathrm{FD}}^{(m)}(E) = Q_{\mathrm{FD}}^{(1)}(E) = \sum_{l=\mathrm{odd}} (2l+1)\sin^2\delta_l \,, m \text{ odd} \tag{2.181b}$$

hold, while for even values of m, the components of $Q_l^{(m)}(E)$ for $m = 2$, 4 are defined by the expressions given, respectively, in eqn (2.180b) and eqn (2.180d). Finally, when spin statistics enter, the summations must also be modified (Hirschfelder *et al.*, 1964): should the atoms be spin-I fermions, for example, the sums over all l must be replaced by

$$\frac{I}{2I+1}\sum_{l \text{ even}} Q_l^{(m)} + \frac{I+1}{2I+1}\sum_{l \text{ odd}} Q_l^{(m)} \tag{2.182a}$$

and, should the atoms be spin-I bosons, these sums must be replaced by

$$\frac{I+1}{2I+1}\sum_{l \text{ even}} Q_l^{(m)} + \frac{I}{2I+1}\sum_{l \text{ odd}} Q_l^{(m)} \,. \tag{2.182b}$$

In each case the scattering cross-section $Q_l^{(m)}$ is defined as in eqn (2.180), but with the factor 4π replaced by 8π.

Because transport properties can currently be obtained with experimental accuracies that are of the order of 0.10% or smaller, it is then necessary to be able to calculate transport coefficients, such as η and λ minimally to an accuracy that is better than 0.10%. Gases with constituents that involve binary collisions in which the reduced mass of the colliding atoms is smaller than approximately twenty will require a quantum-mechanical treatment in order to obtain highly accurate values of the transport coefficients. This is especially the case for pure helium (^{3}He or ^{4}He), binary mixtures of ^{3}He and ^{4}He, and for any binary mixture of helium and one of the other noble gases (or, for that matter, with any other chemical species) at any temperature. A quantum-mechanical treatement will also likely be needed for pure neon and its binary mixtures with Ar, Kr, or Xe at temperatures below ambient temperature. Because the quantum-mechanical expressions for the relevant (monatomic) effective cross-sections can be expressed in terms of omega integrals that in turn depend upon energy-dependent cross-sections $Q^{(m)}(E)$ of eqns (2.180), this means that it is important to be able to compute highly accurate values of the $Q^{(m)}(E)$ quantum-mechanically.

It can be seen from eqns (2.180) that the quantum-mechanical cross-sections $Q^{(m)}(E)$ depend upon weighted sums of squares of the sines of differences between phase shifts associated with the partial-wave components of the radial part of the scattered-state wavefunction. Relative phase shifts are traditionally evaluated by computing the differences between the asymptotic limiting values of the relative phases of the corresponding perturbed and unperturbed wavefunctions as, in this limit, the phase shift is independent of the node number: a node-to-node numerical integration process is thus terminated once the change in the phase shift $\delta_{l,\mathrm{z}}$ between two successive nodes (zeroes), say z and z + 1, is determined to be smaller than a preselected tolerance. The change in phase shift is traditionally given in terms of Bessel and Neumann functions evaluated at the location R_z of the node as

$$\delta_{l,\mathrm{z}} \equiv \arctan\left\{\frac{j_l(E, R_\mathrm{z})}{n_l(E, R_\mathrm{z})}\right\}.$$

However, as the range of the inverse tangent function is restricted to the interval $(-\frac{\pi}{2}, \frac{\pi}{2})$, the absolute phase shift still requires a determination of the multiple, n_π, of π (radians) by which the full phase shift $\delta_l(E)$ has grown in passing through the $n-1$ previous nodes. This is often achieved using eqn (9.2.29) given by Abramowitz and Stegun (1972). Wei and Le Roy (2006) accurately and effectively proposed a promising alternative procedure for calculating absolute scattering phase shifts.

2.6 Dynamical Models for Binary Atomic Interactions

Dynamical models are based upon simplifying assumptions made about the interaction between pairs of atoms, thereby simplifying the description of the collision process. Simple dynamical (or collision) models can be useful in establishing broad connections among physical properties and in delineating the effects of various approximation schemes introduced in order to solve the Boltzmann equation. Moreover, the analytical solutions that simple dynamical models provide often lead to new insights into the interpretation of the results of more complex collision calculations.

Two important and rather simple, but ultimately unrealistic, models frequently considered are the interaction between a pair of hard spheres, for which the interaction energy $V(R)$ is represented in terms of the radius R_0 of a hard sphere by

$$V(R) = \begin{cases} 0, & R > 2R_0 \\ \infty, & R \leq 2R_0 \end{cases}, \tag{2.183a}$$

and atoms that interact via a simple inverse-power-law force via a potential energy function (PEF) that takes the form

$$V(R) = A_\nu R^{-\nu}. \tag{2.183b}$$

The power-law interaction (2.183b) is termed repulsive or attractive depending upon whether the sign of A_ν is positive or negative.

Each of these dynamical models has proven to be quite useful. The hard-sphere interaction, for example, is often employed as a simple reference model to determine which calculations can readily be made. It is also a common practice (Hirschfelder *et al.*, 1964; Chapman and Cowling, 1970; Maitland *et al.*, 1980) to relate the omega-integrals of Sections 5.1, 5.2 to hard-sphere values in order to help visualize how realistic interatomic interactions deviate from hard-sphere behavior. Over 150 years ago, Maxwell (1867) showed how atoms interacting via a repulsive inverse fourth-power of R led to especially simple expressions for λ and η. As a consequence, this interaction is commonly referred to as the Maxwell model interaction. It was nearly one hundred years later that it was realized (Wang Chang and Uhlenbeck, 1970*a*; Waldmann, 1958*c*) that this particular power-law model is actually rather special, as the eigenfunctions of the linearized collision operator $\mathcal{R}$ can be obtained for it. For this reason (see Section 6.3), the first Chapman–Cowling approximation expressions for λ and η are exact for Maxwell molecules, and hence, the eigenfunctions of the linearized collision operator $\mathcal{R}$ for the Maxwell model interaction provide a sound basis for expansions of the distribution function when more general interactions are being considered. This (fortuitous) choice of basis functions made up from symmetric traceless tensors in $\mathbf{W}$ multiplied by associated Laguerre polynomials in W_{r}^2 is also related to the rapid convergence that has been obtained for the transport coefficients of the noble gases (Chapman and Cowling, 1970) (see also Chapter 4). Finally, it is also worth noting that generalized kinetic modeling procedures (see Section 2.7) have evolved from a knowledge of the eigenfunction properties of Maxwell molecules.

Because model potentials like the hard-sphere and power-law interactions often have a number of useful analytical properties that may be exploited, they are discussed in greater detail in Section 2.6.1 (the hard-sphere model), Section 2.6.2 (the general power-law interaction), and Section 2.6.3 (the Maxwell model).

2.6.1 The hard-sphere model interaction

Perhaps the single most successful collision model is the single-parameter hard-sphere model, which is often used to check new theories. Any theory that fails to give a sensible result for the hard-sphere limit may normally be rejected on this basis alone. The interparticle potential is given by eqn (2.183a) [see also Fig. 2.2, which illustrates the relevant collision dynamics].

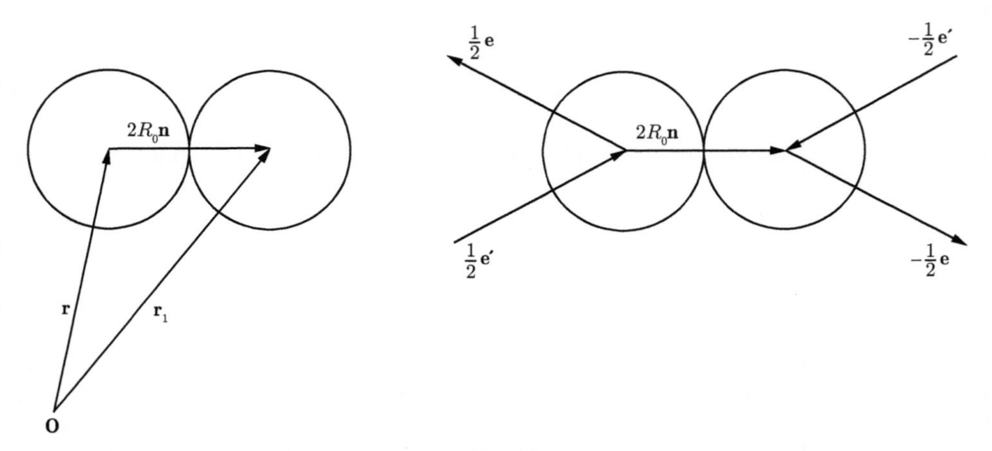

Fig. 2.2 Hard-sphere collision in the laboratory (left) and center-of-mass systems.

Upon substituting $R_c = 2R_0$ for the lower limit of the integral in eqn (2.156b), it can immediately be evaluated to yield

$$\tfrac{1}{2}(\pi - \chi) = \arcsin\left(\frac{b}{2R_0}\right) \tag{2.184}$$

for the scattering angle χ. The angle dependence of the impact parameter b is thus given by

$$b(\chi) = 2R_0 \sin\left\{\tfrac{1}{2}(\pi - \chi)\right\} = 2R_0 \cos(\chi/2)\,, \tag{2.185}$$

while according to eqn (2.159) the differential cross-section becomes

$$\sigma(\chi, c_r) = \left|\frac{b\mathrm{d}b}{\sin\chi\mathrm{d}\chi}\right| = R_0^2\,, \tag{2.186}$$

which is independent of the scattering angle χ and relative speed c_r. The scattering of hard spheres is thus isotropic in the center-of-mass coordinates. The integral cross-section, σ_{int}, is therefore given, according to eqn (2.167), simply as

$$\sigma_{\text{int}}(\text{hs}) = 4\pi R_0^2\,.$$

The advantage of the hard-sphere model now becomes obvious: both the angular and relative velocity integrations in the collision integrals can be performed analytically to give

$$\Omega_{\text{hs}}^{(m,n)} = \tfrac{1}{4}\bar{c}_r \pi R_0^2 \left[2 - \frac{1 + (-1)^m}{m+1}\right](n+1)! \tag{2.187}$$

as the general result for the hard-sphere omega integrals (Waldmann, 1958*d*). This result may be utilized to define dimensionless, or reduced, omega integrals via

$$\Omega^{(m,n)\star} \equiv \frac{\Omega^{(m,n)}}{\Omega_{\text{hs}}^{(m,n)}}\,, \tag{2.188}$$

in which $\Omega^{(m,n)}$ is the omega integral computed with a specified realistic potential energy function $V(R)$. Values of such reduced omega integrals have been tabulated as functions of temperature for a number of potential functions (Hirschfelder *et al.*, 1964; Maitland *et al.*, 1980; Klein and Smith, 1968).

From eqns (2.133), with $\Omega_{\text{hs}}^{(2,2)} = 2\bar{c}_r \pi R_0^2$, the viscosity and thermal conductivity hard-sphere effective cross-sections become

$$\mathfrak{S}_\eta(\text{hs}) = \tfrac{3}{2}\mathfrak{S}_\lambda(\text{hs}) = \tfrac{16\pi}{5}R_0^2 = \tfrac{4}{5}\sigma_{\text{int}}\,. \tag{2.189}$$

These results can then be fed into eqns (2.70) and (2.75) to obtain the first approximations for the viscosity and thermal conductivity coefficients of a hard-sphere gas, namely,

$$\eta(\text{hs}) = \frac{5k_B T}{4\bar{c}_r \sigma_{\text{int}}}\,, \qquad \lambda(\text{hs}) = \frac{75{k_B}^2 T}{16 m \bar{c}_r \sigma_{\text{int}}}\,. \tag{2.190}$$

It is useful to consider the concept of the mean free path, defined as the ratio

$$l = \overline{c}/\overline{\nu}$$

of the mean thermal speed $\overline{c}$ and the mean hard-sphere collision rate $\overline{\nu}$, for a hard-sphere gas. The mean collision rate $\overline{\nu}$ for a hard-sphere gas may be obtained from eqn (1.103) as

$$\overline{\nu} = n^{-1} \iint \mathrm{d}\mathbf{c}\mathrm{d}\mathbf{c}_1 \, \sigma_{\mathrm{int}} c_{\mathrm{r}} \, f_0(\mathbf{c}) f_0(\mathbf{c}_1) = n\overline{c}_{\mathrm{r}}\sigma_{\mathrm{int}} \,, \tag{2.191}$$

in which case the mean free path in a hard-sphere gas is given by

$$l = \frac{\overline{c}}{n\overline{c}_{\mathrm{r}}\sigma_{\mathrm{int}}} = \frac{1}{n\sqrt{2}\sigma_{\mathrm{int}}} \,. \tag{2.192}$$

Upon employing eqn (2.192) to replace σ_{int} in expressions (2.190), $\eta(\mathrm{hs})$ and $\lambda(\mathrm{hs})$ can be obtained in terms of the mean free path l as

$$\eta(\mathrm{hs}) = \frac{5lp}{4\overline{c}} \,, \qquad \lambda(\mathrm{hs}) = \frac{75k_{\mathrm{B}}lp}{16m\overline{c}} \,. \tag{2.193}$$

It has become customary to invert the first of these expressions to provide a general definition of the mean free path for an arbitrary gas, namely,

$$l \equiv \frac{4\eta\overline{c}}{5p} \,. \tag{2.194}$$

The first expression in (2.190) may also be employed to assign an effective hard-sphere integral (collision) cross-section $\sigma_{\mathrm{int}} = 4\pi R_0^2$ between pairs of gas particles, thereby assigning an effective hard-sphere diameter $2R_0$ to each gas atom using its experimental viscosity. A somewhat less artificial choice would be to compare the gas with a more realistic model PEF, such as a Mie/Lennard-Jones (MLJ) interaction [see eqns (2.214)] and to assign the distance σ at which the (fitted) MLJ model PEF vanishes to be $2R_0$. Of course, it will be clear that a hard-sphere model cannot provide the correct temperature dependence of a transport coefficient, especially at low temperatures, as it entirely disregards the attractive part of an interatomic interaction; moreover, even for temperatures at which collisions are dominated by the repulsive short-range interaction, the hard-sphere model cannot be expected to give quantitative comparisons with experimental results, as the repulsive part of a realistic potential function is never infinitely steep.

The collision rate $\nu(c)$ of eqn (1.103) has been shown (Williams, 1972) to be given for a hard-sphere gas by

$$\begin{aligned} \nu(c) &= \int \mathrm{d}\mathbf{c}_1 \, c_{\mathrm{r}}\sigma_{\mathrm{int}} f_0(\mathbf{c}_1) \\ &= \tfrac{1}{2} n\overline{c}_{\mathrm{r}}\sigma_{\mathrm{int}} \left[\left(W + \frac{1}{2W} \right) \sqrt{\pi} \, \mathrm{erf}(W) + \exp(-W^2) \right] , \end{aligned} \tag{2.195}$$

in which $W = c/c_0$, with $c_0 = (2k_{\mathrm{B}}T/m)^{\frac{1}{2}}$, while $\mathrm{erf}(W)$ denotes the error function. The collision rate thus has the limiting behaviors

$$\nu(0) = n\bar{c}\sigma_{\text{int}} \quad \text{and} \quad \nu(c) \rightarrow nc\sigma_{\text{int}}\,, \quad \text{for} \quad c \rightarrow \infty\,. \tag{2.196}$$

If a hard-sphere gas in equilibrium and within it a test particle of equal mass and size is considered, then the scattering kernel for the test particle is obtained from eqn (1.153) by a straightforward, though tedious, calculation as (Grad, 1963)

$$\mathcal{K}_1(\mathbf{c} \rightarrow \mathbf{c}') = \frac{n\sigma_{\text{int}}}{\pi^{\frac{3}{2}} c_0^2 |\mathbf{W} - \mathbf{W}'|} \exp\left[-\frac{1}{4}\left(|\mathbf{W} - \mathbf{W}'| - \frac{W^2 - W'^2}{|\mathbf{W} - \mathbf{W}'|}\right)^2\right]. \tag{2.197}$$

A much shorter derivation is possible using the Fourier transform of a time-dependent correlation function (Williams, 1972). Along the same lines, it is also possible to obtain the scattering kernel for a pure hard-sphere gas, cf. eqns (1.104) and (1.154), as (Grad, 1963)

$$\mathcal{K}(\mathbf{c} \rightarrow \mathbf{c}') = 2\mathcal{K}_1(\mathbf{c} \rightarrow \mathbf{c}') - \frac{n\sigma_{\text{int}}}{\pi^{\frac{3}{2}} c_0^2} |\mathbf{W} - \mathbf{W}'| \mathrm{e}^{-W'^2}\,, \tag{2.198}$$

in terms of $\mathcal{K}_1$.

Notice in Fig. 2.2 that a binary collision between two hard spheres does not correspond to the centers of the colliding atom-pair being coincident, that is, a hard-sphere collision is not a point-like, or local, event. This property enabled Enskog (1922) to formulate a more generalized kinetic equation for the hard-sphere interaction. The kinetic equation that he proposed will facilitate a discussion of some of the limitations of the Boltzmann equation. In particular, one deficiency of the Boltzmann equation is that the factors f', f_1', and so on, in the integrand must be evaluated at the same location $\mathbf{r}$, thus effectively localizing all collision events. This, in turn, requires that spatial variations of f must be sufficiently slow that differences over a distance $2R_0$ may be neglected.

Enskog proposed that, instead of evaluating both f' and f_1' at $\mathbf{r}$, f_1' should be evaluated at $\mathbf{r}_1 = \mathbf{r} + 2R_0\mathbf{n}$ (see Fig.2.2), with $\mathbf{n}$ defined as

$$\mathbf{n} \equiv \frac{\mathbf{e}' - \mathbf{e}}{|\mathbf{e}' - \mathbf{e}|}\,,$$

if f' is evaluated at $\mathbf{r}$. In addition, $\mathbf{n}$ should be replaced by $-\mathbf{n}$ for the inverse collision, so that the non local Enskog Ansatz for the collision term becomes

$$\left(\frac{\delta f(\mathbf{c}, \mathbf{r}, t)}{\delta t}\right)_{\text{coll}} = \iint \mathrm{d}\mathbf{c}_1 \mathrm{d}\mathbf{e}'\ \sigma_{\text{int}} c_{\text{r}} \left[f(\mathbf{c}', \mathbf{r}, t) f(\mathbf{c}_1', \mathbf{r} + 2R_0\mathbf{n}, t) - f(\mathbf{c}, \mathbf{r}, t) f(\mathbf{c}_1, \mathbf{r} - 2R_0\mathbf{n}, t)\right].$$

Because the Enskog collision term accounts for collisional transfer effects in transport phenomena (an example is the instantaneous transfer of momentum over a distance $2R_0$ between two colliding hard spheres), the Enskog equation should thus be applicable at somewhat higher gas densities. At higher densities, the relative volume, φ, occupied by the particles, namely,

$$\varphi = \tfrac{4}{3} n\pi R_0^3 \,,$$

(which is $n/4$ times the van der Waals constant b per particle) is no longer sufficiently small to be ignored. It would be incorrect, however, to assume that the particles are uncorrelated prior to collision, so that it is minimally necessary to introduce a factor having the value of the equilibrium pair-distribution function $g(R, n)$ at $R = 2R_0$ and for a density n at the point $\mathbf{r}$, so as to account for the increased collision rate that can be associated with correlations. A reasonable approximation for the pair distribution function is (Velarde, 1974)

$$g(2R_0, n) = 1 + \tfrac{5}{2}\varphi + 4.59\varphi^2 + \mathcal{O}(\varphi^3) \,,$$

by which the right-hand side of the collision term at page 89 must be multiplied.

Application of either the Chapman–Enskog or moment method to the linearized Enskog equation results in a system of thermo-hydrodynamical equations that go beyond ordinary kinetic theory (Grad, 1963; Schmidt *et al.*, 1981). The final results for the transport coefficients can be given in the form (Velarde, 1974)

$$\frac{\eta(\varphi)}{\eta(0)} = \tfrac{1}{g} + \tfrac{16}{5}\varphi + 12.18 g\varphi^2 + \mathcal{O}(\varphi^3) \,,$$

$$\frac{\lambda(\varphi)}{\lambda(0)} = \tfrac{1}{g} + \tfrac{24}{5}\varphi + 12.12 g\varphi^2 + \mathcal{O}(\varphi^3) \,,$$

in which g stands for $g(2R_0, n)$. The ratios on the left refer to the limits at $\varphi \to 0$, that is, to the values deduced from the Boltzmann equation. A volume viscosity coefficient, given by

$$\frac{\eta_V(\varphi)}{\eta(0)} = 16.03 g\varphi^2 + \mathcal{O}(\varphi^3) \,,$$

also appears. Up to $\varphi \approx 0.1$, the measured density dependence of η and λ for compressed noble gases agrees well with these expressions (Velarde, 1974).

2.6.2 Inverse power-law model potential energy functions

The repulsive inverse power-law potential, eqn (2.183b), can be treated classically in a straightforward fashion, starting from eqn (2.156b). It is convenient to change variables from R to $s = b/R$ so that

$$V(s) = A_\nu \left(\frac{s}{b}\right)^\nu \,, \tag{2.199}$$

while $V(R)/E$ in eqn (2.156b) takes the form

$$\frac{V(s)}{E} = \frac{A_\nu}{k_B T W_r^2} \left(\frac{s}{b}\right)^\nu \equiv \left(\frac{s}{s_0}\right)^\nu \,, \tag{2.200}$$

in which

$$s_0 = s_0(W_\mathrm{r}) = b\left(\frac{k_\mathrm{B}TW_\mathrm{r}^2}{A_\nu}\right)^{1/\nu}. \tag{2.201}$$

With these results, the deflection angle can be written as

$$\chi = \chi(s_0) = \pi - 2\int_0^{s_\mathrm{c}} \mathrm{d}s \left[1 - s^2 - \left(\frac{s}{s_0}\right)^\nu\right]^{-\frac{1}{2}}, \tag{2.202}$$

with $s_\mathrm{c} = b/R_\mathrm{c}$ the (unique) positive root of

$$1 - s_\mathrm{c}^2 - \left(\frac{s_\mathrm{c}}{s_0}\right)^\nu = 0. \tag{2.203}$$

For the inverse power-law then, eqn (2.131) may be employed, together with eqn (2.201), to rewrite the transport cross-section $Q^{(m)}(W_\mathrm{r})$ as

$$Q^{(m)}(W_\mathrm{r}) = 2\pi\left(\frac{A_\nu}{k_\mathrm{B}TW_\mathrm{r}^2}\right)^{2/\nu}\int_0^\infty s_0\mathrm{d}s_0(1-\cos^m\chi) = 2\pi\left(\frac{A_\nu}{k_\mathrm{B}TW_\mathrm{r}^2}\right)^{2/\nu} F_m(\nu), \tag{2.204}$$

with $F_m(\nu)$, defined as

$$F_m(\nu) \equiv \int_0^\infty s_0\mathrm{d}s_0\,(1-\cos^m\chi),$$

determined by numerical quadrature (Chapman and Cowling, 1970). The omega integrals are then obtained from eqn (2.132), and may be expressed in the form

$$\begin{aligned}\Omega^{(m,n)} &= \frac{\pi}{2}\bar{c}_\mathrm{r}\left(\frac{A_\nu}{k_\mathrm{B}T}\right)^{2/\nu} F_m(\nu)\int_0^\infty \mathrm{d}W_\mathrm{r}\mathrm{e}^{-W_\mathrm{r}^2}W_\mathrm{r}^{2n+3-(4/\nu)}\\ &= \frac{\pi}{4}\bar{c}_\mathrm{r}\left(\frac{A_\nu}{k_\mathrm{B}T}\right)^{2/\nu} F_m(\nu)\Gamma(n+2-2/\nu),\end{aligned} \tag{2.205}$$

in which $\Gamma(x)$ is a gamma function (see Appendix A.2.3).

From eqn (2.133) and the result (2.205) for the Ω-integrals, $\mathfrak{S}_\lambda$ and $\mathfrak{S}_\eta$ are obtained as

$$\mathfrak{S}_\lambda = \tfrac{2}{3}\mathfrak{S}_\eta = \frac{4\pi}{15}\left(\frac{A_\nu}{k_\mathrm{B}T}\right)^{2/\nu} F_2(\nu)\Gamma(4-2/\nu) \tag{2.206}$$

for a simple repulsive power-law interaction. The viscosity and thermal conductivity are then given from eqns (2.70, 2.75) by

$$\eta = \frac{5k_\mathrm{B}T}{2\pi\bar{c}_\mathrm{r}F_2(\nu)\Gamma(4-2/\nu)}\left(\frac{k_\mathrm{B}T}{A_\nu}\right)^{2/\nu}, \tag{2.207}$$

and

$$\lambda = \frac{75k_\mathrm{B}^2T}{8\pi m\bar{c}_\mathrm{r}F_2(\nu)\Gamma(4-2/\nu)}\left(\frac{k_\mathrm{B}T}{A_\nu}\right)^{2/\nu}. \tag{2.208}$$

Corresponding formulae may be obtained for an attractive power-law model PEF by reversing the signs of $(s/s_0)^\nu$ in eqns (2.202, 2.203). For such a case, s_c becomes the least positive root of

$$1 - s^2 + \left(\frac{s}{s_0}\right)^\nu = 0\,.$$

A more realistic model PEF must contain both repulsive and attractive branches corresponding, respectively, to separations between the colliding atoms at which their electron charge distributions may overlap and to separations at which the atoms are separated by many atomic diameters. The simplest such model PEF is that introduced independently by Mie (1903) and by Lennard–Jones (1924*a*; 1924*b*), and now commonly referred to as the MLJ model PEF. A family of MLJ model potential energy functions is defined via the general formula

$$V_{\text{MLJ}}(R) = \mathfrak{D}_\text{e}\left[\left(\frac{m}{n-m}\right)\left(\frac{R_{\min}}{R}\right)^n - \left(\frac{n}{n-m}\right)\left(\frac{R_{\min}}{R}\right)^m\right], \qquad n > m\,, \tag{2.209a}$$

in which $\mathfrak{D}_\text{e}$ is the depth of the potential and $R_{\min}$ is the atomic separation at which the minimum, $-\mathfrak{D}_\text{e}$, occurs. We refer to these functions as MLJ(n,m) potential energy functions. The most frequently encountered member of this family, the MLJ(12,6) model PEF [often referred to as the Lennard–Jones (12, 6) PEF], is illustrated in Fig. 2.3. These MLJ potential energy functions may also be represented as

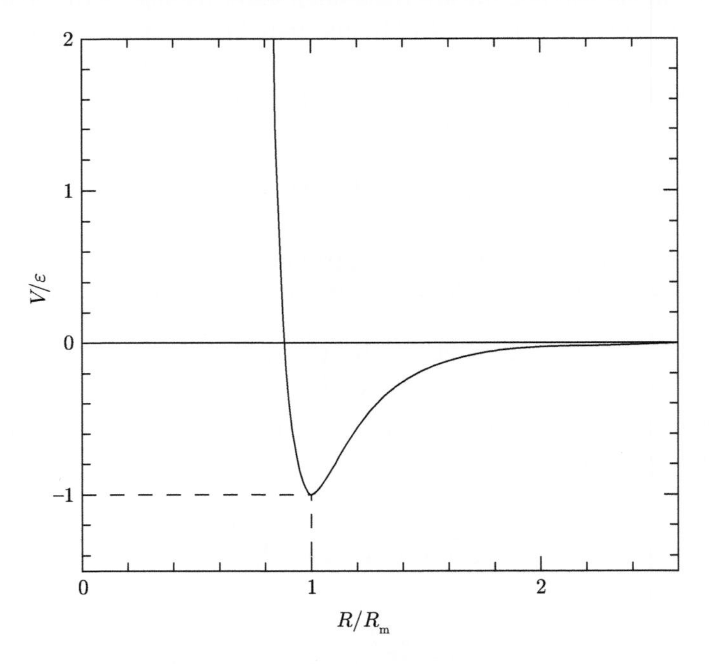

Fig. 2.3 The MLJ(12,6) model PEF in reduced form.

$$V_{\text{MLJ}}(R) = \mathfrak{D}_{\text{e}} \frac{m}{n-m} \left(\frac{m}{n}\right)^{-n/(n-m)} \left[\left(\frac{\sigma}{R}\right)^{n} - \left(\frac{\sigma}{R}\right)^{m}\right]$$

$$= \frac{\mathfrak{D}_{\text{e}}}{\left(\frac{m}{n}\right)^{m/(n-m)} - \left(\frac{m}{n}\right)^{n/(n-m)}} \left[\left(\frac{\sigma}{R}\right)^{n} - \left(\frac{\sigma}{R}\right)^{m}\right], \tag{2.209b}$$

in which σ is the distance at which $V_{\text{MLJ}}(R)$ vanishes, that is, the distance at which the attractive and repulsive forces between the atoms precisely cancel one another. Equivalently, σ may be said to be the distance of closest approach between two atoms whose initial relative translational energy is zero. Note in passing that the position, R_{min}, of the minimum in $V_{\text{MLJ}}(R)$ and σ is related by

$$R_{\text{min}} = \left(\frac{n}{m}\right)^{1/(n-m)} \sigma .$$

Unfortunately, even a simple, more realistic model PEF, such as the MLJ model, with both attractive and repulsive branches, no longer provides the advantage of analytical solutions for the differential cross-section and the omega-integrals that is enjoyed by the simple power-law and hard-sphere models, and they must be evaluated numerically. It is, however, worth noting that extensive tabulations in the form of the reduced omega-integrals defined in eqn (2.188) have been provided (Hirschfelder *et al.*, 1964; Maitland *et al.*, 1980) for a number of MLJ model potential energy functions. It is clear, however, that for comparison with highly accurate measurements extending over an extensive temperature domain, MLJ potential energy functions will also prove insufficient, as no single PEF (with fixed m, n) will be able to provide good agreement with experiment over the entire temperature domain. This means that truly realistic interatomic interactions will necessarily require fairly complicated analytic expressions to represent them accurately enough to provide good agreement with precision experimental measurements over an extended temperature domain.

Finally, note that interatomic potential energy functions of infinite range, such as the power-law interaction (2.183b) or the MLJ model (2.214) will, according to eqn (2.157), give rise to an integral cross-section, σ_{int}, that is infinite, and a differential cross-section, $\sigma(b, \chi)$, that diverges at glancing angles (corresponding to $\chi \to 0$, $b \to \infty$). Despite this behavior, so long as $\nu > 1$, the integral over b in eqn (2.161a) converges to give finite omega integrals, and hence, finite effective cross-sections $\mathfrak{S}\binom{ps}{ps'}$. The Coulomb force (with $\nu = 1$), for which the integral diverges, is an exception. However, the Boltzmann equation in its original form would not be applicable for this case anyway, since in view of the very long range of the Coulomb force, it would make no sense to neglect correlations.

2.6.3 Maxwell molecules

As mentioned in Section 2.6.2, Maxwell molecules constitute an important special model of the interatomic interaction, specifically, that of a repulsive inverse fourth power potential. It is especially important because eigenfunctions and eigenvalues for the linearized collision operator $\mathcal{R}$ have only been found so far for this specific interaction. Maxwell (1867), who first proposed this form of interaction, found that he could develop expressions for the shear viscosity and thermal conductivity in this case without having to determine the nonequilibrium distribution function f.

The simplification for this special power-law potential originates with the observation that the combination $c_{\rm r}\sigma\,\mathrm{d}\mathbf{e}$, given by

$$c_{\rm r}\sigma\mathrm{d}\mathbf{e} \propto W_{\rm r}b\mathrm{d}b = \left(\frac{A_\nu}{k_{\rm B}T}\right)^{\frac{2}{\nu}} W_{\rm r}^{1-4/\nu} s_0 \mathrm{d}s_0\,, \tag{2.210a}$$

is independent of $W_{\rm r}$ for $\nu = 4$, so that evaluation of the omega (or collision) integrals (2.161a) becomes considerably simpler than for other values of ν. By combining this particular result with eqn (2.158), $W_{\rm r}b\,\mathrm{d}b$ is given as

$$W_{\rm r}b\mathrm{d}b = W_{\rm r}\sigma(W_{\rm r},\chi)\sin\chi\,\mathrm{d}\chi = -\gamma(\chi)\cos(\chi/2)\mathrm{d}\chi\,, \tag{2.210b}$$

with $\gamma(\chi)$, given by

$$\gamma(\chi) = 4\sin(\chi/2)\sigma(W_{\rm r},\chi)W_{\rm r} = -\left(\frac{A_4}{k_{\rm B}T}\right)^{\frac{1}{2}} \frac{s_0\mathrm{d}s_0}{\cos(\chi/2)\mathrm{d}\chi}\,, \tag{2.211}$$

a function of the deflection angle χ only. Written explicitly, $\gamma(\chi)$ contains elliptic integrals of the first and second kinds (Maxwell, 1867).

It has been shown by Wang Chang and Uhlenbeck (1970*a*), and later independently by Waldmann (1958*c*), that the velocity independence of $W_{\rm r}\sigma$ leads to an exact treatment of the linearized Boltzmann eqn (1.106) for the homogeneous Maxwell gas. With the Ansatz

$$\Phi(\mathbf{c},t) = \mathrm{e}^{-\omega t}\Phi(\mathbf{c})\,, \tag{2.212}$$

the equation

$$\frac{\partial\mathbf{\Phi}}{\partial t} = -\mathcal{R}\mathbf{\Phi}$$

leads to the eigenvalue equation

$$\mathcal{R}\mathbf{\Phi}(\mathbf{c}) = \omega\mathbf{\Phi}(\mathbf{c})\,. \tag{2.213}$$

The solutions to this eigenvalue equation are the basis functions $\{\mathbf{\Phi}^{(ps)}(\mathbf{W})\}$ that have typically been employed for the expansion of the distribution function. They have associated with them the $(2p+1)$-fold degenerate eigenvalues (Wang Chang and Uhlenbeck, 1970*a*)

$$\begin{aligned} n\bar{c}_{\rm r}\mathfrak{S}(ps) &\equiv \omega_{ps} \\ &= \pi\int_0^\pi \cos(\chi/2)\mathrm{d}\chi\gamma(\chi)\left[(1-\delta_{p0}\delta_{s0}) - \sin^{p+2s}(\chi/2)P_p(\sin(\chi/2))\right. \\ &\qquad \left. - \cos^{p+2s}(\chi/2)P_p(\cos(\chi/2))\right]\,. \end{aligned} \tag{2.214}$$

These eigenvalues must be computed numerically. They are obviously independent of the temperature, so that the corresponding effective cross-sections $\mathfrak{S}(ps)$ all show a $T^{-\frac{1}{2}}$ temperature dependence. The lowest eigenvalues were already determined by

Maxwell (1867). The spectrum of relaxation frequencies of the Maxwell gas has thus been established (Wang Chang and Uhlenbeck, 1970*a*) to be discrete, the relaxation frequencies are positive (except for the fivefold degenerate zero eigenvalue for the hydrodynamic basis functions), and the spectrum has no upper bound. The eigenvalues for $p > 0$ can generally be written as linear combinations of the eigenvalues for $p = 0$, as can be seen from eqn (2.214), by writing down the explicit formulae for the Legendre polynomials. Thus, for example, the eigenvalues for $p = 1$, 2, and 3 may be obtained in terms of the eigenvalues $\omega_{0,s'}$ as

$$\omega_{1s} = \omega_{0,s+1} \,, \tag{2.215a}$$

$$\omega_{2s} = \tfrac{3}{2}\omega_{0,s+2} - \tfrac{1}{2}\omega_{0,s+1} \,, \tag{2.215b}$$

$$\omega_{3s} = \tfrac{5}{2}\omega_{0,s+3} - \tfrac{3}{2}\omega_{0,s+2} \,. \tag{2.215c}$$

Both the viscosity and the thermal conductivity are determined by ω_{20}, as they are related by the Eucken factor $\mathrm{f_E} = \lambda m/\eta c_V = 5/2$. The shear viscosity for a Maxwell gas is given by eqn (2.207) as

$$\eta = \frac{5k_\mathrm{B}T}{8\Omega^{(2,2)}} = \frac{5k_\mathrm{B}T}{2\pi\bar{c}_\mathrm{r}F_2(4)\Gamma(7/2)}\left(\frac{k_\mathrm{B}T}{A_4}\right)^{\frac{1}{2}} \,, \tag{2.216}$$

with $F_2(4) = 0.436$ (Waldmann, 1958*e*).

Because the basis functions $\boldsymbol{\Phi}^{(ps)}(\mathbf{W})$ are eigenfunctions of the linearized collision operator, the first-approximation expressions for the transport coefficients become exact for the Maxwell model. The thermal diffusion coefficient, because it depends for its existence upon the collisional coupling between $\boldsymbol{\Phi}^{10|\mathrm{k}}$ and $\boldsymbol{\Phi}^{11|\mathrm{k}}$, vanishes identically for the Maxwell model. The fact that the basis functions $\boldsymbol{\Phi}^{(ps)}$ are eigenfunctions of $\mathcal{R}$ will be utilized in Section 2.8.

2.7 The Moment Method

The moment method of solving the Boltzmann equation for a pure monatomic gas is conceptually due to Maxwell (1860), but its rigorous mathematical formulation was developed by Grad (1949, 1958). The moment method has seen extensive application in the theory of sound absorption (Goldman and Sirovich, 1967; McCormack and Craven, 1974; Moraal and McCourt, 1972) and Rayleigh–Brillouin light scattering (Yip and Nelkin, 1964; Nelkin and Yip, 1966; McCormack, 1973) in monatomic gases and their mixtures.

In contrast to the Chapman–Enskog treatment, in which an expansion is made about a local Maxwellian, the moment method is based upon the linearization (1.105) about a global Maxwellian for the gas at rest, that is,

$$f(\mathbf{c},\mathbf{r},t) = f_0(\mathbf{c})[1 + \Phi(\mathbf{c},\mathbf{r},t)] \,, \tag{2.217}$$

with the equilibrium distribution function f_0 characterized by parameters n, T, and a vanishing flow velocity. Note that an essential difference between the expansions used in the Chapman–Enskog treatment and in the moment method is that the perturbation function $\Phi(\mathbf{c},\mathbf{r},t)$ in the moment method contains all temporal and spatial dependence.

The number density of the nonequilibrium gas will be denoted as $n + \delta n$, whence

$$n + \delta n(\mathbf{r}, t) = \int \mathrm{d}\mathbf{c}\, f = \int \mathrm{d}\mathbf{c}\ f_0[1 + \Phi] = n[1 + \langle 1|\Phi\rangle]\,, \tag{2.218}$$

so that the relative nonequilibrium deviation of the number density is given by

$$\frac{\delta n}{n} = \langle 1|\Phi\rangle\,. \tag{2.219}$$

Apart from the weight function being the global Maxwellian (which is independent of $\mathbf{r}$ and t), the inner product is as in eqn (1.85). In a similar fashion, the flow velocity $\mathbf{v}$ may be obtained as

$$\begin{aligned} \mathbf{v} = \langle \mathbf{c}\rangle_{\mathrm{ne}} &= \frac{1}{n + \delta n} \int \mathrm{d}\mathbf{c}\, \mathbf{c} f = \frac{1}{n + \delta n} \int \mathrm{d}\mathbf{c}\, \mathbf{c} f_0 [1 + \Phi] \\ &= \left(1 + \frac{\delta n}{n}\right)^{-1} \langle \mathbf{c}|\Phi\rangle\,. \end{aligned} \tag{2.220}$$

In the spirit of linearization, $\delta n/n$, $\mathbf{v}$, and any other quantities describing deviations from equilibrium shall be regarded as small, so that all higher powers or products of such quantities may therefore be discarded systematically. In such a spirit, the flow velocity is given simply by

$$\mathbf{v} = \langle \mathbf{c}|\Phi\rangle\,. \tag{2.221}$$

In order to study the deviations from equilibrium in greater detail, the function $\Phi(\mathbf{c}, \mathbf{r}, t)$ is expanded in terms of the complete orthonormal set of irreducible expansion tensors $\boldsymbol{\Phi}^{ps}(\mathbf{W})$ from Table 2.1. Specifically, $\Phi(\mathbf{c}, \mathbf{r}, t)$ is expanded as

$$\Phi(\mathbf{c}, \mathbf{r}, t) = \sum_{p=0}^{\infty} \sum_{s=0}^{\infty} \mathsf{a}^{ps}(\mathbf{r}, t) \odot \boldsymbol{\Phi}^{ps}(\mathbf{W})\,. \tag{2.222}$$

Because the Maxwellian equilibrium distribution corresponds to a global equilibrium in which the flow velocity vanishes, the dimensionless variable $\mathbf{W}$ becomes $\mathbf{W} = \mathbf{c}/c_0$ [with $c_0 = (2k_\mathrm{B}T/m)^{\frac{1}{2}}$], rather than $\mathbf{C}/c_0$, as obtained for an expansion about a local equilibrium.

In calculating the nonequilibrium average value of the expansion tensor $\boldsymbol{\Phi}^{ps}(\mathbf{c})$ according to definition (1.3), the vanishing of all equilibrium averages $\langle \boldsymbol{\Phi}^{ps}\rangle$ apart from $\langle \Phi^{00}\rangle = \langle 1\rangle = 1$, is utilized. Hence, $\langle \boldsymbol{\Phi}^{ps}\rangle_{\mathrm{ne}}$ takes the form

$$\begin{aligned} \langle \boldsymbol{\Phi}^{ps}\rangle_{\mathrm{ne}} &= \frac{1}{n} \int \mathrm{d}\mathbf{c}\ f \boldsymbol{\Phi}^{ps}(\mathbf{c}, \mathbf{r}, t) = \langle \boldsymbol{\Phi}^{ps}|\Phi\rangle \\ &\equiv \mathsf{a}^{ps}(\mathbf{r}, t)\,, \qquad (p, s) \neq (0, 0)\,, \end{aligned} \tag{2.223}$$

while $\langle \Phi^{00}\rangle_{\mathrm{ne}}$ is given by

$$\langle \Phi^{00}\rangle_{\mathrm{ne}} = 1 + \frac{\delta n}{n} = 1 + a^{00}\,, \tag{2.224}$$

in accord with eqn (2.219). In this way, the tensorial expansion coefficients a^{ps} represent dimensionless moments of the nonequilibrium distribution function.

The above set of moments is particularly useful in that the lowest-order moments represent physical quantities such as density, temperature, velocity, the heat flux vector, and the viscous pressure tensor. The flow velocity,

$$\begin{aligned}\mathbf{v} &= \langle \mathbf{c}|\Phi\rangle = c_0\langle \mathbf{W}|\Phi\rangle = (c_0/\sqrt{2})\langle \mathbf{\Phi}^{10}|\Phi\rangle \\ &= (c_0/\sqrt{2})\langle \mathbf{\Phi}^{10}\rangle_{\rm ne} = (c_0/\sqrt{2})\mathbf{a}^{10}\,, \end{aligned} \tag{2.225}$$

has already been encountered in this context. This point is further illustrated by considering the interpretation of the expansion coefficient a^{01}: for $(p,s) = (0{,}1)$, eqn (2.223) gives $a^{01}(\mathbf{r},t)$ as

$$a^{01}(\mathbf{r},t) = \left(\tfrac{2}{3}\right)^{\frac{1}{2}} \langle W^2 - \tfrac{3}{2}\rangle_{\rm ne} = \left(\frac{2}{3}\right)^{\frac{1}{2}} \left[\frac{1}{k_{\rm B}T}\langle \tfrac{1}{2}mc^2\rangle_{\rm ne} - \frac{3}{2}\right], \tag{2.226}$$

so that should the average translational energy be employed to define a nonequilibrium temperature via $\frac{3}{2}k_{\rm B}(T+\delta T) = \langle \frac{1}{2}mc^2\rangle_{\rm ne}$, then eqn (2.226) readily reduces to

$$a^{01} = \left(\frac{3}{2}\right)^{\frac{1}{2}} \frac{\delta T}{T}, \tag{2.227}$$

so that a^{01} thus represents the relative deviation of the temperature from its equilibrium value.

The moment method transforms the linearized Boltzmann integro-differential eqn (1.106) into an infinite set of coupled partial differential equations for the moments: it is often referred to as the system of moment or transport-relaxation equations (TRE). To obtain the general form for this system of equations, first substitute expansion (2.222) into the linearized Boltzmann equation, then multiply both sides of the resultant equation by an expansion tensor $\mathbf{\Phi}^{ps}$, form the equilibrium average on both sides of the equation and simplify the result using the normalization and orthogonality properties of the expansion tensors, with the end result of this process being the TRE

$$\frac{\partial \mathsf{a}^{ps}}{\partial t} + \sum_{p's'} \langle \mathbf{\Phi}^{ps}|\mathbf{\Phi}^{p's'}\mathbf{c}\rangle \odot \nabla \mathsf{a}^{p's'} + \sum_{p's'} \langle \mathbf{\Phi}^{ps}|\mathcal{R}\mathbf{\Phi}^{p's'}\rangle \odot \mathsf{a}^{p's'} = 0\,. \tag{2.228}$$

As usual, the symbol $\odot$ denotes a full scalar product, with the convention that the innermost Cartesian indices are contracted first.

The moment eqns (2.228) may be simplified by making use of the rotational invariance of the linearized collision operator $\mathcal{R}$. An important consequence of rotational invariance is that if $\mathcal{R}$ is applied to a symmetric traceless p$^{\rm th}$-rank tensor, a tensor of the same kind, and with the same orientation, is produced. As in Section 2.3, the Wigner–Eckart theorem may be utilized to introduce the reduced matrix elements

$$\langle \mathbf{\Phi}^{ps}|\mathcal{R}\mathbf{\Phi}^{p's'}\rangle = \delta_{pp'}\langle ps\|\mathcal{R}\|ps'\rangle \mathbf{\Delta}^{(p)} = \delta_{pp'} n\bar{c}_{\rm r}\mathfrak{S}\left(\begin{smallmatrix} ps \\ ps' \end{smallmatrix}\right)\mathbf{\Delta}^{(p)} \tag{2.229}$$

[see also eqn (1.95a)]. A shorthand notation, in which the labels ps are substituted for $\mathbf{\Phi}^{ps}$, is used here to represent the reduced matrix elements of $\mathcal{R}$.

The streaming term in eqn (2.228) may be simplified using similar arguments: according to the coupling rules of irreducible tensors, a symmetric traceless pth-rank tensor can only be coupled with a vector to symmetric traceless tensors of ranks $p' = p-1, p, p+1$. Because $\mathbf{c}$ has odd parity, $p' = p$ must be ruled out. The flow term thus takes on the form (see Appendix A.3) (Hess, 2015)

$$\langle \mathbf{\Phi}^{ps} | \mathbf{\Phi}^{p's'} \mathbf{c} \rangle = \delta_{p',p+1} \langle ps \| \mathbf{c} \| p+1, s' \rangle \cdot \mathbf{\Delta}^{(p+1)} + \delta_{p',p-1} \langle ps \| \mathbf{c} \| p-1, s' \rangle \mathbf{\Delta}^{(p)} \,. \quad (2.230)$$

Again, the reduced matrix elements are obtained by contraction, so that

$$(2p+3)\langle ps \| \mathbf{c} \| p+1, s' \rangle = \langle \mathbf{\Phi}^{ps} \mathbf{c} \odot | \mathbf{\Phi}^{p+1,s'} \rangle \quad (2.231)$$

and

$$(2p+1)\langle ps \| \mathbf{c} \| p-1, s' \rangle = \langle \mathbf{\Phi}^{(ps)} \odot | \mathbf{c} \mathbf{\Phi}^{p-1,s'} \rangle \,. \quad (2.232)$$

When the above equations are utilized, the system (2.228) of transport-relaxation equations reduces to

$$\frac{\partial \mathsf{a}^{ps}}{\partial t} + \sum_{s'=0}^{\infty} \{ \langle ps \| \mathbf{c} \| p+1, s' \rangle \nabla \cdot \mathsf{a}^{p+1,s'} + \langle ps \| \mathbf{c} \| p-1, s' \rangle \overline{\nabla \mathsf{a}^{p-1,s'}} \} + \sum_{s'=0}^{\infty} n \bar{c}_{\mathrm{r}} \mathfrak{S}\left({}^{ps}_{ps'} \right) \mathsf{a}^{ps'} = 0 \,. \quad (2.233)$$

Overhooks, of the form $\overline{\nabla \mathsf{a}^{p-1,s'}}$, indicate that the tensor has been made symmetric traceless (see Appendix A.3) (Hess, 2015).

The system of coupled partial differential equations for the moments a^{ps} is the starting point for the treatment of transport and relaxation phenomena in a monatomic gas. For practical purposes, the infinite system (2.233) must obviously be truncated in some way. Many applications are adequately described by the thirteen-moment approximation, based upon the moments

$$a^{00} = \frac{\delta n}{n} \,, \quad (2.234\text{a})$$

$$a^{01} = \left(\tfrac{2}{3}\right)^{\frac{1}{2}} \langle W^2 - \tfrac{3}{2} \rangle_{\text{ne}} = \left(\tfrac{3}{2}\right)^{\frac{1}{2}} \frac{\delta T}{T} \,, \quad (2.234\text{b})$$

$$\mathbf{a}^{10} = \sqrt{2} \langle \mathbf{W} \rangle_{\text{ne}} = \sqrt{2} \frac{\mathbf{v}}{c_0} \,, \quad (2.234\text{c})$$

$$\mathbf{a}^{11} = \left(\tfrac{4}{5}\right)^{\frac{1}{2}} \langle (W^2 - \tfrac{5}{2}) \mathbf{W} \rangle_{\text{ne}} = \left(\tfrac{4}{5}\right)^{\frac{1}{2}} (p c_0)^{-1} \mathbf{q} \,, \quad (2.234\text{d})$$

$$\mathsf{a}^{20} = \sqrt{2} \langle \overline{\mathbf{W}\mathbf{W}} \rangle_{\text{ne}} = (\sqrt{2}\, p)^{-1} \mathbf{\Pi} \,, \quad (2.234\text{e})$$

with $p = nk_{\mathrm{B}}T$ being the equilibrium ideal gas pressure.

The set of transport-relaxation equations associated with the thirteen-moment approximation is thus given by

$$\frac{\partial a^{00}}{\partial t} + \tfrac{1}{\sqrt{2}} c_0 \nabla \cdot \mathbf{a}^{10} = 0 \,, \quad (2.235\text{a})$$

$$\frac{\partial a^{01}}{\partial t} + \tfrac{1}{\sqrt{3}} c_0 \nabla \cdot [\mathbf{a}^{10} + (\tfrac{5}{2})^{\frac{1}{2}} \mathbf{a}^{11}] = 0\,, \tag{2.235b}$$

$$\frac{\partial \mathbf{a}^{10}}{\partial t} + c_0 \left[\nabla \left(\tfrac{1}{\sqrt{2}} a^{00} + \tfrac{1}{\sqrt{3}} a^{01}\right) + \nabla \cdot \mathsf{a}^{20}\right] = 0\,, \tag{2.235c}$$

$$\frac{\partial \mathbf{a}^{11}}{\partial t} + \left(\tfrac{5}{6}\right)^{\frac{1}{2}} c_0 \nabla a^{01} + \left(\tfrac{2}{5}\right)^{\frac{1}{2}} c_0 \nabla \cdot \mathsf{a}^{20} + n\bar{c}_{\mathrm{r}} \mathfrak{S}(11) \mathbf{a}^{11} = 0\,, \tag{2.236a}$$

$$\frac{\partial \mathsf{a}^{20}}{\partial t} + c_0 \overline{\nabla \mathbf{a}^{10}} + \left(\tfrac{2}{5}\right)^{\frac{1}{2}} c_0 \overline{\nabla \mathbf{a}^{11}} + n\bar{c}_{\mathrm{r}} \mathfrak{S}(20) \mathsf{a}^{20} = 0\,. \tag{2.236b}$$

Equations (2.235a, 2.235b, 2.235c) represent the (linearized) conservation equations for number density, energy, and linear momentum, respectively, while eqns (2.236a, 2.236b) can be employed to infer the thermal conductivity and shear viscosity coefficients. To this end, consider the stationary versions of the latter two equations, and invoke the hydrodynamical approximation, in which the gradients and divergences of the nonconserved quantities $\mathbf{a}^{11}$ and a^{20} are neglected. The simplified equations

$$\tfrac{5}{4} c_0^2 p T^{-1} \nabla T + n\bar{c}_{\mathrm{r}} \mathfrak{S}(11) \mathbf{q} = 0 \tag{2.237}$$

and

$$2p\overline{\nabla \mathbf{v}} + n\bar{c}_{\mathrm{r}} \mathfrak{S}(20) \mathbf{\Pi} = 0 \tag{2.238}$$

thereby obtained agree with the familiar phenomenological equations. Comparison of these two equations with their phenomenological forms

$$\mathbf{q} = -\lambda \nabla T\,, \tag{2.239}$$

$$\mathbf{\Pi} = -2\eta \overline{\nabla \mathbf{v}}\,, \tag{2.240}$$

results in the transport coefficients η and λ being given by eqns (2.70, 2.75), respectively, identical to the results obtained rather more readily from the simplest inverse-matrix approximation.

Employment of the exact relation

$$\mathfrak{S}(11) = \tfrac{2}{3} \mathfrak{S}(20)$$

between the effective cross-sections $\mathfrak{S}(20)$ and $\mathfrak{S}(11)$ [see eqn (2.133)] in eqn (2.70) for η, followed by elimination of the effective cross-section $\mathfrak{S}(11)$ between η and eqn (2.75) for λ, shows that these two transport coefficients are related (at this level of description) by

$$\frac{\lambda}{\eta} = \frac{15 k_{\mathrm{B}}}{4m} = \frac{5 c_V}{2m}\,, \tag{2.241}$$

a result referred to as the Eucken relation. This relation is approximately satisfied for noble gases. The Eucken relation can be rearranged into a characteristic dimensionless ratio

$$\mathrm{f}_{\mathrm{Eu}} = \frac{\lambda m}{\eta c_V}\,, \tag{2.242}$$

which is now called the Eucken number: in the present approximation, it has the value $\frac{5}{2}$. A related dimensionless ratio that is popular in hydrodynamics, called the Prandtl number, is given by $\mathrm{Pr} = \eta c_p/(\lambda m) = (c_p/c_V)\mathrm{f}_{\mathrm{E}}^{-1}$.

2.7.1 Extension to mixtures

The starting point for the derivation of the system of moment equations for a noble gas mixture is the Ansatz

$$f_{\mathrm{k}} = f_{0\mathrm{k}}(1 + \Phi_{\mathrm{k}})\,, \tag{2.243}$$

which defines the deviation Φ_{k} of f_{k} from global equilibrium. A system of linearized Boltzmann equations

$$\frac{\partial \Phi_{\mathrm{k}}}{\partial t} + \mathbf{c}\cdot\nabla\Phi_{\mathrm{k}} + \sum_{\mathrm{k}'} \mathcal{R}_{\mathrm{kk}'}\Phi_{\mathrm{k}'} = 0 \tag{2.244}$$

is obtained in this way.

An expansion of each Φ_{k} into a complete set of orthonormalized irreducible tensors is carried out as described in Section 2.6.3 on pure gases, that is, Φ_{k} is expanded as

$$\Phi_{\mathrm{k}}(\mathbf{c},\mathbf{r},t) = \sum_{ps} \boldsymbol{\Phi}^{ps|\mathrm{k}}(\mathbf{c}) \odot \mathsf{a}^{ps|\mathrm{k}}(\mathbf{r},t)\,, \tag{2.245}$$

with the normalization property (2.57) for the expansion tensors. For the pair of indices $(p,s) \neq (0,0)$, the tensors $\mathsf{a}^{ps|\mathrm{k}}$ are simply the moments (i.e. nonequilibrium averages) of $\boldsymbol{\Phi}^{ps|\mathrm{k}}$, that is,

$$\mathsf{a}^{ps|\mathrm{k}} = \langle \boldsymbol{\Phi}^{ps|\mathrm{k}} \rangle_{\mathrm{ne}}\,. \tag{2.246}$$

These moments then satisfy the system of transport-relaxation equations given by

$$\frac{\partial \mathsf{a}^{ps|\mathrm{k}}}{\partial t} + \sum_{s'} \langle (ps)_{\mathrm{k}} \|\mathbf{c}\| (p-1,s')_{\mathrm{k}} \rangle \overline{\nabla \mathsf{a}^{p-1,s'|\mathrm{k}}}$$
$$+ \langle (ps)_{\mathrm{k}} \|\mathbf{c}\| ((p+1,s')_{\mathrm{k}} \rangle \nabla \cdot \mathsf{a}^{p+1,s'|\mathrm{k}} + \sum_{\mathrm{k}'=1}^{\mathrm{N}} n x_{\mathrm{k}}^{-1} S\big(\begin{smallmatrix} ps \\ ps' \end{smallmatrix}\big|\begin{smallmatrix} \mathrm{k} \\ \mathrm{k}' \end{smallmatrix}\big) \mathsf{a}^{ps'|\mathrm{k}'} = 0\,. \tag{2.247}$$

The reduced matrix elements of the streaming operator are defined as in the pure gas case, but with a factor x_{k} on the right-hand side in the equation corresponding to eqn (2.230). The definition (2.59) has been adopted for the linearized collision operator.

The infinite set of moment equations generated in this way must be truncated. In lowest order, the 13–moment equations for a mixture are ($\mathrm{k} = 1, 2, \cdots, \mathrm{N}$):

$$a^{00|\mathrm{k}} = \frac{\delta n_{\mathrm{k}}}{n_{\mathrm{k}}}\,, \tag{2.248a}$$

$$a^{01|\mathrm{k}} = \left(\frac{3}{2}\right)^{\frac{1}{2}} \frac{T_{\mathrm{k}} - T}{T}\,, \tag{2.248b}$$

$$\mathbf{a}^{10|\mathrm{k}} = \sqrt{2}\,\mathbf{v}_{\mathrm{k}}/c_0^{\mathrm{k}}\,, \tag{2.248c}$$

$$\mathbf{a}^{11|\mathrm{k}} = (n_\mathrm{k} k_\mathrm{B} T)^{-1} \left(\frac{2m_\mathrm{k}}{5kT}\right)^{\frac{1}{2}} \mathbf{q}_\mathrm{k} \,, \tag{2.249a}$$

$$\mathsf{a}^{20|\mathrm{k}} = (n_\mathrm{k} k_\mathrm{B} T)^{-1} \sqrt{2}\, \mathbf{\Pi}_\mathrm{k} \,, \tag{2.249b}$$

In these equations, $T \equiv \sum_\mathrm{k} x_\mathrm{k} T_\mathrm{k}$ is the equilibrium temperature of the multicomponent mixture and $c_0^\mathrm{k} = (2k_\mathrm{B}T/m_\mathrm{k})^{\frac{1}{2}}$ is the most probable speed for atoms of component k. The moments represent, respectively, density deviation, temperature deviation, flow velocity, heat flux, and viscous pressure tensor associated with each species k. This set of moments has a corresponding set of transport-relaxation equations for the mixture components. At the 13-moment level described above for the pure gas, the transport-relaxation equations are, upon utilizing eqn (2.59), given as

$$\frac{\partial a^{00|\mathrm{k}}}{\partial t} + \tfrac{1}{\sqrt{2}} c_0^\mathrm{k} \nabla \cdot \mathbf{a}^{10|\mathrm{k}} = 0 \,, \tag{2.250a}$$

$$\frac{\partial a^{01|\mathrm{k}}}{\partial t} + \tfrac{1}{\sqrt{3}} c_0^\mathrm{k} \nabla \cdot \left[\mathbf{a}^{10|\mathrm{k}} + \left(\tfrac{5}{2}\right)^{\frac{1}{2}} \mathbf{a}^{11|\mathrm{k}}\right] + n \sum_{\mathrm{k}'} x_\mathrm{k}^{-1} S\!\left(\begin{smallmatrix}01\\01\end{smallmatrix}\middle|\begin{smallmatrix}\mathrm{k}\\\mathrm{k}'\end{smallmatrix}\right) a^{01|\mathrm{k}'} = 0 \,, \tag{2.250b}$$

$$\begin{aligned} \frac{\partial \mathbf{a}^{10|\mathrm{k}}}{\partial t} &+ \tfrac{1}{\sqrt{2}} c_0^\mathrm{k} \left[\nabla a^{00|\mathrm{k}} + \left(\tfrac{2}{3}\right)^{\frac{1}{2}} \nabla a^{01|\mathrm{k}} + \sqrt{2}\, \nabla \cdot \mathsf{a}^{20|\mathrm{k}}\right] \\ &+ n \sum_{\mathrm{k}'} \sum_r x_\mathrm{k}^{-1} S\!\left(\begin{smallmatrix}10\\1r\end{smallmatrix}\middle|\begin{smallmatrix}\mathrm{k}\\\mathrm{k}'\end{smallmatrix}\right) \mathbf{a}^{1r|\mathrm{k}'} = 0 \,, \end{aligned} \tag{2.250c}$$

$$\frac{\partial \mathbf{a}^{11|\mathrm{k}}}{\partial t} + \left(\tfrac{5}{6}\right)^{\frac{1}{2}} c_0^\mathrm{k} \nabla a^{01|\mathrm{k}} + \left(\tfrac{2}{5}\right)^{\frac{1}{2}} c_0^\mathrm{k} \nabla \cdot \mathsf{a}^{20|\mathrm{k}} + n \sum_{\mathrm{k}'} \sum_r x_\mathrm{k}^{-1} S\!\left(\begin{smallmatrix}11\\1r\end{smallmatrix}\middle|\begin{smallmatrix}\mathrm{k}\\\mathrm{k}'\end{smallmatrix}\right) \mathbf{a}^{1r|\mathrm{k}'} = 0 \,, \tag{2.251a}$$

$$\frac{\partial \mathsf{a}^{20|\mathrm{k}}}{\partial t} + c_0^\mathrm{k} \left[\overline{\nabla \mathbf{a}^{10|\mathrm{k}}} + \left(\tfrac{2}{5}\right)^{\frac{1}{2}} \overline{\nabla \mathbf{a}^{11|\mathrm{k}}}\right] + n \sum_{\mathrm{k}'} \sum_r x_\mathrm{k}^{-1} S\!\left(\begin{smallmatrix}20\\2r\end{smallmatrix}\middle|\begin{smallmatrix}\mathrm{k}\\\mathrm{k}'\end{smallmatrix}\right) \mathsf{a}^{2r|\mathrm{k}'} = 0 \,. \tag{2.251b}$$

The system of transport-relaxation eqns (2.250, 2.251) differs from the corresponding moment eqn (2.234) for the pure gas in two respects: firstly, different species are coupled via the collision term and secondly, a concentration gradient $\nabla a^{00|\mathrm{k}}$ occurs in eqn (2.250c) as a new driving force. The treatment of thermal conductivity, viscosity, diffusion, and thermal diffusion with the moment equations leads for $N = 2$ to precisely the same expressions for λ, η, D, and D_T that have been obtained using the inverse operator technique, provided gradients of nonconserved quantities are omitted (the hydrodynamical approximation). For this reason, the moment method for mixtures are not pursued further.

2.8 Kinetic Models

The dynamic models of the Boltzmann equation discussed in Section 2.7 were based on an ability to evaluate binary collision dynamics exactly for a number of specific simple interaction models. A more mathematical approach is to simplify the operator appearing in the equation (with little regard for the physics of the scattering process), rather than the inter particle interaction. Such modifications are normally carried out for the linearized collision operator. Kinetic models obtained in this manner are often simple enough to permit efficient and accurate procedures of solution, so that there is no need to employ truncated Chapman–Enskog or moment expansions. Solutions obtained from suitably chosen models may result in rather accurate descriptions of observed phenomena.

2.8.1 Modeling for Maxwell molecules

It was pointed out in Section 2.6.3 that the basis tensors $\mathbf{\Phi}^{(ps)}$ are eigenfunctions of the collision operator for Maxwell molecules, that is,

$$\mathcal{R}\mathbf{\Phi}^{(ps)}(\mathbf{c}) = \omega_{ps}\mathbf{\Phi}^{(ps)}(\mathbf{c}) \,. \tag{2.252}$$

Moreover, the spectrum, $\omega_{ps} = n\bar{c}_{\mathrm{r}}\mathfrak{S}(ps)$, of eigenvalues of $\mathcal{R}$ consists entirely of discrete eigenvalues of finite multiplicity and, apart from the fivefold degenerate zero eigenvalue corresponding to the fundamental collisional invariants, is positive and has no upper bound (Wang Chang and Uhlenbeck, 1970*b*). As the eigenfunctions $\mathbf{\Phi}^{(ps)}$ form a complete set of functions spanning the Hilbert space, an arbitrary function $\phi(\mathbf{c})$ may thus be expanded in terms of them, so that

$$\phi(\mathbf{c}) = \sum_{\alpha} \mathbf{\Phi}^{\alpha}(\mathbf{c}) \odot \langle \mathbf{\Phi}^{\alpha} | \phi \rangle \,, \tag{2.253}$$

and

$$\mathcal{R}\phi(\mathbf{c}) = \sum_{\alpha} \omega_{\alpha} \mathbf{\Phi}^{\alpha}(\mathbf{c}) \odot \langle \mathbf{\Phi}^{\alpha} | \phi \rangle \,, \tag{2.254}$$

upon using a collective label α for ps. The coefficients $\langle \mathbf{\Phi}^{\alpha} | \phi \rangle$ are identical to the moments determined in Section 2.5. Should ϕ depend upon $\mathbf{r}$ and t, that dependence will be transferred to the moments. It is also understood that the linearized collision operator $\mathcal{R}$ has been expanded in terms of projectors. Although this expression is exact for Maxwell molecules, it remains of little use as it stands.

A systematic approximation scheme for the mathematical modeling of the linearized Boltzmann equation based upon expansion (2.254) was introduced by Gross and Jackson (1959). A simplification of the spectrum of $\mathcal{R}$ is made by collapsing all eigenvalues beyond a certain point into a single eigenvalue $\nu > 0$, so that eqn (2.254) can be rewritten in the form

$$\mathcal{R}\phi(\mathbf{c}) = \sum_{\alpha=0}^{N} \omega_{\alpha} \mathbf{\Phi}^{\alpha}(\mathbf{c}) \odot \langle \mathbf{\Phi}^{\alpha} | \phi \rangle + \nu \sum_{\alpha=N+1}^{\infty} \mathbf{\Phi}^{\alpha}(\mathbf{c}) \odot \langle \mathbf{\Phi}^{\alpha} | \phi \rangle \,. \tag{2.255}$$

The collective label α may then be employed to indicate an ordering index that labels the tensors in a sequential order. Completeness of the $\mathbf{\Phi}^\alpha(\mathbf{c})$ functions is then employed to replace the second (infinite) sum by

$$\sum_{\alpha=N+1}^{\infty} \mathbf{\Phi}^\alpha(\mathbf{c}) \odot \langle \mathbf{\Phi}^\alpha | \phi \rangle = \sum_{\alpha=0}^{\infty} \mathbf{\Phi}^\alpha(\mathbf{c}) \odot \langle \mathbf{\Phi}^\alpha | \phi \rangle - \sum_{\alpha=0}^{N} \mathbf{\Phi}^\alpha(\mathbf{c}) \odot \langle \mathbf{\Phi}^\alpha | \phi \rangle$$
$$= \phi(\mathbf{c}) - \sum_{\alpha=0}^{N} \mathbf{\Phi}^\alpha(\mathbf{c}) \odot \langle \mathbf{\Phi}^\alpha | \phi \rangle \,. \tag{2.256}$$

An interpretation of ν as a collision rate enables $\mathcal{R}$ to be split as in eqn (1.102) into

$$\mathcal{R}\phi(\mathbf{c}) = \nu\phi(\mathbf{c}) - \sum_{\alpha=0}^{N} (\nu - \omega_\alpha) \mathbf{\Phi}^\alpha(\mathbf{c}) \odot \langle \mathbf{\Phi}^\alpha | \phi \rangle \,. \tag{2.257}$$

The Gross–Jackson procedure thus defines a class of kinetic model operators for which N is the order of the model. The larger the chosen value of N, the more details of the spectrum of $\mathcal{R}$ are included in the description.

The best-known example of such modeling, the linearized Bhatnagar–Gross–Krook (1954) (or BGK) model, treats only the hydrodynamic eigenfunctions faithfully, and replaces the entire positive spectrum of $\mathcal{R}$ by a single value, ν. Thus, the BGK model replaces $\mathcal{R}\phi$ by

$$\mathcal{R}\phi = \nu \left\{ \phi - \left[\langle \phi \rangle + 2\mathbf{W} \cdot \langle \mathbf{W} | \phi \rangle + \tfrac{2}{3}(W^2 - \tfrac{3}{2}) \langle (W^2 - \tfrac{3}{2}) | \phi \rangle \right] \right\} . \tag{2.258}$$

Note that the model was originally designed for the nonlinear Boltzmann equation. The effect of collisions was pictured as a relaxation toward the local Maxwellian with a single relaxation rate ν (Williams, 1972).

Because the hydrodynamic eigenfunctions have been correctly taken into account, the BGK model preserves the conservation laws. One of the transport coefficients, say the viscosity, is reproduced if the collision rate is adjusted to $\nu = nkT/\eta$. It cannot be hoped, however, to obtain simultaneously also the correct value for the thermal conductivity. Indeed, instead of $\frac{5}{2}$ for the Eucken number (2.241), $f_\mathrm{E} = \lambda m/(\eta c_V)$, which is exact for Maxwell molecules, the value $\frac{5}{3}$ [equivalently, 1 instead of $\frac{2}{3}$ for the Prandtl number $\mathrm{Pr} = \eta c_p/(\lambda m)$] is obtained for f_E. The remedy is to switch to a higher-order model, thereby refining the expansion of the last term in eqn (2.257).

For the sake of the conservation laws, of course, any extension must leave the terms in (2.258) intact. Which further terms should be incorporated may then become a somewhat delicate question. Two procedures were suggested by Sirovich (1963), one based upon a polynomial ordering of the eigenfunctions and the other based upon an ordering in terms of increasing eigenvalues. These two procedures are known as P-ordering and E-ordering and have been shown to be practically equivalent for 15 or more terms (Sugawara *et al.*, 1968; Ranganathan and Yip, 1966). When only a few terms are kept, the two procedures are not equivalent, and care must be exercised in their use. Detailed discussions may be found in the original literature (Sugawara *et al.*, 1968; Ranganathan and Yip, 1966; Tenti *et al.*, 1974).

2.8.2 Generalized kinetic models

Sirovich (1962) established that the modelling procedure developed for the Maxwell gas by Gross and Jackson (1959) can be generalized to deal with arbitrary spherically symmetric interactions. Although the functions $\mathbf{\Phi}^\alpha$ are now no longer eigenfunctions of the collision operator, they can still be used to expand ϕ, as they are complete. The action of $\mathcal{R}$ upon expansion (2.253) gives

$$\mathcal{R}\phi(\mathbf{c}) = \sum_{ps} \mathcal{R}\mathbf{\Phi}^{ps}(\mathbf{c}) \odot \langle \mathbf{\Phi}^{ps}|\phi\rangle\,, \tag{2.259}$$

which may again be expanded in terms of the set $\{\mathbf{\Phi}^{ps}\}$ to obtain

$$\begin{aligned}\mathcal{R}\phi(\mathbf{c}) &= \sum_{p's'ps} \mathbf{\Phi}^{p's'} \odot \langle \mathbf{\Phi}^{p's'}|\mathcal{R}\mathbf{\Phi}^{ps}\rangle \odot \langle \mathbf{\Phi}^{ps}|\phi\rangle \\ &= \sum_{pss'} n\bar{c}_{\mathrm{r}}\mathfrak{S}\big({}^{ps}_{ps'}\big)\,\mathbf{\Phi}^{p's'} \odot \langle \mathbf{\Phi}^{ps}|\phi\rangle\,. \end{aligned} \tag{2.260}$$

The proportionality of the matrix element to $\delta_{pp'}\mathbf{\Delta}^{(p)}$ has been utilized in the last step. Though the choice of the basis set $\{\mathbf{\Phi}^{ps}\}$ is convenient, it is by no means essential, since any complete set of functions would do.

Models for $\mathcal{R}$ may be obtained by truncating a suitably ordered expansion (2.260) and replacing the remainder, as was done for Maxwell molecules. Thus, upon setting

$$\langle \mathbf{\Phi}^{\beta}\mathcal{R}\mathbf{\Phi}^{\alpha}\rangle = \nu\delta_{\alpha\beta}\mathbf{\Delta}^{(p)}\,, \qquad \text{for } \alpha \text{ or } \beta > N\,, \tag{2.261}$$

with α and β ordering labels replacing ps, and following rearrangement of the sum, the model becomes

$$\mathcal{R}\phi = \nu\phi - \sum_{\alpha,\beta=0}^{N} \Big[\nu\delta_{\alpha\beta} - n\bar{c}_{\mathrm{r}}\mathfrak{S}\big({}^{\beta}_{\alpha}\big)\Big]\,\mathbf{\Phi}^{\beta} \odot \langle \mathbf{\Phi}^{\alpha}|\phi\rangle\,. \tag{2.262}$$

The remaining problem is then the calculation of the matrix elements retained in the model. Even though for a chosen interparticle potential this is a straightforward procedure with modern computational facilities, it is also possible to sidestep this aspect by selecting the order of the model sufficiently low so that all parameters may be determined in terms of measured transport coefficients. The advantage of doing this is that no adjustable parameters appear in the final results.

The major difference between the kinetic modeling procedures and the moment method discussed in Section 2.5 lies in the assumption (2.261). The moment method truncates the infinite expansion (2.260) at $\alpha, \beta = N$ as if

$$\langle \mathbf{\Phi}^{\alpha}|\mathcal{R}\phi\rangle = 0\,, \qquad \text{for } \alpha > N\,,$$

which is tantamount to introducing conserved quantities other than mass, energy and momentum into the description. Assumption (2.261) guarantees that kinetic models retain some effect of the infinite set of expansion functions with only the weightings of some of them changed.

A basic flaw of the kinetic models is that they imply a speed-independent collision rate ν. However, a variable $\nu(c)$ may readily be taken into account if the basis set $\{\boldsymbol{\Phi}^{ps}\}$ is replaced by a basis set that is orthonormal with respect to the modified weight $\nu(c)f^{(0)}(\mathbf{c})/n$. Inevitably, such a generalization increases the computational effort required for the solution, though not unduly if a low-order model is chosen. It suffices here to give the appropriate generalization of the linearized BGK model (Williams, 1972),

$$\mathcal{R}\phi(\mathbf{c}) = \nu(c)\left\{\phi(\mathbf{c}) - \left[\frac{\langle\nu\phi\rangle}{\nu_0} + \frac{3\mathbf{W}\cdot\langle\nu\mathbf{W}|\phi\rangle}{\nu_2}\right.\right.$$
$$\left.\left. + \frac{\nu_0}{\nu_4\nu_0 - \nu_2^2}\left(W^2 - \frac{\nu_2}{\nu_0}\right)\langle\nu|\left(W^2 - \frac{\nu_2}{\nu_0}\right)|\phi\rangle\right]\right\}, \tag{2.263}$$

in which the moments $\nu_n = \langle W^n|\nu\rangle$ of the collision rate are employed. The coefficients have been modified in such a way that the conservation laws are still obeyed, that is,

$$\mathcal{R}1 = 0\,, \qquad \mathcal{R}\mathbf{W} = 0\,, \qquad \mathcal{R}W^2 = 0\,.$$

A generalization to mixtures is relatively straightforward, and is not described here in detail. Both ϕ and the basis tensors are replaced by arrays, and the collision operator is replaced by a matrix of operators, as explained in Section 1.6. There is a considerable literature on this topic, and interested readers are referred there for further details (Sirovich and Thurber, 1961; Boley and Yip, 1972; McCormack, 1973; Gross and Krook, 1956; Oppenheim, 1965; Holway Jr., 1966; Hamel, 1965).

3
Realistic Interatomic Potential Energy Functions

3.1 The Need for Realistic Potential Energy Functions

Chapter 2 focused on the relationship between effective cross-sections and the microscopic scattering cross-sections in both classical mechanical and quantum mechanical terms. It discussed the relatively simple final expressions obtained for the viscosity η and thermal conductivity λ of a pure monatomic gas when an inverse-power model or a hard-sphere model potential energy function (PEF) is employed to represent the interaction energy between a pair of atoms as well as the first realistic model PEF of (Mie, 1903; Lennard-Jones 1924*a*, 1924*b*), which includes both short-range repulsive and long-range attractive components of the interaction energy. This chapter examines the nature of atom–atom interaction energies.

This section discusses the need for realistic representations of atom–atom interaction energies and Section 3.2 describes the Mie/Lennard-Jones (MLJ) empirical model (Mie, 1903; Lennard–Jones, 1924*a*, 1924*b*) for the interaction energies. Section 3.3 introduces and develops the concept of semi-empirical modeling of the interaction energy. More specifically, Section 3.3.1 discusses in detail both the basic Hartree-Fock plus damped dispersion (or HFD) model of Scoles and co-workers (Ahlrichs *et al.*, 1977; Douketis and Scoles, 1982; Hepburn *et al.*, 1975) for the interaction energy between a pair of like closed-shell atoms, as well as a similar model introduced by Tang and Toennies (1977, 1984, 1986) (referred to as the TT model), while Section 3.3.2 addresses extensions of the HFD and TT models to provide descriptions of the interaction energies between pairs of unlike closed-shell atoms via combining rules. Section 3.3.3 concerns an application of the TT model to describe the interaction energies between noble gas atom pairs (six homonuclear, fifteen heteronuclear), while Section 3.3.4 deals with an extension of the TT model to chemically bonded atom pairs. Section 3.4 introduces and discusses the exchange-Coulomb (or XC) semi-empirical models of Meath and co-workers (Knowles and Meath, 1986; Koide *et al.*, 1981; Kreek and Meath, 1969; Wheatley and Meath, 1993*a*; Wheatley and Meath, 1993*b*).

Section 3.5 examines the development by Le Roy and co-workers (2007, 2011, 2009, 2011) of a model PEF that is capable of representing sets of spectroscopic transitions in chemically bound diatomic species with typical spectroscopic accuracy. Chemically bound heteronuclear molecules typically have very deep potential energy wells that are capable of supporting hundreds of energy levels between which

Transport Properties and Potential Energy Models for Monatomic Gases. Hui Li and Frederick R.W. McCourt, Oxford University Press.
 DOI: 10.1093/oso/9780198888253.003.0003

spectroscopic transitions are possible. A typical data set for such molecules enables the accurate determination of a sufficient number of potential parameters to provide a spectrscopically-accurate fit to the input data. This process may also be considered in the context of an inversion of the experimental spectroscopic data (transition frequencies) to obtain the PEF for a particular molecule.

Section 3.6 is concerned with the various means available for direct inversion of experimental data (other than spectroscopic) to obtain the relevant interaction energy dependence upon the atom–atom separation R. In general, to determine the interaction (potential) energy between a pair of atoms as a function of the pair separation requires the inversion of a number of layers of averaging. Atomic (or molecular) beam scattering cross-section data involves the least averaging, while bulk property (BP) measurements (transport properties, for example) require the most averaging. Section 3.6.1 discusses the inversion of microscopic data, such as atomic beam integral or differential scattering measurements, while Section 3.6.2. focuses upon the direct inversion of experimental data based upon an iterative multiparameter fitting procedure that minimizes the statistical deviation between sets of calculated and measured transport properties.

Section 3.7 focuses on the *ab initio* calculation of PEFs for the interaction between the five homonuclear pairs formed by the ground-state noble gas atoms He, Ne, Ar, Kr, Xe, and their fitting to the Morse/long-range (MLR) analytical form. Following summaries of the supermolecule and perturbation approaches in Sections 3.7.1 and 3.7.2, the analytical representation of *ab initio* atom–atom interaction energies is described in some detail in Section 3.7.3. Section 3.8 describes the results obtained for each of the homonuclear interactions in Section 3.8.1 [He–He ($X\Sigma_g^+$) interaction] through 3.8.5 [Xe_2 ($X\Sigma_g^+$) and $Rn_2(X^1\Sigma_g^+)$ interactions], while Section 3.8.6 is devoted to results obtained for ten of the fifteen noble gas heteronuclear interaction energies. Finally, Section 3.9 extends the employment of the MLR form to obtain analytical representations (Dattani and Le Roy, 2011; Le Roy and Henderson, 2007; Le Roy *et al.*, 2011; Le Roy *et al.*, 2009) of the interaction energies between pairs of ground-state alkali atoms.

3.2 The Mie/Lennard-Jones Potential Energy Functions

In 1860, Maxwell (1860) introduced the hard-sphere model to describe the interaction between two atoms. The expression for the viscosity η obtained from the kinetic theory of gases could then be evaluated for the hard-sphere model, see eqn (2.224), and was found to have a $T^{\frac{1}{2}}$ temperature dependence. Once it became possible to access temperatures well below room temperature, it became clear that the viscosity coefficient clearly did not behave even approximately as $T^{\frac{1}{2}}$. This led Sutherland (1893, 1909) to propose in 1893 that a weak attractive force field, proportional to $R^{-(m+1)}$ surrounds the hard sphere, leading to a modification of the hard-sphere model from eqn (2.183a) to

$$V_S(R) = \begin{cases} -C_m R^{-m}\,, & R > 2R_0, \\ \infty\,, & R \leq 2R_0\,, \end{cases} \tag{3.1a}$$

with m an integer. The first Chapman–Cowling approximation to η for the Sutherland model function may be expressed in terms of the hard-sphere result (2.224) for $[\eta(\text{hs})]_1$ as (Chapman and Cowling, 1970)

$$[\eta(\text{S})]_1 = \frac{k_\text{B}T}{k_\text{B}T + S(m)}\,[\eta(\text{hs})]_1\,, \tag{3.1b}$$

in which $S(m) \equiv I_2(m)C_m/\sigma^m$ is proportional to the potential energy of the mutual attraction between the two atoms evaluated at $\sigma \equiv 2R_0$. The constant of proportionality, $I_2(m)$, was shown by Enskog (1917) to vary between 0.2337 for $m = 2$ and 0.1556 for $m = 8$. Although the Sutherland model produced better agreement between calculated and measured viscosity coefficients at low temperatures than did the simple hard-sphere model, the differences for Ar, for example, at temperatures near 100 K were still of the order of 8%, which was significantly poorer than the approximately 1% agreement achievable at high temperatures.[1] Just as the repulsive power-law interaction represented an improvement over the hard-sphere interaction, it made sense to introduce an attractive component to the repulsive power-law model, and this is what Mie in 1903, and Lennard–Jones in 1924, independently set out to do.

The general MLJ model PEF, traditionally expressed in the form of eqns (2.209), may also be written conveniently as

$$V_\text{MLJ}(R) = \frac{C_n}{R^n} - \frac{C_m}{R^m}\,, \qquad n > m\,, \tag{3.2a}$$

in which $C_n \equiv \mathfrak{D}_\text{e}R_\text{e}^n$ and $C_m \equiv \mathfrak{D}_\text{e}R_\text{e}^m$ represent, respectively, the strengths associated with the repulsive and attractive branches of the PEF. Lennard–Jones (1924*a*, 1924*b*) initially fixed m at the value 2 (for mathematical convenience), and examined cases for which $n = \infty$, 20, $\frac{27}{2}$, 10, $\frac{20}{3}$. For the purpose of fitting to viscosity data, he was able to write $[\eta(T)]_1$ in the form

$$[\eta(T)]_1 = [\eta(T_0)]_1 \left(\frac{T}{T_0}\right)^{\frac{3}{2}} \frac{T_0^{(n-2)/n} + S}{T^{(n-2)/n} + S}\,, \tag{3.2b}$$

in which $\eta(T_0)$ is the viscosity for a standard temperature (chosen to be $T_0 = 273.15$ K), and S is referred to as the Sutherland constant. The fitting procedure then involved the determination of $[\eta(T_0)]_1$, n, and S from the experimental data. As Lennard–Jones was able to obtain equally good fits (i.e. within the experimental uncertainties) to the viscosity data for Ar for a number of different combinations of n and $[\eta(T_0)]_1$ for a given value of m, he concluded that it was not possible to determine the interatomic force-field of Ar uniquely based upon viscosity data alone. Although a limited amount of data was available for the only other transport property accessible for pure monatomic gases, namely, the coefficient of thermal conductivity, it was at that time both less accurate than the viscosity data and limited to a much smaller temperature range. More important, however, is the fact that the first Chapman–Cowling approximations

[1] It should be understood that such a statement merely represents the *goodness of fit* of an assumed potential model to the experimental data.

to the viscosity and thermal conductivity coefficients for atomic systems essentially depend upon the same effective cross-section [see eqns (2.70, 2.75) and (2.133)], so that thermal conductivity measurements for monatomic gases do not really provide independent data, but merely additional data of the same type.

Lennard–Jones then turned to equation of state data and measurements of the second coefficient in the virial expansion in powers of the number density proposed by Kamerlingh Onnes (1901). The virial equation of state for a gas may be written in the form of a series expansion of the compressibility factor $Z \equiv pV/(Nk_{\rm B}T)$ in terms of the number density $n \equiv N/V$ as

$$\frac{pV}{Nk_{\rm B}T} = 1 + B_2(T)n + B_3(T)n^2 + \cdots, \tag{3.3a}$$

in which the leading term corresponds to the ideal gas, and the other terms represent deviations from ideal gas behavior associated with increasing number density. The coefficients $B_2(T)$, $B_3(T)$, $\cdots$, referred to as the second, third, $\cdots$, virial coefficients, are functions only of temperature. The second (virial) coefficient, $B_2(T)$, is related to the interaction potential energy $V(R)$ via (Mason and Spurling, 1969; McCourt, 2003)

$$B_2(T) = -2\pi \int_0^\infty [{\rm e}^{-\beta V(R)} - 1]R^2\,{\rm d}R, \tag{3.3b}$$

with β defined as $\beta \equiv (k_{\rm B}T)^{-1}$.[2] From an analysis of the temperature dependence of the second virial coefficient data for Ar, Lennard–Jones (1924*b*) obtained consistency between the two sets of data using his model PEF with $n = 40/3$ and $m = 5$. Note that he used $m = 5$ rather than $m = 2$ for his comparison, as it is necessary that m be greater than 3 for $B_2(T)$ to remain finite.

It might appear that it should be easier to improve the form for the model PEF by employing binary mixture data, given that in addition to the mixture second virial coefficient, $B_{2,\rm mix}$, the number of transport phenomena that can be measured for a binary mixture increases by two, with the addition of the binary diffusion coefficient, D, and the thermal diffusion factor, $\alpha_{\rm T}$, to the mixture viscosity and thermal conductivity coefficients, $\eta_{\rm mix}$ and $\lambda_{\rm mix}$. Moreover, $\eta_{\rm mix}$ and $\lambda_{\rm mix}$ are somewhat less correlated for binary mixtures than they are for pure gases. The advantage provided by the addition of the two new transport phenomena was, however, more than offset by the fact that extraction of the contributions to $\eta_{\rm mix}$, $\lambda_{\rm mix}$, and $B_{2,\rm mix}(T)$ arising from the unlike atom interaction required greatly improved accuracy in the measurement of these phenomena both for the binary mixture itself and for the two pure gases.

Evolution of the MLJ (12–6) PEF into *the* characteristic representative PEF for the interaction between two noble gas atoms may be attributed to the establishment by London (Eisenschitz and London, 1930; London, 1930; London, 1937) that the leading long-range interaction between two closed-shell atoms is attractive, and varies as R^{-6}. The value 12 for n, however, has no physical significance, and was selected

[2] Some authors express the virial equation of state as an expansion in powers of the reciprocal $\widetilde{\rho}$ of the molar volume $\widetilde{v} \equiv N_A V/N$ as $p\widetilde{v}/(RT) = 1 + \widetilde{B}_2(T)\widetilde{\rho} + \widetilde{B}_3(T)\widetilde{\rho}^2 + \cdots$, with N_A the Avogadro number and $R = N_A k_{\rm B}$ the universal gas constant. Upon substituting $\widetilde{\rho} = n/N_A$ into this version of the virial equation of state, $B_2(T)$ and $\widetilde{B}_2(T)$ may be seen to be related by $\widetilde{B}_2(T) = N_A B_2(T)$.

essentially for mathematical convenience. The temperature dependence of the viscosity coefficient, together with that of the second virial coefficient were essentially the sole sources of data utilized for the determination of the depth, $\mathfrak{D}_e$, and position, R_e, associated with the MLJ (12,6) model PEF until the middle of the twentieth century. Largely because the experimental values for these properties typically had uncertainties ranging between ± 1% and ± 10% (especially for those data obtained at temperatures below 273 K), it was often possible to obtain statistically good agreement between experiment and calculations based upon MLJ (n,m) model functions.

3.2.1 Improving upon the simple MLJ empirical models

In order to proceed beyond the functional form of the MLJ models toward truly realistic PEFs, it became necessary to be able to obtain much more accurate viscosity and second virial coefficient data over a more extended temperature range. At the same time, as any realistic PEF would necessarily be considerably more complicated than the MLJ (n,m) model functions, it also became necessary to have access to faster computers and more accurate computational procedures. Further development of realistic PEFs was therefore stalled for nearly thirty years while waiting for these needs to be met.

With the advent of modern digital computers and modern programming languages on the calculational side, together with oscillating-body and vibrating viscometers and transient hot-wire thermal conductivity cells (Assael *et al.*, 1991; Nieuwoudt and Shankland, 1991), all of which were capable of giving experimental data with greatly reduced systematic experimental errors plus uncertainties as small as ± 0.1%, the stage was set to pursue a number of avenues toward the determination of truly realistic PEFs. It is also important to note that at about the same time (mid-1960s to mid-1970s), accurate molecular beam integral and differential scattering cross-section data (Scoles, 1988), Van der Waals ro-vibrational spectra (Colbourn and Douglas, 1976; Freeman *et al.*, 1974; Herman *et al.*, 1988; LaRocque *et al.*, 1986; Tanaka and Yoshino, 1970; Tanaka and Yoshino, 1972; Tanaka *et al.*, 1973), and dielectric and refractivity virial data (Achtermann *et al.*, 1993; Buckingham and Graham, 1974; Orcutt and Cole, 1967; Buckingham *et al.*, 1970) became available, thereby providing additional independent phenomena needed for the determination of more realistic PEFs. Shortly after these experimental developments, improved theoretical expressions for the PEF were introduced (Ahlrichs *et al.*, 1977; Hepburn *et al.*, 1975; Tang and Toennies, 1977) and tested for the noble gases.

Two different approaches to the provision of more realistic potential energy model functions were proposed in the early 1970s. One approach, taken by Maitland and Smith (1973) and essentially empirical in nature, was to generalize the relatively successful MLJ $(n,6)$ model functions by making the repulsive power a function of the separation R. Specifically, they proposed replacing the two-parameter MLJ functional form eqn (2.248a) with an improved three-parameter form,

$$V_{\text{iMLJ}}(R) = \mathfrak{D}_e \left[\frac{m}{n(R) - m} \left(\frac{R_e}{R} \right)^{n(R)} - \frac{n(R)}{n(R) - m} \left(\frac{R_e}{R} \right)^{m} \right], \tag{3.4a}$$

in which $m = 6$, $n(R)$ is defined by

$$n(R) \equiv 13 + \gamma \frac{R - R_e}{R}, \tag{3.4b}$$

and $\mathfrak{D}_e$, R_e, and γ are adjustable parameters. For Ar, $\eta(T)$ values computed from eqns (3.4) with specific parameter values $\mathfrak{D}_e = 98.77\,\text{cm}^{-1}$, $R_e = 3.76\,\text{Å}$, and $\gamma = 9$ gave significantly better agreement with the recommended experimental values than did $\eta(T)$ values computed from the 13–parameter Barker–Bobetic–Maitland–Smith (BBMS) potential energy function obtained (Maitland and Smith, 1971) by adapting an 11–parameter (Barker *et al.*, 1971) PEF to agree with the widths of the potential well obtained from a Rydberg–Klein–Rees (Le Roy, 2010; Kaplan, 2003) (RKR) inversion of spectroscopic data obtained for the Ar_2 dimer (Colbourn and Douglas, 1976; Freeman *et al.*, 1974; Herman *et al.*, 1988; LaRocque *et al.*, 1986; Tanaka and Yoshino, 1970; Tanaka and Yoshino, 1972; Tanaka *et al.*, 1973). The agreement between the experimental temperature dependence of $B_2(T)$ and that computed using the $V_{\text{iMLJ}}(R)$ PEF was comparable to that computed using the much more elaborate 13–parameter BBMS PEF. Due in part to its mathematical simplicity and in part to its flexibility, an extension of eqn (3.1a) has been introduced by Pirani *et al.* (2008) both to include interactions involving ionic species and as a recommended simple yet quite realistic model PEF for use in molecular dynamics (MD) simulations. Their generalization involves replacement of expression (3.4b) for the repulsive exponent $n(R)$ by

$$n(R) = \beta + 4\left(\frac{R}{R_e}\right)^2, \tag{3.4c}$$

with β a parameter correlated with the polarizabilities of the interacting atoms/ions and expected to vary slowly, between 8 for strongly polarizable species (like alkali atoms) or for interactions in which the long-range attraction is dominated by electrostatic interactions (such as ion-permanent-dipole and ion-ion interactions), to 10 for more weakly polarizable atoms (such as noble gas atoms) (Cambi *et al.*, 1991; Pirani *et al.*, 2004).

In determining the efficacy of using the iMLJ functional form to generate the dependence of the differential scattering cross-section $\sigma(\theta; E)$ on the laboratory scattering angle θ and of the speed dependence of the integral scattering cross-section, $\sigma_{\text{int}}(v)$, Pirani *et al.* (2008) compared results obtained from iMLJ functions with those obtained from the corresponding MLJ (12,6) function, for which $n(R) \equiv 12$, and a multiproperty-fit PEF based in part upon fitting to integral scattering cross-section data. For the Ne–Ar, Kr, Xe interactions, for example, the three PEFs are essentially indistinguishable for $R \leq R_e$ and for $R \gtrsim 0.7\,\text{nm}$, and differ slightly, but significantly, for $R_e < R < 0.7\,\text{nm}$. The iMLJ and multiproperty-fit PEFs for the Ne–Ne interaction are nearly indistinguishable over the entire range of R. This near-indistinguishability is reflected in an inability to differentiate, within the experimental uncertainties, between the calculated differential and integral Ne–Ne scattering cross-sections. The slightly greater attraction of the MLJ (12,6) PEF at intermediate separations (in the vicinity of $R = 0.45\,\text{nm}$, however, is sufficient to cause the calculated speed dependence of the integral cross-section to lie nearly 20% above the experimental curve, and well outside the experimental uncertainties. Moreover, although plots of the product

$\sigma(\theta;E)\theta^{\frac{7}{3}}$ of the differential cross-section versus θ (done to emphasize the rainbow structure that is superimposed upon the diffraction oscillations) for low collision energies ($E < 800$ cm^{-1}) were unable to distinguish between the iMLJ and multiproperty-fit PEFs, the simple MLJ (12,6) function was clearly unable to reproduce the experimental behavior of the oscillations. Much the same conclusions were drawn (Pirani *et al.*, 2008) for the relative accuracy of the Ne–Ar, Kr, Xe interactions with regard to the integral cross-section data: it was also possible for Ne–Kr and Ne–Ar scattering to distinguish between the fits to the experimental differential cross-section data provided by the iMLJ and multiproperty-fit PEFs.

3.3 Hartree–Fock Plus Damped Dispersion Semi-Empirical Models

The basic concept driving this approach to the construction of improved potential energy model functions was the observation by Slater and others (Born and Mayer, 1932; Buckingham, 1938; Slater and Kirkwood, 1931) that the nature of the small-separation repulsion between two atoms is essentially exponential. Based upon this observation, Buckingham and Corner (1947) proposed a model function that can be expressed in the form (Thakkar, 2001)

$$V(R) = A\mathrm{e}^{-bR} - \left(\frac{C_6}{R^6} + \frac{C_8}{R^8}\right) f(R)\,, \tag{3.5a}$$

in which the Born–Mayer repulsion $A\mathrm{e}^{-bR}$ is coupled with a long-range dispersion series damped by an empirical function $f(R)$ given by

$$f(R) = \begin{cases} \mathrm{e}^{-4(R_\mathrm{e}-R)^3/R_\mathrm{e}^3}\,, & R < \frac{5}{4}R_\mathrm{e} \\ 1 & R \geq \frac{5}{4}R_\mathrm{e} \end{cases}\,. \tag{3.5b}$$

This model for the potential energy is now referred to as the Buckingham–Corner (6,8), or BC (6,8), model.

3.3.1 The basic HFD model

Nearly thirty years after the introduction of the BC (6,8) function, Scoles and coworkers (1975, 1977), inspired by the observation by Toennies (1973) that a simple combination of the Born–Mayer repulsion and the long-range disperions series truncated at $-C_{10}R^{-10}$ provided surprisingly good agreement with experimental values of σ, R_e, and $\epsilon_{\mathrm{min}} \equiv \mathfrak{D}_\mathrm{e}$ for the noble gases, introduced the Hartree–Fock (HF) plus damped dispersion, or HFD, model

$$V_{\mathrm{HFD}}(R) \equiv \Delta E_{\mathrm{SCF}} - E_{\mathrm{disp}} \tag{3.6a}$$

$$= A\mathrm{e}^{-bR} - \left(\frac{C_6}{R^6} + \frac{C_8}{R^8} + \frac{C_{10}}{R^{10}}\right) f(R)\,. \tag{3.6b}$$

The main differences between the $V_{\mathrm{HFD}}(R)$ and BC (6,8) models are the addition of a C_{10} dispersion term and the introduction of a more physically based form for the damping function. The form for $f(R)$ is assumed to apply to all atom–atom interactions, and is determined by requiring that the high-accuracy *ab initio* interaction energies for $H_2(a\,^3\Sigma_u^+)$ computed by Kołos and Wolniewicz (1974) match those obtained via the HFD model function (3.6b) by employing accurate Hartree–Fock SCF energies[3] and *ab initio* long-range dispersion coefficient calculations as input data. The specific form determined for $f(R)$ in this way could be represented by

$$f(R) = \begin{cases} e^{-[(\frac{5}{4}R_e - R)/R]^{1.9}}, & R < \frac{5}{4}R_e \\ 1, & R \geq \frac{5}{4}R_e \end{cases}. \tag{3.6c}$$

Model HFD PEFs generated for the He–He, Ne_2, and Ar_2 interactions via eqns (3.5) were seen to be in very good agreement with experimental PEFs extracted from He–He and Ne–Ne crossed-beam differential scattering cross-section data and from a multi-property fit to BP of argon. However, less satisfactory agreement was obtained between the HFD model PEFs generated for the He–Ar interaction (based upon ΔE_{SCF} values obtained via combining rules (Olson and Smith, 1972) for the A and b coefficients in the Born–Mayer repulsion term) and PEFs obtained either by inversion of beam scattering data or via an empirical fit to mixture viscosity data. Because the disagreement in the case of the He–Ar mixed interaction was thought to be due to the use of combining rules to obtain the Born–Mayer repulsion energies, Ahlrichs *et al.* (1977) carried out new HF–SCF calculations for the He–Ar and Ne–Ar repulsion energies in order to enable a direct test of the SCF combining rules to be carried out. Comparison of their directly calculated HF–SCF He–Ar and Ne–Ar repulsion energies with those generated from accurate He–He, Ne_2, and Ar_2 HF–SCF data via combination rules established that the Smith combining rules (Olson and Smith, 1972) did indeed provide reliable results for the He–Ne and He–Ar interactions.

Ahlrichs *et al.* (1977) also generated final HFD model functions using a slightly simplified version of the original HFD model of eqn (3.6a), in which eqn (3.6c) for the damping function $f(R)$ was replaced by

$$f(R) = \begin{cases} e^{-[(1.28R_e - R)/R]^2}, & R < 1.28R_e \\ 1, & R \geq 1.28R_e \end{cases}. \tag{3.6d}$$

Input data employed for the like interactions were the Born–Mayer fit-parameters A and b obtained from HF–SCF repulsion-energy calculations and *ab initio* long-range dispersion coefficients C_6, C_8, and C_{10}. For the unlike interactions, however, all parameters were obtained via the Smith combining rules (Olson and Smith, 1972) for HF–SCF repulsion energies and the Moelwyn–Hughes (Kramer and Herschbach, 1970; Moelwyn-Hughes, 1957; Thakkar, 2001) combining rules for long-range dispersion coefficients. The R-dependences of the HFD model functions for the He–He and Ne_2 interactions were essentially indistinguishable from those of the respective experimental PEFs determined from inversions of crossed-beam differential scattering cross-section

[3]In the context of the HFD model, the Hartree–Fock calculations are such as to *always* ensure appropriate dissociation of the diatomic complex into neutral atoms at infinite separation; see, for example, Bowman, Hirschfelder, and Wahl, (1970).

data. Similar comparisons for the He–Ne and He–Ar interactions revealed differences in the well-positions, R_e, and well-depths $\mathfrak{D}_e \equiv -V(R_e)$, of 0.19 Å and 5.1 cm^{-1}, respectively, for the He–Ne interaction, and 0.06 Å and 4.8 cm^{-1}, respectively, for the He–Ar interaction between the HFD model functions and PEFs obtained by Chen *et al.* (1973) through inversion of their differential scattering data. However, the HFD well-depth parameters for these interactions showed substantive agreement with the well-depth parameters characterizing PEFs obtained from fittings to bulk data (specifically, binary diffusion and interaction second virial coefficient data). A further comparison of the dependence of newer experimental He–Ne differential scattering cross-section data of Smith *et al.* (1977) upon the laboratory-frame scattering angle with that computed from both the ESMSV (exponential-spline-Morse-spline) PEF of Chen *et al.* (1973) and the HFD model PEF showed that the HFD He–Ne PEF provided a considerably better resolution of the experimental He–Ne second differential scattering minimum than did the ESMSV PEF.

As pointed out by Tang and Toennies (1977), even though the HFD model functions of Scoles and coworkers (Ahlrichs *et al.*, 1977; Hepburn *et al.*, 1975) provided very good to excellent agreement with experimentally derived PEFs for rare gas atom–atom interactions, the HFD model itself has no real physical basis, as it fails dramatically, for example, in the prediction of unrealistic purely attractive interactions between rare gas atoms and alkali atoms. An examination of the nature of the long-range dispersion series

$$V_{\text{disp}}(R) \sim -\sum_{n\geq 3} \frac{C_{2n}}{R^{2n}}\,, \tag{3.7a}$$

taking particular note (as first pointed out by Dalgarno and Lewis, 1955) that the truncation error for this series at a given value of R is of the order of magnitude of the smallest term in the truncated series, thus requires that the number of terms retained should actually increase with increasing R. The normal potential energy model practice of employing a fixed number of dispersion terms at all separations can thus be, at best, correct only for a limited range of R values, with the consequence that the error in the vicinity of $R = R_e$ can be appreciable (should the dispersion expansion actually be valid there).

Examination of the dispersion series (3.7a) shows (Rae, 1975; Tang and Toennies, 1977) that for the interaction between a noble gas atom and another noble gas or alkali atom $-C_{2n}R^{-2n}$ is (algebraically) smallest within a characteristic range $R_{n-1,n} < R \leq R_{n,n+1}$ for $n = 4, 5, \cdots$, with $R_{n',n}$ defined in terms of the dispersion coefficients as $R_{n',n} \equiv (C_{2n}/C_{2n'})^{\frac{1}{2}}$, $n' < n$. The existence of these characteristic ranges for the dispersion terms suggests a means of incorporating the asymptotic dispersion series (3.7a) into realistic and more physically based models for the interaction energy between a pair of atoms. One mechanism for achieving this end is to apply switching or damping functions to the individual terms of the dispersion series, thereby providing an additive dispersion PEF that applies over the entire range of separation. Thus, the model dispersion energy component of $V(R)$ may be represented as

$$V_{\text{disp}}(R) = -\sum_{n\geq 3} f_{2n}(R)\,\frac{C_{2n}}{R^{2n}}\,. \tag{3.7b}$$

in which the $f_{2n}(R)$ are switching or damping functions for the individual dispersion terms. For $n \geq 4$, Tang and Toennies (1977) initially chose to employ linear switching functions that may be represented in the form

$$f_{2n}(R) = \begin{cases} 1\,, & R \geq R_{n,n+1} \\ \dfrac{R - R_{n-1,n}}{R_{n,n+1} - R_{n-1,n}}\,, & R_{n-1,n} \leq R < R_{n,n+1} \qquad n \geq 3\,, \\ 0\,, & R < R_{n-1,n} \end{cases} \tag{3.7c}$$

while for $n = 3$, they chose a linear switching function

$$f_6(R) = \frac{R - R_{\rm i}}{R_{3,4} - R_{\rm i}}\,, \tag{3.7d}$$

with $R_{\rm i} < \sigma$ a separation *inside* the repulsive wall of $V(R)$ chosen so as to convert the unrealistic negative pole of the $-C_6R^{-6}$ dispersion term at $R = 0$ into a realistic positive pole there. They also found that for many rare gas atom interactions it is important to include dispersion contributions up to and including the C_{14} term in order to provide a proper description of dispersion forces in the neighborhood of $R_{\rm e}$. For the Ar_2 interaction, for example, the C_{12} and C_{14} terms were each found to contribute an additional lowering of $V(R_{\rm e})$ of approximately 5%. Although the HFD-type models of Scoles *et al.* (Hepburn *et al.*, 1975; Ahlrichs *et al.*, 1977) and of Tang and Toennies (1977) introduced some physicality through the introduction of damping/switching functions into their models, thereby removing the clearly unphysical divergence of the dispersion series (3.7a), the somewhat ad hoc nature of the corrections remained less than fully satisfactory.

The long-range dispersion energy expression (3.7a) is valid when both electron exchange and charge overlap between a pair of interacting atoms can be neglected. It is, however, possible to avoid the neglect of charge overlap by evaluating the perturbation energy expression obtained following the expansion of the energy eigenstates of the combined atom–atom system using second-order perturbation theory in which the perturbation is the interaction Hamiltonian (Longuet-Higgins, 1956). Hence, if the second mathematical expansion employed in London's original derivation of expressions for the long-range dispersion coefficients, namely, an expansion of the matrix elements of the perturbation in inverse powers of the internuclear separation, is not carried out, direct evaluation of the perturbation terms will provide contributions to the attractive component of the interaction energy that converge at all interatomic separations R. In two important studies, Meath and coworkers (Kreek and Meath, 1969; Koide *et al.*, 1981) carried out variational evaluations of the second order nonexpanded perturbation energy terms that tend asmptotically to R^{-2n}, $n = 3, 4, \cdots 10$, in order to examine the roles of charge overlap contributions in determining the attractive nature of the Born–Oppenheimer potential energy for the $H_2(a^3\Sigma_u^+)$ prototypical van der Waals molecule. A comparison between their nonexpanded results for a given interatomic separation R and the corresponding values obtained from eqn (3.7a) enables the determination of the individual damping functions $f_{2n}(R)$, thereby giving the dispersion energy for $H_2(a^3\Sigma_u^+)$ in the form of the series (3.7b). The earlier study (Kreek and Meath, 1969) focused in part only upon the first three members of the dispersion series for $H_2(a^3\Sigma_u^+)$, while the later study (Koide *et al.*, 1981) employed a much larger basis-set and a different method for computing the nonexpanded energies. Their results therefore represented an improvement over the Kreek and Meath (1969) results for the $n = 8, 10$ nonexpanded dispersion energies, thereby providing the basis for any sensible damping function developed for HFD-type model PEFs.

Scoles and coworkers (1982) extended their preliminary (Ahlrichs *et al.*, 1977) HFD model, eqn (3.6a), firstly by modifying their representation of the HF–SCF repulsion term to[4]

$$\Delta E_{\mathrm{SCF}}(R) = A R^{\beta} \mathrm{e}^{-\alpha R}\,, \tag{3.8a}$$

and secondly, by improving the model dispersion component through the addition of C_{12} and C_{14} terms in the long-range dispersion series and by the introduction both of individual damping for each dispersion term and of an overall damping for the dispersion series itself, so that $E_{\mathrm{disp}}(R)$ is represented by

$$E_{\mathrm{disp}}(R) = -\left[\sum_{n\geq 3}^{7} f_{2n}(\rho R)\,\frac{C_{2n}}{R^{2n}}\right] g(\rho R)\,, \tag{3.8b}$$

in which ρ is a scaling parameter introduced to consider systematic changes in the characteristic ranges over which the dispersion energy terms are important for different atom pairs. Note that because very few accurate literature values are available for the C_{12} and C_{14} dispersion coefficients, it is often necessary to employ approximations like (Douketis and Scoles, 1982)

$$C_{12} \approx 1.028\left(\frac{C_{10}}{C_8}\right)^2 C_6, \qquad C_{14} \approx 0.975\left(\frac{C_{12}}{C_{10}}\right)^2 C_8, \tag{3.8c}$$

to generate reasonable values for them.

The forms for the individual damping functions $f_{2n}(R)$ were determined as

$$f_{2n}(R) = \left[1 - \mathrm{e}^{-(2.1R - 0.109R^2\sqrt{n})/n}\right]^n\,, \tag{3.8d}$$

by fitting them to *ab initio* values extracted from the Kreek and Meath (1969) $H_2(a^3\Sigma_u^+)$ unexpanded dispersion energy data (for which ρ is, by definition, 1). Similarly, the form for the overall long-range dispersion energy damping function $g(\rho R)$, which was introduced to correct both for the neglect of charge exchange between the partners and higher-order effects that were neglected in the construction of the model, was determined by requiring that the HFD model PEF for $H_2(a^3\Sigma_u^+)$ coincides with the

[4]For interactions involving alkali atoms, the HF–SCF repulsion is represented by $\Delta E_{\mathrm{SCF}}(R) = A_1 R^{\beta_1} \mathrm{e}^{-\alpha_1 R} + A_2 R^{\beta_2} \mathrm{e}^{-\alpha_2 R}$, rather than by eqn (3.8a). This revised form for the HFD model is then referred to as the HFD-C model. Unfortunately, no systematic naming convention for, or logical characterization of, the various HFD-type models that have been employed has been developed and, as a consequence, it is necessary to check the defining relations for each case.

ab initio results of Kołos and Wolniewicz (1974). The functional form for $g(R)$ thereby determined is

$$g(R) = 1 - R^{1.68}\mathrm{e}^{-0.78R}. \tag{3.8e}$$

Scoles and coworkers (1982) proposed an initial expression for the dimensionless scaling parameter ρ based upon the physically reasonable assumption that the asymptotic decay constant appearing in the exponential decay of electron charge density with distance from the nucleus for a ground-state atom (Hoffmann–Ostenhof and Hoffmann–Ostenhof, 1977; Tal, 1978) may be employed to characterize the range of interaction between a pair of like atoms. Since this decay constant is given by $\sqrt{2I_\mathrm{p}}$, with I_p the atomic ionization potential, a reasonable first approximation for ρ is simply $\rho \equiv [I_\mathrm{p}(\mathrm{X})/I_\mathrm{p}(\mathrm{H})]^{\frac{1}{2}}$. However, because this simple square-root scaling did not work well for interactions involving alkali atoms, the initial expression for ρ was revised empirically to

$$\rho = \left[\frac{I_\mathrm{p}(\mathrm{X})}{I_\mathrm{p}(\mathrm{H})}\right]^{\frac{2}{3}}, \tag{3.8f}$$

which not only worked well for interactions involving alkali atoms but also, surprisingly, resulted in slightly improved model functions for rare gas atom interactions. The development described via eqns (3.8a–f), may be summarized as

$$V_\mathrm{HFD}(R) = AR^{\beta}\mathrm{e}^{-\alpha R} - \left[\sum_{n\geq 3}^{7} f_{2n}(\rho R)\,\frac{C_{2n}}{R^{2n}}\right] g(\rho R) \tag{3.9}$$

for the HFD model PEF. This expression is normally referred to as the HFD-A form for $V_\mathrm{HFD}(R)$. A modified version, in which the exponential term is replaced by $\mathrm{e}^{-(\alpha R+\beta R^2)}$, is often referred to as the HFD-B form.

A model that has a similar format to that of the HFD model of Scoles and coworkers (Douketis and Scoles, 1982), and that was introduced at about the same time by Tang and Toennies (1984), is given by the expression

$$\begin{aligned} V_\mathrm{TT}(R) &= A\mathrm{e}^{-\alpha R} - \sum_{n\geq 3} f_{2n}(\alpha R)\,\frac{C_{2n}}{R^{2n}} \\ &\equiv V_\mathrm{rep}(R) + V_\mathrm{disp}(R)\,, \end{aligned} \tag{3.10a}$$

in which the damping functions $f_{2n}(z)$, given by

$$f_{2n}(z) = 1 - \mathrm{e}^{-z}\sum_{k=0}^{2n}\frac{z^k}{k!} \equiv 1 - \frac{\Gamma(2n+1, z)}{(2n)!}, \tag{3.10b}$$

with $\Gamma(2n+1, z)$ an incomplete gamma function (Abramowitz and Stegun, 1972; Lide, 2001) of order $2n+1$. Expressions (3.10) are collectively referred to as the TT model.

The model function $V_\mathrm{TT}(R)$ defined via eqns (3.10) was originally intended (Tang and Toennies, 1984) to be constructed from fitting the A and α parameters to *ab initio* high-level HF–SCF data to provide the repulsion term, then combining it with the damped dispersion series in which the form for the (universal) damping functions is determined from "exact" calculations of the dispersion energies for $H_2(a^3\Sigma_u^+)$. Tang and Toennies (1984) found that addition of C_{12} and C_{14} dispersion terms was required in order to obtain good agreement with the experimental $\mathfrak{D}_e$ and R_e values extracted from accurate beam scattering data or from PEFs derived from (multiproperty) fits to bulk gas data. It is fortunate, since accurate values for the C_{12} and C_{14} dispersion coefficients are not generally available, that the C_{12} and C_{14} dispersion terms play only a relatively limited role in improving the representation of $V_{TT}(R)$ in the vicinity of $R = R_e$, so that reasonable estimates for their values will suffice, and they may be estimated with sufficient accuracy by using either approximation (3.8c) or the semi-empirical recursion relation (Tang and Toennies, 1984; Thakkar, 1988)

$$C_{2n+4} = \left(\frac{C_{2n+2}}{C_{2n}}\right)^3 C_{2n-2} \, . \tag{3.11}$$

An alternative approach (Tang and Toennies, 1984) is to assume that the functional form (3.10a) provides a suitable scaffold upon which to build accurate PEFs for a large number of atom–atom interactions, retain only the generally accurately known C_6, C_8, and C_{10} dispersion coefficients in the model, and then determine repulsion parameters A and α by requiring that they conform to the experimentally determined position, R_e, and depth, $\mathfrak{D}_e$, for the interaction. Values of A obtained in this manner (Tang and Toennies, 1984) for the He–He and Ar_2 interactions, for example, were of the order of 13% and 17% larger than the corresponding A_{SCF} values, while the values of α so obtained differed insignificantly from the corresponding α_{SCF} values. The much less-repulsive nature of the $V_{TT}(R)$ model functions obtained by employing the HF–SCF values A_{SCF} and α_{SCF} in expressions (3.10) has tentatively been attributed (Tang and Toennies, 1984) to the existence of non-negligible repulsive contributions arising from terms in the expansion of the potential energy that were neglected in the development of the original HFD and TT model forms. Note also that the value for A obtained via the constraint method depends rather sensitively upon the accuracy of R_e, as an experimental uncertainty of the order of 0.1% in R_e may lead to an uncertainty as large as 10% for the final fitted value.

Thus, even though the form for the TT model expression (3.10a) seems very similar to the HFD-B expression (3.9), there are three important differences between these two models. Firstly, the parameters A and α determining the repulsion term in the TT model, unlike those determining the repulsion term in the HFD model, are no longer determined via a Born–Mayer fit to the R-dependence of the repulsive interaction calculated at the HF–SCF level. Rather, they are determined by fitting to previously determined experimental values of the position and depth of the minimum in the interaction. Secondly, the manner in which systematic differences in the ranges of R over which the damping functions will act for different like-atom interactions X–X is accounted for differently in the two models. The HFD model inserts an empirical scaling factor ρ, defined as a fractional power of the ratio between the first ionization potential of atom X and that of H, into the argument of the damping functions. By contrast, the damping functions in the TT model are effectively driven by the parameter α that appears in the repulsive component of the interaction, thereby providing a type of natural scaling in terms of the range of charge overlap between the interacting atoms. Finally, the HFD model is intended to serve more as a predictive tool, while the TT model has actually evolved into a fitting function intended to contain built-in features of the interaction. As such, however, it does require that quite accurate values be available for $\mathfrak{D}_e$ and R_e, either from experiment or from high-level *ab initio* quantum calculations, before an appropriate $V_{TT}(R)$ model function can be constructed.

If the TT model potential function $V_{TT}(T)$ is written in reduced form $U \equiv V(R)/\mathfrak{D}_e$ as a function of $x \equiv R/R_e$ as

$$U_{TT}(x) = A^* e^{\alpha^* x} - \sum_{n \geq 3} \left[1 - \sum_{k=0}^{2n} \frac{(\alpha^* x)^k}{k!} e^{\alpha^* x} \right] \frac{C_{2n}^*}{x^{2n}} , \tag{3.12}$$

then by utilizing $\mathfrak{D}_e = -V(R_e)$, and the definitions $A^* \equiv A/\mathfrak{D}_e$, $\alpha^* \equiv \alpha R_e$, and $C_{2n}^* \equiv C_{2n}/(\mathfrak{D}_e R_e^{2n})$, the derivative $U'_{TT}(x)$ of U_{TT} with respect to x becomes

$$U'(x) = -A^* \alpha^* e^{-\alpha^* x} - \sum_{n \geq 3} \alpha^* e^{-\alpha^* x} \frac{(\alpha^*)^{2n}}{(2n)!} C_{2n}^* + \sum_{n \geq 3} \left[1 - e^{\alpha^* x} \sum_{k=0}^{2n} \frac{(\alpha^* x)^k}{k!} \right] \frac{2n C_{2n}^*}{x^{2n+1}} .$$

By requiring that $U'_{TT}(1)$ vanish, A^* may be obtained as

$$A^* = \sum_{n \geq 3} \left[e^{\alpha^*} - \sum_{k=0}^{2n} \frac{(\alpha^*)^k}{k!} \right] \frac{2n C_{2n}^*}{\alpha^*} - \sum_{n \geq 3} \frac{(\alpha^*)^{2n}}{(2n)!} C_{2n}^* . \tag{3.13a}$$

Substitution of expression (3.13a) for A^* into $U_{TT}(1) = -1$ then gives a transcendental equation for α^*, namely,

$$\sum_{n \geq 3} \left[1 - e^{-\alpha^*} \sum_{k=0}^{2n} \frac{(\alpha^*)^k}{k!} \right] \left(\frac{2n}{\alpha^*} - 1 \right) C_{2n}^* - \sum_{n \geq 3} e^{-\alpha^*} \frac{(\alpha^*)^{2n}}{(2n)!} C_{2n}^* = -1 . \tag{3.13b}$$

Equations (3.13) may be solved iteratively from an input data set consisting of values for C_6, C_8, C_{10}, R_e, and $\mathfrak{D}_e$ (Tang and Toennies, 1986).

It has been suggested (Tang and Toennies, 1984) on the basis of comparisons made between the model damping functions (3.10b) and damping functions obtained directly from *ab initio* calculations for $H_2(a^3\Sigma_u^+)$ (Koide *et al.*, 1981) and the He–He interaction (Jacobi and Csanak, 1975), as well as with damping functions constructed from *ab initio* calculations of individual multipole damping functions (Krauss and Neumann, 1979; Krauss and Stevens, 1982; Krauss *et al.*, 1980) for the He–He, Ar_2, and Xe_2 interactions, and with damping functions generated from the HFD and Feltgen (1981) models, that these damping functions have a "universal" character. It seems plausible that the presence of the factor $e^{-\alpha R}$ in expression (3.10b) may be responsible for their apparent universal character: a more elaborate exposition of the TT model format and justification for the form taken by these damping functions comes from Tang, Toennies, and Yiu (1993, 1994, 1998).

3.3.2 Combining rules and hetero-atomic interactions

Commonly, a hetero-atomic PEF is not available, so it becomes necessary to construct one or more components of the PEF from a knowledge of the components of the two corresponding homo-atomic PEFs. The Smith (Olson and Smith, 1972) combining rule mentioned earlier applies when the short-range repulsion energy between pairs of atoms can be adequately represented by the simple Born–Mayer form $Ae^{-\alpha R}$, and expresses the repulsive component of the unlike-atom interaction energy in terms of the A and α parameter values that characterize the repulsion energies between the homo-atomic pairs A_2 and B_2. More specifically, if the repulsive component of the interaction energy, $V_{AB}(R)$, between atoms A and B can be represented by $V_{AB}^{rep}(R) = A_{AB}e^{-\alpha_{AB}R}$, then A_{AB} and α_{AB} can be obtained from the parameters A_{AA}, α_{AA} from $V_{AA}^{rep}(R)$ and A_{BB}, α_{BB} from $V_{BB}(R)$ via the combining rules

$$\alpha_{AB} = \frac{2\alpha_{AA}\alpha_{BB}}{\alpha_{AA} + \alpha_{BB}} \tag{3.14a}$$

and

$$\frac{1}{\alpha_{AB}} \ln[\alpha_{AB}A_{AB}] = \frac{1}{2}\left\{\frac{1}{\alpha_{AA}} \ln[\alpha_{AA}A_{AA}] + \frac{1}{\alpha_{BB}} \ln[\alpha_{BB}A_{BB}]\right\}. \tag{3.14b}$$

Note that these specific representations of the Smith combining rules stem from the fact that α has the form of a reciprocal length.

The dispersion component of the attraction between unlike atoms A, B often requires the determination of the three leading dispersion coefficients C_6^{AB}, C_8^{AB}, C_{10}^{AB} in terms of the C_6^{kk}, C_8^{kk}, C_{10}^{kk} (k = A, B) like-interaction coefficients. However, in some cases, it may be sufficient to obtain only an estimate of C_{10}^{AB} from the empirical relation (Thakkar, 1988)

$$C_{10}^{AB} \simeq 1.21\frac{(C_8^{AB})^2}{C_6^{AB}}. \tag{3.15}$$

Higher-order dispersion coefficients C_{2n}^{AB} ($n = 6,\ 7,\ \cdots$), if needed, can then be approximated via eqn (3.8c). Each coefficient C_{2n} (more explicitly, C_{2n}^{ij} for atoms i, j) in the asymptotic expression (3.7a) for the long-range dispersion energy can be broken down (Tang and Toennies, 1986) into sums of components $C^{ij}(l, l')$ giving rise to the same overall inverse power $2n$ of R, with l and l' satisfying the conditions $l, l' \geq 1$, $l + l' + 1 = n$. In this way, the three leading dispersion coefficients C_6^{ij}, C_8^{ij}, and C_{10}^{ij} are given as

$$C_6^{ij} = C^{ij}(1,1)\,, \tag{3.16a}$$

$$C_8^{ij} = C^{ij}(1,2) + C^{ij}(2,1)\,, \tag{3.16b}$$

$$C_{10}^{ij} = C^{ij}(1,3) + C^{ij}(3,1) + C^{ij}(2,2)\,, \tag{3.16c}$$

in terms of their multipole–multipole components, with $l = 1, 2, 3$ denoting electronic dipole, quadrupole, and octupole moments, respectively.

Tang and Toennies Tang and Toennies (1986) derived combining rules for the $C^{AB}(l, l')$, given ultimately in terms of the homo-atomic dispersion coefficients C_{2n}^{kk}

(k = A, B) and the multipole polarizabilities $\alpha_l(\mathrm{k})$. Their expression for the induced-dipole–induced-dipole dispersion coefficient C_6^{AB} is, as may be expected, the same as the well known Moelwyn–Hughes combining rule (Kramer and Herschbach, 1970; Moelwyn-Hughes, 1957), namely,

$$C_6^{\mathrm{AB}} = \frac{2\alpha_1(\mathrm{A})\alpha_1(\mathrm{B})C_6^{\mathrm{AA}}C_6^{\mathrm{BB}}}{C_6^{\mathrm{AA}}\alpha_1^2(\mathrm{B}) + C_6^{\mathrm{BB}}\alpha_1^2(\mathrm{A})}, \tag{3.17a}$$

in which $\alpha_1(\mathrm{k})$ is the static (dipole) polarizability of atom k. The dipole–quadrupole contribution, $C^{\mathrm{AB}}(1,2)$, to C_8^{AB} can be expressed in terms of the C_6^{kk} and C_8^{kk}, together with the various electronic polarizabilities α_l^{k}, as

$$C^{\mathrm{AB}}(1,2) = \frac{5\alpha_1(\mathrm{A})\alpha_2(\mathrm{B})C_6^{\mathrm{AA}}C_6^{\mathrm{BB}}C_8^{\mathrm{BB}}}{C_6^{\mathrm{AA}}[10\alpha_1(\mathrm{B})\alpha_2(\mathrm{B})C_6^{\mathrm{BB}} - \alpha_1^2(\mathrm{B})C_8^{\mathrm{BB}}] + \alpha_1^2(\mathrm{A})C_6^{\mathrm{BB}}C_8^{\mathrm{BB}}}, \tag{3.17b}$$

with a similar result for $C^{\mathrm{AB}}(2,1)$ obtained from eqn (3.17b) by interchanging the atom labels A and B, that is, $C^{\mathrm{AB}}(2,1) = C^{\mathrm{BA}}(1,2)$. A final expression for C_8^{AB} may be obtained by substituting the results for $C^{\mathrm{AB}}(1,2)$ and $C^{\mathrm{AB}}(2,1)$ into eqn (3.16b). The hetero-atomic dipole–octupole dispersion coefficient can be obtained as

$$C^{\mathrm{AB}}(1,3) = \frac{28\alpha_1(\mathrm{A})\alpha_3(\mathrm{B})C_6^{\mathrm{AA}}C_6^{\mathrm{BB}}C_8^{\mathrm{BB}}}{C_6^{\mathrm{AA}}[28\alpha_1(\mathrm{A})\alpha_3(\mathrm{B})C_6^{\mathrm{BB}} - 3\alpha_1^2(\mathrm{A})C_6^{\mathrm{BB}}C^{\mathrm{BB}}(1,3)] + 3\alpha_1^2(\mathrm{A})C_6^{\mathrm{BB}}C^{\mathrm{BB}}(1,3)}, \tag{3.17c}$$

while $C^{\mathrm{AB}}(3,1)$ is related to $C^{\mathrm{AB}}(1,3)$ in the same manner that $C^{\mathrm{AB}}(2,1)$ is related to $C^{\mathrm{AB}}(1,2)$, that is, $C^{\mathrm{AB}}(3,1) = C^{\mathrm{BA}}(1,3)$. Finally, according to eqn (3.16c), in order to obtain the full expression for C_{10}^{AB}, an expression for the quadrupole–quadrupole dispersion coefficient is still needed: it is given by

$$C^{\mathrm{AB}}(2,2) = \frac{70}{3}\,\frac{\alpha_2(\mathrm{A})\alpha_2(\mathrm{B})C_6^{\mathrm{AA}}C_6^{\mathrm{BB}}C_8^{\mathrm{AA}}C_8^{\mathrm{BB}}}{C_{qq}(\mathrm{A},\mathrm{B}) + C_{qq}(\mathrm{B},\mathrm{A})}, \tag{3.17d}$$

with $C_{qq}(\mathrm{k},\mathrm{k}')$ defined, for convenience, as

$$C_{qq}(\mathrm{k},\mathrm{k}') \equiv C_6^{\mathrm{kk}}C_8^{\mathrm{kk}}[10\alpha_1(\mathrm{k}')\alpha_2(\mathrm{k}')C_6^{\mathrm{k'k'}} - \alpha_1^2(\mathrm{k}')C_8^{\mathrm{k'k'}}].$$

Higher dispersion terms may then be obtained, if needed, by employing the approximate bootstrap expressions of eqn (3.8c) or the recursion relations (3.11). Finally, the simple empirical combining rule (Douketis and Scoles, 1982),

$$\rho_{\mathrm{AB}} = \frac{2\rho_{\mathrm{A}}\rho_{\mathrm{B}}}{\rho_{\mathrm{A}} + \rho_{\mathrm{B}}}, \tag{3.18}$$

may be employed for the scaling factor ρ_{AB} appearing in the HFD model damping function of eqn (3.9).

3.3.3 Application of the TT model to noble gas interactions

Tang and Toennies (2003) use a version of their model PEF, eqns (3.10), given by the three leading long-range dispersion coefficients (C_{2n}, $n = 3, 4, 5$), the pre-exponential factor A, and the repulsion exponent α. For the six homo-atomic rare gas interactions, they employed *ab initio* values for C_6, C_8, and C_{10}, together with values of A and α obtained from iterative fits to eqns (3.13) using the *ab initio* values of the three long-range dispersion coefficients and "best experimental" values of $\mathfrak{D}_e$ and R_e as input data.[5] Values for $\mathfrak{D}_e$ and R_e were determined for the homo-atomic series of interactions either from *ab initio* calculations (Anderson, 2001; Runeberg and Pyykkö, 1998) (He, Rn) or from multiproperty-fit empirical PEFs (Aziz, 1993; Aziz and Slaman, 1986*b*; Aziz and Slaman, 1989*b*; Dham *et al.*, 1989; Dham *et al.*, 1990) (Ne, Ar, Kr, Xe). Although such TT03 PEFs may be employed to illustrate the trend in the depth of the minimum interaction energy and the equilibrium separation between the nuclei anticipated qualitatively in terms of increases in the number of electrons and atomic size in passing from He to Rn atoms, they are not useful for accurate calculations (i.e. to better than 1%) of either spectroscopic or BP data. See also the more quantitative statement in the final paragraph of Section 3.8.3 on the Ar–Ar interaction.

It is more challenging to generate appropriate values of A and α for many of the hetero-atomic interactions, however, because there are relatively few spectroscopic studies from which sufficiently accurate determinations of experimental $\mathfrak{D}_e$, and especially R_e, values can be made. As beam scattering and bulk-propery-fit results for $\mathfrak{D}_e$ and for R_e may often differ by as much as 1%, Tang and Toennies (1986) instead utilized a pair of combining rules, namely,

$$\mathfrak{D}_e^{AB} = [\mathfrak{D}_e^{A}\mathfrak{D}_e^{B}]^{\frac{1}{2}} \left(\frac{[R_e^{A}R_e^{B}]}{[R_e^{AB}]^2}\right)^3 \frac{2\alpha(\mathrm{A})\alpha(\mathrm{B})[C_6^{AA}C_6^{BB}]^{\frac{1}{2}}}{C_6^{AA}\alpha^2(\mathrm{B}) + C_6^{BB}\alpha^2(\mathrm{A})}$$

for the magnitude of the potential energy minimum and

$$[R_e^{AB}]^6 = \frac{(\frac{1}{2}[\mathfrak{D}_e^{A}]^{\frac{1}{13}}[R_e^{A}]^{\frac{12}{13}} + \frac{1}{2}[\mathfrak{D}_e^{B}]^{\frac{1}{13}}[R_e^{B}]^{\frac{12}{13}})^{13}}{[\mathfrak{D}_e^{A}\mathfrak{D}_e^{B}]^{\frac{1}{2}}[R_e^{A}R_e^{B}]^3}$$

for the position of the minimum. Employment of $\mathfrak{D}_e^{AB}$ and R_e^{AB} values obtained from combining rules was found to give a more systematic and consistent derivation of the repulsive parameters A and α for the hetero-atomic rare gas interactions.

Two broad reviews of the history behind the development of damping function models are available (Feltgen, 1981; Tang and Toennies, 1984), including a detailed examination (Feltgen, 1981) of the nature of the various contributions to interatomic interactions between pairs of atoms. The combination of these two studies provides a thorough analysis of the appropriateness of the various approximations employed in developing the HFD model forms.

[5] A word of caution is necessary here for those who may wish to construct PEFs using more recent values of $\mathfrak{D}_e$ and R_e since, as explained by Tang and Toennies (2003), values obtained for the repulsion parameters A and α are quite sensitive to the input values $\mathfrak{D}_e$ and R_e for a chosen set of dispersion coefficients C_{2n} $(n = 3, 4, \cdots, N)$.

At the lowest level of approximation, the interatomic interaction energy can be expressed in terms of the Hartree–Fock and electron correlation energies as

$$V(R) = \Delta E_{\mathrm{HF}}(R) + E_{\mathrm{corr}}(R)\,,$$

with $\Delta E_{\mathrm{HF}}(R)$ representing the net Coulombic and inductive contributions to the interaction associated with (at most) single excitations on one of the atoms. The electron correlation energy may be split formally into two components, one a contribution to the Hartree–Fock energy that represents *intra*-atomic electron correlation (which vanishes for the H atom) resulting from the inclusion of double and/or multiple excitations on a single atom, the other representing the *inter*atomic electron correlation energy plus *intra*-atomic electron correlation effects arising from simultaneous single or multiple excitations on both atoms. From the viewpoint of HFD models, the former component may be combined with $\Delta E_{\mathrm{HF}}(R)$ and labeled $\Delta E_{\mathrm{SCF}}(R)$, while the latter component is referred to as the dispersion energy, $E_{\mathrm{disp}}(R)$. Thus, at the second level of approximation, the interaction may be represented as

$$V(R) = \Delta E_{\mathrm{SCF}} + E_{\mathrm{disp}}(R)\,.$$

For atomic separations R sufficiently large for there to be negligible charge overlap and electron exchange, $E_{\mathrm{disp}}(R)$ may be obtained by (Rayleigh–Schrödinger) perturbation theory calculations in the form of the nonexpanded dispersion energies of Meath and coworkers (Koide *et al.*, 1981; Kreek and Meath, 1969), a result that is commonly referred to as the polarization approximation (Hirschfelder, 1967). Such calculations include charge overlap effects, but neglect electron exchange effects. Expansion of the terms obtained in this manner into power series in R^{-1} gives rise to the well-known London asymptotic long-range dispersion series. Fortunately, the electron exchange contribution to the dispersion energy, which is rather difficult to compute, represents only a very small fraction of the total interaction energy for $R < R_{\mathrm{e}}$ and dies off exponentially with R for $R > R_{\mathrm{e}}$, so that its neglect in HFD-type models would thus appear to be a reasonable approximation (Moszynski, 2007). The third-order perturbation calculation of the polarization dispersion energy for the H(1s)–H(1s) interaction, for example, gives positive C_{11}, C_{13}, and C_{15} contributions that have magnitudes of the order of 50–100 times smaller than the corresponding second-order (negative) contributions (Feltgen, 1981). Because there is no particular reason to anticipate that third-order contributions to the polarization dispersion energies of other atom–atom interactions should behave differently relative to their corresponding second-order contributions, the odd inverse powers are often ignored (Douketis and Scoles 1982; Tang and Toennies, 1977, 1984, 2003) in the construction of HFD-type models. The TT and HFD model PEFs have the advantage over many of the more frequently employed semi-empirical piecewise continuous PEFs of being continuous everywhere and having derivatives to all orders: this feature has no doubt enhanced their use in numerous applications.

3.3.4 Extension of the TT model to chemically bonded species

Under the assumption that the damped dispersion series, $V_{\mathrm{disp}}(R)$ of eqn (3.10a), representing the non-exchange part of the dispersion energy remains valid even for separations R in the region of the diatomic chemical bond, while the exchange component, $V_{\mathrm{X,disp}}(R)$, of the dispersion energy can no longer be neglected, the TT potential energy model has been extended (Lau *et al.*, 2016; Tang *et al.*, 1991) by assuming that the interaction potential energy, $V(^1\Sigma_g^+;R)$, for the ground electronic term of a chemically bound closed-shell diatomic molecule may be represented as

$$V(^1\Sigma_g^+;R) = V_{\mathrm{SCF}}(^1\Sigma_g^+;R) + V_{\mathrm{disp}}^{(2)}(R) - V_{\mathrm{X,disp}}(R) + V_{\mathrm{disp}}^{(3)}(R)\,, \tag{3.19}$$

in which $V_{\mathrm{SCF}}(^1\Sigma_g^+;R)$ is obtained from two-configuration SCF calculations carried out in such a manner that the correct neutral-atom dissociation limits are maintained, $V_{\mathrm{disp}}^{(2)}(R)$ is the second-order dispersion energy, $V_{\mathrm{X,disp}}(R)$ is the electron exchange contribution to the dispersion energy, and $V_{\mathrm{disp}}^{(3)}(R)$, given by

$$V_{\mathrm{disp}}^{(3)}(R) = \sum_{n=5} f_{2n+1}(R)\,\frac{C_{2n+1}}{R^{2n+1}}\,, \tag{3.20a}$$

with the $f_{2n+1}(R)$ damping functions assumed to take the same form as the $f_{2n}(R)$ functions defined in eqn (3.10b), is the third-order perturbation contribution to the dispersion energy.

The exchange–dispersion contribution to $V(^1\Sigma_g^+;R)$ is given for H_2 and homonuclear alkali molecules in terms of the asymptotic exchange energy, $V_{\mathrm{aX}}(R)$, encountered in the theory of magnetism (Herring and Flicker, 1964; Smirnov and Chibisov, 1965), that is,

$$V_{\mathrm{aX}}(R) = R^{(3.5-\alpha)/\alpha}\,B\mathrm{e}^{-2\alpha R}\,, \tag{3.20b}$$

with $\alpha \equiv \sqrt{2I_p^{\mathrm{A}}}$, and I_p^{A} the first ionization potential of atom A. The SCF exchange energy is given by one-half the difference between the triplet and singlet two-configuration SCF energies, that is,

$$V_{\mathrm{X,SCF}}(R) = \tfrac{1}{2}[V_{\mathrm{SCF}}(a^3\Sigma_u^+;R) - V_{\mathrm{SCF}}(X^1\Sigma_g^+;R)]\,.$$

Although the asymptotic exchange–dispersion energy properly accounts for correlation contributions to the exchange energy (missing from the SCF calculations) for sufficiently large separations R, it will grossly overestimate $V_{\mathrm{X,disp}}(R)$ for separations corresponding to the bonding region. The TT model for the $X^1\Sigma_g^+$ interaction energy includes the ad hoc assumption that $V_{\mathrm{X,disp}}(R)$ can be represented by the damped exchange–dispersion energy given by

$$V_{\mathrm{X,disp}}(R) = f_6(R)V_{\mathrm{aX,disp}}(T) = f_6(R)[V_{\mathrm{aX}}(R) - V_{\mathrm{X,SCF}}(R)]\,. \tag{3.20c}$$

This approximation has been found to give results for $V(R)$ that compare reasonably well (Tang *et al.*, 1991; Lau *et al.*, 2016) with accurate *ab initio* (Kolos and Wolniewicz, 1974) and two-configuration SCF (Bowman Jr. *et al.*, 1970) calculations for the ground electronic state of $H_2(^1\Sigma_g^+)$.

By writing the Coulomb and exchange energies in terms of the singlet- and triplet-term energies as

$$V_C = \tfrac{1}{2}[V(a^3\Sigma_u^+; R) + V(X^1\Sigma_g^+; R)]\,, \tag{3.21a}$$

and

$$V_X(R) = \tfrac{1}{2}[V(a^3\Sigma_u^+; R) - V(X^1\Sigma_g^+; R)]\,, \tag{3.21b}$$

the interaction energy of the ground term singlet is given by

$$V(X^1\Sigma_g^+; R) = V_C(R) - V_X(R)\,, \tag{3.21c}$$

and a corresponding expression,

$$V(a^3\Sigma_u^+; R) = V_C(R) + V_X(R)\,, \tag{3.21d}$$

is obtained for the $a^3\Sigma_u^+$ (first) excited electronic term having the same free atomic dissociation limit as the $n\,^1S_0$ ground electronic term, with n the value of the atomic principal quantum number.

If the same splitting for the SCF energy is made, that is,

$$V_{SCF}(X^1\Sigma_g^+; R) = V_C^{SCF}(R) - V_X^{SCF}(R)\,,$$

the result

$$\begin{aligned} &V(X^1\Sigma_g^+; R) \\ &= \left[V_C^{SCF}(R) + V_{disp}^{(2)}(R) + V_{disp}^{(3)}(R)\right] - [[1 - f_6(R)]V_X^{SCF}(R) + f_6(R)V_{aX}(R)] \end{aligned} \tag{3.22}$$

is obtained as the final expression for $V(X^1\Sigma_g^+; R)$. The first term in this expression represents the Coulomb energy, the second term the exchange energy. In this model for the chemically bound electronic ground term for $H_2(X^1\Sigma_g^+)$ or for a homonuclear diatomic alkali molecule, the sum $V_{disp}^{(2)}(R) + V_{disp}^{(3)}(R)$ accounts for the correlation energy that is neglected in the calculation of $V_C^{SCF}(R)$. The second term gives the asymptotic exchange energy for separations sufficiently large for $f_6(R)$ to be unity, and serves as a switching function that gradually turns the SCF exchange energy on as the asymptotic exchange energy is gradually turned off. The only difference between this version of the TT model and that described earlier [see eqns (3.10)] for nonbonded atomic interactions is the inclusion of the third-order dispersion and exchange dispersion contributions, which have typically been found to be negligible for nonbonded interactions.

The principal experimental data associated with chemically bound diatomic molecules is a set of spectroscopic transition frequencies corresponding to energy differences between (internal) ro-vibrational states of the molecule. As the frequencies of spectroscopic transitions can be determined with high accuracy (often to better than 0.1%), it may thus be prudent to introduce a word of caution with respect to the development of a relatively simple model for the PEF for such interactions. As shown by Tang *et al.* (Lau *et al.*, 2016; Tang *et al.*, 1991) for $H_2(^1\Sigma_g^+)$ and $Li_2(^1\Sigma_g^+)$, these model functions are useful for illustrating qualitative differences in the relative shapes of the attractive wells of chemically bound diatomic interaction potentials. Indeed, these model PEFs work fairly well should the experimental accuracies be of the order of a few percent, but will likely turn out to be much less useful for experimental accuracies that are of the order of $\pm 0.1\%$ or better, as is typical for spectroscopic frequency measurements.

3.4 Exchange-Coulomb (XC) Semi-Empirical Models

The XC semi-empirical model for the interaction energy $V(R)$ between two closed electronic shell chemical species (here, atoms) separated by a distance R is based upon a partitioning of the overall interaction energy into the sum of a repulsive energy term, to be represented by the interatom electron exchange energy, $E_{\rm X}(R)$, and an attractive term, to be obtained from the expression for the total interatomic Coulomb energy for the A–B pair.

An expression for the Coulomb energy between atoms A and B is typically obtained from Rayleigh–Schrödinger perturbation theory based upon the separated atom zeroth-order product wavefunction $\Psi_{\rm AB}^{(0)} \equiv \phi_{\rm A}^{(0)}\phi_{\rm B}^{(0)}$ of high-level SCF monomer ground electronic state basis functions $\phi_{\rm A}^{(0)}$ and $\phi_{\rm B}^{(0)}$. The Coulomb energy may thus be expressed in the form of a perturbation series as

$$E_{\rm C} = E_{\rm C}^{(1)} + E_{\rm C}^{(2)} + \cdots ,$$

in which $E_{\rm C}^{(n)}$ represents the $n^{\rm th}$-order perturbation contribution. The first-order Coulomb energy, $E_{\rm C}^{(1)}(R)$, represents the net Coulombic interaction energy between the electrons and nucleus of atom A with those of atom B, and may be computed from the basic expression

$$E_{\rm C}^{(1)}(R) = \langle \Psi_{\rm AB}^{(0)} | \mathcal{V}_{\rm e} | \Psi_{\rm AB}^{(0)} \rangle ,$$

in which $\mathcal{V}_{\rm e}$ is the Coulombic interaction Hamiltonian in the absence of interatomic electron exchange, given by

$$\mathcal{V}_{\rm e} = \frac{1}{4\pi\varepsilon_0} \sum_{i,j} \frac{q_i q_j}{|\mathbf{R} + \mathbf{r}_j - \mathbf{r}_i|} ,$$

with $\mathbf{R}$ the position vector of atom B relative to atom A, $\mathbf{r}_i$ the position vector of constituent particle (electron, nucleus) i of atom A, $\mathbf{r}_j$ the position vector of constituent particle j of atom B, and q_i, q_j the (signed) charges on particles i, j. Typically, $E_{\rm C}^{(1)}(R)$ is evaluated by representing the ground-state isolated atom wavefunctions $\phi_{\rm A}^{(0)}$ and $\phi_{\rm B}^{(0)}$ in terms of reliable nonrelativistic SCF wavefunctions. The numerical output obtained from such a computation is often least squares fitted to the functional form (Aziz *et al.*, 1992; Meath and Koulis, 1991; Ng *et al.*, 1978; Ng *et al.*, 1979)

$$E_{\rm C}^{(1)}(R) = -\exp\left\{-a_0 R + a_1 + a_2 R^{-1} + a_3 R^{-2}\right\} . \tag{3.23a}$$

The second, and all higher-order, perturbation contributions to the Coulomb energy may be represented in terms of series of damped inverse-power contributions to the interatomic interaction energy.

To express $V(R)$ as the sum of a repulsive electron exchange component plus a net attractive Coulomb-based component, it is first necessary to separate out the repulsive exchange component of the total interaction energy. This task is accomplished within the XC model through approximating the exchange energy, $E_{\mathrm{X}}(R)$, by the difference, $E_{\mathrm{X}}^{(1)}(R)$, between the computed first-order Heitler–London energy, $E_{\mathrm{HL}}^{(1)}(R)$, and the first-order Coulomb energy, $E_{\mathrm{C}}^{(1)}(R)$. The first-order Heitler–London energy is obtained from the expression

$$E_{\mathrm{HL}}^{(1)}(R) = \frac{\langle \mathcal{A}\phi_{\mathrm{A}}\phi_{\mathrm{B}}|\mathcal{H}|\phi_{\mathrm{A}}\phi_{\mathrm{B}}\rangle}{\langle \mathcal{A}\phi_{\mathrm{A}}\phi_{\mathrm{B}}|\phi_{\mathrm{A}}\phi_{\mathrm{B}}\rangle} - (E_{\mathrm{A}}^{0} + E_{\mathrm{B}}^{0}),$$

in which $\mathcal{H}$ is the full Hamiltonian for the A–B dimer, $\mathcal{A}$ is the interaction antisymmetrizer, and $E_{\mathrm{A}}^{0} + E_{\mathrm{B}}^{0}$ is the sum of the separated atom/monomer energies. The essence of the XC semi-empirical model is contained in two approximations. The first approximation is to express $E_{\mathrm{X}}^{(1)}(R)$ in terms of the first-order Coulomb energy $E_{\mathrm{C}}^{(1)}(R)$ via (Aziz *et al.*, 1992; Meath and Koulis, 1991; Ng *et al.*, 1978; Ng *et al.*, 1979)

$$E_{\mathrm{X}}^{(1)}(R) = -\gamma(1 + aR)E_{\mathrm{C}}^{(1)}(R), \tag{3.23b}$$

in which γ, a are adjustable parameters. The second approximation is to drop terms of order three or higher in the perturbation expansion of the Coulomb energy, and to represent the second-order perturbation term, $E_{\mathrm{C}}^{(2)}(R)$, as a series of damped reciprocal powers of the separation R, referred to as the damped dispersion series. The parameter a in eqn (3.23b) is often set to 0.1, in which case $E_{\mathrm{X}}^{(1)}(R)$, and thus $V(R)$, may be determined in principle by fitting the single remaining parameter γ either to a fixed "exact" value of $V(R)$ or to an experimental property determined by $V(R)$.

Once faster computers with large active storage capacities were available, it was possible to develop extensive high-level sets of basis functions and to use them to carry out very accurate first-order Heitler–London and Coulomb energy calculations for the interactions between many-electron atoms, including Ar and Kr, thereby enabling a test of the accuracy of the fundamental approximation eqn (3.23b) to be made for van der Waals dimers other than $H_2(a^3\Sigma_u^+)$. In particular, calculations of $E_{\mathrm{X}}^{(1)}$ and $E_{\mathrm{C}}^{(1)}$ carried out (Wheatley and Meath, 1993*b*) for the He–He, He–Ne, He–Ar, Ne_2, Ne–Ar, Ar_2, and Kr_2 interactions established that the basic linear relation between $E_{\mathrm{X}}^{(1)}$ and $E_{\mathrm{C}}^{(1)}$ is accurate to better than 1% for the He–He, He–Ar, Ne–Ar, Ar_2, and Kr_2 interactions, and to within 0.2% for the He–He and He–Ar interactions when both γ and a are allowed to vary. For a fixed at the value 0.1, the quality of the approximation was slightly compromised, with deviations from linearity generally exceeding 2%. Note, however, that even when both γ and a are free parameters, deviations still occur at sufficiently short range (for the He–He interaction (Komasa and Thakkar, 1995), for example, for $R < 2$ Å) and long range (typically for the interactions mentioned earlier, for $R \gtrsim 3.5$ Å). To distinguish between these one and two free-parameter possibilities, the XC models so constructed are labeled as XC-1 and XC-2, respectively. Significantly larger deviations obtained for interactions involving the Ne atom, especially the Ne_2 interaction, were attributed to the poor quality of the Ne basis-set (Wheatley and Meath, 1993*b*).

For the initial versions of the XC model (Aziz *et al.*, 1992; Meath and Koulis, 1991; Ng *et al.*, 1978; Ng *et al.*, 1979), the second-order damped dispersion series was represented by the overall-damped expression

$$E_{\mathrm{C}}^{(2)}(R) \simeq -g(R)\left[\frac{C_6}{R^6}+\frac{C_8}{R^8}+\frac{C_{10}}{R^{10}}\right],$$

in which the damping function $g(R)$ was a modified version of the HFD-A model damping function, namely

$$g(R)=\begin{cases} \mathrm{e}^{-0.4(1.28R_{\mathrm{e}}R^{-1}-1)^2}, & R<1.28R_{\mathrm{e}} \\ 1, & R\geq 1.28R_{\mathrm{e}} \end{cases}.$$

Several variants of the XC model exist: in the main, they differ with respect to the manner in which the Coulomb energy beyond $E_{\mathrm{C}}^{(1)}(R)$ is represented.

The most general version of the XC model can be represented by the sum of a repulsive exchange energy term, $V_{\mathrm{rep}}(R)$, having the form of eqn (3.23b), and an attractive term, $V_{\mathrm{att}}(R)$, given by the sum of the first-order Coulomb energy and the product of the damped dispersion energy, $V_{\mathrm{disp}}(R)$, and an overall damping function, $g_N(R)$, that depends upon the highest inverse power, $2N$, included in the approximation to $V_{\mathrm{disp}}(R)$. Thus, the XC model function may be written as

$$\begin{aligned} V_{\mathrm{XC}}(R) &= V_{\mathrm{rep}}(R)+V_{\mathrm{att}}(R) \\ &= -\gamma(1+aR)E_{\mathrm{C}}^{(1)}(R)+E_{\mathrm{C}}^{(1)}(R)-g_N(R)\sum_{n=3}^{N} f_{2n}(R)\,\frac{C_{2n}}{R^{2n}}, \end{aligned} \tag{3.24}$$

in which the overall damping function $g_N(R)$ accounts for those components of the Coulomb energy that have been neglected in the determination of the approximation for $V_{\mathrm{disp}}(R)$, such as (Aziz *et al.*, 1992) nonexpanded dispersion terms with $n>N$, second-order "spherical" induction and dispersion energies, and third- and higher-order Coulomb energies.

The only prototypical van der Waals molecule for which essentially exact calculations had been achieved (Kolos and Wolniewicz, 1974) at the time of development of the XC model (similarly, HFD-type models) was the $\mathrm{H_2}(a^3\Sigma_u^+)$ interaction. The XC model could therefore be tested by fitting the free parameter γ in the model to give agreement with a selected "exact" value (Kolos and Wolniewicz, 1974) for $V(R)$. The value $\gamma=4.971$ obtained (Ng *et al.*, 1978) by fitting the XC model expression for $V(R)$ to the "exact" computed interaction energy (Kolos and Wolniewicz, 1974) for $R=3.7042$ Å ($7a_0$) agreed to better than 4% with the full set of "exact" energies for $R\geq 1.3229$ Å ($2.5a_0$), and agreed to better than 0.4% with the "exact" energies for values of R corresponding to the attractive region of $V(R)$ (i.e. $R>\sigma\approx 3.65$ Å).

The availablity of new extensive basis-sets coupled with improved and expanded computational facilities also made it possible to examine the nature of the damping functions for atomic interactions other than the H(1s)–H(1s) interaction.

Thakkar (2001) showed that the dispersion energy between a pair of S-term atoms A, B can be expressed either in what is called the nonexpanded form,

$$E_{\text{disp}}(\text{AB}; R) = \sum_{l_{\text{A}}=1} \sum_{l_{\text{B}}=1} E^{\text{disp}}_{l_{\text{A}},l_{\text{B}}}(\text{AB}; R)\,, \tag{3.25a}$$

obtained when interatomic electron exchange, but not charge overlap, effects are neglected, or more commonly, in the long-range asymptotic limit, as

$$E_{\text{disp}}(\text{AB}; R) \sim -\sum_{l_{\text{A}}=1} \sum_{l_{\text{B}}=1} C^{\text{AB}}_{l_{\text{A}},l_{\text{B}}} R^{-2(l_{\text{A}}+l_{\text{B}}+1)}\,. \tag{3.25b}$$

The dispersion coefficients $C^{\text{AB}}_{l_{\text{A}},l_{\text{B}}}$ appearing in this particular asymptotic form are associated with individual ($2^{l_{\text{A}}}$-pole)–($2^{l_{\text{B}}}$-pole) partial wave interaction terms. The individual $E^{\text{disp}}_{l_{\text{A}},l_{\text{B}}}(\text{AB}; R)$ terms in expression (3.25a) can be related to their long-range asymptotic limits in eqn (3.25b) through the introduction of individual damping functions defined by the relation

$$f_{l_{\text{A}},l_{\text{B}}}(\text{AB}; R) = -E^{\text{disp}}_{l_{\text{A}},l_{\text{B}}}(\text{AB}; R)\,\frac{R^{2(l_{\text{A}}+l_{\text{B}}+1)}}{C^{\text{AB}}_{l_{\text{A}},l_{\text{B}}}}\,. \tag{3.26a}$$

Upon collecting together all terms leading to the same inverse powers of R in eqn (3.25b), the nonexpanded dispersion energy may be approximated by the truncated expression

$$E_{\text{disp}}(\text{AB}; R) = -\sum_{n\geq 3}^{5} C^{\text{AB}}_{2n} f_{2n}(\text{AB}; R) R^{-2n}\,, \tag{3.26b}$$

in which the dispersion coefficients C^{AB}_{2n} are given by

$$C^{\text{AB}}_{2n} = \sum_{l_{\text{A}}=1} \sum_{l_{\text{B}}=1} C^{\text{AB}}_{l_{\text{A}},l_{\text{B}}}\,, \qquad l_{\text{A}} + l_{\text{B}} + 1 = n\,, \tag{3.26c}$$

and the individual multipole–multipole damping functions are given as

$$f_{2n}(\text{AB}; R) = \frac{1}{C^{\text{AB}}_{2n}} \sum_{l_{\text{A}}=1} \sum_{l_{\text{B}}=1} C^{\text{AB}}_{l_{\text{A}},l_{\text{B}}} f_{l_{\text{A}},l_{\text{B}}}(\text{AB}; R)\,, \quad l_{\text{A}} + l_{\text{B}} + 1 = n\,. \tag{3.26d}$$

From the restriction $l_{\text{A}}+l_{\text{B}}+1 = n$, together with eqn (3.26c) the C_6, C_8, and C_{10} long-range dispersion coefficients are given in terms of the multipolar dispersion coefficients $C^{\text{AB}}_{l_{\text{A}},l_{\text{B}}}(R)$ by $C^{\text{AB}}_6 \equiv C^{\text{AB}}_{1,1}$, $C^{\text{AB}}_8 = C^{\text{AB}}_{2,1} + C^{\text{AB}}_{1,2}$, and $C^{\text{AB}}_{10} = C^{\text{AB}}_{3,1} + C^{\text{AB}}_{1,3} + C^{\text{AB}}_{2,2}$.

A rather thorough study of the dispersion damping functions for the H–He, H–Li, He–He, He–Li, $Li_2(a^3\Sigma^+_u)$, and Ar_2 interactions has been carried out by Meath and coworkers (Knowles and Meath, 1986; Wheatley and Meath, 1993*a*). To maintain internal consistency, they also computed values for the C_6, C_8, and C_{10} long-range dispersion coefficients for each interaction, obtaining results that were in excellent agreement with previous high-level calculations for all six interactions. By combining the C_{2n} values with their nonexpanded dispersion energy results, obtained at a series of separations R lying between 1.25 Å and 10.5 Å in the same manner as had been employed previously (Kreek and Meath, 1969) for $H_2(a^3\Sigma_u^+)$, they were able to back out not only the $n = 3, 4, 5$ damping functions $f_{2n}(R)$, but also the corresponding dipole–quadrupole, $f_{1,2}$, and quadrupole–dipole, $f_{2,1}$, damping functions contributing to f_8, and the dipole–octupole, $f_{1,3}$, octupole–dipole, $f_{3,1}$, and quadrupole–quadrupole, $f_{2,2}$, damping functions contributing to f_{10}. As a check on their new calculational procedure, the $H_2(a^3\Sigma_u^+)$ dispersion damping functions $f_6(R)$, $f_8(R)$, and $f_{10}(R)$ were recomputed for a denser grid of R-values (specifically, $1.5\,\text{Å} < R < 6\,\text{Å}$), while the $f_{1,3}(R) = f_{3,1}(R)$ and $f_{2,2}(R)$ multipole (or partial-wave) component damping functions that make up $f_{10}(R)$ were reported separately. Precise agreement was obtained with the earlier results of Koide *et al.* (1981) for values of the separation R that were common to the two studies. Their computed results for the dispersion damping functions $f_6(R)$, $f_8(R)$, and $f_{10}(R)$ plus multipolar contributions to the homo-atomic He–He ($1.25\,\text{Å} < R < 4.5\,\text{Å}$), $Li_2(a^3\Sigma_u^+)$ ($2.5\,\text{Å} < R < 10.5\,\text{Å}$), and Ar_2 ($2.0\,\text{Å} < R < 5\,\text{Å}$) interactions, and for $f_6(R)$, $f_8(R)$, and $f_{10}(R)$, together with the corresponding multipolar contributions for the hetero-atomic H–He ($1.5\,\text{Å} < R < 5.3\,\text{Å}$), H–Li and He–Li ($2\,\text{Å} < R < 8.5\,\text{Å}$) interactions provide high-quality data for use in the testing of both proposed forms for the damping functions for use in more general atom–atom interactions and the scaling factors that had been proposed for establishing the proper ranges of R over which the damping functions apply.

The $f_{2n}(\text{AB}; R)$ damping functions obtained from the direct calculations for the five A–B interactions all behave as functions of R for given values of n qualitatively as expected on the basis of the forms for the already-known damping functions for $H_2(a^3\Sigma_u^+)$ (Koide *et al.*, 1981; Kreek and Meath, 1969), namely, the $f_{2n}(\text{AB}; R)$ tend to 0 as $R^{2n+\delta}$, increase monotonically with R, and tend to 1 as R becomes infinite. The approach of the f_{2n} damping functions to their limiting values 0 and 1 does depend upon n, with f_{2n} approaching 0 more rapidly and 1 more slowly with increasing n, in agreement with the more rapid divergence for a fixed value R of the individual terms of the multipolar expansion (3.5a) from the corresponding nonexpanded dispersion energies of eqn (3.25a) as n increases.

For homo-atomic interactions, the $f_{l_A,l_B}(\text{AB}; R)$ multipolar damping functions corresponding to fixed values of n are seen to be independent of l_A and l_B, as may be expected from an asymptotic series in inverse powers of R for which the charge overlap effects that determine the damping functions will be much the same for all dispersion contributions to a given overall (Knowles and Meath, 1986) inverse power $2n$ at any separation R. Deviations from this behavior found in calculated results obtained for homo-atomic interactions may therefore be attributed to the employment of insufficiently complete basis-sets (CBS). This argument does not hold, however, for smaller separations between hetero-atoms, as the effects of charge overlap between two unlike chemical species can be quite different for the various multipole interactions contributing to a specific R^{-2n} overall dispersion coefficient. Of course, as the separation R increases, such differences will decrease, and for R sufficiently large, all differences will become negligible. Differences between the multipolar contributions to $f_{2n}(\text{AB}; R)$ for small and intermediate separations are reflected in the structure of eqn (3.4a–4d) defining the overall damping functions. The overall damping function $f_{2n}(\text{AB}; R)$ for the R^{-2n} dispersion energy most strongly resembles the multipolar damping function associated with the dominant partial-wave contribution which, for the Li–He interaction for example, is the ($l_{\mathrm{He}} = 1, l_{\mathrm{Li}} = 3$) dipole–octupole multipole interaction (Knowles and Meath, 1986; Wheatley and Meath, 1993*a*).

The procedure used to obtain dispersion damping functions typically employed in assembling HFD and XC model PEFs requires further consideration. As $\mathrm{H}_2(a^3\Sigma_u^+)$ was the only van der Waals dimer for which both "exact" interaction energies (Kolos and Wolniewicz, 1974) and precisely determined damping functions (Koide *et al.*, 1981; Kreek and Meath, 1969) were available, all damping functions for other atomic interaction energies were necessarily predicated upon generalizations based upon the H(1*s*)–H(1*s*) damping function forms. While it might be thought that these damping functions should provide excellent precursors for the damping functions for alkali dimer interactions, it is also possible that they would not necessarily provide the optimal precursor damping functions for interactions between closed electronic shell species, such as noble gas atoms, given that the H–H interaction lacks contributions from intra-atomic electron correlation effects. While it was not possible to employ damping functions based upon those for the more natural He–He prototype interaction at the time that these models were initially developed, the introduction of highly accurate software program suites, together with vastly expanded hardware capabilities, has made it both feasible and practical to compute accurate values for many binary interaction energies, especially those between closed-shell multielectron ground-state atoms. These developments have also enabled the computation of accurate first-order Coulomb and exchange energies for atom–atom interactions (Knowles and Meath, 1986; Wheatley and Meath, 1993*b*; Wheatley and Meath, 1993*a*), thereby also making it possible to test the usefulness of the HFD and XC models.

An especially attractive feature in the construction of HFD- and XC-like models is the estimation of the damping functions $f_{2n}(\mathrm{AB}; R)$ for the A–B interaction via a scaling of the separation R in the $f_{2n}(\mathrm{HH}; R)$ damping functions for $\mathrm{H}_2(a^3\Sigma_u^+)$. This scaling is achieved by a recipe of the type

$$f_{2n}(\mathrm{AB}; R) = f_{2n}(\mathrm{HH}; \rho_{\mathrm{AB}} R)\,, \qquad \rho_{\mathrm{AB}} \equiv X_{\mathrm{HH}}/X_{\mathrm{AB}}\,, \tag{3.27}$$

in which the scaling parameter ρ_{AB} is expressed in terms of physically intuitive size-related arguments X_{AB}. Four types of scaling have been proposed for the A–B interaction:

- $X_{\mathrm{AB}} = R_{\mathrm{e}}^{\mathrm{AB}}$, the position of the minimum in $V(R_{\mathrm{AB}})$,
- $X_{\mathrm{AA}} = [I_{\mathrm{p}}(\mathrm{A})]^{-\frac{1}{2}}$ for homo-atomic interactions, taken together with the combination rule $X_{\mathrm{AB}} = 2X_{\mathrm{AA}}X_{\mathrm{BB}}/[X_{\mathrm{AA}} + X_{\mathrm{BB}}]$ for the A–B interaction (Douketis and Scoles, 1982),
- $X_{\mathrm{AB}} = \sqrt{C_8^{\mathrm{AB}}/C_6^{\mathrm{AB}}}$ of asymptotic dispersion coefficients (Aziz *et al.*, 1992; Fuchs *et al.*, 1984), and
- $X_{\mathrm{AB}} = b_{\mathrm{AB}}^{-1}$ (Tang and Toennies, 1984, 1986), in which b_{AB} is the fitted Born–Mayer exponential repulsion parameter.

Wheatley and Meath (1993*a*) examined the possible dependence of ρ_{AB} upon n and R in addition to the chemical species A and B involved in the interaction: they found in general, based upon their *ab initio* calculations of damping functions for eight different binary interactions, that the scaling parameters needed to satisfy eqn (3.4–5) were, apart from relatively small deviations of order 3% for six of the eight atom pairs studied and up to 10% for the remaining two, essentially independent of both n and R. They found also that scaling based upon the ratio of $[C_8^{\mathrm{AB}}/C_6^{\mathrm{AB}}]^{\frac{1}{2}}$ to $[C_8^{\mathrm{HH}}/C_6^{\mathrm{HH}}]^{\frac{1}{2}}$ for $\mathrm{H}_2(a^3\Sigma_u^+)$ yields the best overall results, providing excellent agreement especially for the He–He and $\mathrm{Li}_2(a^3\Sigma_u^+)$ interactions: the overall success of this scaling method can be related to the reflection of the correction of the divergent nature of the long-range expansion by the damping functions as R decreases. They proposed an improved, though nonetheless semi-empirical, combining rule

$$X_{\mathrm{AB}} = \frac{\sqrt{C^{\mathrm{AB}}(1,2)} + \sqrt{C^{\mathrm{AB}}(2,1)}}{\sqrt{2C^{\mathrm{AB}}(1,1)}}\,, \tag{3.28}$$

for obtaining the optimal characteristic scaling variable for hetero-atomic interactions. Note that X_{AB} reduces to $\sqrt{C_8^{\mathrm{AA}}/C_6^{\mathrm{AA}}}$ for the A $\equiv$ B homo-atomic interaction. This semi-empirical combining rule resulted in excellent agreement between their scaled H–H damping functions and their *ab initio* results for the hetero-atomic H–He, H–Li, He–Li, and He–Ne interactions.

XC model PEFs of high quality have been obtained for all six homo-atomic noble gas interactions (Aziz *et al.*, 1992; Meath and Koulis, 1991; Ng *et al.*, 1978; Ng *et al.*, 1979), while the exchange and Coulomb interaction energies for the hetero-atomic species (Wheatley and Meath, 1993*b*) He–Ne, He–Ar, and He–Ar, and dispersion energy damping functions have been determined (Wheatley and Meath, 1993*a*) for the six binary interactions involving the H, He, and Li atoms.

Note that PEFs determined from the HFD, TT, and XC models can be extremely useful for the determination of van der Waals interaction energies between pairs of atoms (or, in principle, molecules) with sufficient accuracy to allow accurate calculations of bulk gas properties, such as virial coefficients and transport properties, for which the experimental uncertainties are normally not better than $\pm 0.1\%$. These models were not intended, however, to provide PEFs capable of producing bound-state energy differences with a precision of the order of $0.001\ \mathrm{cm}^{-1}$ or better, which is typical of the precision that is needed for the computation of spectroscopic transition frequencies having accuracies sufficient for comparison with modern spectroscopic measurements.

3.5 Modern Empirical Multiproperty-Fit Potential Energy Functions

Diatomic PEFs have long been of interest to the molecular spectroscopy community, and numerous model representations of the interatomic interactions between ground and excited terms of diatomic molecular species exist. As most formulations of the functional form were concerned primarily with being able to reproduce accurate experimental data, they were not overly concerned with the behavior of the PEF outside the experimental data range. This was especially true for separations for which the interaction energy is positive, and hence, corresponding to unbound (continuum) states.

An example of a potential energy model that both relates to the models discussed in Section 3.4 and provides the flexibility necessary to produce the best possible agreement with all available spectroscopic data is the MLR potential form developed by Le Roy and coworkers (Kaplan, 2003; Le Roy and Henderson, 2007; Le Roy *et al.*, 2011; Le Roy, 2010). The particular empirical form for the PEF was originally developed for use with direct potential fit (DPF) methods (Kaplan, 2003; Le Roy, 2010) of spectroscopic data analysis. This empirical model incorporates the correct limiting long-range behavior and is generally well behaved for both large and small atomic separations lying outside the range spanned by the available spectroscopic experimental data. Moreover, inclusion of non-Born–Oppenheimer correction terms in the MLR fitting function allows for (small) atomic mass effects that differentiate molecular isotopologues.

The MLR empirical PEF was developed (Le Roy and Henderson, 2007; Le Roy *et al.*, 2011) to provide a flexible, yet compact, minimally parameterized functional form that is primarily characterized in terms of the physically important parameters $\mathfrak{D}_{\mathrm{e}}$ and R_{e} that represent the depth and equilibrium separation, respectively, of the binary interaction energy. The MLR PEF has the form (Le Roy and Henderson, 2007; Le Roy *et al.*, 2011)

$$V_{\mathrm{MLR}}(R) = \mathfrak{D}_{\mathrm{e}} \left[1 - \frac{u_{\mathrm{LR}}(R)}{u_{\mathrm{LR}}(R_{\mathrm{e}})} \, \mathrm{e}^{-\beta(R) y_p^{\mathrm{eq}}(R)} \right]^2 , \tag{3.29}$$

with R the internuclear separation of the interacting atom pair, $\mathfrak{D}_{\mathrm{e}}$ the equilibrium well-depth, R_{e} the equilibrium internuclear separation, and $u_{\mathrm{LR}}(R)$ the long-range tail function, which is related to $V(R)$ by

$$V(R) \equiv \mathfrak{D}_{\mathrm{e}} - u_{\mathrm{LR}}(R) + \cdots \tag{3.30}$$

with the ellipsis representing the nonconstant nonlong-range components of $V(R)$. The long-range interaction energy, $u_{\mathrm{LR}}(R)$, is typically defined as a sum of simple inverse-power terms, namely,

$$u_{\mathrm{LR}}(R) = \sum_{i=1}^{\mathrm{last}} \frac{C_{m_i}}{R^{m_i}} , \tag{3.31}$$

and the radial variable $y_p^{\mathrm{eq}}(R)$ appearing in the exponent of $V(R)$ is defined as

$$y_p^{\mathrm{eq}}(R) \equiv \frac{R^p - R_{\mathrm{e}}^p}{R^p + R_{\mathrm{e}}^p} , \tag{3.32}$$

while $\beta(R)$ is given by

$$\beta(R) \equiv y_p^{\mathrm{ref}}(R)\beta_\infty + \left[1 - y_p^{\mathrm{ref}}(R)\right] \sum_{i=0}^{N} \beta_i \left[y_q^{\mathrm{ref}}(R)\right]^i , \tag{3.33}$$

in terms of the radial variables

$$y_p^{\text{ref}}(R) \equiv \frac{R^p - R_{\text{ref}}^p}{R^p + R_{\text{ref}}^p} \quad \text{and} \quad y_q^{\text{ref}}(R) \equiv \frac{R^q - R_{\text{ref}}^q}{R^q + R_{\text{ref}}^q} . \tag{3.34}$$

Although the structures of these two variables are similar to that of $y_p^{\text{eq}}(R)$, they are defined relative to a different expansion center, R_{ref}, and employ in general (Dattani and Le Roy, 2011; Le Roy *et al.*, 2009; Le Roy *et al.*, 2011) different powers p and q. Early versions of the MLR model form for the PEF for a pair of interacting atoms employed only the radial variables $y_p^{\text{eq}}(R)$ of eqn (3.4) with p set to 1: this early version (Le Roy and Henderson, 2007) of the MLR PEF led to $\beta(R)$ becoming strongly negative in the repulsive region ($R \ll R_{\text{e}}$) of the PEF, so that $V(R)$ passed through a maximum in this region, and ultimately became strongly negative, to cause a turnover of the PEF.

Le Roy and colleagues 2011 argued that the MLR model PEF of eqn (3.29) provides perhaps the most useful representation of a single-minimum PEF, as it is flexible, compact, contains physically interesting parameters (such as the depth, $\mathfrak{D}_{\text{e}}$, and location, R_{e}, of the equilibrium minimum), provides physically realistic extrapolations to both very large and very small separations, and incorporates the physically correct inverse-power long-range behavior. In addition, it is an analytic function, so that it is continuous everywhere and is infinitely differentiable.

The long-range interaction between a pair of atoms is typically defined as the sum of simple inverse powers, namely eqn (3.31). By choosing the long-range interaction to have the form of eqn (3.31), the exponent coefficient $\beta(R)$ is required to assume the long-range limiting form eqn (3.35). This limiting form for $\beta(R)$ then requires (Dattani and Le Roy, 2011; Le Roy *et al.*, 2009) that the power p that appears in eqn (3.32) must exceed $m_{\text{last}} - m_1$ in order to maintain the long-range nature of eqn (3.31). In particular, this condition requires that the value p in $y_p^{\text{ref}}(R)$ must be greater than $m_{\text{last}} - m_1$ in order that the correct long-range behavior of $u_{\text{LR}}(R)$ be maintained. There is, however, no similar requirement imposed upon the value of q in $y_q^{\text{ref}}(R)$, although experience (Dattani and Le Roy, 2011; Le Roy and Henderson, 2007; Le Roy *et al.*, 2009; Le Roy *et al.*, 2011) suggests that its value should lie in the range $3 \lesssim q \lesssim p$.

A number of problems associated with early versions (Le Roy and Henderson, 2007; Le Roy *et al.*, 2009) of the MLR form were eliminated via the introduction of the two-center expansion of eqn (3.11) and the use of damping functions (Dattani and Le Roy, 2011; Le Roy *et al.*, 2011). However, because the original damping function forms introduced by Douketis *et al.* (1982) and by Tang and Toennies (1984) possess unphysical short-range behavior that prevents them from being employed within the MLR format, two new families of damping functions were introduced (Le Roy *et al.*, 2011), specifically a generalized Douketis–Scoles damping function given by

$$D_m^{\text{DS}(s)}(r) = \left[1 - \exp\left(-\frac{b^{\text{DS}}(s)(\rho R)}{m} - \frac{c^{\text{DS}}(s)(\rho R)^2}{\sqrt{m}}\right)\right]^{m+s} , \tag{3.35}$$

and a generalized Tang–Toennies (TT) damping function given by

$$D_m^{\mathrm{TT}(s)}(r) = 1 - \exp\left[-b^{\mathrm{TT}}(s)(\rho R)\right] \sum_{k=0}^{m-1+s} \frac{\left[b^{\mathrm{TT}}(s)(\rho R)\right]^k}{k!} . \tag{3.36}$$

The b and c parameters appearing in eqns (3.5–3.9) are system independent, and are obtained by optimizing agreement with the *ab initio* $m = 6, 8, 10$ damping function behavior of a pair of ground-state H atoms (for which $p = 1$) determined by Kreek and Meath (1969). Damping functions were originally intended to account for the weakening of the inverse-power terms that are associated with the net dispersion energy when the overlap between "electron clouds" on the interacting atoms increases as the internuclear separation decreases. Because the sizes of these electron clouds differs from atom to atom, and the distance at which the overlap becomes important will vary, Douketis *et al.* (1982) introduced a system-dependent range parameter ρ_{AB} having the form

$$\rho_{AB} = \frac{2\rho_A \rho_B}{\rho_A + \rho_B} , \tag{3.37}$$

with ρ_A (and similarly ρ_B) defined in terms of the ratio of the ionization potential I_p for atom A relative to I_p^H for a ground-state hydrogen atom: thus ρ_A, for example, is given by

$$\rho_A = \left(\frac{I_\mathrm{p}^A}{I_\mathrm{p}^\mathrm{H}}\right)^{\frac{2}{3}} . \tag{3.38}$$

If expression (3.29) is expanded as

$$V_{\mathrm{MLR}}(R) = \mathfrak{D}_\mathrm{e} \left[1 + \left(\frac{u_{\mathrm{LR}}(R)}{u_{\mathrm{LR}}(R_\mathrm{e})}\right)^2 \exp\left[-2\beta(R) y_p^{\mathrm{eq}}(R)\right] -2\frac{u_{\mathrm{LR}}(R)}{u_{\mathrm{LR}}(R_\mathrm{e})} \exp\left[-\beta(R) y_p^{\mathrm{eq}}(R)\right]\right] ,$$

the dependence of $V_{\mathrm{MLR}}(R)$ on the internuclear separation R can be seen to be determined by the sum of a repulsive term and an attractive term. Moreover, as all three expansion variables $y_p^{\mathrm{eq}}(R)$, $y_p^{\mathrm{ref}}(R)$, and $y_q^{\mathrm{ref}}(R)$ approach the value -1 as R approaches 0, the exponential factor in the repulsive term approaches a value of $e^{2\beta(0)}$ and has zero limiting slope, so that the limiting short-R behavior of the repulsive term in $V_{\mathrm{MLR}}(R)$ is governed by the factor $[u_{\mathrm{LR}}(R)]^2$. At short internuclear separation, this factor approaches the square of the highest-order inverse-power term in eqn (3.35), namely,

$$V(R) \propto R^{-2m_{\mathrm{last}}} . \tag{3.39}$$

When the long-range potential is represented by the leading dispersion energy terms with $n = 6, 8, 10$, this means that the PEF would have a singularity of order 20 for $R = 0$. This behavior is, of course, nonphysical, as the limiting short-range behavior of any interatomic PEF will be governed by the interatomic repulsion term $Z_A Z_B e^2/(4\pi\varepsilon_0 R)$ of the united atom, which has a singularity of order 1. Note, however, that the incipient high-order singularity of a MLR PEF does not really present a problem in the well-region that generates much of the spectroscopic data employed for the fitting procedure.

Kreek and Meath (1969) first examined the nature of the expanded and non expanded interaction energies of a pair of ground-state H atoms, thereby giving rise to the concept of damping functions, which are defined as the ratio of the nonexpanded to expanded interaction energies. This means that the correct long-range (attractive) interaction energy is then given by eqn (3.40),

$$u_{\mathrm{LR}}(R) = \sum_{i=1}^{\mathrm{last}} D_{m_i}(R)\frac{C_{m_i}}{R^{m_i}}. \tag{3.40}$$

Because calculations of nonexpanded interaction energies have only been reported for a small number of few-electron systems, it is customary to employ the ground-state H atom interaction energies as a representation of a universal behavior that may be scaled radially for other atom–atom systems. The damping behavior for all terms arising in eqn (3.40) may then be mapped onto the behavior of ground-state H atoms as determined by Kreek and Meath.

The damping function (3.36) employed by Le Roy *et al.* (2011) to obtain their MLR model PEF for the ArXe($X^1\Sigma_u^+$) dimer provides a physically reasonable parameterization of the interaction energy that is also convenient, as accurate values of the ionization potentials of Ar and Xe are readily available (as they are for all atomic species). The original versions (Le Roy and Henderson, 2007) of these damping functions correspond to $s = 0$ ($\overline{dd} = 0.008$) for eqn (3.36) and to $s = 1$ ($\overline{dd} = 0.018$) for eqn (3.37). A later examination (Le Roy *et al.*, 2011) of the quality of fits provided by these damping functions for $-2 \lesssim s \leq +2$ found, in general, that the damping function (3.36) provided significantly better fits ($0.0064 \lesssim \overline{dd} \leq 0.0094$) than did the damping functions (3.37), for which $0.018 \lesssim \overline{dd} \leq 0.086$, except for $s = 2$, for which $\overline{dd} = 0.0096$.

Le Roy *et al.* (2011) also examined the quality of fits to the *ab initio* H-atom data (Koide *et al.*, 1981; Kreek and Meath, 1969) provided by damping functions for s in the range $-2 \lesssim s \leq 2$ and found that, in general, the damping function forms proposed by Douketis *et al.* (1982) provided significantly better fits ($0.0064 \leq \overline{dd} \leq 0.0094$) to the (original spectroscopic) data than did the damping functions proposed by Tang and Toennies (1984), with the former damping function providing root-mean-square deviation (RMSD) values $\overline{dd}$ that are typically about an order of magnitude smaller than the $\overline{dd}$ values obtained using eqn (3.13). Note also that within the MLR model for the PEF, any damping function form for which $s > 0$ is physically unacceptable, as $V_{\mathrm{MLR}}(R)$ behaves as R^{2s} for very small values of R.

A thorough test (Le Roy *et al.*, 2011) of the employment of the MLR model PEF as a tool for the inversion of spectroscopic data sets was provided by a reanalysis of the sets of spectroscopic line frequencies for two distinctly different diatomic species MgH($X^1\Sigma^+$) and Li_2($X^1\Sigma_g^+$) based, respectively, upon 7453 (Le Roy *et al.*, 2011) and 17,477 (Le Roy *et al.*, 2009) spectroscopic transition frequencies. In addition, the very weakly bound ArXe($X^1\Sigma^+$) dimer PEF obtained originally (Dattani and Le Roy, 2011) from a mix of microwave data for the $v = 0$ levels of five zero-spin isotopologues, VUV emission lines into the $v = 0, 1$ levels of $Ar^{132}Xe$ and $Ar^{136}Xe$, plus interaction virial coefficient data was reanalyzed, as the essentially physical bonding in rare gas dimers is significantly different from that in chemically bonded dimers. This study clearly illustrated the versatility of the MLR PEF with damping functions. A total of three different fits were carried out for the MgH dimer: a basic 18-parameter fit ($\overline{dd} = 0.755$) using parameter values determined by Shayesteh at al. (Le Roy and Henderson, 2007), a 14-parameter fit without damping, but including two additional Born–Oppenheimer breakdown (BOB) corrections that had been overlooked in the original fit (giving $\overline{dd} = 0.076$), and a 12-parameter fit with damping ($\overline{dd} = 0.757$). A 16-parameter fit (Le Roy *et al.*, 2011) using the same set of parameters as those employed previously (Le Roy and Henderson, 2007) but with the addition of damping, was carried out for the Li_2 dimer. Three sets of damping functions with $s = 0, -(1/2), -1$ and, for comparison, two sets of damping functions with $s = 0, -1$ were employed for the ground-state ArXe dimer.

Were $u_{LR}(R)$ to behave as a simple sum of inverse powers as in eqn (3.31), then for very short separations R, $V_{MLR}(R)$ would behave as $R^{-2M_{last}}$. Thus, in the most common case, in which the leading terms have $m_i = 6,\ 8,\ 10$, $V_{MLR}(R)$ would then give rise to a positive singularity of order 20 at the origin. Such a singularity is clearly unphysical, as the leading short-range behavior for any realistic interatomic PEF has a positive singularity of order 1 at the origin associated with the physically required Coulombic internuclear repulsion term, $Z_A Z_B e^2/(4\pi\varepsilon_0 R)$, in which Z_A, Z_B are the atomic numbers for atoms A, B, and ε_0 is the vacuum permittivity. Of course, a higher-order singularity does not necessarily affect the determination of the fitting parameters in the attractive well-region of $V_{MLR}(R)$, as the behavior of the component coefficient function $\beta(R)$ will ensure proper behavior of $V_{MLR}(R)$ in the well-region.

The unphysically steep small-separation behavior of $V_{MLR}(R)$ may be alleviated by employing damping functions very much like those employed in the HFD, TT, and XC models discussed previously. Two generalized damping functions, given by eqns (3.35, 3.36), have been proposed (Le Roy *et al.*, 2011), corresponding respectively, to the HFD and TT models with $s = 0$ and $s = 1$. Note that the parameters $b^{DS}(s)$, $c^{DS}(s)$ of the Douketis–Scoles damping function and the $b^{TT}(s)$ of the Tang–Toennies damping function are system-independent functions that are optimized with the $m_i = 6, 8, 10$ *ab initio* damping functions (Koide *et al.*, 1981; Kreek and Meath, 1969; Wheatley and Meath, 1993*a*) for $H_2(a^3\Sigma_u^+)$. The introduction of damping functions is automatically accommodated in the MLR potential model (Le Roy *et al.*, 2011) by changes in $\beta(R)$.

The level of agreement between computed and experimentally determined values of a physical attribute, such as a set of emission or absorption line frequencies or the temperature dependence of a transport coefficient is often stated in terms of a statistical measure referred to as the dimensionless root-mean-square deviation (DRMSD), denoted here by the symbol $\overline{dd}$, and defined as

$$\overline{dd} \equiv \sqrt{\frac{1}{N}\sum_{i=1}^{N}\left(\frac{y_i^{calc} - y_i^{obs}}{u_i}\right)^2}. \tag{3.41}$$

In expression (3.41), the calculated and experimental values, y^{calc} and y^{obs} are, respectively, the model-predicted and experimentally observed values of the datum y_i, N is the number of observed data values, and u_i is the experimental uncertainty associated with y_i^{obs}. By definition, a $\overline{dd}$ value 0 means that the fitting function provides a perfect fit to the set of data being fitted, while a value 1 means that the results obtained from the fitting function all fall just within the given uncertainties of the input data being fitted, while $\overline{dd}$ values greater than 1 indicate that the fit to the input data is of a poorer quality. Typically, the greater the $\overline{dd}$ value, the poorer the quality of the fit. Similarly, the smaller the $\overline{dd}$ value, the better the quality of the fit.

It is worth illustrating the role of the parameter s that appears in the two damping functions in terms of the quality of fit provided by fitting the *ab initio* damping functions for the $H_2(a^3\Sigma_u^+)$ state of molecular hydrogen for the $-2 \leq s \leq +2$ members of the generalized HFD and TT model damping function families given in eqns (3.35, 3.36). For convenience, a uniform uncertainty u_i = 1will be employed for this comparison (giving RMSD values rather than $\overline{dd}$ values). See Table 3.1 for the results obtained in this fashion.

That the *ab initio* results for the unexpanded dispersion energy for $H_2(a^3\Sigma_u^+)$ increase in magnitude monotonically and approach finite limits as R approaches 0 suggests that the physically most appropriate damping function models may be those having $s = 0$. Note that as the repulsive component of $V_{MLR}(R)$ now behaves as R^{2s} for R very small, any damping function that corresponds to a positive value of s will automatically lead to a turnover in the repulsive wall of $V_{MLR}(R)$ for very small separations R: this is clearly a physically unacceptable behavior for any atomic interaction. Note also that, although the TT $s = 1$ damping function is physically unacceptable for use in the MLR model, in which $u_{LR}(R)$ appears multiplicatively, it is a physically acceptable damping function for a potential model in which $u_{LR}(R)$ appears as an additive attractive term or as a multiplicative factor. No effort has been made to force $V_{MLR}(R)$ to satisfy the united atom (Pathak and Thakkar, 1987) asymptotic form for $R \to 0$, as this limiting behavior occurs only for separations that are typically much smaller than a_0. Moreover, it was found by Dattani and Le Roy (2011) to be difficult to impose such a constraint. The introduction of a damping function whose short-range behavior is given by R^s, with s small and negative (such as

Table 3.1 Effect of parameter s on HFD and TT models damping function of for the $H_2(a^3\Sigma_u^+)$

	TT-type		HFD-type		
s	RMSD	$b^{TT}(s)$	RMSD	$b^{DS}(s)$	$c^{DS}(s)$
2	0.0091	3.47(2)	0.0066	4.99(12)	0.34(1)
1	0.018	3.13(3)	0.0064	4.53(12)	0.36(1)
0	0.034	2.78(6)	0.0068	3.95(12)	0.39(1)
−1	0.055	2.44(8)	0.0078	3.30(14)	0.423(16)
−2	0.081	2.10(10)	0.0094	2.50(16)	0.468(20)

$s = -1, -2$) thus simultaneously improves the long-range behavior of the interatomic potential energy and prevents the short-range repulsive wall of an MLR PEF from becoming excessively steep for very small interatomic separations.

The MLR model was originally developed to represent the potential energy of chemically bound diatomic molecules. However, it is (in principle) equally valid for the much more weakly bound van der Waals molecules, for which a significant reduction in the number of fitting parameters occurs. This may be illustrated by the Li_2 molecule, for example, for which accurate PEFs for both the ground $X^1\Sigma_g^+$ electronic term (Dattani and Le Roy, 2011; Le Roy *et al.*, 2009) ($D_e = 8516.709$ cm^{-1}, $R_e = 2.67299$) and the $a^3\Sigma_u^+$ related electronic term (Dattani and Le Roy, 2011) ($D_e = 333.758$ cm^{-1}, $R_e = 4.17005$) have been obtained from fits to 17,477 and 5584 spectroscopic line positions, respectively. Twenty-four free parameters were required to obtain a minimum $\overline{dd}$ value of 1.0059 for the ground-state, but instances of where only nine parameters were needed to obtain a minimal $\overline{dd}$ value of 0.7099 for the $Li_2(a^3\Sigma_u^+)$ state have also been reported (Douketis and Scoles, 1982; Lau *et al.*, 2016).

The original TT model PEF has been extended to represent the lowest spin-aligned $a^3\Sigma_u^+$ state of a homonuclear alkali dimer via the inclusion of an additional term in the exponent of the Born–Mayer repulsion term (Runeberg and Pyykkö, 1998). The resultant LTT model PEF,

$$V_{\rm LTT}(R) = Ae^{-bR-cR^2} - \sum_{n=3}^{N} f_{2n}(bR + 2cR^2)C_{2n}/R^{2n}, \tag{3.42}$$

with parameters A, b, c determined numerically from experimental values of the location R_e, depth D_e, and harmonic vibrational frequency ω_e, provided good agreement with a number of experimental spectroscopic features for the X_2 spin-aligned dimers, with X = Na, K, Rb, Cs. The level of agreement between predictions employing a LTT PEF for $Li_2(a^3\Sigma_u^+)$ and experiment, although better than that achievable using the original TT model form, still fell outside the 0.1 cm^{-1} experimental resolution for the spectral data (Linton *et al.*, 1999) and was unable to match the level of agreement achieved by an earlier multi-parameter MLR model fit carried out by Dattani and Le Roy (2011).

Both multi-parameter MLR and two-parameter TT fitting functions have been used in carrying out direct potential fits (Piticco *et al.*, 2010) to a data set consisting of 130 rotationally resolved resonance-enhanced vacuum ultraviolet (VUV) two-photon ionization spectral transitions, 26 microwave transitions (Jäger and Gerry, 1993), and 26 interaction second virial coefficient data (Dymond *et al.*, 2003; Rentschler and Schram, 1977; Schramm *et al.*, 1977) obtained for the Ar–Xe Van der Waals dimer, with the aim of obtaining an optimal PEF for the Ar–Xe interaction. The dominant long-range dispersion coefficients C_6, C_8, C_{10} were set to the values recommended by Tang and Toennies (Kleinekathöfer *et al.*, 1997; Pack *et al.*, 1982; Tang and Toennies, 1984), while both the damping functions of eqn (3.10) for the TT model PEF and of eqn (3.31) for the MLR model PEF were employed. Three additional dispersion terms employing values of C_{12}, C_{14}, C_{16} that were generated from the C_6, C_8, C_{10} coefficients by employing the Thakkar recursion relations (Dalgarno and Lewis, 1955) were also incorporated into the TT two-parameter fitting function. By expressing the pre-exponential Born–Mayer factor A in terms of R_e and the Born–Mayer exponential parameter b, the usual A and b free parameters for the TT fitting function were replaced by R_e and b, in which case, the depth, $\mathfrak{D}_e$, of the PEF minimum is obtained from

$$-\mathfrak{D}_e \equiv V(R_e) = Ae^{-bR_e} - \frac{e^{bR_e}}{b}\frac{dV_{att}(R)}{dR} \tag{3.43}$$

(the derivative is evaluated at $R = R_e$), so that the usual MLR parameter $\mathfrak{D}_e$ (including an estimate of its uncertainty) can readily be obtained from the values obtained for R_e, b, and their associated uncertainties. The TT fitting function thus represents a compromise between the overall quality of fit to an extensive set of experimental data (with attendant uncertainties) and a minimization of the number of adjustable parameters required to fit the full set of data within an overall acceptable level of accuracy. Finally, the argument of the damping functions for the MLR fitting function employs the scale factor ρ_{AB} of eqn (3.34). The flexible multi-parameter exponent fitting factor $\beta(R)$ defined in eqn (3.29) could be simplified considerably in this case because none of the high-resolution spectroscopic data for the Ar–Xe dimer corresponds to an outer vibrational turning point R_I that is significantly larger than R_e, and consequently there is no need to employ the additional separation $R_{ref} > R_e$ (similarly, $q < p$) typically utilized in the determination of $V_{MLR}(R)$ functions for chemically bound molecules.

Five pairs of $\mathfrak{D}_e$, R_e values characterizing the minimum in the Ar–Xe interaction energy, together with their representative dd-values, are collected in Table 3.2.

For purposes of the present discussion, a number of fit-versions of the MLR model, designated by MLR(N), with N representing the number of free β_i parameters employed in the expression for $\beta(R)$, and the TT fitting function have been considered. The two-parameter TT model fitting function actually fares remarkably well, as it provides a better overall fit to the combined VUV and virial data ($\overline{dd} = 2.32$) than does the corresponding two-parameter fitting function MLR(0) (with $\overline{dd} = 4.20$). The values for $\mathfrak{D}_e$, R_e for the TT fitting function are also remarkably close to the best-fit values given for the final recommended (Piticco *et al.*, 2010) MLR(2) PEF, for which $\overline{dd}$

Table 3.2 Values characterizing the minimum in the Ar–Xe interaction energy for different PEFs

Acronym	Data-types fitted	$\mathfrak{D}_e$ cm^{-1}	R_e Å	Ref.	$\overline{dd}$
MLR(2)	VUV, MW, Virial	129.81	4.09577_4	a	0.554
TT-fit	VUV, Virial	131.30	4.09577_3	b	2.32
M3SV	DSC, Virial, Diffusion	131.24	4.03762	c	0.54
HFD-C*	Virial, Viscosity, Diffusion	131.10	4.0668	d	-
TT	-	130.37	4.09054	e	-

* The HFD-C PEF differs from the HFD PEF defined in Douketis and Scoles, 1982.
a, Piticco *et al.*, 2010; b, Piticco *et al.*, 2010; c, Pack *et al.*, 1982; d, Aziz and van Dalen, 1983*b*; e, Tang and Toennies, 2003.

values of 0.574, 0.372, 0.539 were obtained for the 130 VUV transitions, 26 MW transitions, and 26 virial data values, respectively, and an overall value $\overline{dd} = 0.554$ obtained for the full set of 182 experimental data employed for the fitting. The larger value of $\mathfrak{D}_e$ obtained for the TT overall interaction (equivalently, the monotone increase in the value of $\mathfrak{D}_e$) that typically results from the inclusion of successive additional terms in the dispersion series. However, as pointed out by Feltgen Feltgen (1981), and reiterated by Kleinekathoefer *et al.* (1997), termination of this series at the R^{-16} contribution may be justified by the observation that the overall (net negative) contribution to the second-order dispersion term with $n > 8$ appears in many cases to be canceled by the (positive) third-order dispersion terms that are normally neglected in HFD-type models. Perhaps surprisingly, the value $\mathfrak{D}_e = 130.37$ cm^{-1} predicted (Tang and Toennies, 2003) on the basis of combining rules is in excellent agreement with the final MLR(2) best-fit value for $\mathfrak{D}_e$, differing from it by only 0.43%, while the predicted value 4.0905_4 Å differs by only 0.13% from the MLR(2) best-fit value. This extraordinary agreement between the model predictions and the best semi-empirical values for R_e and $\mathfrak{D}_e$ for the Ar–Xe van der Waals interaction is a tribute to the combining rules developed in Tang and Toennies Tang and Toennies (1986).

Two empirical multiproperty-fit PEFs have been obtained for the ArXe interaction, one by Pack et al. (1982) from fitting a Morse-Morse-spline-van der Waals (M3SV) PEF to a combination of crossed molecular beam scattering data (for collision energy 516.2 cm^{-1}), mixture viscosity data, and interaction virial coefficient data (with the BP data available over a temperature range of about 500 K), the other by Aziz and Van Dalen (1983*b*) from a fitting to the temperature dependence of the interaction second virial coefficient, interaction viscosity, and diffusion data. The high quality of the Pack *et al.* $\mathfrak{D}_e$, R_e values is reflected in the corresponding $\overline{dd}$ 0.26, 0.56, and 0.54 for the fits to the virial data, mixture shear viscosity data, and differential scattering data, respectively, plus the $\overline{dd}$ value of 0.47 for the fit to the full set of experimental data. The relatively close agreement between the respective M3SV and MLR(3) values for the depth (0.43%) and position (1.42%) of the minimum in the Ar–Xe interaction energy suggests that even though neither the beam scattering nor binary mixture viscosity data were included in the MLR(2) fit procedure, a forward calculation of these quantities using the MLR(2) PEF should result in rather good agreement also with the molecular beam scattering and viscosity data. It is clear, however, from the results that have been obtained for the Ar–Xe interaction that no simple two-parameter fit function, such as the TT or HFD model function, will be sufficiently flexible to provide fits either to extensive sets of spectral data spanning a large portion of the full interaction range or to the very precise data associated with MW transitions among low lying states in the vicinity of a potential energy minimum.

It was possible to conclude from the Le Roy *et al.* (2011) study that the incorporation of damping functions into the long-range tail $u_{\mathrm{LR}}(R)$ of the MLR model PEF provided both a more physically correct description of the long-range tail of the PEF itself and a more physically realistic description of the repulsive wall of the PEF. The MLR model PEF often turns out to be more compact, as it requires fewer parameters in order to achieve a specific level of agreement with the experimental data (for example, 12 vs. 18 parameters for MgH). Section 3.8 shows that the MLR PEF has proved to be rather useful for fitting to *ab initio* interaction energies in a constructive manner. A complex case involving the coupling between the strongly bonded $Li_2(1^2\Sigma_g^+)$ state ($\mathfrak{D}_e = 7093.44\,cm^{-1}$) and the relatively weakly bonded $Li_2(a^2\Sigma_u^-)$ state ($\mathfrak{D}_e = 333.758\,cm^{-1}$) was treated by Dattani and Le Roy Dattani and Le Roy (2011), who were able to employ the MLR form to explain the spectroscopic behaviors of all states of the ^{6}Li and ^{7}Li homonuclear isotopologues.

3.6 Direct Inversions of Experimental Data

Previous sections have focused on various empirical and semi-empirical models describing the potential energy, $V(R)$, between pairs of interacting atoms. The most realistic (and successful) of these models has incorporated physical intuition, the known functional behaviors of repulsion and attraction associated with small, respectively, very large separations R, and a means of spanning the intermediate separations via damping functions derived from knowledge of the $H_2(a^3\Sigma_u^+)$ interaction energy over the full range of H–H separations. A direct potential fit model that is primarily based upon an indirect inversion of spectroscopic data via an iterative multiparameter fitting procedure that minimizes the DRMSD between the sets of calculated and observed spectroscopic transition frequencies has also been examined.

3.6.1 Inversion of microscopic data

Direct inversion of experimental data acquired from measurements of microscopic phenomena, such as bound-state spectroscopic transition frequencies and molecular beam differential and integral scattering cross-sections, is inherently easier to accomplish than is the inversion of experimental data acquired from measurements of macroscopic phenomena, also referred to as BPs. Indeed, because very accurate wavelength and frequency measurements have been possible for many years, direct inversion of spectroscopic transition frequency data to give numerical representations of PEFs of both chemically bound and van der Waals diatomic molecules, has been carried out routinely for several decades via the well-known (Kaplan, 2003; Le Roy, 2010) RKR inversion method. Note, however, that the (indirect) inversion of spectroscopic data that is implicit in the MLR fitting procedure is inherently superior to the RKR (direct) inversion procedure, as it results in a closed functional form that is capable of producing turning points whose values are more accurate than those generated directly from the first-order semi-classical RKR inversion equations. Successful procedures for the inversion of molecular beam scattering data took much longer to develop, as in most cases, two steps, rather than one, as for the RKR method, are required to obtain $V(R)$ from scattering data.

From the earliest days of atomic beam scattering we know that crossed atomic beams provide an almost collision-free environment for the study of interatomic (similarly intermolecular) interactions via microscopic single-collision scattering events occurring under carefully controlled conditions. Unlike the direct (RKR) inversion of spectroscopic data, which allows the determination of at most the attractive well of the PEF, the inversion of beam scattering data enables the accurate determination, in principle, of both the repulsive and attractive components of the PEF.

Although it can be shown in principle that atom–atom scattering phase-shift data, $\delta_\ell(E)$, may be inverted to obtain the PEF, $V(R)$, either from a knowledge of the complete energy dependence of the phase-shift for a fixed value of orbital angular momentum ℓ, or from the complete set of phase-shifts for a fixed scattering energy E (Newton, 1966), it appears that uniqueness of the resultant $V(R)$ cannot be assured. Because in practice neither phase-shifts nor deflection function values can be directly accessed experimentally, a two-step inversion procedure is required. The first step of this procedure involves the determination of either the classical deflection angle as a function of impact parameter or a full set of phase-shifts for a fixed collision energy from the scattering data. The second step proceeds via inversion either of the classical deflection function $\chi(b)$ to obtain $V(R)$ via a method due initially to Firsov (1953) or of the phase-shift results via extensions (Miller, 1969; Sabatier, 1965; Vollmer, 1969) of the original Firsov method. Strictly speaking, the original Firsov method was developed for high-energy (keV-range) atomic scattering that may be described in terms of a monotonic repulsive PEF, so that classical mechanics is applicable and the deflection function $\chi(b)$ is also a monotonic function of the impact parameter b, in which case uniqueness of the inversion process is assured (Kaplan, 2003). Thermal energy atomic scattering (typically in the meV range) is in general sensitive to the full interatomic PEF, in which case the deflection function, with the exception of that portion obtained from backward scattering data, will be multi-valued in accordance with both the low-frequency supernumerary rainbow scattering oscillations and the superimposed high-frequency diffraction oscillations that typically characterize the behavior of experimental thermal energy differential scattering.

A combination of eqn (2.174) for the partial wave expansion of the scattering amplitude and eqn (2.175b) relating the partial wave scattering amplitude $a_l(E)$ to the scattering S-matrix element $S_l(E)$ gives the relation

$$a(E,\chi) = \frac{1}{2ik}\sum_{l'}(2l'+1)[S_{l'}(E)-1]P_{l'}(\cos\chi)\,, \tag{3.44}$$

in which the wavenumber $k \equiv \sqrt{2m_\mathrm{r}E/\hbar^2}$ depends upon the reduced mass m_r. By multiplying the result (3.44) for the scattering amplitude by $P_l(\cos\chi)$, followed by integration over χ and employment of the Legendre orthogonality relation [eqn (A.9a)], the $S_l(E)$ partial-wave S-matrix elements at collision energy E can be related to the scattering amplitude via

$$S_l(E) - 1 = ik\int_0^\pi a(E,\chi)P_l(\cos\chi)\sin\chi\,\mathrm{d}\chi\,. \tag{3.45}$$

Now, upon recalling that $S_l(E)$ is related to the partial-wave phase-shift $\delta_l(E)$ $S_l(E)$ is equivalent either to the determination of the phase-shifts $\delta_l(E)$ or of the full scattering amplitude $a(E,\chi)$.

One procedure (Miller, 1969; Vollmer, 1969) by which the scattering phase-shift function $\delta_\ell(E)$ can be inverted to give $V(R)$ has been elegantly and succinctly presented by Buck (1974, 1976, 1986): his description of this procedure has been adopted here. Within the semi-classical, or Wentzell–Kramers–Brillouin (WKB) approximation

(Child, 1974, 1991), the classical deflection function $\chi(E, b)$ is related to the semi-classical phase-shift $\delta_l(E)$ via

$$2\frac{\partial \delta_l(E)}{\partial l} = \chi(E, b), \qquad b = (l + \tfrac{1}{2})/k,$$

with the phase-shift given by

$$\delta_\ell(E) = k\left\{\int_{R_c}^{\infty}\left[1 - \frac{V(R)}{E} - \frac{(\ell+\frac{1}{2})^2}{k^2R^2}\right]^{\frac{1}{2}} \mathrm{d}R - \int_{(\ell+\frac{1}{2})^2/k}^{\infty}\left[1 - \frac{(\ell+\frac{1}{2})^2}{k^2R^2}\right]^{\frac{1}{2}} \mathrm{d}R\right\}, \tag{3.46}$$

in which R_c is the classical turning point and E is the scattering energy, respectively, associated with the collision pair. By making the transformation

$$s^2 \equiv R^2\left[1 - \frac{V(R)}{E}\right], \tag{3.47}$$

$\delta_\ell(E)$ can be re-expressed in terms of the impact parameter $b \equiv (\ell + \frac{1}{2})/k$ as

$$\delta(b, E) = \int_b^{\infty} \sqrt{s^2 - b^2}\left[\frac{\mathrm{d}\ln(R/s)}{\mathrm{d}s}\right] \mathrm{d}s, \tag{3.48}$$

provided only that eqn (3.47) can be inverted uniquely to give $R(s)$. Integration by parts of eqn (3.48), which is allowed for any PEF for which the condition $\lim_{R\to\infty} RV(R) = 0$ is satisfied, then gives $\delta(E, b)$ as

$$\delta(b, E) = -k\int_b^{\infty} bI(s, E)\frac{s}{\sqrt{s^2 - b^2}} \mathrm{d}s, \tag{3.49}$$

with $I(s, E)$ defined as $I(s, E) \equiv \ln[R(s, E)/s]$. Multiplication of eqn (3.49) by $2b(b^2 - t^2)^{-\frac{1}{2}}$, then integrating over b from t to ∞, followed by an interchange of the order of the s and b integrations, gives

$$\begin{aligned}\int_t^{\infty} \frac{2b\delta(b, E)}{k\sqrt{b^2 - t^2}} \mathrm{d}b &= -\int_t^{\infty} sI(s, E)\int_t^{s} \frac{2b}{\sqrt{(b^2 - t^2)(s^2 - b^2)}} \mathrm{d}b\mathrm{d}s \\ &= -\pi\int_t^{\infty} sI(s, E)\,\mathrm{d}s\,.\end{aligned}$$

This result allows $I(s, E)$ to be obtained in terms of the derivative of the phase-shift as

$$I(s, E) = \frac{2}{k\pi}\int_s^{\infty} \frac{\partial\delta(b, E)}{\partial b}\frac{1}{\sqrt{b^2 - s^2}} \mathrm{d}b\,. \tag{3.50}$$

As the semi-classical expression corresponding to the classical deflection function $\chi(b, E)$ is given in terms of $\delta(b, E)$ by

$$\chi(b, E) = \frac{2}{k}\frac{\partial\delta(b, E)}{\partial b},$$

$I(s, E)$ may be re-expressed in terms of $\chi(b, E)$ as

$$I(s, E) = \frac{1}{\pi} \int_s^{\infty} \frac{\chi(b, E)}{\sqrt{b^2 - s^2}} \, db \,, \tag{3.51}$$

which is the Firsov result (Firsov, 1953; Miller, 1969). Because the defining relation for $I(s, E)$ may also be expressed in the form

$$\frac{s}{R(s, E)} = e^{-I(s,E)} \,,$$

eqn (3.45) may also be employed to obtain $V(R)$ in terms of $I(s, E)$ as

$$V(R) = E \left[1 - e^{-2I(s,E)} \right] . \tag{3.52}$$

The inversion of scattering data to obtain the interaction PEF involves, as has been pointed out by Buck (1974, 1976), a relatively complicated procedure. In particular, the following experimental inputs are required in order to carry out a successful inversion of scattering data:

(a) from differential scattering cross-section measurements:
- well-defined locations of the rainbow oscillations,
- separations between the rapid (or diffraction) oscillations that are superimposed upon the rainbow oscillations,
- backward scattering data,
- the ratio of the amplitude of the first rainbow oscillation to the backward scattering;

(b) from integral scattering cross-section measurements, the speeds associated with the glory oscillation extrema;

(c) from other sources, for example, the most accurate available value of the C_6 long-range dispersion constant.

The first two items in (a) are particularly important because, (i) they are unaffected by instrumental averaging occurring in typical high-resolution beam scattering experiments, (ii) they are readily extracted from high-resolution data and, most importantly, (iii) they are very sensitive to both the absolute value and the form of the PEF. All of the listed information, apart from the backward scattering cross-section data, is required to construct the nonmonotonic attractive branch of the deflection function $\chi(b)$ by an iterative process of minimization of the differences between calculated and experimentally measured values of $\sigma(E, \chi)$. It also helps if differential scattering measurments can be carried out for a series of scattering energies, rather than just for a single scattering energy. Finally, the repulsive component of $V(R)$ can be determined from the backward scattering associated with angles larger than the classical rainbow angle.

A detailed description of a specific procedure whereby differential scattering data containing at least three well-resolved rainbow oscillations can be inverted in a piecewise manner using (modified) Firsov inversions either of deflection functions or, equivalently, of semi-classical WKB phase-functions, to obtain a nonmonotonic PEF was given by Buck (1971): its validity was established via an in-principle inversion of a set of differential cross-section pseudo-data generated from a MLJ (12,6) PEF. The first practical application of this procedure was to obtain $V(R)$ for the Na–Hg interaction from the inversion of high-resolution differential scattering cross-section data obtained at five scattering energies lying between $1400\,\text{cm}^{-1}$ and $2500\,\text{cm}^{-1}$ (Buck and Pauly, 1971), supplemented by the positions of glory oscillation extrema in the speed dependence of the integral scattering cross-section (Buck *et al.*, 1971), backward differential scattering data, and the value for the Na–Hg C_6 long-range dispersion coefficient. The inversion gave values of $V(R)$ for Na–Hg separations lying between 3.05 Å and 9.5 Å: as the minimum $\mathfrak{D}_\text{e} \simeq 442.5\,\text{cm}^{-1}$ in $V(R)$ occurs for separation $R_\text{e} \simeq 4.72$ Å, the inversion of the thermal energy scattering data also enabled a portion of the low repulsive wall (up to approximately $1100\,\text{cm}^{-1}$) of the PEF corresponding to separations $3\,\text{Å} \lesssim R \lesssim 3.85\,\text{Å}$ to be determined. The occurrence of significant overlap between the domains of separation covered by the separate inversions of the five sets of thermal energy scattering data also enabled a check to be made of the practical uniqueness of the PEF obtained from the inversion procedure. The values obtained for $V(R)$ in the overlapping inversion domains were, to within the precision allowed by the experimental data, the same. Similar inversions were also carried out for atomic scattering data acquired for the (K, Cs)–Hg (Buck *et al.*, 1972) and Li–Hg (Buck *et al.*, 1974) alkali–mercury interactions. The shape of the PEF obtained from the inversion of alkali–mercury differential scattering data differs significantly from that for the traditional MLJ (12,6) PEF, as it is much less repulsive at short range and possesses a much broader attractive well than that of a MLJ (12,6) PEF having the same $(\mathfrak{D}_\text{e}, R_\text{e})$ values. It is perhaps interesting that although the Na, K, Cs–Hg interactions satisfy the principle of corresponding states within the mutual experimental uncertainties, the Li–Hg interaction does not.

A stringent test of the quality of a PEF obtained from the inversion of differential scattering cross-section data can be carried out for the Ar–Kr interaction, for which a PEF based primarily upon the inversion of atomic beam scattering data (Aziz *et al.*, 1979; Buck *et al.*, 1978), three PEFs determined from BP data (Gough *et al.*, 1975; Lee *et al.*, 1975; Maitland and Wakeham, 1978*a*), a multi-property fit PEF (Aziz and van Dalen, 1983*b*) and a model PEF (Tang and Toennies, 2003) are available. The beam-scattering inversion PEF was obtained (Buck *et al.*, 1978) from a combination of differential scattering data (including two well-resolved rainbow oscillations) acquired over scattering angles lying between 3° and 70° at a collision energy of $523.5\ \text{cm}^{-1}$, coupled with additional higher-resolution measurements carried out to identify the positions of the diffraction oscillations plus integral glory oscillation data (Linse *et al.*, 1979; Stokvis *et al.*, 1980; van den Biesen *et al.*, 1982). This set of scattering data enabled a determination to be made of the Ar–Kr attractive well-region of the PEF. Two (slightly) different inversion procedures applied to the same beam scattering data gave essentially the same attractive well. Minor adjustments were made to the inverted PEF data to improve agreement with the amplitudes of the rainbow oscillations so as to ensure the uniqueness of the attractive component of the inversion PEF. The uniqueness of the repulsive component of the PEF was assured, as it was derived from large-angle scattering data, supplemented by high-temperature second virial coefficient data. The entire set of numerical values defining the inversion PEF was then fitted, for calculational convenience, to a MSMSV functional form with a maximum of 16 adjustable parameters. Only a few relatively minor parameter changes (Aziz *et al.*, 1979) were necessary to modify the original MSMSV fitted PEF in order to bring the calculated viscosity and diffusion coefficient values into acceptable agreement with experiment.

Two of the four Ar–Kr PEFs based either fully or largely upon bulk data were obtained from direct inversions of mixture viscosity data (Gough *et al.*, 1975; Maitland and Wakeham, 1978*a*), while the other two resulted from multi-property fits involving diffusion, mixture viscosity, and interaction second virial coefficient data in the first (Lee *et al.*, 1975) case, and both BP (Arora *et al.*, 1978; Gough *et al.*, 1976; Hogervorst, 1971; Kestin *et al.*, 1970; Maitland and Smith, 1974; van Heijningen *et al.*, 1968) and beam scattering data (Linse *et al.*, 1979; Stokvis *et al.*, 1980; van den Biesen *et al.*, 1982) in the second (Aziz and van Dalen, 1983*b*) case. Comparisons between the attractive well-portion of the reduced PEF, $U(x)$, for the Buck *et al.* (1978) PEF and the attractive well portions of the reduced forms of the BP and model PEFs show that there is consistency among the six Ar–Kr PEFs. The DRMSD values, $\overline{dd}$, obtained from comparisons between forward calculations of $\sigma(\chi)$, $\sigma_{\text{int}}(v)$, and $B_2(T)$ and the corresponding experimental values carried out for six ArKr PEFs are given in Table 3.3.

Table 3.3 shows, for example, that the $\overline{dd}$ values for $\sigma(\chi)$ are significantly larger for the three PEFs derived exclusively from BP data than those obtained from PEFs

Table 3.3 "Goodness-of-fit" as represented by DRMSD values for comparisons between calculated and experimental Ar–Kr scattering and bulk virial coefficient data

Property	Data Source	$\overline{dd}$ value for PEF of Ref.					
		a	b	c	d	e	f
$\sigma(\chi)$	Buck *et al.* 1978	4.94	3.53	4.32	9.01	8.86	13.3
$\sigma_{\text{int}}(v)$	Linse *et al.* 1979	0.69	1.40	1.83	3.36	2.50	2.18
$B_2(T)$	Schramm *et al.* 1977	0.29	0.32	0.30	0.53	0.73	0.32
	Dymond *et al.* 2003	0.40	0.50	0.35	2.53	4.43	0.75
$\eta_{\text{AB}}(T)$	Kestin *et al.* 1970	0.55	0.75	0.79	0.39	2.00	0.70
	Maitland and Smith 1974	0.64	1.61	0.54	0.87	1.71	1.14
$D_{\text{AB}}(T)$	Hogervorst 1971	3.20	4.94	2.82	3.61	2.12	4.83
	van Heijningen *et al.* 1968	0.51	1.73	0.44	0.66	1.53	1.13
	Arora *et al.* 1978	0.30	14.6	0.50	1.00	15.3	8.0
overall $\overline{dd}$		2.45	4.66	2.34	4.47	5.87	6.40

a, Aziz and van Dalen, 1983*b*; b, Buck *et al.*, 1978; c, Aziz *et al.*, 1979; d, Gough *et al.*, 1975; e, Maitland and Wakeham, 1978*a*; f, Lee *et al.*, 1975.

based upon beam scattering data. It can also be seen that the PEFs derived from BPs fare relatively better in describing the integral scattering cross-section data than they do in describing the differential scattering data. Although the $\overline{dd}$ values reported for the virial data of Schramm *et al.* Schramm *et al.* (1977) and Rentschler and Schram Rentschler and Schram (1977) are generally smaller than those for the virial data of Dymond *et al.* Dymond *et al.* (2003)b, the difference is due primarily to the typically five times larger experimental uncertainties associated with the former data. This clearly illustrates the potential danger of placing too much weight upon the raw $\overline{dd}$ values for individual sets of data, especially in comparisons between $\overline{dd}$ values from one set of experimental measurements to another: in this case, the smaller $\overline{dd}$ values obtained for the data of Schramm *et al.* Schramm *et al.* (1977) and Rentschler and Schram Rentschler and Schram (1977) are mainly due to the large experimental uncertainties in these particular virial measurements in comparison with the tighter uncertainties assigned to the virial data of Dymond *et al.* Dymond *et al.* (2003).

Section 2.5.4 showed that the quantum mechanical expression for the differential cross-section is given by [see eqn (2.166)] as

$$\sigma(E,\chi) = |a(E,\chi)|^2 \,,$$

in terms of the scattering amplitude $a(E,\chi)$. Chapter 2 also showed that $a(E,\chi)$ may, in turn, be expressed in terms of partial-wave phase-shifts, $\delta_\ell(E)$, associated with the solutions of the Schrödinger eqns (2.170) governing the radial components $Z_\ell(R)$ of the quantum mechanical wavefunction $\psi(\mathbf{R})$ of eqn (2.169) via eqns (2.174, 2.175) as

$$\sigma(E,\chi) = \frac{1}{k^2}\sum_{\ell,\ell'}^{\infty}(2\ell+1)(2\ell'+1)\mathrm{e}^{i(\delta_l-\delta_{l'})}\sin\delta_\ell\sin\delta_{l'}P_\ell(\cos\chi)P_{\ell'}(\cos\chi)\,. \qquad (3.53)$$

It will be evident from this expression for $\sigma(E,\chi)$ that a great deal of structure may be associated with the χ-dependence of the differential scattering cross-section measured at a given collision energy E.

Thus, to evaluate the differential scattering cross-section (3.53) for given (E,χ) and a known $V(R)$, it is necessary to solve the (uncoupled) set of Schrödinger eqns (2.170), one for each value of the (conserved) orbital angular momentum ℓ, and from the solutions extract the associated scattering partial-wave phase-shifts $\delta_\ell(E)$, and substitute the results into eqn (3.53). It is clear from this description of the forward calculations of the differential cross-section from a known PEF that the relationship between $V(R)$ and $\sigma(E,\chi)$ is rather complicated even for atom–atom scattering, and therefore that the task of extracting $V(R)$ from a set of experimental scattering data is fairly daunting.

A method (Gerber and Karplus, 1970; Martin, 1969; Newton, 1968) for carrying out the first step of the direct inversion process begins with the generalized optical, or unitarity, theorem (Newton, 1966)

$$\operatorname{Im} a(E,\chi) = \frac{k}{4\pi}\int_0^{2\pi}\int_0^{\pi} a^*(E,\cos\chi')a(E,\cos\chi'')\sin\chi'\mathrm{d}\chi'\mathrm{d}\phi'\,,$$

in which $\cos\chi''$ is given by $\cos\chi'' = \cos\chi'\cos\chi + \sin\chi\sin\chi'\cos\phi'$. If the differential cross-section at collision energy E has been determined for all scattering angles χ, then $a(E,\chi)$ becomes

$$a(E,\cos\chi) = \sqrt{\sigma(E,\cos\chi)}\,\mathrm{e}^{i\alpha(E,\cos\chi)}\,, \tag{3.54}$$

with $\alpha(E,\cos\chi)$ the phase of the scattering amplitude at scattering angle χ for fixed collision energy E. The resulting governing equation for $\alpha(E,\cos\chi)$ is then obtained as

$$\sqrt{\sigma(E,\chi)}\,\sin\alpha = \frac{k}{4\pi}\int_0^{2\pi}\int_0^{\pi}\sqrt{\sigma(E,\chi')\sigma(E,\chi'')}\cos(\alpha'-\alpha'')\sin\chi'\mathrm{d}\chi'\mathrm{d}\phi'\,, \tag{3.55a}$$

with α, α', α'' representing $\alpha(E,\cos\chi)$, $\alpha(E,\cos\chi')$, and $\alpha(E,\cos\chi'')$, respectively, and $\sigma(E,\theta)$ understood to stand for $\sigma(E,\cos\theta)$. As this procedure requires knowledge of $\sigma(E,\chi)$ at fixed scattering energy for $0 \leq \chi \leq \pi$, measurements over as large an angular range as possible are required, including minimally all oscillations, in the event that extrapolation to large angles needs to be made. Moreover, instrumental averaging effects associated with the finite spreads of beam velocity distributions and the finite sizes of detectors must be removed by deconvolution. These effects are smaller and more readily deconvoluted when there are relatively few, though pronounced and well separated, quantum oscillations. Finally, experimental relative intensities must also be converted into the absolute differential cross-section values required by the unitarity inversion method. Thus, when a fairly dense set of experimental values for $\sigma(E,\cos\chi)$ is available, eqn (3.55a) becomes an equation from which the phase $\alpha(E,\cos\chi)$ of the scattering amplitude $a(E,\cos\chi)$ may be determined by employing an appropriate iterative procedure.

A mathematically sufficient condition for a sequence of iterates of eqn (3.55a) to converge uniformly to an unique solution (apart from the absolute sign of $\cos\alpha(\chi)$, which the unitarity relation cannot provide) has been shown by Gerber and Karplus (1970) to be

$$P\sqrt{\sigma(E,\chi)} \leq \frac{k}{4\pi}\int_0^{2\pi}\int_0^{\pi}\sqrt{\sigma(E,\chi')\sigma(E,\chi'')}\cos(\alpha'-\alpha'')\sin\chi'\mathrm{d}\chi'\mathrm{d}\phi'\,,$$

with $P \leq 0.79$. They further showed that the sequence of iterations

$$\sin\alpha_{N+1}(E,\chi) = \frac{k}{4\pi}\int_0^{2\pi}\int_0^{\pi}\left[\frac{\sigma(E,\chi')\sigma(E,\chi'')}{\sigma(E,\chi)}\right]^{\frac{1}{2}}\cos[\alpha_N(\chi')-\alpha_N(\chi'')]\sin\chi'\mathrm{d}\chi'\mathrm{d}\phi'\,, \tag{3.55b}$$

for $N = 0,1,2,\cdots$, initialized by $\alpha_0(\chi) = 0$, and with $\cos\alpha_N$ defined via $\cos\alpha_N \equiv +(1-\sin^2\alpha_N)^{\frac{1}{2}}$, converges uniformly to the correct solution, with the associated error for the Nth iterant bounded above by

$$|\sin\alpha(\chi) - \sin\alpha_N(\chi)| \leq \frac{PJ^N}{1-J}\,, \qquad J \equiv 2P\left[1 + \frac{P}{\sqrt{1-P^2}}\right]. \tag{3.55c}$$

The sufficiency condition will be violated should $\sigma(E,\chi)$ become very small (or, in particular, vanish) for some value of the scattering angle. Violation of this condition, which is manifested as a divergence of the Nth iterate such that $|\sin\alpha_N(\chi)| > 1$ (Gerber and Shapiro, 1976), may occur in the inversion of differential cross-section data at scattering angles corresponding to the positions of the minima of pronounced diffraction oscillations. Such violations may be avoided by modifying the iteration procedure to (Gerber and Shapiro, 1976):

- employ a starting phase function $\alpha_0(\chi)$ whose behavior corresponds to general features, such as attractive at long range, repulsive at short range, that are typical of an atom–atom interaction;
- include a mechanism to determine, should $\sin\alpha_N(\chi)$ approach ± 1 at some particular scattering angle, whether $\cos\alpha_N(\chi)$ will change sign or not via analytical continuation arguments.

The absolute sign of $\cos\alpha(0)$ is determined from knowing that the long-range interatomic forces responsible for the scattering of large-l partial waves are generally attractive, and give rise to positive phase-shifts $\delta_l(E)$.

The iteration scheme (3.55b) was tested (Gerber and Shapiro, 1976) with two sets of simulated partial-wave phase-shifts, one based upon twenty values selected in an essentially random manner, the other employing phase-shifts calculated from a MLJ(12,6) PEF. Although these sets of simulated input data do not represent truly realistic sets of experimental atomic scattering data, they did provide clear violations of some of the mathematical conditions required for an unique inversion result and, in so doing, enabled the proposed modifications to the iterative numerical inversion procedure to be tested. Nonetheless, this iteration scheme was found to converge to the correct phase within relatively few iterations: for example, neither of the two trial simulations (Gerber and Shapiro, 1976) required more than 20 iterations for reasonably realistic (though not necessarily close to the true behavior) nonzero nonconstant initial phase functions $\alpha_0(\chi)$. Convergence was achieved more rapidly for small scattering angles than for large scattering angles, and was not achievable for physically unrealistic starting phase functions $\alpha_0(\chi)$.

The final step in the first stage of inversion of experimental beam scattering data is accomplished by employing the set of scattering amplitude values obtained from the differential cross-section data to determine the partial-wave phase-shifts $\delta_l(E)$ via eqns (3.45) and (2.175c). Because these partial-wave phase-shifts will be modulo π, the correct absolute phase-shifts are obtained through analytic continuation based upon the known general form of the l-dependence of the phase function (Buck, 1986).

The second stage of the quantum mechanical inversion procedure employs the scattering amplitude information extracted from the differential scattering cross-section data via the iterative scheme described above, to obtain $V(R)$ from the phase-shifts. Shapiro and Gerber (Shapiro and Gerber, 1976) developed a process, evocatively termed "inversion by peeling," for achieving this goal. Their method is based upon the screening of the shorter-range components of the potential energy by the centrifugal potential, so that only the longest-range components of the potential energy determine the large-l partial-wave contributions to the scattering amplitude. The potential energy $V(R)$ is represented by a finite sum of terms $v_i(R)$ for this procedure, with the domain of $v_{i+1}(R)$ smaller than that of $v_i(R)$. Thus,

$$V(R) = \sum_{i=1}^{N} v_i(R)\,, \tag{3.56a}$$

with individual component functions $v_i(R)$ given by one of

$$v_i(R) = \begin{cases} c_i e^{-d_i R} \\ c_i R^{-d_i} \end{cases}, \tag{3.56b}$$

in which c_i and d_i are constants to be determined: the number of terms, N, in eqn (3.56a) is also determined during the inversion process.

The peeling procedure is initiated by choosing the largest value of l, l_{last}, for which the phase-shift $\delta_{l_{\mathrm{last}}}(E)$ is nonzero. Only the longest-range term in $V(R)$ which, according to eqn (3.56b) is $v_1(R)$, will contribute to $\delta_{l_{\mathrm{last}}}$. The asymptotically longest-range dispersion interaction between ground-term atoms has the form $V(R) \sim -C_n R^{-n}$, with C_n an asymptotic dispersion energy coefficient, $n = 6$, 5, for scattering in which one, respectively, neither, of the atoms is an S–term atom. If electronically excited atoms are involved, the longest-range dispersion interaction may have a power other than 5 or 6. Moreover, if an ionic species interacts with a ground-term atom, the relevant longest-range power becomes $n = 4$. Note that in many cases, quite accurate *ab initio* values of the long-range dispersion coefficients, C_n, are available, especially for the C_6 coefficient for closed-shell ground-term atom–atom interactions. Thus, for $v_1(R) = c_1 R^{-d_1}$, an initial value for c_1 may be taken to be $c_1 = C_n$, and d_1 may be set to the value of n appropriate to the interacting atoms. As $v_1(R)$ will be weak (relative to the centrifugal potential term), the Born approximation (Newton, 1966) should give $\delta_{l_{\mathrm{last}}}$ rather accurately as

$$\delta^{\mathrm{calc}}_{l_{\mathrm{last}}} = \frac{2m_{\mathrm{r}}k}{\hbar^2} C_n \int_0^\infty \left[j_{l_{\mathrm{last}}}(kR)\right]^2 R^{-n}\,\mathrm{d}R\,, \tag{3.57a}$$

in which $j_l(x)$ is a spherical Bessel function (Lide, 2001). If the difference between $\delta_{l_{\mathrm{last}}}$ and its experimental counterpart lies outside the assigned experimental uncertainty, the value of C_n may be adjusted within its assigned error bounds in order to improve the level of agreement between the calculated and experimental values. Once good agreement has been attained in this manner to provide $v_1(R)$, the phase-shifts δ_l may be determined for a decreasing sequence of l-values until the difference between the calculated and experimental phase-shifts for a particular value l_{k-1} of l exceeds the experimental uncertainty. Thus, when the partial-wave phase-shifts for $l_k \le l \le l_{\mathrm{last}}$ are consistent with $V(R) = v_1(R)$, but that for l_{k-1} is not, a new term, $v_2(R)$ may then be added to $v_1(R)$, so that $V(R) = v_1(R) + v_2(R)$, and $\delta^{\mathrm{calc}}_{l_k-1}$ is given, still within the Born approximation, by

$$\delta^{\mathrm{calc}}_{l_k-1} = \delta^{\mathrm{calc}}_{l_k-1}(c_1, d_1) + \frac{2m_{\mathrm{r}}k}{\hbar^2}\, c_2 \int_0^\infty \left[j_{l_k-1}(kR)\right]^2 R^{-d_2}\,\mathrm{d}R\,, \tag{3.57b}$$

with $\delta^{\mathrm{calc}}_{l_k-1}(c_1, d_1)$ the known contribution obtained from $v_1(R)$. Carrying out the same calculation for $l = l_k - 2$ then gives two equations from which c_2 and d_2 can be determined. Once c_2 and d_2 have been determined, all phase-shifts δ_l for $l_k - 2 \leq l \leq l_{\text{last}}$ must be recalculated to ensure that they are still within the experimental uncertainties. This procedure is continued for ever-decreasing values of l until, for $\delta_{l_{k'}}$, the Born approximation is no longer sufficient to determine $v_{N_1}(R)$, $1 < N_1 < N$, properly. The distorted-wave-Born approximation (DWBA) may then be employed to calculate the remaining partial-wave phase-shifts $\delta_{l_{k'}}$ to δ_0, as each individual $v_j(R)$ term in $V(R)$ may be treated as a perturbation on the potential given by the sum of the first $N_1{-}1$ terms of eqn (3.56a) already determined in previous steps of the procedure. Buck Buck (1986) and Shapiro and Gerber Shapiro and Gerber (1976) should be consulted for a more detailed and in-depth description of the peeling method.

Numerical tests of the inversion by peeling procedure were carried out for two sets of partial-wave phase-shift pseudodata (Shapiro and Gerber, 1976): one set of 81 phase-shifts ($l = 0, \cdots, 80$) was generated from a MLJ (12,6) PEF, the other set of 53 phase-shifts ($l = 0, \cdots, 52$) generated from a Morse PEF. Each PEF corresponds to a chemically bound diatomic species with an equilibrium separation of 7.9377 nm and a well-depth of 10,974 cm^{-1}. The scattering calculations were carried out at a collision energy $E = 2853\,\text{cm}^{-1}$. For strict comparison purposes, the inverse power form for the $v_i(R)$ of $V(R)$ in eqn (3.56b) was used for the inversion calculations carried out for the partial-wave phase-shifts obtained from the MLJ (12,6) PEF, while the exponential form was used for inversion of data generated from the Morse potential. Excellent agreement was obtained for both trial inversions: Table 3.4 compares the level of agreement attained in these two cases.

No constraint was placed upon the values to be assumed by the c_i and d_i coefficients in the representations of $V(R)$ given in eqns (3.56), so that the inversion of the MLJ (12,6) pseudodata using the expansion in inverse powers of R works extremely well, giving exponent values very near to 12 and 6 (within 2.46% and 0.04%, respectively). The relatively larger deviation of nearly 2.5% in the value of the repulsive exponent is partly compensated for by the approximately 13.8% smaller value obtained for the c_2 coefficient, so that the final "inversion" PEF is in remarkably good agreement with the "input" PEF. Even though the original Morse example PEF used in (Gerber and Karplus, 1970; Newton, 1968; Martin, 1969) was a sum of two exponentials, it ultimately needed four exponential $v_i(R)$ terms to provide an inversion PEF that was sufficiently accurate to reproduce all of the original input partial-wave phase-shifts. This example thus provides a concrete demonstration that even though the interaction potential energy may be unique, there are many ways of representing it within the experimental uncertainties of properties that depend upon it.

The inversion-by-peeling method has been employed (Gerber *et al.*, 1978) to obtain a He–Ne PEF from high-resolution experimental differential scattering cross-section data acquired using crossed He and Ne atomic beams. The scattering measurements covered an angular range $3.7° \leq \chi \leq 112.1°$ in the center-of-mass system. Extrapolation of the experimental cross-section data to small angles was carried out using the known R^{-6} long-range component of $V(R)$, while extrapolation to large angles was based upon the known behavior of large-angle scattering. A deconvolution procedure (Shapiro *et al.*, 1977) applicable to differential scattering cross-sections dominated by diffraction oscillations was applied to the experimental data lying between 3.7° and

Table 3.4 Convergence of the peeling method using partial-wave phase-shift pseudodata obtained from MLJ (12,6) and Morse PEFs†

MLJ (12,6) PEF				Morse PEF			
Range of l-values	i	d_i	c_i $/E_{\mathrm{h}}$	Range of l-values	i	d_i $/a_0^{-1}$	c_i $/E_{\mathrm{h}}$
$l = 80$	1	6.0023	−0.2861	$l = 52$	1	1.50414	−0.97929
$23 \leq l \leq 80$	1	6.0023	−0.2683	$30 \leq l \leq 52$	1	1.50417	−0.96950
	2	11.7049	0.5286		2	2.87030	3.70380
$22 \leq l \leq 80$	1	6.0023	−0.2853	$28 \leq l \leq 52$	1	1.50417	−0.96950
	2	11.7049	1.3496		2	2.87030	3.70380
					3	3.44360	−0.42598
$17 \leq l \leq 80$	1	6.0023	−0.2869	$27 \leq l \leq 52$	1	1.50417	−0.96950
	2	11.7049	1.4548		2	2.97030	3.70380
					3	3.44360	−0.42598
					4	4.20864	1.71224
Exact	1	6.0	−0.2962	Exact	1	1.5	−0.94877
	2	12.0	1.687		2	3.0	4.50086

† Data from Figs. 1 and 2 of Shapiro and Gerber, Shapiro and Gerber (1976).

40° to correct for instrumental damping to obtain not only accurate magnitudes but also very accurate values for the locations of the diffraction oscillations in order to pin down the separation $R = \sigma$ at which the PEF vanishes. Convergence of the inversion-by-peeling iteration scheme of eqn (3.55b) was obtained after only 28 iterations, and the partial-wave phase-shifts were then extracted from the S-matrix elements obtained via eqn (3.45). The PEF was then determined by applying the Firsov method using eqns (3.50)–(3.52).

3.6.2 Inversion of bulk property data

Inversion methods for BPs tend to fall into two classes. The first class, which is typified by the second virial coefficient, involves in principle a single layer of integration, as can be seen in eqn (3.3b), between the PEF, $V(R)$, and the BP. The second class of BP data, typified by the viscosity coefficient $\eta = [\eta]_1 f_\eta$ (with f_η a factor of order 1 that is relatively insensitive to the specific form for $V(R)$), is separated from $V(R)$ by three layers of integration. Both classes are based on classic backgrounds.

3.6.2.1 Inversion of second virial coefficient data

Integration by parts in eqn (3.3b) gives an equivalent form for the second virial coefficient $B_2(T)$ that will prove useful for the development of a procedure that allows $V(R)$ to be determined from the inversion of its temperature dependence. This procedure gives

$$B_2(T) = -\tfrac{2}{3}\pi\beta \int_0^\infty \mathrm{e}^{-\beta V(R)} \frac{\mathrm{d}V}{\mathrm{d}R} R^3 \, \mathrm{d}R\,, \tag{3.58}$$

with $\beta \equiv (k_\mathrm{B}T)^{-1}$. Upon recalling that force and potential energy are related by $F(R) = -\mathrm{d}V/\mathrm{d}R$, expression (3.58) can be seen to provide the connection between the interatomic force operative during the collision of a pair of atoms and the second virial coefficient. By defining a new function $\phi(R)$ as $\phi(R) = V(R) + \mathfrak{D}_\mathrm{e}$, using $\phi(R_\mathrm{e}) = 0$ to define a new energy origin, and by expressing $\phi(R)$ as the sum of two monotonic functions $\phi(R) = \phi_+(R) + \phi_-(R)$, in which $\phi_+(R)$, $\phi_-(R)$ corresponds to the branch of $\phi(R)$ that gives rise to a repulsive and attractive force, respectively, between the interacting atoms due to its monotone decreasing and increasing nature, respectively, the second virial coefficient $B_2(T)$ may be rewritten as

$$B_2(T) = \frac{2\pi \mathrm{e}^{\beta\mathfrak{D}_\mathrm{e}}}{3k_\mathrm{B}T}\left[-\int_0^{R_\mathrm{e}} R^3\left(\frac{\mathrm{d}\phi_+}{\mathrm{d}R}\right)\mathrm{e}^{-\beta\phi_+(R)}\mathrm{d}R - \int_{R_\mathrm{e}}^\infty R^3\left(\frac{\mathrm{d}\phi_-}{\mathrm{d}R}\right)\mathrm{e}^{-\beta\phi_-(R)}\mathrm{d}R\right].$$

Because $\phi_+(R)$ and $\phi_-(R)$ are monotonic functions, they may be inverted to give monotonic inverse functions $R_+(\phi)$ and $R_-(\phi)$, thereby enabling the variable of integration in each integral to be replaced by ϕ to give $B_2(T)$ as

$$B_2(T) = \frac{2\pi \mathrm{e}^{\beta\mathfrak{D}_\mathrm{e}}}{3k_\mathrm{B}T}\left[\int_0^\infty R_+^3(\phi)\mathrm{e}^{-\beta\phi}\,\mathrm{d}\phi - \int_0^{\mathfrak{D}_\mathrm{e}} R_-^3(\phi)\mathrm{e}^{-\beta\phi}\,\mathrm{d}\phi\right].$$

By extending the integration limit in the second term from $\mathfrak{D}_\mathrm{e}$ to ∞ through a redefinition of $R_-(\phi)$ as

$$R_-(\phi) = \begin{cases} R_-(\phi)\,, & 0 \le \phi \le \mathfrak{D}_\mathrm{e} \\ 0\,, & \phi > \mathfrak{D}_\mathrm{e} \end{cases},$$

the two integrals may be recombined in order to express $B_2(T)$ in the alternative form (Frisch and Helfand, 1960; Maitland *et al.*, 1981)

$$B_2(T) = \frac{2\pi \mathrm{e}^{\beta\mathfrak{D}_\mathrm{e}}}{3k_\mathrm{B}T}\int_0^\infty \Delta(\phi)\mathrm{e}^{-\beta\phi}\,\mathrm{d}\phi\,, \tag{3.59a}$$

with $\Delta(\phi)$, called the well-width function (Maitland *et al.*, 1981), defined as

$$\Delta(\phi) \equiv \begin{cases} R_+^3(\phi) - R_-^3(\phi)\,, & \phi \le \mathfrak{D}_\mathrm{e} \\ R_+^3(\phi)\,, & \phi > \mathfrak{D}_\mathrm{e} \end{cases}. \tag{3.59b}$$

For energies that correspond to bound states of $V(R)$, $R_+(\phi)$ and $R_-(\phi)$, with $R_+(\phi) < R_-(\phi)$, represent the vibrational turning points for the relative vibrational motions of the atom pair.

As the integral in eqn (3.59a) represents a Laplace transform of the width-function $\Delta(\phi)$, it follows that the inverse Laplace transform of $2\pi\beta e^{\beta\mathfrak{D}_e}B_2(\beta)/3$ should, in principle, give $\phi(R)$, and hence, $V(R)$. Smith and coworkers (Cox *et al.*, 1980; Maitland and Smith, 1972; Smith *et al.*, 1980; Smith *et al.*, 1981) showed that unfortunately, evaluation of the inverse Laplace transform for nonmonotonic PEFs is numerically ill-conditioned (Maitland *et al.*, 1985) because, although $B_2(T)$, which is equivalent to the Laplace transform $\widetilde{\Delta}$ of $\Delta(\phi)$, is relatively insensitive to small changes in ϕ, the potential energy $V(R)$ obtained via an inverse Laplace transform is inordinately sensitive to small variations in $B_2(T)$, so that unless virial coefficient data accurate to 1 ppm are available, convergence of iterates to give accurate values of $\Delta(\phi)$ will be very difficult to achieve.

The introduction of a length scale $\overline{R}$ directly related to the BP whose temperature dependence is to be inverted is an important initial step for any such inversion procedure. Because $B_2(T)$ has dimensions of volume, it is tempting to employ a length scale given by $\overline{R} = [\frac{3}{2}B_2(T)/\pi]^{\frac{1}{3}}$ for the inversion of second virial coefficient data. This scale works well for the inversion of a monotonic repulsive PEF, but becomes problematic for realistic atomic interactions that have both repulsive and attractive branches because $B_2(T)$ becomes negative below the Boyle temperature. To overcome this problem Smith and coworkers (Cox *et al.*, 1980; Maitland and Smith, 1972; Smith *et al.*, 1980; Smith *et al.*, 1981) suggested defining the characteristic length $\overline{R}$ in terms of the positive-definite combination of $B_2(T)$ and its first temperature derivative given by

$$\mathcal{B}(T) \equiv B_2(T) + T\frac{\mathrm{d}B_2}{\mathrm{d}T}\,, \tag{3.60a}$$

rather than in terms of $B_2(T)$ alone, that is, to define $\overline{R}$ as

$$\overline{R}(T) \equiv \left[\frac{3}{2\pi}\mathcal{B}(T)\right]^{\frac{1}{3}}\,. \tag{3.60b}$$

One clear advantage of such a definition for $\overline{R}$ is that $\mathcal{B}(T)$ is always positive: it enabled quite successful inversions to be carried out both for a set of $\mathcal{B}(T)$ pseudodata generated over the temperature range 85 K $\leq T \leq$ 650 K from a known Ar_2 PEF (with and without assigned random uncertainties) (Cox *et al.*, 1980; Smith *et al.*, 1981), and for experimental virial coefficient data for Kr over the temperature range 100 K $\leq T \leq$ 600 K (Clancy *et al.*, 1975; Gough *et al.*, 1974), although a significant lateral displacement (of approximately 1.5%) was obtained for points on the Kr_2 repulsive wall.

It may be shown that $\mathcal{B}(T)$ takes the form

$$\mathcal{B}(T) = \tfrac{2}{3}\pi\beta^2 e^{\beta\mathfrak{D}_e}\int_0^\infty (\phi - \mathfrak{D}_e)\Delta(\phi)\,e^{-\beta\phi}\,\mathrm{d}\phi\,, \tag{3.61a}$$

from which it will be clear that $3e^{-\beta\mathfrak{D}_e}\mathcal{B}(T)/(2\pi\beta^2)$ is the Laplace transform of $(\phi - \mathfrak{D}_e)\Delta(\phi) = V(R)\Delta(\phi)$. The corresponding inverse Laplace transform gives $V(R)\Delta(\phi)$ as

$$V(R)\Delta(\phi) \equiv \frac{1}{2\pi i}\int_{d-i\infty}^{d+i\infty} \frac{3e^{-\beta\mathfrak{D}_e}\,\mathcal{B}(T)}{2\pi\beta^2}\, e^{\beta\phi}\, d\beta$$

$$= -\frac{k_B}{2\pi i}\int_{c-i\infty}^{c+i\infty} \frac{3e^{-\beta\mathfrak{D}_e}\mathcal{B}(T)}{2\pi}\, e^{\beta\phi}\, dT\,, \tag{3.61b}$$

so that $V(R)$ may be expressed formally as

$$V(R) = \frac{3k_B e^{-\beta\mathfrak{D}_e}}{2\pi(R_-^3 - R_+^3)}\,\frac{1}{2\pi i}\int_{c-i\infty}^{c+i\infty} e^{\beta\phi}\,\mathcal{B}(T)\, dT\,. \tag{3.62}$$

This expression provides a formal basis for the inversion procedure developed by Smith and coworkers, in which the inversion function $G_{\mathcal{B}}(T)$ is defined via eqn (3.62) as

$$G_{\mathcal{B}}(T) \equiv \frac{V(\overline{R})}{k_B T} = \frac{3e^{-\beta\mathfrak{D}_e}}{2\pi(\overline{R}_-^3 - \overline{R}_+^3)T}\,\frac{1}{2\pi i}\int_{c-i\infty}^{c+i\infty} e^{\beta\phi}\,\mathcal{B}(T)\, dT\,. \tag{3.63}$$

Expression (3.63) provides a formal justification of the validity of the earlier ad hoc procedure employed for the direct inversion of the temperature dependence of second virial coefficient data to obtain PEFs for atomic interactions (Cox *et al.*, 1980; Smith *et al.*, 1980; Smith *et al.*, 1981). As noted by Maitland *et al.* (1985), while it would be feasible in principle to employ eqn (3.63) to obtain the initial iterate of $G_{\mathcal{B}}(T)$ from a starting PEF $V_0(\overline{R})$, the encumbrance of evaluating the inverse Laplace transform makes such a method impractical. The practical inversion procedure consists of four steps:

(i) Obtain an initial approximation $V_0\overline{R})$ to $V(R)$, such as a previous MLJ (12,6) PEF determined from fitting to experimental data (see also Maitland and Wakeham 1978*b*) and generate an initial inversion function $G_{\mathcal{B}}^{(0)}(T) = \beta V_0(\overline{R}_0)$, with $\overline{R}_0$ given by $\overline{R}_0(T) = [3\mathcal{B}_0(T)/(2\pi)]^{\frac{1}{3}}$ in terms of $\mathcal{B}_0(T)$ values computed from $V_0(R)$.

(ii) Generate a better estimate of $V(\overline{R})$ as $V_1(\overline{R}) = kTG_{\mathcal{B}}^{(0)}(T)$, but with $\overline{R}$ now given by eqn (3.60b) in terms of the experimental second virial coefficient data.

(iii) Generate an improved inversion function $G_{\mathcal{B}}^{(1)}(T)$ from $V_1(\overline{R})$.

(iv) Repeat steps (ii)–(iii), to obtain new values $V_{i+1}(\overline{R})$ and $G_{\mathcal{B}}^{(i)}(T)$ until a predefined acceptable level of convergence has been achieved, that is, until $V_{i+1}(\overline{R})$ and $V_i(\overline{R})$ are essentially indistinguishable.

In almost all cases, convergence of this iteration scheme is typically achieved within three to four iterations, as shown both by tests carried out using simulated data (pseudodata) and by inversion of actual experimental data.

3.6.2.2 Inversion of transport property data

Although the focus in the following is upon inversion of viscosity data, the same principles and procedures will apply to any transport property (Maitland and Wakeham, 1978*b*). The viscosity coefficient $\eta(T)$ is represented as

$$\eta(T) = [\eta]_1 f_\eta\,, \tag{3.64a}$$

in which $[\eta]_1 \equiv [\eta]_1(T)$ is the first Chapman–Cowling approximation to η while f_η, which corrects for Chapman–Cowling contributions to η from higher-order terms, has a value that is near unity, depends only weakly upon temperature, and is (as shown in Chapter 2) relatively insensitive to the actual form of the pair interaction. The idea that the temperature dependence of a transport property like the viscosity could be inverted to generate the interatomic interaction as a function of separation goes back to a remark by Hirschfelder and Eliason (1957) that for monotonic simple inverse-power repulsive (or attractive) interactions the product $\overline{c}_r\Omega^{(2,2)}(T)$ at a particular temperature T has the value $\pi\overline{R}^2$ for separation $\overline{R}$ corresponding to $V(\overline{R}) = Ck_BT$, with C a constant. As pointed out by Maitland *et al.* (1978*b*), this comment implies that it is possible to find a function $G(T)$, with the property $V(\overline{R}) = k_BTG(T)$ when $\overline{c}_r\Omega^{(2)}(T) = \pi\overline{R}^2$, that serves as an inversion function. It will be instructive to follow the derivation of the inversion process outlined in Maitland *et al.* 1985.

Upon employing eqn (2.70) for $[\eta]_1$, the viscosity is given in terms of the effective cross section $\mathfrak{S}_\eta(T) \equiv \mathfrak{S}(20)$ as

$$\eta(T) = \frac{k_BTf_\eta}{\overline{c}_r(T)\mathfrak{S}_\eta(T)}\,, \tag{3.64b}$$

and by combining eqns (2.197) and (2.206), with $m = n = 2$, $\mathfrak{S}_\eta(T)$ may be obtained as

$$\mathfrak{S}_\eta(T) = \frac{1}{5}\,\beta \int_0^\infty (\beta E)^3 \mathrm{e}^{-\beta E}\, Q^{(2)}(E)\,\mathrm{d}E\,, \tag{3.65}$$

with $\beta \equiv (k_BT)^{-1}$. The temperature dependence of $\eta(T)$ is thus related to the interaction potential energy $V(R)$, through eqns (3.65) and (2.202, 2.206), which involves three layers of integration between $V(R)$ and $\eta(T)$, namely, over separation R, impact parameter b (equivalently scattering angle, χ), and collision energy E. Inversion of the temperature dependence of the viscosity coefficient may thus be said to be equivalent to the problem of determining $V(R)$ from a set of values of its functional $\mathfrak{S}(20)$ (Maitland *et al.*, 1985). Because three layers of integration must be inverted, at least in principle, in order to obtain $V(R)$ from the temperature dependence of $\mathfrak{S}(20)$, it makes sense to proceed in three steps, beginning with the connection between $V(R)$ and the deflection (scattering) angle χ.

Expression (2.202) for the deflection angle $\chi(b, E)$, namely,

$$\chi(b, E) = \pi - 2b \int_{R_c}^\infty \frac{1}{R^2}\left[1 - \frac{b^2}{R^2} - \frac{V(R)}{E}\right]^{-\frac{1}{2}} \mathrm{d}R\,, \tag{3.66a}$$

may be rewritten in terms of $\alpha(R, E) \equiv R^2[1 - V(R)/E]$ as

$$\chi(b, E) = \pi - 2b \int_{R_c}^\infty [\alpha(R, E) - b^2]^{-\frac{1}{2}} R^{-1}\mathrm{d}R\,. \tag{3.66b}$$

By defining $\gamma(R, E)$ as $\gamma(R, E) \equiv \ln[\alpha(R, E)/R^2]$, $R^{-1}\,\mathrm{d}R$ may be replaced by $\frac{1}{2}(\mathrm{d}\ln\alpha - \mathrm{d}\gamma)$, so that $\chi(b, E)$ takes the form

$$\chi(b,E) = \pi - \int_{b^2}^{\infty} \left(\frac{\mathrm{d}\ln\alpha}{\mathrm{d}\alpha}\right) \frac{b}{\sqrt{\alpha - b^2}}\,\mathrm{d}\alpha + \int_{b^2}^{\infty} \left(\frac{\mathrm{d}\gamma}{\mathrm{d}\alpha}\right) \frac{b}{\sqrt{\alpha - b^2}}\,\mathrm{d}\alpha$$

$$= \int_{b^2}^{\infty} \left(\frac{\mathrm{d}\gamma}{\mathrm{d}\alpha}\right) \frac{b}{\sqrt{\alpha - b^2}}\,\mathrm{d}\alpha\,. \tag{3.66c}$$

The final step then follows from recognition that the first integral has the value π so that eqn (3.66c) is an Abelian integral equation whose inverse is

$$\gamma(z^2) = -\frac{2}{\pi}\int_{z}^{\infty} \frac{\chi(b,E)}{\sqrt{b^2 - z^2}}\,\mathrm{d}b\,. \tag{3.67}$$

This result plays the same role in the inversion of experimental viscosity data that eqn (3.51) plays in the Firsov inversion of beam scattering data. Note also that in this expression $z(E) \equiv z > b$ is real, but otherwise arbitrary.

A physical basis for the viscosity inversion procedure described by Maitland *et al.* (1978*b*), is provided by three key observations:

1. The dominant contribution to the scattering angle $\chi(b,E)$ results from $V(R)$ when the separation R between the two atoms is close to their distance of closest approach, R_{c}.
2. The nature of the factor $1 - \cos^2\chi(b,E)$ in the integral that determines $Q^{(2)}(E)$ suppresses the small-angle scattering contribution to the integral, and allows the introduction of the random-phase approximation (RPA) for $\cos^2\chi(b,E)$ together with an appropriately chosen cut-off value $b_{\mathrm{c}}(E)$ of b.
3. The strongly peaked energy weight factor $E^3\mathrm{e}^{-\beta E}$ in the integrand of eqn (3.65) effectively carries the role of $b_{\mathrm{c}}(E)$ over into $\mathfrak{S}_{\eta}(T)$.

These observations are also used for the present discussion.

The distance of closest approach, $\overline{R} \equiv R_{\mathrm{c}}$, between a pair of atoms participating in a collision at energy E characterized by impact parameter z must satisfy the relation

$$z^2 = \overline{R}^2\left[1 - \frac{V(\overline{R})}{E}\right], \tag{3.68}$$

from which $V(\overline{R})$ may be obtained as

$$V(\overline{R}) = E\left(1 - \frac{z^2}{\overline{R}^2}\right) \tag{3.69}$$

once z has been determined. Note, however, that in order for z to be real, eqn (3.68) requires $\overline{R}$ at fixed collision energy E to be assigned only values greater than the value for which $V(\overline{R}) = E$. Moreover, the definition of $\gamma(R,E)$ allows $\alpha(R,E)$ to be expressed as $\alpha(R,E) = R^2\mathrm{e}^{\gamma(R,E)}$, which may then be employed in eqn (3.68) to obtain $\overline{R}(E)$ as

$$\overline{R}(E) = z\exp\left\{\frac{1}{\pi}\int_{z}^{\infty}\frac{\chi(b,E)}{\sqrt{b^2 - z^2}}\,\mathrm{d}b\right\}. \tag{3.70}$$

Upon combining eqns (3.69, 3.70), $V(\overline{R})$ may be obtained as

$$V(\overline{R}) = E\left[1 - \exp\left\{-\frac{2}{\pi}\int_z^\infty \frac{\chi(b,E)}{\sqrt{b^2 - z^2}}\,\mathrm{d}b\right\}\right]. \tag{3.71}$$

This last result is the key inversion equation for the final stage of the full inversion process. Note that eqn (3.68) plays the same role for the inversion of viscosity data that eqn (3.47) plays for the inversion of atomic beam scattering data. Moreover, should the Firsov expression (3.51) for $I(z, E)$ be employed, then eqn (3.71) is identical to eqn (3.52).

Because $\mathfrak{S}_\eta(T)$ is a monotonic function of temperature determined from the monotonic function $Q^{(2)}(E)$ of the collision energy through eqn (3.65), it is possible to define a collision energy $\overline{E}^{(2)}$ such that $\mathfrak{S}_\eta(T) \equiv Q^{(2)}(\overline{E}^{(2)})$. As the weight function $(\beta E)^3 \mathrm{e}^{-\beta E}$ in eqn (3.65) is sharply peaked around $E = 3k_\mathrm{B}T$, application of the mean value theorem of integral calculus, in which $Q^{(2)}(E)$ is replaced by $Q^{(2)}(3k_\mathrm{B}T)$, gives $\mathfrak{S}_\eta(T)$ approximately as

$$\mathfrak{S}_\eta(T) \simeq \frac{6}{5}\, Q^{(2)}(3k_\mathrm{B}T). \tag{3.72}$$

This approximation has been found for a MLJ (12,6) PEF to be quite accurate over an extended range of temperature (Maitland *et al.*, 1985): moreover, as mentioned at the start of this section, it effects an approximation of the required inversion of the integration over collision energy. Because $\mathfrak{S}_\eta(T)$ has the dimensions of a cross-section, as does $Q^{(2)}(3k_\mathrm{B}T)$, $\mathfrak{S}_\eta(T)$ may be written formally as

$$\mathfrak{S}_\eta(T) = g_2 \overline{R}^2(3k_\mathrm{B}T), \tag{3.73}$$

with g_2 an appropriately chosen proportionality factor (that has yet to be determined). Thus, $\overline{R}(3k_\mathrm{B}T)$ is given by

$$\overline{R}(3k_\mathrm{B}T) = \sqrt{\frac{\mathfrak{S}_\eta(T)}{g_2}}. \tag{3.74}$$

Alternatively, upon substitution of $\overline{R}(3k_\mathrm{B}T)$ from eqns (3.6–3.27) into eqn (3.73), $\mathfrak{S}_\eta(T)$ may be obtained as

$$\mathfrak{S}_\eta(T) = g_2 z^2 \exp\left\{\frac{2}{\pi}\int_z^\infty \frac{\chi(b, 3k_\mathrm{B}T)}{\sqrt{b^2 - z^2}}\,\mathrm{d}b\right\}. \tag{3.75}$$

Note that this expression may be solved to give a value $z(3k_\mathrm{B}T)$ associated with each particular value of $\mathfrak{S}_\eta(T)$. Finally, upon substituting eqn (3.74) into eqn (3.69) for $V(\overline{R})$, the result

$$\frac{V[\overline{R}(3k_\mathrm{B}T)]}{k_\mathrm{B}T} = 3\left[1 - \frac{g_2 z^2(3k_\mathrm{B}T)}{\mathfrak{S}_\eta(T)}\right] \tag{3.76a}$$

$$\equiv G_\eta(T), \tag{3.76b}$$

thereby providing an implicit definition of the inversion function $G_\eta(T)$. To proceed further, however, it is necessary to determine g_2.

The integration over impact parameter may be circumvented (Maitland *et al.*, 1978) by utilizing the RPA for $\cos^2\chi$, namely, $\cos^2\chi \overset{\text{RPA}}{=} \frac{1}{2}$, from which $Q^{(2)}(E)$ becomes

$$Q^{(2)}(E) \simeq \tfrac{1}{2}\,\pi b_{\text{c}}^2(E)\,, \tag{3.77}$$

in which $b_{\text{c}}(E)$ corresponds to the impact parameter at which the deflection angle $\chi(b, E)$ falls to π^{-1}. This critical value $b_{\text{c}}(E)$ will be relatively large, and the corresponding deflection angle sufficiently small ($< \pi^{-1}$) that the distance of closest approach, $\overline{R}(E)$, will approximately equal $b_{\text{c}}(E)$, in which case $\mathfrak{S}_\eta(T)$ becomes

$$\mathfrak{S}_\eta(T) = g_2\overline{R}^2(3k_{\text{B}}T) \simeq \frac{3\pi}{5}b_{\text{c}}^2(3k_{\text{B}}T) \approx \frac{3\pi}{5}\overline{R}^2(3k_{\text{B}}T)\,, \tag{3.78}$$

which suggests that an appropriate choice of g_2 is $g_2 = 3\pi/5$.

For a monotonic repulsive (or attractive) PEF, the scattering angle $\chi(b_{\text{c}}, E) \approx \pi^{-1}$ associated with $\overline{R}(E)$ [and approximately equal to $b_{\text{c}}(E)$] can be approximated by the small-angle scattering result

$$\begin{aligned}\chi(b_{\text{c}}, E) &= \left|\frac{\overline{R}}{E}\int_{\overline{R}}^{\infty}\frac{1}{\sqrt{R^2 - \overline{R}^2}}\frac{\mathrm{d}V(R)}{\mathrm{d}R}\,\mathrm{d}R\right| \\ &= \frac{\overline{R}}{E}\int_{\overline{R}}^{\infty}\frac{1}{\sqrt{R^2 - \overline{R}^2}}\frac{\mathrm{d}|V(R)|}{\mathrm{d}R}\,\mathrm{d}R\,.\end{aligned}$$

The approximate result $\chi(b_{\text{c}}, E) \simeq \pi^{-1}$ gives rise to an expression for $|V(R)|$ that may be written as (Maitland *et al.*, 1978)

$$|V(\overline{R})| = k_{\text{B}}TG_\eta(T)\,, \tag{3.79a}$$

with $\overline{R}$ given in terms of $\mathfrak{S}_\eta(T)$ by eqn (3.78), and $G_\eta(T)$ given by

$$G_\eta(T) = \frac{6}{\pi^2}\int_0^1\left\{\ln\sqrt{\frac{\mathfrak{S}_\eta(T)}{\mathfrak{S}_\eta(sT)}} - \ln\left[1 + \sqrt{\frac{\mathfrak{S}_\eta(T)}{\mathfrak{S}_\eta(sT)}}\right]\right\}\mathrm{d}s\,. \tag{3.79b}$$

Note that because $|V(R)|$, rather than $V(R)$ itself, appears in eqn (3.6–36a), it is still necessary to know whether $V(R)$ is monotone positive (repulsive) or monotone negative (attractive) before the inversion can be completed.

For a PEF having a single minimum at separation $R = R_{\text{e}}$ and derivative $V'(R) < 0$ for $R < R_{\text{e}}$, $V'(R) > 0$ for $R > R_{\text{e}}$, $\chi(b_{\text{c}}, E)$ can be represented by

$$\chi(\overline{R}, E) = \left|\frac{\overline{R}}{E}\int_{\overline{R}}^{\infty}\frac{1 - 2H(R_{\text{e}})}{\sqrt{R^2 - \overline{R}^2}}\left|\frac{\mathrm{d}V(R)}{\mathrm{d}R}\right|\mathrm{d}R\right| \simeq \frac{1}{\pi}\,,$$

in which $H(R_{\text{e}})$ is the Heaviside function. For collision energies E sufficiently low that $\overline{R} \geq R_{\text{e}}$ or sufficiently high for $\overline{R} < R_{\text{e}}$ but still sufficiently low for the integral to remain dominated by contributions from the attractive branch of the PEF, the solution to this equation gives rise to $V(\overline{R}) = k_B T G_\eta(T)$ which, because $G_\eta(T)$ is, as may be seen from eqn (3.79b), always negative, determines the attractive branch of the PEF. However, for collision energies E high enough that the attractive branch plays only a minor role in determining the integral, the solution takes the form $V(\overline{R}) = -k_B T G_\eta(T)$, and gives rise to the repulsive branch of the PEF. The inversion function $G_\eta(T)$ has been found to be relatively insensitive to the detailed nature of the PEF and to depend almost solely upon the reduced temperature $T^* \equiv k_B T/\mathfrak{D}_e$ (Maitland *et al.*, 1985), with $\mathfrak{D}_e$ the depth of the PEF, thereby enhancing the rate of convergence of the iterative scheme employed to perform the inversion of the experimental data.

Equation (3.64b), together with expressions (3.74)–(3.76) form a set of equations that provides a means for carrying out the inversion of a set of temperature-dependent viscosity measurements for a pure gas to give the interaction PEF between a pair of its constituent atoms. The iteration process is similar to that already described for the inversion of second virial coefficient data, and may be itemized as follows:

(i) An initial PEF $V_0(R)$ is assumed, and employed to compute the deflection function of eqn (3.66a) for $E = 3k_B T$ for each experimental temperature T.

(ii) The impact parameter $z(3k_B T)$ is computed from eqn (3.6-32) for each experimental temperature T.

(iii) $V_1(\overline{R})/(k_B T)$ is evaluated from eqn (3.6-33a), with $\overline{R}$ determined from eqn (3.74).

(iv) Steps (i)–(iii) are repeated until $V_{i+1}(\overline{R})$ and $V_i(\overline{R})$ are essentially indistinguishable.

As for the iterative scheme developed for second virial coefficient data, rather few iterations, typically three to four only, are necessary to achieve a converged experimentally determined PEF. A practical set of three criteria for monitoring the convergence of the iterands has been given by Gough *et al.* (1972):

(i) The degree to which $V_i(\overline{R})$ is unchanged upon the next iteration to give $V_{i+1}(\overline{R})$.

(ii) The degree to which the values of $\mathfrak{S}_\eta(T)$ calculated from $V_i(\overline{R}$ correspond to the experimental values.

(iii) The degree to which $V_i(\overline{R})$ obtained from different initial trial functions $V_0(R)$ are in accord with one another.

Inversion of experimental viscosity data was first carried out successfully for Ar (Gough *et al.*, 1972), both because an extensive set of very accurate experimental viscosity data was available and because an accurate PEF determined (Maitland and Smith, 1971) from a multi-property fit to an extensive set of bulk data already existed to serve as a benchmark PEF. A similar successful inversion of viscosity data was also carried out for Kr (Cox *et al.*, 1980; Smith *et al.*, 1980; Smith *et al.*, 1981) using an accurate estimate of the well depth obtained from spectroscopy (Freeman *et al.*, 1974; Tanaka and Yoshino, 1970; Tanaka and Yoshino, 1972; Tanaka *et al.*, 1973). The need in the earliest versions of the viscosity data inversion procedure to procure a good prior estimate of the well-depth $\mathfrak{D}_e$ from an external source was resolved by Maitland and Wakeham (1978*b*), who showed that a good initial value for $\mathfrak{D}_e$ may be obtained by studying the rate and degree of convergence of the iteration scheme.

Note that the procedure described for inversion of the temperature dependence of viscosity (equivalently, thermal conductivity) data may readily be extended to the inversion of the temperature dependence of an arbitrary $\Omega^{(m,n)}(T)/\bar{c}_{\mathrm{r}}(T)$ group (Maitland *et al.*, 1985). However, the only distinct case that arises naturally for pure gases is the group $\Omega^{2,2)}(T)/\bar{c}_{\mathrm{r}}(T)$ associated with viscosity: all other pure gas omega-integrals appear only in the correction factors f_η and f_λ for the viscosity and thermal conductivity coefficients.

The inversion procedure has been straightforwardly extended (Gough *et al.*, 1975) to include the inversion of interaction viscosities extracted (Kestin *et al.*, 1970) from binary mixture viscosity data via eqns (2.82), as it depends upon the mixture group $\Omega^{2,2)}_{\mathrm{AB}}(T)/\bar{c}_{\mathrm{AB}}(T)$, which is the unlike-interaction equivalent of the group associated with the viscosity of pure gases. Moreover, for binary mixtures of species A and B, a new group, $\Omega^{1,1)}_{\mathrm{AB}}(T)/\bar{c}_{\mathrm{AB}}(T)$, appears naturally in the description of the first Chapman–Cowling approximation for the diffusion coefficient $[D_{\mathrm{AB}}]_1(T)$. The mixture inversion procedure was tested successfully for the inversion of Ar–Kr interaction viscosity data extracted from mixture viscosity measurements over the temperature range 120–1600 K and employing an initial estimate of the well-depth obtained from second virial coefficient data (Gough *et al.*, 1975). An extensive follow-up study reported "experimental" inversion PEFs for all unlike noble gas atom binary interactions (Maitland and Wakeham, 1978*a*).

The ion mobility coefficient, $K_{\mathrm{AB}}(T_{\mathrm{eff}})$, is defined as the ratio of the drift speed, v_{d}, of a positively charged ion of chemical species A (with mass m_{A}) through a dilute gas of a neutral single-component chemical species B (mass m_{B}), to the strength, E, of an electric field applied across the sample chamber. The ions are characterized by an effective temperature, T_{eff}, that is related to the bulk gas equilibrium temperature T in lowest approximation by (Mason and McDaniel, 1988)

$$\tfrac{3}{2}k_{\mathrm{B}}T_{\mathrm{eff}} = \tfrac{3}{2}k_{\mathrm{B}}T + \tfrac{1}{2}m_{\mathrm{B}}v_{\mathrm{d}}^2 \,.$$

At very low field strengths the drift speed will be small and the second term may be ignored relative to the first one, so that $T_{\mathrm{eff}} \equiv T$: for higher electric field strengths, T_{eff} is higher than the thermal temperature of the neutral species, essentially because some of the energy imparted to the ions by the electric field can be thermalized via collisions with the neutral species. The ionic mobility coefficient, $K_{\mathrm{AB}}(T_{\mathrm{eff}})$, for a singly charged cation A^+ is related to its diffusion coefficient, $D_{\mathrm{AB}}(T_{\mathrm{eff}})$, via (Mason and McDaniel, 1988)

$$K_{\mathrm{AB}}(T_{\mathrm{eff}}) = \frac{e[D_{\mathrm{AB}}]_1(T_{\mathrm{eff}})f_D(T_{\mathrm{eff}})}{k_{\mathrm{B}}T_{\mathrm{eff}}} \,,$$

and hence, depends upon the diffusion effective cross-section $\mathfrak{S}(10|\mathrm{A})_{\mathrm{AB}}$, or equivalently, the group $\Omega^{(1,1)}_{\mathrm{AB}}(T_{\mathrm{eff}})/\bar{c}_{\mathrm{AB}}(T_{\mathrm{eff}})$, as

$$K_{\mathrm{AB}}(T_{\mathrm{eff}}) = \frac{e}{n_{\mathrm{B}}}\,\frac{f_D}{m_{\mathrm{A}}\bar{c}_{\mathrm{AB}}\mathfrak{S}(10|\mathrm{A})_{\mathrm{AB}}} \,.$$

An inversion scheme along the lines of those employed for the inversion of neutral gas transport properties was developed by Viehland *et al.* (1976) for the inversion of ion mobility data. Additional steps are required for this inversion scheme, both because the Chapman–Cowling approximation scheme does not converge as rapidly for the ion mobility coefficient as it does for the transport coefficients in neutral gases and because the scheme includes a means for determining the depth, $\mathfrak{D}_e$, of the PEF based upon the experimental input data. An extensive set of simulated experimental data (pseudodata) was generated using a previously determined PEF for the Li^+–He interaction and employed to test the efficacy of the inversion procedure. Excellent agreement between the numerical PEF obtained from inversion of the pseudodata and the original PEF used to generate the pseudodata was achieved with only two iterations. A thorough discussion of the program INVERT for inversion of the temperature dependence of ion mobility data, its application to trace alkali ion mobility measurements in rare gases, and the resultant ion atom interaction PEFs has been given by Viehland (1983, 1984).

3.7 *Ab Initio* Calculation of Potential Energy Functions

A great deal of attention has thus far been paid to the development of various means of representing the interaction energies between atomic species and to the relationship between experimentally measured microscopic (e.g. beam scattering cross-section and spectroscopic transitions) and bulk gas, or macroscopic (e.g. second virial and transport coefficients) phenomena. There are two important reasons for doing so. The most obvious reason is historical and is based upon the relative ease of carrying out experimental dilute gas measurements, then inferring the dependence of the interaction energy between two atoms on their separation $\mathbf{R}$, in comparison to performing quantitatively accurate theoretical calculations of the interaction energies as a function of $\mathbf{R}$. As the present focus is upon atom–atom interactions, the only relevant variable is the distance of separation, R, between the two atoms. Moreover, with the availability of high-speed digital computers with extremely large memories and massive parallelization capabilities coupled with versatile and highly efficient software packages, it has become relatively straightforward to carry out highly accurate *ab initio* electronic structure calculations of the interaction energy between an atom pair.

3.7.1 Quantum mechanical background

The Schrödinger equation for a system consisting of two atoms, A and B, having nuclei with masses m_α and charges $Z_\alpha e$ plus N_α electrons (α = A, B) must be considered. Thus, consider a space-fixed coordinate system in which the nuclear coordinates are $\{\mathbf{R}'_A, \mathbf{R}'_B\}$ and the electronic coordinates are $\{\mathbf{r}'_i\}$ $(i = 1, \cdots, N_A + N_B)$. This system of nuclei and electrons is governed, in the absence of external electric and magnetic fields, simply by Coulombic interactions, so that the total Hamiltonian, expressed in terms of space-fixed coordinates, may be written as the sum of nuclear and electronic Hamiltonians as

$$\mathcal{H}_{AB} = \mathcal{H}_{nuc} + \mathcal{H}_{el} = -\frac{\hbar^2}{2}\left[\frac{1}{m_A}\nabla^2_{\mathbf{R}'_A} + \frac{1}{m_B}\nabla^2_{\mathbf{R}'_B}\right] + \mathcal{H}_{el}\,, \tag{3.80a}$$

with $\mathcal{H}_{\rm el}$ given in terms of the internuclear separation R as

$$\mathcal{H}_{\rm el} = -\frac{\hbar^2}{2m_{\rm e}}\sum_{i=1}^{N}\nabla^2_{\mathbf{r}'_i} + V_{\rm el}(R,\{r_{{\rm A}i}\},\{r_{{\rm B}i}\},\{r_{ij}\})\,. \tag{3.80b}$$

The potential energy operator $V_{\rm el}(R,\{r_{{\rm A}i}\},\{r_{{\rm B}i}\},\{r_{ij}\})$ is given explicitly by

$$V_{\rm el}(R,\{r_{{\rm A}i}\},\{r_{{\rm B}i}\},\{r_{ij}\}) = \frac{1}{4\pi\varepsilon_0}\left[\frac{Z_{\rm A}Z_{\rm B}e^2}{R} - \sum_{i=1}^{N}\left(\frac{Z_{\rm A}e^2}{r_{{\rm A}i}} + \frac{Z_{\rm B}e^2}{r_{{\rm B}i}}\right) + \sum_{i=1}^{N-1}\sum_{j=i+1}^{N}\frac{e^2}{r_{ij}}\right], \tag{3.80c}$$

in which R, $r_{{\rm A}i}$, $r_{{\rm B}i}$, r_{ij} are the distances between the elementary (charged) constituent particles making up atoms A and B. Because $V_{\rm el}$ depends only upon the magnitudes of the relative separations between pairs of the constituent particles, the center-of-mass (CM) motion of the nuclear-electronic system can be separated out from the relative motions of the constituent particles, thereby reducing the set of $3N{+}6$ coordinates to $3N{+}3$ coordinates defined relative to the CM motion. Even though a transformation from space-fixed coordinates to coordinates relative to the CM position would be the most logical motion choice to make in order to separate out the CM motion, the difference of approximately a factor 2000 between individual nucleon and electron masses makes it more convenient[6] to work with the set of coordinates relative to the CM, $\mathbf{R}'_{\rm CMN}$, of the two nuclei (Hirschfelder and Meath, 1967; Pack and Hirschfelder, 1968; Pack and Hirschfelder, 1970*a*; Pack and Hirschfelder, 1970*b*).

The position of nucleus B relative to nucleus A is given by $\mathbf{R} = \mathbf{R}'_{\rm B} - \mathbf{R}'_{\rm A}$, while the location of the CM of the A and B nuclei, $\mathbf{R}'_{\rm CMN}$, is given by

$$\mathbf{R}'_{\rm CMN} = y^2_{\rm A}\mathbf{R}'_{\rm A} + y^2_{\rm B}\mathbf{R}'_{\rm B}\,, \tag{3.81a}$$

in terms of mass fractions $y^2_\alpha \equiv m_\alpha/M$, $\alpha =$ A, B, but with $M \equiv m_{\rm A} + m_{\rm B}$ the total mass of the nuclei [rather than the total mass of atoms A and B, as defined in eqn (2.3)–(2.29)]. Similarly, the CM of the entire A–B system in space-fixed coordinates is

$$\mathbf{R}'_{\rm CM} = \frac{m_{\rm A}}{M+Nm_{\rm e}}\mathbf{R}'_{\rm A} + \frac{m_{\rm B}}{M+Nm_{\rm e}}\mathbf{R}'_{\rm B} + \frac{1}{M+Nm_{\rm e}}\sum_{i=1}^{N}\mathbf{r}'_i\,, \tag{3.81b}$$

while the gradient operators become

[6] Note that although the specific choice of relative coordinates to employ for separating out the CM motion does not affect the present discussion, it does affect the reduced mass that appears in the Schrödinger equation determining the relative vibrational motion of the nuclei, and also has ramifications for approximations yet to be discussed.

$$\nabla_{\mathbf{R}'_\mathrm{A}} = \frac{m_\mathrm{A}}{M + Nm_\mathrm{e}} \nabla_{\mathbf{R}'_\mathrm{CM}} - \nabla_\mathbf{R} - y_\mathrm{A}^2 \sum_{i=1}^{N} \nabla_{\mathbf{r}_i} ,$$

$$\nabla_{\mathbf{R}'_\mathrm{B}} = \frac{m_\mathrm{B}}{M + Nm_\mathrm{e}} \nabla_{\mathbf{R}'_\mathrm{CM}} + \nabla_\mathbf{R} - y_\mathrm{B}^2 \sum_{i=1}^{N} \nabla_{\mathbf{r}_i} , \qquad (3.81c)$$

$$\nabla_{\mathbf{r}'_i} = \nabla_{\mathbf{r}_i} + \frac{m_\mathrm{e}}{M + Nm_\mathrm{e}} \nabla_{\mathbf{R}'_\mathrm{CM}} .$$

Upon substitution of eqns (3.81) into eqns (3.80a, 3.80b), the total Hamiltonian for the A–B nucleus–electron system is given by

$$\mathcal{H}_\mathrm{total} = -\frac{\hbar^2}{2(M + Nm_\mathrm{e})} \nabla^2_{\mathbf{R}'_\mathrm{CM}} - \frac{\hbar^2}{2m_\mathrm{r}} \nabla^2_\mathbf{R} - \frac{\hbar^2}{2m_\mathrm{e}} \sum_{i=1}^{N} \nabla^2_{\mathbf{r}_i} + V_\mathrm{el} + \mathcal{H}_\mathrm{pol} , \qquad (3.82a)$$

with V_el given by eqn (3.80c), and $\mathcal{H}_\mathrm{pol}$, given by

$$\mathcal{H}_\mathrm{pol} = -\frac{\hbar^2}{2(M + Nm_\mathrm{e})} \left[\nabla^2_{\mathbf{r}_i} + 2 \sum_{i=1}^{N-1} \sum_{j=i+1}^{N} \nabla_{\mathbf{r}_i} \cdot \nabla_{\mathbf{r}_j} \right] , \qquad (3.82b)$$

is referred to as the mass-polarization Hamiltonian. The specific form (3.82b) arises from choosing to express the electron positions relative to $\mathbf{R}'_\mathrm{CMN}$. Note also that $\mathbf{R}'_\mathrm{CM}$ and $\mathbf{R}'_\mathrm{CMN}$ are related by

$$\mathbf{R}'_\mathrm{CM} = \mathbf{R}'_\mathrm{CMN} + \frac{m_\mathrm{e}}{M + Nm_\mathrm{e}} \sum_{i=1}^{N} \mathbf{r}_i , \qquad \mathbf{r}_i \equiv \mathbf{r}'_i - \mathbf{R}'_\mathrm{CMN} .$$

If the first term in expression (3.82a) is identified with the Hamiltonian governing the translational motion of the CM of the A–B system, while the remaining terms are identified with the Hamiltonian $\mathcal{H}_\mathrm{rel}(\mathbf{R}, \{\mathbf{r}_i\})$ governing the relative motion, then the Schrödinger equation

$$\mathcal{H}\Psi(\mathbf{R}'_\mathrm{CM}, \mathbf{R}, \{\mathbf{r}_i\}) = E\Psi(\mathbf{R}'_\mathrm{CM}, \mathbf{R}, \{\mathbf{r}_i\}) ,$$

becomes separable. Thus, the wavefunction $\Psi(\mathbf{R}'_\mathrm{CM}, \mathbf{R}, \{\mathbf{r}_i\})$ may be written as a product function

$$\Psi(\mathbf{R}'_\mathrm{CM}, \mathbf{R}, \{\mathbf{r}_i\}) = \psi(\mathbf{R}'_\mathrm{CM})\Psi_\mathrm{rel}(\mathbf{R}, \{\mathbf{r}_i\}) ,$$

with $\Psi_\mathrm{rel}(\mathbf{R}, \{\mathbf{r}_i\})$ obeying the Schrödinger equation

$$\mathcal{H}_\mathrm{rel}(\mathbf{R}, \{\mathbf{r}_i\})\Psi_\mathrm{rel}(\mathbf{R}, \{\mathbf{r}_i\}) = E_\mathrm{rel}\Psi_\mathrm{rel}(\mathbf{R}, \{\mathbf{r}_i\}) , \qquad (3.83a)$$

in which the Hamiltonian $\mathcal{H}_\mathrm{rel}(\mathbf{R}, \{\mathbf{r}_i\})$ is defined as

$$\mathcal{H}_\mathrm{rel}(\mathbf{R}, \{\mathbf{r}_i\}) = -\frac{\hbar^2}{2m_\mathrm{r}} \nabla^2_\mathbf{R} - \frac{\hbar^2}{2m_\mathrm{e}} \sum_{i=1}^{N} \nabla^2_{\mathbf{r}_i} + V_\mathrm{el} + \mathcal{H}_\mathrm{pol}$$

$$\equiv -\frac{\hbar^2}{2m_\mathrm{r}} \nabla^2_\mathbf{R} + \mathcal{H}_\mathrm{el} . \qquad (3.83b)$$

The presence of the Coulombic nuclear repulsion term in V_{el} means that the Schrödinger eqn (3.83a) cannot be further decoupled. However, should the separation $\mathbf{R}$ between the A and B nuclei be fixed, then the nuclear translational term contributes zero, and the Schrödinger equation reduces to an electronic Schrödinger equation that determines the motions of the N electrons when the two nuclei have the fixed distance R separating them. This is the "clamped-nuclei" approximation. The energy eigenfunctions $\Psi_k^{\text{el}}(\{\mathbf{r}_i\};\mathbf{R})$ and eignenvalues $E_k(\mathbf{R})$ obtained upon solution of this electronic Schrödinger equation will thus depend parametrically upon the separation R between the nuclei. As the set of functions $\{\Psi_k^{\text{el}}(\{\mathbf{r}_i\};\mathbf{R})\}$ forms a complete basis for the electronic coordinates of the A–B system at each fixed internuclear separation R, the general solutions of the electronic Schrödinger equation may be expressed as linear combinations of these basis functions.

Following Moszynski (2007), $\Psi_{\text{rel}}(\mathbf{R},\{\mathbf{r}_i\})$ may be expressed in the form of a Born–Huang expansion

$$\Psi_{\text{rel}}(\mathbf{R},\{\mathbf{r}_i\}) = \sum_l \Psi_l^{\text{el}}(\{\mathbf{r}_i\};\mathbf{R})\chi_l(\mathbf{R}) \tag{3.84a}$$

in terms of the eigenfunctions $\Psi_l^{\text{el}}(\{\mathbf{r}_i\};R)$ of the electronic Schrödinger equation

$$\mathcal{H}^{\text{el}}\Psi_l^{\text{el}}(\{\mathbf{r}_i\};R) = E_l(R)\Psi_l^{\text{el}}(\{\mathbf{r}_i\};R)\,, \tag{3.84b}$$

and the wavefunctions $\chi_l(\mathbf{R})$ that describe the nuclear motions. The eigensolutions $\{\Psi_l^{\text{el}}(\{\mathbf{r}_i\};\mathbf{R})\}$ depend only parametrically upon the nuclear coordinates $\mathbf{R}$, while the energy eigenvalues, $E_l(R)$, are functions only of R: note also that the index l labels the electronic states of the van der Waals dimer.

Upon substituting the Born–Huang expansion (3.84a) for $\Psi_{\text{rel}}(\mathbf{R},\{\mathbf{r}_i\})$ into eqns (3.83), then forming inner products with $\Psi_k^{\text{el}}(\mathbf{R},\{\mathbf{r}_i\})$, a set of coupled differential equations that determine the $\chi_k(\mathbf{R})$ is obtained. These equations may be expressed in the form

$$[-\frac{\hbar^2}{2m_{\text{r}}}\nabla_{\mathbf{R}}^2 + C_{kk}(\mathbf{R}) + E_k(R) - \mathcal{E}_k]\chi_k(\mathbf{R}) = -\sum_{l\neq k}[C_{kl}(\mathbf{R}) + 2\mathbf{B}_{kl}(\mathbf{R})\cdot\nabla_{\mathbf{R}}]\chi_l(\mathbf{R})\,, \tag{3.85}$$

in which $C_{kl}(\mathbf{R})$ and $\mathbf{B}_{kl}(\mathbf{R})$ are defined as

$$C_{kl}(\mathbf{R}) \equiv -\frac{\hbar^2}{2m_{\text{r}}}\langle\Psi_k^{\text{el}}|\nabla_{\mathbf{R}}^2|\Psi_l^{\text{el}}\rangle_{\{\mathbf{r}\}}\,, \tag{3.86a}$$

and

$$\mathbf{B}_{kl}(\mathbf{R}) \equiv -\frac{\hbar^2}{2m_{\text{r}}}\langle\Psi_k^{\text{el}}|\nabla_{\mathbf{R}}|\Psi_l^{\text{el}}\rangle_{\{\mathbf{r}\}}\,, \tag{3.86b}$$

while $\mathcal{E}_k$ is the energy eigenvalue characterizing the relative motion of the two nuclei, and the notation $\langle\Psi_k^{\text{el}}|\cdots|\Psi_l^{\text{el}}\rangle_{\{\mathbf{r}\}}$ indicates that integration has only been carried out over the electronic coordinates. The diagonal elements of $\mathbf{B}_{kl}$ vanish for real $\Psi_k^{\text{el}}(\{\mathbf{r}_i\};\mathbf{R})$, and the diagonal elements of $C_{kl}(\mathbf{R})$ depend only upon the magnitude R. The $C_{kk}(R)$ are referred to as adiabatic corrections, while the $C_{kl}(\mathbf{R})$ are known as diabatic corrections: as may be seen from eqns (3.86), both the diabatic and adiabatic terms are inversely proportional to the nuclear reduced mass, m_r, and hence, they play increasingly less-important roles as the reduced-mass of the dimer increases.

Neglect of the nondiagonal terms in (i.e. the RHS of) eqn (3.85) gives the Schrödinger equation (Born and Oppenheimer, 1927)

$$\left[-\frac{\hbar^2}{2m_\mathrm{r}}\nabla^2_\mathbf{R} + U_k(R)\right]\chi_k(\mathbf{R}) = \mathcal{E}_k\chi_k(\mathbf{R})\,, \tag{3.87a}$$

in which the potential energy operator $U_k(R)$ is given by the sum of the electronic energy and the adiabatic correction term, that is,

$$U_k(R) = E_k(R) + C_{kk}(R)\,. \tag{3.87b}$$

Note that this assumption is tantamount to saying that $\Psi^\mathrm{ad}_{\mathrm{rel},k}(\{\mathbf{r}_i\},\mathbf{R})$ is given by

$$\Psi^\mathrm{ad}_{\mathrm{rel},k}(\{\mathbf{r}_i\},\mathbf{R}) = \Psi^\mathrm{el}_k(\{\mathbf{r}_i\})\chi_k(\mathbf{R})\,. \tag{3.87c}$$

If, in addition to neglecting the diabatic terms in eqn (3.85), the adiabatic correction term on the left-hand side is also neglected, then the "clamped-nuclei" approximation is obtained. This approximation is also commonly referred to as the Born–Oppenheimer (Born and Oppenheimer, 1927) approximation to the full Schrödinger equation for the relative nuclear motion. In the Born–Oppenheimer approximation $U_k(\mathbf{R}) \equiv E_k(\mathbf{R})$.

If the interaction energy, $E_\mathrm{int}(R)$, between two atoms A and B is defined as the difference between the energy $E_\mathrm{AB}(R)$ of the dimer AB and the sum of the energies, E_A, E_B, of the isolated atoms, that is, as

$$V(R) \equiv E_\mathrm{int}(R) = E_\mathrm{AB}(R) - E_\mathrm{A} - E_\mathrm{B}\,, \tag{3.88}$$

then $E_\mathrm{AB}(R)$ is precisely the electronic energy $E_k(R)$ obtained for the AB dimer in the Born–Oppenheimer (adiabatic and clamped-nuclei) approximations. The potential energies $U(R)$ and $V(R)$ are thus related by a shift of the energy origin, namely,

$$V(R) = U(R) - E_\mathrm{A} - E_\mathrm{B}\,.$$

Equations (3.87) apply to all electronic states $\{k\}$ of the AB dimer. Here, however, concern is primarily with the ground electronic state of the AB dimer formed by two closed-shell atoms A, B, in which case, the electronic state subscript k is no longer needed. This is commonly done, so that the particular PEF is simply identified by specifying the appropriate electronic term symbol associated with the electronic state of the AB dimer. Thus, for dimers formed from open shell atoms, such as H or Li, for example, $\mathrm{H}_2(X\,^1\Sigma_g^+)$ denotes H_2 in its ground (X) electronic state, while $\mathrm{Li}_2(a\,^3\Sigma_u^+)$ denotes Li_2 in its lowest-lying (first) excited electronic (a) state. For noble gas dimers, however, even this convention is unnecessary, as the $^1\Sigma^+$ ground electronic state has an energy that is so much lower than that associated with any excited electronic state that no source of confusion can be possible.

It has long been known (Eisenschitz and London, 1930; London, 1930; London, 1937; Stone, 1996) that van der Waals interaction energies between two monomers can in general be broken down into four basic components whose origins involve electrostatic, inductive, dispersive, and electron exchange interactions. The first three components are determined by the natures of the interacting monomers: electrostatic terms may be associated with interactions between permanent electric dipole and/or higher (multipole) moments possessed by the two monomers, while induction terms are associated with interactions between permanent electric moments on one monomer and electric moments induced in the other monomer by them, and dipsersion terms are associated with interactions between instantaneous electric multipole moments arising from the dynamic multipole polarizabilities possessed by the two monomers. These three components of the interaction energy are all attractive. The intermonomer electron exchange component, however, is repulsive as a consequence of the role of the Pauli exclusion principle in preventing penetration of the electrons from one monomer into the space occupied by the electrons of the other monomer. Thus, van der Waals interactions between closed-shell atoms, which possess no permanent electric moments, will have only dispersion and electron exchange components, while van der Waals interactions between two open-shell atoms or between a neutral atom and an atomic ion may have contributions of all four types.

3.7.2 Quantum methodology

Ab initio computation may be carried out at two levels: one, designated by the acronym HF for Hartree–Fock, is based essentially upon the independent particle model for electrons; the other (referred to as the post-HF or electron-correlated level), also includes the effects of electron correlation. Post-HF calculations are considerably more difficult to carry out, and are often fraught with basis-set effects that must be overcome in order to obtain accurate values of the interaction energy are to be obtained. Calculations carried out fully at the HF level provide contributions to all four components of the interaction energy via the HF monomer charge densities, permanent electric moments, static, and dynamic polarizabilities. Although such HF calculations do include inter-monomer correlation effects, otherwise known as the HF dispersion energies, they exclude all intramonomer correlation effects, which can only be obtained at the post-HF level. Such calculations have nonetheless provided a good starting point for the development of realistic potential energy model functions, as discussed in Sections 3.3 and 3.4. There are two main approaches to post-HF *ab initio* determination of the interaction energy as a function of the internuclear separation R between a pair of atoms A and B. These approaches are commonly referred to as the supermolecule and perturbative approaches.

3.7.2.1 The supermolecule approach

The supermolecule approach to intermonomer interactions is based upon expression (3.88) together with the clamped-nuclei approximation. The A–B complex for each fixed internuclear separation is therefore treated as a rigid diatomic molecule governed by the Hamiltonian (3.80a). The eigenvalues $E_{\mathrm{AB}}(R)$, E_{A}, and E_{B} of $\mathcal{H}_{\mathrm{AB}}$, $\mathcal{H}_{\mathrm{A}}$, and $\mathcal{H}_{\mathrm{B}}$, respectively, are then computed using one of the high-level size-consistent computational methods of electronic structure theory, and the relevant interaction energies $V_{\mathrm{AB}}(R)$ are obtained as appropriate differences. However, not only is $V_{\mathrm{AB}}(R)$

typically several orders of magnitude smaller than $E_{\rm AB}(R)$, $E_{\rm A}$, $E_{\rm B}$, it is also often much smaller than the differences between their respective exact and approximate values (even those obtained using the most accurate aproximations). By carrying out calculations of $E_{\rm AB}(R)$, $E_{\rm A}$, and $E_{\rm B}$ using the *same* basis-sets and applying the basis-set superposition error (BSSE) counterpoise correction (Boys and Bernardi, 1970), net errors in the values of $E_{\rm AB}(R)$ may be minimized. This approach has the advantages that the energies of the A–B "supermolecule" dimer and the A and B monomer atoms are all treated uniformly, possible convergence problems associated with the perturbation expansion of the interaction energy are avoided, and interatomic exchange effects are incorporated automatically (Chałasiński and Gutowski, 1988). It does, however, have the unfortunate disadvantage of ultimately providing a single value for the interaction energy at each A–B separation R, with little or no information as to its physical origins. The size-extensive methods most commonly employed for supermolecule calculations are the coupled-cluster (CC) method (Shavitt and Bartlett, 2009) and many-body perturbation theory (Jeziorski and Szalewicz, 2003)[7] (MBPT).

3.7.2.2 The perturbation approach

The basic concept hinges upon being able to obtain $V_{\rm AB}$ directly, that is, without prior knowledge of $E_{\rm AB}(R)$, $E_{\rm A}$, or $E_{\rm B}$, from the expression

$$V_{\rm Ab} = \frac{\langle \Phi_{\rm A}\Phi_{\rm B}|\mathcal{V}|\Psi_{\rm AB}\rangle}{\langle \Phi_{\rm A}\Phi_{\rm B}|\Psi_{\rm AB}\rangle}\,, \tag{3.89a}$$

in which $\Phi_{\rm A}$, $\Phi_{\rm B}$, $\Psi_{\rm AB}$ are eigenfunctions of $\mathcal{H}_{\rm A}$ $\mathcal{H}_{\rm B}$, and $\mathcal{H}_{\rm AB}$, respectively, and $\mathcal{V}$ is given for the diatomic AB by an expression similar to eqn (3.80b), namely,

$$\mathcal{V} = \frac{e^2}{4\pi\varepsilon_0}\left[\frac{Z_{\rm A}Z_{\rm B}}{R} - \sum_{j\in{\rm B}}\frac{Z_{\rm A}}{r_{j{\rm A}}} - \sum_{i\in{\rm A}}\frac{Z_{\rm B}}{r_{i{\rm B}}} + \sum_{i\in{\rm A}}\sum_{j\in{\rm B}}\frac{1}{r_{ij}}\right], \tag{3.89b}$$

with $r_{j{\rm A}} \equiv |\mathbf{r}'_j - {\rm R}'_{\rm A}|$, $r_{i{\rm B}} \equiv |\mathbf{r}'_i - {\rm R}'_{\rm B}|$, and $r_{ij} \equiv |\mathbf{r}'_i - \mathbf{r}'_j|$. The unperturbed Hamiltonian for this procedure is $\mathcal{H}_0 \equiv \mathcal{H}_{\rm A} + \mathcal{H}_{\rm B}$. Even though the energy shift $V_{\rm AB}$ associated with the perturbation operator $\mathcal{V}$ is typically several orders of magnitude smaller than the eigenenergies $E_{\rm AB}$, $E_{\rm A}$, and $E_{\rm B}$, $\mathcal{V}$ unfortunately still cannot be viewed as a small operator in the mathematical sense, because changes in the wavefunctions are almost never small, and the spectra of $\mathcal{H}_0$ and $\mathcal{H} \equiv \mathcal{H}_0 + \mathcal{V}$ are quite different (Jeziorski and Szalewicz, 2003). For such a perturbation, conventional Rayleigh–Schrödinger (RS) perturbation theory breaks down (Kutzelnigg, 1980), and alternative perturbation formulations are required. The most succssful such formulation is based upon proper electron antisymmetrization of the many-body wavefunctions, and is referred to by its developers (Korona *et al.*, 1997; Rybak *et al.*, 1991; Szalewicz and Jeziorski, 1979; Williams *et al.*, 1993) as symmetry-adapted perturbation theory (SAPT), which represents a double perturbation procedure, in terms of the intermonomer interaction

[7]MBPT is also often referred to as Møller–Plesset (MP) perturbation theory (Chałasiński and Gutowski, 1988), hence the terms MP2, MP4, respectively, for second- and fourth-order perturbation calculations.

and intramonomer correlation energies, and calculations based upon it are able to provide all four components of the inter-monomer interaction energy up to, and including, the complete treatment of electron correlation via CC theory (Bartlett, 1981; Bartlett, 1989). Moreover, the SAPT interaction energy is represented as a sum of well-defined and physically meaningful contributions corresponding, more generally, to electrostatic, induction, dispersion, and electron exchange interactions (Jeziorski *et al.*, 1994).

The supermolecule and perturbation theory (particularly SAPT) approaches are complementary. On the one hand, the absolute accuracies of the outcomes of supermolecule calculations decrease as the atom–atom separation increases because the interaction energy (which is the difference between the total electronic energy of the supermolecule at separation R and the total electronic energy of the separated atoms) becomes an increasingly smaller fraction of the total supermolecule electronic energy. On the other hand, the absolute accuracies of SAPT calculations improves as the atom–atom separation increases because the perturbation represented by the interaction energy becomes smaller with increasing separation. Thus, it likely makes sense in general to employ a combination of the supermolecule and perturbation approaches in order to compute atom–atom interaction energies having the highest possible accuracy.

The theory of intermolecular interactions has been developed both to enable the generation of interaction energies between specific monomer pairs and to illucidate, as far as possible, the physical mechanisms contributing to the interactions. Perturbation theory has provided a framework for the achievement of a detailed understanding of the nature of intermonomer interactions in terms of electrostatic, electron exhange, induction, and dispersion contributions. The asymptotic behavior of the perturbation expansion has provided large-intermonomer separation constraints that must be satisfied by any PEF (more generally, potential energy surface, PES) determined either from experiment or from theory. Moreover, as pointed out by Chałasiński and Szczeńiak (1994, 2000), perturbation theory not only helps in the interpretation of the results of supermolecule calculations, but also it is essential for understanding and controlling their accuracy.

3.7.3 Analytical representation of *ab initio* interaction energies

The potential energy literature is rife with analytic functional forms that have been introduced for compact representations both of models for and *ab initio* computations of interaction potential energies. In addition, numerous variants of the individual forms have been proposed, in many cases because the already existing functional forms cannot readily accommodate more accurate determinations of the interaction energies. The most successful analytic functions have typically used the partition of the overall interaction into a modified exponential repulsive component plus an attractive damped dispersion compnent having a form similar to the HFD and XC type model functions introduced in Sections 3.3 and 3.4.

All analytic functions employed as fit-functions representing the R-dependence of *ab initio* calculations of atomic interaction energies can be represented by the generic functional form

$$V(R) = V_{\mathrm{rep}}(R) + V_{\mathrm{att}}(R)\,, \tag{3.90a}$$

in which the repulsive and attractive components $V_{\text{rep}}(R)$ and $V_{\text{att}}(R)$, respectively, are given in general by the functional forms

$$V_{\text{rep}}(R) = \left(\sum_{i=-m}^{m'} c_i R^i\right) \exp\left\{-\sum_{k=1}^{p}(a_k R^k + a_{-k} R^{-k})\right\} + \left(\sum_{j=-m''}^{m'''} d_j R^j\right) \mathrm{e}^{-\beta R - \gamma R^2}, \tag{3.90b}$$

and

$$V_{\text{att}}(R) = -\sum_{n=3}^{q} f_{2n}(\rho R)\frac{C_{2n}}{R^{2n}} - \sum_{n'=1}^{q'} f_{2n'+1}(\rho R)\frac{C_{2n'+1}}{R^{2n'+1}}, \tag{3.90c}$$

in which A, the a_k, c_i, d_j, α, β, γ, and ρ often serve as free parameters in the fitting procedure, while the long-range dispersion coefficients C'_n with n' an odd integer and C_{2n} with $n = 3, 4, 5$ often have fixed values or, for $n \geq 6$, may be varied, but normally only within predetermined bounds. The damping functions $f_{2n}(z)$ and $f_{2n+1}(z)$ are given by eqn (3.10b).

On occasion, it will be convenient to express $V(R)$ as the sum of the dominant clamped-nuclei (or Born–Oppenheimer) PEF, $V_{\text{BO}}(R)$, plus a series of correction terms, namely (Cencek *et al.*, 2012; Przybytek *et al.*, 2010; Przybytek *et al.*, 2012),

$$V(R) = V_{\text{BO}}(R) + \Delta V_{\text{ad}}(R) + \Delta V_{\text{rel}}(R) + \Delta V_{\text{QED}}(R), \tag{3.91a}$$

in which $\Delta V_{\text{ad}}(R)$ is the (diagonal) adiabatic correction to the clamped-nuclei approximation, $\Delta V_{\text{rel}}(R)$ is the total relativistic correction, and $V_{\text{QED}}(R)$ is the correction arising from quantum electrodynamics. In addition, the Casimir–Polder (CP) retardation correction (Casimir and Polder, 1948) to $V(R)$ may be introduced by adding $\Delta V_{\text{ret}}(R)$, defined as

$$\Delta V_{\text{ret}}(R) \equiv V_{\text{CP}}(R) + \frac{C_6^{\text{BO}}}{R^6} + \frac{C_4}{R^4} + \frac{C_3}{R^3}, \tag{3.91b}$$

to the right-hand side of eqn (3.91a) (Cencek *et al.*, 2012; Przybytek *et al.*, 2010; Przybytek *et al.*, 2012). The retardation correction has been written in this form because the CP long-range asymptotic behavior is given to order α^3 [α is the atomic fine-structure constant, with value $(137.03599679)^{-1}$] by (Meath and Hirschfelder, 1966)

$$V_{\text{CP}}(R) \sim -\frac{C_6^{\text{BO}}}{R^6} - \frac{C_4}{R^4} - \frac{C_3}{R^3} + \mathcal{O}(\alpha^4), \tag{3.91c}$$

in which C_6^{BO} is the usual Born–Oppenheimer asymptotic long-range dipolar contribution, while C_4 and C_3 are the asymptotic constants determining the long-range behavior of the $\Delta V_{\text{rel}}(R)$ and $\Delta V_{\text{QED}}(R)$ correction terms, respectively.

3.8 Interactions between Pairs of Ground-Term Noble Gas Atoms

This section provides both a critical survey (up to 2020) of available *ab initio* PEFs for the interactions between atom pairs and to provide a common functional form representing the interaction energies between noble gas atom pairs. The MLR form (Le Roy and Henderson, 2007) of Section 3.5 has been selected for this purpose, as it is continuous and differentiable everywhere and, by selecting an appropriate damping function for the dispersion forces, provides an ultra-short-range behavior that qualitatively has the limiting (shielded) Coulomb-like repulsion that may (in principle) be anticipated for any atom–atom interaction at very small internuclear separations. Naturally, the principal focus will be upon interactions between pairs of noble gas atoms in their ground electronic (closed-shell) states.

3.8.1 The $He_2(X^1\Sigma_g^+)$ interaction

It is, of course, no surprise that the most intensively studied van der Waals interaction is that between two helium atoms, as it gives rise to the simplest and smallest non-chemically bound dimer. Moreover, both microscopic properties, such as high-quality atomic beam differential scattering data at 40.247 cm^{-1} and at 508.13 cm^{-1} and integral scattering data over the speed range 80 $m/s^{-1} \lesssim v \lesssim 4000$ m/s^{-1}, and macroscopic (bulk) property data, such as helium virial and transport coefficient measurements over an extensive temperature range (5 K $\lesssim T \lesssim$ 2000 K), are available to be utilized for the vetting of any proposed He–He PEF. As the methodology evolves, the uncertainties in the transport properties computed from the PEF are now smaller than those obtained from experiment. Because of the small number of electrons involved, the interaction energy may (in principle) be determined with great accuracy via *ab initio* computation. Accurate determination of the He–He interaction energy has nonetheless been a very challenging task, as the weakness of the interaction necessitates extraordinary precision in the computations, and is consequently very sensitive to the quality and completeness of the basis-sets employed in the *ab initio* calculations.

The earliest relatively successful computation of the He–He PEF was that of Liu and McLean in 1973, who employed quantum Monte Carlo calculations (Liu and McLean, 1973). Their calculations led to a well-depth of approximately 6.6 cm^{-1}, which was too shallow to support even a single He–He bound state. By 1989 more elaborate quantum Monte Carlo calculations (Liu and McLean, 1989) gave an improved He–He PEF with a well-depth of 7.60 cm^{-1}, which was in excellent agreement with the then best multi-property-fit PEF of Aziz and coworkers (1987) for separations R between 1.6 Å and 16 Å. Their calculations demonstrated the important role played by basis-set truncation, and clearly established the necessity of including the BSSE counterpoise correction procedure introduced by Boys and Bernardi (1970) in 1970.

Neither the best-empirical (Aziz *et al.*, 1987; Aziz *et al.*, 1995; Janzen and Aziz, 1997) nor the most accurate PEF obtainable using the clamped-nuclei approximation will support a bound state for the interaction of two ^{3}He atoms, and will support only a single bound state (corresponding to vibrational quantum number $v = 0$ and rotational quantum number $j = 0$) at an energy E_{00} of the order of $10^{-3}cm^{-1}$ relative to the ^{4}He–^{4}He dissociation limit $\mathfrak{D}_0$. As the binding energy of a dimer is of the order of the accuracy with which a typical *ab initio* calculation could be carried out prior to 1990, the very existence of a bound $^4He_2$ dimer was controversial: it was, however, detected experimentally by Luo *et al.* (1993, 1996) in 1993 and its existence was independently confirmed by Schöllkopf *et al.* (Grisenti *et al.*, 2000; Schöllkopf and Toennies, 1994; Schöllkopf and Toennies, 1996) in 1994. Moreover, because the average interhelium distance, $\langle R \rangle_0$, of (47.1 ± 0.5) Å computed using the best clamped-nuclei He–He PEF (Cencek *et al.*, 2012; Przybytek *et al.*, 2010; Przybytek *et al.*, 2012) is significantly shorter than the experimentally determined value (Luo *et al.*, 1993; Luo *et al.*, 1996) of (62 ± 10) Å, it is clear that only very high-level *ab initio* computations that incorporate both improved and better-converged basis-sets and that include also higher-order corrections, such as the diagonal Born–Oppenheimer correction, retardation, and other relativistic corrections, and quantum electrodynamic (or radiative) corrections, can resolve differences between theoretically calculated and experimentally determined properties of such a weakly bound dimer. So far, the most accurate potential energy for ^{4}He and ^{3}He comes from the work of Czachorowski *et al.* (2020), which includes both relativistic and quantum electrodynamic components.

Early coupled-cluster single-double-triple (CCSD(T)) *ab initio* calculation of the He–He interaction energy as a function of internuclear separation R were published in 1999 by van Mourik and Dunning (1999) and by Cybulski and Toczyłowski (1999), the former at 35 values of R between 1.85 Å and 7.93 Å, the latter at 13 value of R between 1.75 Å and 5.00 Å. These PEFs were characterized by minima $(R_e, -\mathfrak{D}_e)$ with $R_e = 2.9634$ Å, $\mathfrak{D}_e = 7.6385\,\text{cm}^{-1}$ for the former PEF, and $R_e = 2.9764$ Å, $\mathfrak{D}_e = 7.3897\,\text{cm}^{-1}$ for the latter PEF. Moreover, as reported by Nasrabad *et al.* (2004), the basis-set results calculated by Cybulski and Toczyłowski can be utilized to carry out extrapolations to the CBS limits and obtain a more accurate $\text{Ne}_2(X^1\Sigma_g^+)$ PEF. Although the minimum in the interaction energy for the PEF obtained by Cybulski and Toczyłowski is nearly 4% shallower than that for the SAPT2 PEF of Janzen and Aziz (1997), which is based upon a combination of quantum Monte Carlo calculations for small separations (Ceperley and Partridge, 1986), the high-level SAPT (Rybak *et al.*, 1991; Szalewicz and Jeziorski, 1979; Williams *et al.*, 1993) results of Korona *et al.* (1997), with an added retardation correction (Jamieson *et al.*, 1995) applied to the dipole–dipole long-range dispersion coefficient (Bishop and Pipin, 1993) for separations $R > 3$ Å, the equilibrium separation R_e is only about 0.4% larger than the SAPT2 value $R_e = 2.9643$ Å. Similarly, the minimum for the van Mourik and Dunning PEF (1999), is characterized by $R_e = 2.9634$ Å, which is about 0.04% shorter, and $\mathfrak{D}_e = 7.6385\,\text{cm}^{-1}$, which is 0.64% shallower, than the corresponding R_e, $\mathfrak{D}_e$ values for the SAPT2 PEF. A later CCSD(T) computation (Laschuk *et al.*, 2003) that was carried out for 60 He–He internuclear separations, including CBS extrapolation, resulted in a revised minimum, characterized by $R_e = 2.969$ Å, $\mathfrak{D}_e = 7.6926\,\text{cm}^{-1}$. The value for R_e is 0.004 Å larger than that characterizing the minimum in the *ab initio*-based SAPT2 PEF (Janzen and Aziz, 1997), while the value for $\mathfrak{D}_e$ is $0.0048\,\text{cm}^{-1}$ larger than the corresponding value for the SAPT2 PEF.

Two high-level accurate sets of calculations of the clamped-nuclei He–He PEF, one (Hellmann *et al.*, 2007) based upon the coupled-cluster singles, doubles, and non-iterative triples, or CCSD(T), supermolecule procedure, the other (Cencek and Szalewicz, 2008; Jeziorska *et al.*, 2007) based upon the SAPT procedure, are in excellent agreement, and both PEFs may be considered to be state of the art for the present time. In their definitive study, Cencek *et al.* (2012) reviewed the need for such exacting treatments of the He–He interaction. A He–He PEF that is accurate to a few parts in 10^4 for separations R between 0.5 and 6.5 Å has been constructed (Cencek *et al.*, 2012) by utilizing the clamped-nuclei PEF of Jeziorska *et al.* (2007) for $V_{\mathrm{BO}}(R)$ and carefully evaluating the correction terms listed in eqn (3.91). The value of the clamped-nuclei potential energy, $V_{\mathrm{BO}}(R)$, obtained by Jeziorska *et al.* (2007) at $R = 0.529177$ Å ($1.0\,a_0$) for the He–He interaction is in excellent agreement (differing only by approximately 0.003%) with the variational value obtained by Komasa (1999) using a 1200-term explicitly correlated Gaussian (ECG) basis-set. The Komasa results may therefore be employed to extend the overall PEF into the ultra-short separation range, provided that appropriate error estimates can be obtained to account for the neglect of the corrections to the clamped-nuclei values. Such an extension may prove useful for the fitting of analytical functions to the set of computed values in order that they not be overly repulsive at small He–He separations.

A great deal of attention has been focused upon the full PEF in the vicinity of the interaction energy minimum, as the precise value of the bound state energy close to the dissociation limit will be very sensitive to this region of the PEF. This point can be illustrated by an examination of the contributions to $V(R_{\mathrm{e}})$, the zero-point energy (equivalently, the dissociation energy), and the expectation value $\langle R\rangle_0$ associated with the near-dissociation bound state (Przybytek *et al.*, 2010; Przybytek *et al.*, 2012) shown in Table 3.5. Column 2 gives the values of the well-depth, $\mathfrak{D}_{\mathrm{e}} \equiv -V(R_{\mathrm{e}})$, the dissociation energy, $\mathfrak{D}_0$, and the expectation value, $\langle R\rangle_0$, of R for the single (near-dissociation) bound state obtained when only the clamped-nuclei PEF, $V_{\mathrm{BO}}(R)$, is considered, while while columns 3–6 give the values obtained for these same quantities upon successive additions of the correction terms that define the columns. From the changes in the values for $\mathfrak{D}_{\mathrm{e}}$, it is clear that both the diagonal adiabatic and radiative (QED) corrections to $V(R)$ are negative, while both the net relativistic and retardation corrections are positive in the region around R_{e}.

The values of the full supermolecule/perturbation theory (SMPT) PEF, $V_{\mathrm{SMPT}}(R)$, defined according to eqn (3.90), and computed for 17 He–He separations R lying

Table 3.5 Contributions* from the incremental addition of various corrections to $\mathfrak{D}_{\mathrm{e}}$ and the bound state of the $^4\mathrm{He}_2(X^1\Sigma_g^+)$ dimer†

	$V_{\mathrm{BO}}(R)$	$+\,\Delta V_{\mathrm{ad}}(R)$	$+\,\Delta V_{\mathrm{rel}}(R)$	$+\,\Delta V_{\mathrm{QED}}(R)$	$+\,\Delta V_{\mathrm{ret}}(R)$
$\mathfrak{D}_{\mathrm{e}}$	7.6455	7.6518	7.6411	7.6420	7.6419
$-10^3 E_{00}$	1.194	1.262	1.105	1.126	1.122
$\langle R\rangle_0$	45.77	44.62	47.43	47.02	47.09

* $\mathfrak{D}_{\mathrm{e}}$ and E_{00} have units cm^{-1}, $\langle R\rangle_0$ has units Å.

† Data from Cencek *et al.* 2012; Przybytek *et al.* 2010; Przybytek *et al.* 2012.

between 0.5 Åand 6.5 Å, were fitted (Cencek *et al.*, 2012; Przybytek *et al.*, 2010; Przybytek *et al.*, 2012) to a version of eqns (3.90) in which $A = 0$, $m = 2$, $m' = 0$, $m'' = 1$, $\gamma = 0$, $q = 8$, $q' = 7$, that is, to

$$V^{\text{fit}}_{\text{SMPT}}(R) = (c_0 + c_1 R + c_2 R^2)\mathrm{e}^{-\alpha R} + (d_0 + d_1 R)\mathrm{e}^{-\beta R - \gamma R^2} - \sum_{n=3}^{16} f_n(\rho R)\frac{C_n}{R^n}, \tag{3.92}$$

in which the $f_n(\rho R)$ are TT damping functions, defined in eqn (3.10b). Values for the long-range dispersion coefficients C_n and the various fit-coefficients may be found in Przybytek *et al.* (2010, Przybytek *et al.* (2012): their fitted PEF is characterized by $\mathfrak{D}_\mathrm{e} = (7.642 \pm 0.003)\,\mathrm{cm}^{-1}$ and $R_\mathrm{e} = (2.9687 \pm 0.0001)$ Å. Because the inclusion of retardation effects via the addition of eqn (3.91a) to the analytical PEF of eqn (3.92) could lead to a loss of significant figures from numerical round-off errors incurred during the computation of thermophysical properties of He, Cencek *et al.* (2012, Przybytek *et al.* (2010, Przybytek *et al.* (2012) introduced a means for simulating the incorporation of retardation corrections by modifying their analytically fitted PEF. Specifically, the effect of retardation on the R^{-3} and R^{-4} relativistic contributions to $\Delta V_{\text{rel}}(R)$ were accounted for by replacing the TT damping functions $f_3(\rho R)$ and $f_4(\rho R)$ by the functions $\widetilde{f}_i(\rho R) \equiv f_i(\rho R) - 1$, in order to engender no additional loss of accuracy in thermophysical property calculations. As the fit function does not explicitly contain C_6^{BO}, a slightly more elaborate fix was made by replacing $-f_6(\rho R)C_6 R^{-6}$ by $-[\widetilde{f}_6(\rho R)C_6^{\text{BO}} + f_6(\rho R)\widetilde{C}_6]R^{-6}$, with $\widetilde{C}_6$, defined as $\widetilde{C}_6 \equiv C_6 - C_6^{\text{BO}}$, the sum of all post–BO C_6 asymptotic constants defined in eqn (3.92).

Strictly speaking, the result (3.92) and the specific fit-coefficients given by Cencek *et al.* (2012) apply only to the ^{4}He–^{4}He interaction due to incorporation of $\Delta V_{\text{ad}}(R)$ in the final expression for $V_{\text{SMPT}}(R)$. Diagonal adiabatic corrections computed for all isotopologous He–He interactions, namely ^{4}He–^{4}He, ^{4}He–^{3}He, and ^{3}He–^{3}He, by Komasa *et al.* (1999) using a 1200-term ECG basis-set may be employed to construct the full interaction energies for the other two isotopologues. However, note that other independent computations for the ^{4}He–^{4}He interaction based upon extensive ECG and orbital sets of basis functions are in substantive agreement with one another (Cencek *et al.*, 2012; Hellmann *et al.*, 2007; Przybytek *et al.*, 2010; Przybytek *et al.*, 2012), but differ slightly from the earlier Komasa *et al.* (1999) results, especially for separations corresponding to the attractive region of the He–He PEF. Moreover, the diagonal adiabatic correction is not sufficiently large to affect significantly any conclusions arrived at for ^{4}He–^{3}He or ^{3}He–^{3}He properties computed using $V_{\text{SMPT}}(R)$.

The final (Cencek *et al.*, 2012; Przybytek *et al.*, 2010; Przybytek *et al.*, 2012) *ab initio* SMPT $^4\text{He}(X^1\Sigma_g^+)$ PEF differs from all its predecessors in that it includes all relevant post-BO corrections and provides uncertainty estimates for both the non-relativistic BO predominant contribution and for the full PEF in the form of upper- and lower-limit potentials $V^{\pm}_{\text{SMPT}}(R) \equiv V^{\text{fit}}_{\text{SMPT}}(R) \pm \sigma(R)$, in which the PEF uncertainty function $\sigma(R)$ has been obtained by fitting the individual uncertainties σ_i

(associated with the values of $V_{\mathrm{SMPT}}(R)$ at each of the 17 grid-points R_i employed for its computation) to the functional form

$$\sigma(R) = \sum_{i=1}^{3} \sigma_i \mathrm{e}^{-\eta_i R}, \tag{3.93}$$

subject to the constraint that $\sigma(R) \geq \sigma_i$, $\{i = 1, \cdots 17\}$. Table 3.6 shows the fit-parameters for $\sigma(R)$ obtained using these constraints (Przybytek *et al.*, 2010; Przybytek *et al.*, 2012). The uncertainty function (3.93) was proposed (Cencek *et al.*, 2012; Przybytek *et al.*, 2010; Przybytek *et al.*, 2012) as a means of providing estimates of the computational uncertainties associated with calculations that employ the *ab initio* $V_{\mathrm{SMPT}}(R)$ PEF in order to obtain values for the uncertainties in the physical quantities that are governed by the He–He interatomic interaction. The original publications (Cencek *et al.*, 2012; Przybytek *et al.*, 2010; Przybytek *et al.*, 2012) should be consulted for additional details of both relevant theoretical methodology and computational procedures employed in the determination of thermophysical properties from the $V_{\mathrm{SMPT}}(R)$ PES.

The ^{4}He–^{4}He PEFs obtained by Cybulski and Toczyłowski (1999) and by van Mourik and Dunning (1999) are too shallow to support a bound state, while the $^4\mathrm{He}_2(X^1\Sigma_g^+)$ PEF of Cencek *et al.* (2012, Przybytek *et al.* (2010, Przybytek *et al.* (2012), with a dissociation energy $\mathfrak{D}_0 = 1.722\times10^{-3}\,\mathrm{cm}^{-1}$, supports a single bound ($v = 0,\ j = 0$) ro-vibrational state lying near the top of the potential well. As a

Table 3.6 Fit-parameters for the $\mathrm{Ne}_2(X^1\Sigma_g^+)$ interaction energy‡

Long-range coefficients*				Fit-parameters†	
n	C_n	n	C_n	Coeff.	Value
3	0.126688	10	4.032840×10^7	α	3.65271949356113
4	−7.7620	11	-1.6842×10^7	β	2.36720871471273
5	3.024001×10^4	12	7.241×10^8	ρ	4.09707982651218
6	3.212322×10^5	13	-8.353×10^8	c_0	−25.2616315711638
7	0.0	14	1.872×10^{10}	c_1	269.244425630616
8	3.09975×10^6	15	-3.746×10^{10}	c_2	−56.3879970402079
9	0.0	16	6.28×10^{11}	d_0	38.7957487310071
				d_1	−2.76577136772754
σ_1	36.657	σ_2	0.9929	σ_3	0.004045
η_1	2.456	η_2	1.100	η_3	0.9929

* The C_n coefficients have units $\mathrm{cm}^{-1}\,a_0^n$, the fit-coefficients α, β, ρ, and all η_i have units a_0^{-1}, while c_0, d_0, and all σ_i have units cm^{-1}, c_1, d_1 have units $\mathrm{cm}^{-1}a_0^{-1}$, and d_2 has units $\mathrm{cm}^{-1}a_0^{-2}$.

† The fit-parameter γ was found to have the value 0.0.

‡ Data from Cybulski and Toczyłowski 1999.

consequence, the $^4He_2$ dimer is extremely diffuse, with an average internuclear separation (Cencek *et al.*, 2012) $\langle R \rangle_0 = (47.09 \pm 0.5)$ Å. As noted earlier, this *ab initio* value for $\langle R \rangle_0$ differs significantly from the value (Luo *et al.*, 1993; Luo *et al.*, 1996) (62 ± 10)Å deduced experimentally: the reason for such a large discrepancy between theory and experiment is still not fully understood. The final row of Table 3.7 contains parameters characterizing the SAPT2 PEF determined by Janzen and Aziz (1997) via the fitting of a set of high-level *ab initio* energies spanning internuclear separations $R > 0.5$ Å to a modified TT functional form. The SAPT2 PEF supports a single bound state of the $^4He_2(^1\Sigma_g^+)$ dimer, with $\langle R \rangle_0 = 44.87$ Å and $\mathfrak{D}_0 = 1.247 \times 10^{-3}\,\text{cm}^{-1}$.

3.8.2 The $Ne_2(X^1\Sigma_g^+)$ interaction

The earliest high-level *ab initio* determination of the interaction energy between a pair of ground-term Ne atoms is due to Cybulski and Toczyłowski (1999) (Table 3.6), who carried out CCSD(T) supermolecule calculations that employed a large basis-set (aug-cc-pV5Z{33221}) to develop a set of parametrized functions for the pair-interaction energies to use for rare-gas cluster calculations. They fitted $Ne_2(X^1\Sigma_g^+)$ interaction energies computed at 13 Ne–Ne separations R lying between 2.25 Å and 5 Å to a version of eqns (3.90) in which A, all c_i, all $C_{2n'+1}$, m', and m'' are set to zero, and $q = 8$, that is, to the modified HFD-type PEF

$$V(R) = d_0 \mathrm{e}^{-\alpha R - \beta R^2} - \sum_{n=3}^{8} f_{2n}(\rho R)\, \frac{C_{2n}}{R^{2n}}\,, \tag{3.94a}$$

with d_0, α, β, and ρ as adjustable parameters, and with $f_{2n}(\rho R)$ being the TT damping function of eqn (3.10b). The dispersion coefficients C_{2n} ($n = 3, 4, 5$) were essentially determined by empirical estimates (Tang and Toennies, 1986) and the C_{2n} ($n = 6, 7, 8$) were obtained from extrapolation formulae (Thakkar, 1988). The position and depth of the minimum in $V(R)$, namely $R_e = 3.0988$ Åand $\mathfrak{D}_e = 28.604\,\text{cm}^{-1}$, indicated that a still higher-level basis-set was necessary to achieve parity with the multi-property best-fit empirical PEF of Aziz and Slaman (1989*b*) for which R_e and $\mathfrak{D}_e$ have values 3.091 Å and 29.3657 cm^{-1}, respectively (Table 3.7).

Other theoretical groups have employed a number of high-level basis-sets and computational procedures associated with some of the approximations made in a typical CCSD(T) supermolecule calculation to show that the neglect of core–core and core–valence electron correlations contributes an insignificant error in the neighborhood of the PEF minimum (van Mourik and Dunning, 1999), and (based upon an interaction-optimized set of basis functions) that differences between the *ab initio* (Nasrabad *et al.*, 2004) and the multi-property empirical (Aziz and Slaman, 1989*b*) PEFs are likely due to incomplete consideration of triple and higher excitations, incompleteness of the basis-set, and relativistic effects (van de Bovenkamp and van Duijneveldt, 1999; van Mourik and Dunning, 1999; Woon, 1994). The addition to the CCSD(T) aug-cc-pV5Z{33221} PEF of Cybulski and Toczyłowski (1999) of corrections for incompleteness of the basis-set, lack of full configuration–interaction, omission of core–core and core–valence electron correlation contributions, and the neglect of scalar relativistic effects (often referred to as the Cowan–Griffin approximation) by Gdanitz (2001) gave

Table 3.7 Properties characterizing the $^4He_2(X^1\Sigma_g^+)$ and $Ne_2(X^1\Sigma_g^+)$ van der Waals dimers and their interaction energies

$^4He_2(X^1\Sigma_g^+)$				$Ne_2(X^1\Sigma_g^+)$			
R_e /Å	σ /Å	$\mathfrak{D}_e$ /cm^{-1}	Ref.	R_e /Å	σ /Å	$\mathfrak{D}_e$ /cm^{-1}	Ref.
2.9764		7.3597	a	3.0988		28.604	b
2.9734		7.4320	c	3.10	2.75	28.49	d
2.969	2.6403	7.6926	e	3.1050	2.7686	25.6439	f
2.9679	2.6439	7.6439	g	3.086	2.7582	30.9854	h
2.9676		7.6426	i	3.097		28.740	j
				3.0895	2.7612	29.2980	k
2.9646	2.6390	7.6878	l	3.091	2.759	29.365	m

a, Cybulski and Toczyłowski, 1999; b, Haley and Cybulski, 2003; cm, van Mourik and Dunning, 1999; d, van de Bovenkamp and van Duijneveldt 1999; e, Laschuk *et al.*, 2003; f, Hättig *et al.*, 2003; g, Hellmann *et al.*, 2007; h, Casimir and Polder, 1948; i, Cencek *et al.*, 2012; j, Nasrabad *et al.*, 2004; k, Hellmann *et al.*, 2008*a*; Hellmann *et al.*, 2008*b*; l, Janzen and Aziz, 1997 ; m, Aziz and Slaman, 1989*b*.

only a slight change to the original Ne_2 PEF, resulting, for example, in an increase of 0.262 cm^{-1} in the well-depth to give $\mathfrak{D}_e = 28.866\,\text{cm}^{-1}$. A different method for extrapolating to the CBS limit (Halkier *et al.*, 1998; Klopper, 2001; Patkowski, 2012) resulted in an increase of 0.476 cm^{-1} in the depth of the PEF to give $\mathfrak{D}_e = 29.080\,\text{cm}^{-1}$ for the Ne_2 dimer, which is within 1% of the value $\mathfrak{D}_e \simeq 29.365\,\text{cm}^{-1}$ for the multi-property best-fit empirical PEF of Aziz and Slaman (1989*b*).

Nasrabad *et al.* (2004) found that by using Cybulski and Toczyłowski's (1999) aug-cc-pVQZ{33221} and aug-cc-pV5Z{33221} basis-set results to carry out extrapolations to the CBS limits using Klopper's (2001) X^{-3} method of they were able to obtain a $Ne_2(X^1\Sigma_g^+)$ PEF that closely resembled Aziz and Slaman's (1989*b*) best-fit empirical PEF over the full range of internuclear separations R.

Hellmann *et al.* (2008*a*, 2008*b*) determined a definitive *ab initio* PEF for the Ne_2 dimer from CCSD(T) calculations employing basis-sets up to t-aug-cc-pV6Z quality supplemented with bond functions, corrected for BSSE, and extrapolated to the CBS limit. In addition, correction terms accounting for missing core–core and core–valence electron correlations, scalar relativistic effects, and CC contributions up to coupled-cluster singles doubles triples and quadruples (CCSDT(Q)) were evaluated and added to the CCSD(T) values. However, post-BO corrections were not attempted, as they are difficult to evaluate and were believed to be negligible, nor were CP retardation corrections computed for this interaction. Values for $V_{\text{SM}}(R)$ were obtained at 32 different Ne–Ne separations lying between 1.4 Å and 8 Å, and were fitted to a version of eqns (3.90) in which all c_i, d_j, and $C_{2n'+1}$ are set to zero, that is, to the specific modified TT function

$$V_{\text{SM}}(R) = A \exp\left\{a_1 R + a_2 R^2 + \frac{a_{-1}}{R} + \frac{a_{-2}}{R^2}\right\} - \sum_{n=3}^{8} f_{2n}(\rho R)\,\frac{C_{2n}}{R^{2n}}\,, \tag{3.94b}$$

with A, ρ, the a_i, and the C_{2n} $(n = 3, 4, 5)$ treated as free parameters in the fitting procedure, while the C_{2n} $(n = 6, 7, 8)$ were determined by employing recursion formula (3.10c). Values of the fit-parameters for the Hellmann *et al.* (2008*a*, 2008*b*) *ab initio* PEF are given in Table 3.8. The minimum for the analytical fitted PEF has depth $\mathfrak{D}_e = 29.297\,\text{cm}^{-1}$ for the equilibrium Ne–Ne separation $R_{\text{min}} = 3.1000$ Å, values that are in excellent agreement with $\mathfrak{D}_e = 29.364\,\text{cm}^{-1}$ and $R_e = 3.091$ Å for the empirical multi-property best-fit Ne_2 PEF (Aziz and Slaman, 1989*b*). More than a decade later, they considered the diagonal Born–Oppenheimer correction and retardation of the dispersion interactions, and obtained more accurate potential for the $Ne_2(X^1\Sigma_g^+)$ dimer (Hellmann *et al.*, 2021). Moreover, $V_{\text{SM}}(R)$ is fully consistent with the *ab initio* results obtained by Bytautas and Ruedenberg (2008) in an in-depth study of the Ne_2 correlation energy in general, and of its long-range dispersion component in particular. Their Table VI gives an extensive comparison amongst *ab initio* values of R_e and $\mathfrak{D}_e$ computed for the $Ne_2(X^1\Sigma_g^+)$ dimer upon employing a variety of *ab initio* methods with a large number of basis-sets, including comparisons with experimentally derived values.

The $Ne_2(X^1\Sigma_g^+)$ PEF supports three bound vibrational levels, corresponding to a total of 13 ro-vibrational levels for the $^{20}Ne_2$ isotopologue and 25 ro-vibrational levels for the $^{20}Ne^{22}Ne$ isotopologue. From their experimental data, Wüest and Merkt (2003) obtained energy differences $E_{0,0} - E_{v,j}$ corresponding to $v = 0$, $j = 2, 4, \cdots, 10$ and $v = 1$, $j = 0, 2, 4$ for the $^{20}Ne_2$ isotopologue, and to $v = 0$, $j = 2, 3, 4, 5$ for the $^{20}Ne^{22}Ne$ isotopologue. Values for these energy differences computed from $V_{\text{SM}}(R)$ for the $^{20}Ne_2$ dimer agree on average to within 0.31% with the experimental values (average experimental uncertainty, ±0.60%) for the $^{20}Ne_2$ dimer, and differ by no more than 1.07% on average from the experimental values (average experimental uncertainty, ±4.38%) for the $^{20}Ne^{22}Ne$ dimer. The corresponding percentage differences for the multi-property empirical best-fit PEF (Aziz and Slaman, 1989*b*), which incorporated fits to the earlier VUV data of Tanaka and Yoshino (Freeman *et al.*, 1974; Tanaka and Yoshino, 1970; Tanaka and Yoshino, 1972; Tanaka *et al.*, 1973) (but not to the Wüest and Merkt (2003) data) are 0.45% and 1.09%, respectively. Values for the characteristic features of the $Ne_2(X^1\Sigma_g^+)$ PEF, specifically, the depth, $\mathfrak{D}_e$, the position, R_e, of the minimum interaction energy, and the location σ of the finite-R zero interaction energy for a pair of Ne atoms are given in Table 3.7.

The minima of the *ab initio* $Ne_2(X^1\Sigma_g^+)$ PEFs obtained by van de Bovenkamp and van Duijneveldt (1999) and Cybulski and Toczyłowski (1999) are approximately 2.5% shallower and located about 0.26% further out than that of the best-fit experimental PEF of Aziz and Slaman (1989*b*), and are further characterized by $\langle R \rangle_0 = 3.3433$ Å and $\mathfrak{D}_0 = 16.2631\,\text{cm}^{-1}$. The *ab initio* PEF of Hättig *et al.* (2003) is characterized by values of $\mathfrak{D}_e$ and R_e that are about 13% smaller and 0.44% larger, respectively, than the values of $\mathfrak{D}_e$, R_e that characterize the empirical PEF. The value $\mathfrak{D}_0 = 16.746\,\text{cm}^{-1}$ obtained from the *ab initio* $Ne_2(^1\Sigma_g^+)$ PEF of Hellmann *et al.* (2008*a*, 2008*b*) for the $^{20}Ne_2$ isotopologue is in excellent agreement with the value $\mathfrak{D}_0 = 16.730\,\text{cm}^{-1}$ obtained using the best-fit empirical PEF of Aziz and Slaman (1989*b*).

3.8.3 The $\mathrm{Ar}_2(X^1\Sigma_g^+)$ interaction

An empirical multiproperty-fit PEF (Aziz, 1993) for the $\mathrm{Ar}_2(X\,^1\Sigma_g^+)$ interaction having $f_{2n}(\rho R)$ given by the HFD doubly damped form of eqn (3.8b), was determined via fitting to a large body of available experimental data, including both the temperature dependence of bulk gas phenomena (second pressure virial, viscosity, and thermal conductivity coefficients) and microscopic phenomena (high energy differential and integral glory scattering cross-section data, spectroscopic vibrational spacings, rotational constants).

The adiabatic speed of sound in a gaseous medium may be expressed in terms of a virial expansion in powers of the number density n, having the form (McCourt, 2003)

$$u_{\mathrm{ad}}^2(n;T) = \frac{kT\gamma^\circ}{m}\left[1 + \beta_{2a}(T)n + \beta_{3a}(T)n^2 + \cdots\right],$$

in which $\gamma^\circ \equiv C_p^\circ/C_V^\circ$ is the ratio of the ideal gas isobaric and isochoric heat capacities (with a value 5/3 for monatomic gases), and $\beta_{2\mathrm{a}}(T)$, $\beta_{3\mathrm{a}}(T)$, $\cdots$, are termed the second, third, $\cdots$, acoustic virial coefficients. These acoustic virial coefficients can be related to the pressure virial coefficients and their temperature derivatives: in particular, the second acoustic virial coefficient, $\beta_{2a}(T)$, is obtained in terms of the second pressure virial coefficient, $B_2(T)$, and its temperature derivatives as (McCourt, 2003)

$$\beta_{2a}(T) = 2B_2(T) + 2(\gamma^\circ - 1)T\,\frac{\mathrm{d}B_2}{\mathrm{d}T} + \frac{(\gamma^\circ - 1)^2}{\gamma^\circ}T^2\,\frac{\mathrm{d}^2B_2}{\mathrm{d}^2T}.$$

By developing a means of measuring the adiabatic speed of sound in gases with unprecedented accuracy, Ewing and co-workers (Benedetto *et al.*, 2004; Estrada-Alexanders and Trusler, 1995; Ewing and Goodwin, 1992; Ewing and Trusler, 1992*b*; Ewing *et al.*, 1989; Moldover *et al.*, 1999) were able to determine second acoustic virial coefficients in argon with very high accuracy. Boyes (1994) set out to compare the accurate experimental values of $\beta_{2\mathrm{a}}(T)$ obtained for argon with predictions of the most accurate argon PEF available. However, when he employed the Aziz HFD-ID $\mathrm{Ar}_2(X^1\Sigma_g^+)$ PEF for this purpose, Boyes (1994) found significant disagreement between the predicted and experimental results. As every $\mathrm{Ar}_2(X^1\Sigma_g^+)$ PEF proposed prior to 1994 had failed to reproduce the experimentally observed temperature dependence of β_{2a}, Boyes then carried out a redetermination of the $\mathrm{Ar}_2(X^1\Sigma_g^+)$ PEF using the HFD-ID functional form developed by Aziz (1993), but replaced the set of second pressure virial coefficient data by the considerably more accurate acoustic second virial coefficient data in his fitting routine. His revised Ar_2 PEF differed from the Aziz PEF mainly in the values obtained for R_e and σ [R_e(Boyes) $= 3.76445$ Å vs. R_e(Aziz) $= 3.75700$ Å; σ(Boyes) $= 3.35778$ Å vs. σ(Aziz) $= 3.35336$ Å], with the depth essentially unchanged [$\mathfrak{D}_\mathrm{e}$(Boyes) $= 99.5412\,\mathrm{cm}^{-1}$ vs. $\mathfrak{D}_\mathrm{e}$(Aziz) $= 99.5498\,\mathrm{cm}^{-1}$], and with slightly larger values for the C_{2n} dispersion coefficients (but with the C_6, C_8, and C_{10} values still lying within their *ab initio* computed upper and lower bounds).

A comparison between DRMSD values (Myatt *et al.*, 2018) $\overline{dd}$(Aziz) $= 15.253$ and $\overline{dd}$(Boyes) $= 2.467$ for the available acoustic second virial coefficient experimental data

(Benedetto *et al.*, 2004; Estrada-Alexanders and Trusler, 1995; Ewing and Goodwin, 1992; Ewing and Trusler, 1992*b*; Ewing *et al.*, 1989; Moldover *et al.*, 1999) indicates the inability of the Aziz HFD-ID Ar_2 PEF to provide an accurate representation of the temperature dependence of the acoustic second virial coefficient of argon. Note that although the empirical fit giving the Boyes HFD-ID $Ar_2(X^1\Sigma_g^+)$ PEF did not include the available pressure second virial coefficient data, it did give a slightly better representation of those data, as is clearly indicated by the DRMSD values (Myatt *et al.*, 2018) $\overline{dd}$(Aziz) = 1.764 and $\overline{dd}$(Boyes) = 1.338. The Boyes HFD-ID PEF may be viewed largely as a translation of the Aziz PEF by 0.00745Å toward larger separations coupled with a small change in shape in order to produce acceptable agreement between calculated and experimental $\beta_{2a}(T)$ values. Such an interpretation is supported by the rather close agreement obtained between the sets of computed spectroscopic rotational constants, B_v, and successive level separation values, $\Delta G_{v+\frac{1}{2}}$, determined by the two $Ar_2(X^1\sigma_g^+)$ PEFs, as these two derived spectroscopic constants depend upon the depth and shape of a PEF, but are independent of its location R_e.

Two relevant sets of benchmark calculations have been carried out on the employment of augmented correlation-consistent basis-sets for the determination of accurate interaction energies between pairs of ground-electron-state noble gas atoms (van de Bovenkamp and van Duijneveldt, 1999; van Mourik and Dunning, 1999; Woon, 1994). These studies established the importance of employing counterpoise-corrections for the BSSE and then using successively higher-quality augmented basis-sets to arrive at appropriate CBS limiting values. For the Ar–Ar interaction, in particular, Woon (1994) showed that by proceeding from double-zeta to triple-zeta to quadruple-zeta (DZ → QZ) precision CCSD(T) calculations, then extrapolating to the CBS limit, improved the fraction of the total attractive energy from 76% to 94.7% to 98.8%, with the final 1.2% resulting from the extrapolation to the CBS limit. Extensions by van Mourik *et al.* (1999) of the basis-sets employed for the Ar–Ar calculationss to include full quintuple- and sextuple-zeta basis sets plus core and core–valence electron correlation effects led to $\mathfrak{D}_e$, R_e values $\mathfrak{D}_e(Ar_2) = 96.50\,cm^{-1}$, $R_e(Ar_2) = 3.7836$ Å, thereby further improving the agreement between *ab initio* results and the multi-property empirical values $\mathfrak{D}_e(Ar_2) = 99.55\,cm^{-1}$, $R_e(Ar_2) = 3.7572$ Å. The additional contribution to the correlation energy arising from employment of all-electron correlation-consistent basis-sets was $\Delta E_{corr} = -0.79\,cm^{-1}$. These two benchmark studies clearly established that the use of augmented correlation-consistent basis-sets provides a systematic means for obtaining converged CBS-limit values for the interaction energies between pairs of ground term noble gas atoms.

Fernández and Koch (1998) employed doubly augmented correlation-consistent polarized valence basis-sets, specifically d-aug-cc-pVXZ{33211} (X = T, Q, 5) basis-sets, to compute the $Ar_2(X^1\Sigma_g^+)$ interaction energy for 58 Ar–Ar separations between 2Å and 10.5Å. In general, their computed interaction energies are significantly more repulsive than the corresponding values of the Aziz (1993) multi-property-fit empirical PEF for $R \lesssim 5.0\,a_0$, with the differences lessening as R approaches $R_e = 3.7780$Å, for which the interaction energy has the value $-97.60\,cm^{-1}$—a value that is approximately $2.0\,cm^{-1}$ higher than the minimum for the empirical PEF. For Ar–Ar separations $R > R_e$ their *ab initio* results merge smoothly with the corresponding values for the empirical PEF. Slightly better overall agreement is obtained between the *ab initio* interaction energies and the corresponding values obtained from a refitted empirical PEF obtained by Boyes (1994) upon incorporating acoustic virial coefficient data into the fit procedure: for example, the minimum in the *ab initio* PEF is shifted outward relative to R_e(Boyes) by 0.014Å, compared to an outward shift of 0.021Å relative to R_e(Aziz).

Slavíček *et al.* (2003) took up the challenge of improving the level of agreement between *ab initio* and empirical determinations of the $Ar_2(X^1\Sigma_g^+)$ PEF. They identified the four most likely reasons for the small remaining discrepancy between the Fernández–Koch *ab initio* PEF and the Aziz–Boyes empirical PEF as the incompleteness of the N-electron expansion, inadequacy of the frozen-core approximation, omission of higher-level correlations, and relativistic effects. To determine the importance of the approximations thereby invoked, they employed a series of aug-cc-pVXZ basis-sets, with and without a {33221} set of bond functions, to calculate the interaction energy at 29 Ar–Ar separations R lying between 2Å and 20Å. The level of agreement between interaction energies computed using their highest-level basis-set, aug-cc-pV6Z{33221}, and interaction energies obtained by Fernández and Koch (1998) using a d-aug-cc-pV5Z{33221} basis set strongly suggested that their X = 6 basis-set calculations were very near the CBS limit values, so that finiteness of the basis-set should not be a major source of the remaining discrepancy between their computed *ab initio* PEF and the empirical PEF. To address the errors incurred by employing the frozen-core approximation, they first carried out calculations for $R = 3.80$Å by employing a series of aug-cc-pVXZ basis-sets with X = D, T, Q, 5, and with X = 5 plus a {33221} set of bond functions to examine the difference between the interaction energy calculated with frozen-core orbitals and with full electron correlation: these calculations showed that removal of the frozen-core approximation could indeed account for a significant portion of the difference between *ab initio* and empirical interaction energies, but it was not possible to obtain quantitative results without the availability of appropriately optimized sets of orbital basis functions. The contribution from omitted triples was estimated using CCSD(T) and CCSDT calculations on basis-sets with frozen core orbitals, coupled with extrapolations of results obtained from aug-cc-pVDZ{332} and aug-cc-pVTZ{332} basis sets: a contribution from the omitted triples of $-1.1\,\text{cm}^{-1}$ to $E_{\text{int}}(R)$ at $R = R_e$ was obtained in this way. Upon combining this correction with a relativistic correction estimated by Faas *et al.* (2000) to be $0.7\,\text{cm}^{-1}$ at $R = R_e$ led to a final value $\mathfrak{D}_e = 98.9\,\text{cm}^{-1}$ for the well depth.

Patkowski *et al.* (2005) obtained an improved $Ar_2(X^1\Sigma_g^+)$ PEF by extending 29 $Ar_2(X^1\Sigma_g^+)$ interaction energies (Slavíček *et al.*, 2003) for separations $2\text{Å} \le R \le 20\text{Å}$ via extrapolation of their results obtained for X = 5, 6 using the aug-cc-pVXZ{33221} basis set to their CBS limits, by computing an additional nine interaction energies for separations lying in the interval $0.25\text{Å} \le R \le 1.85\text{Å}$, and by correcting for the frozen-core approximation associated with the orbital basis-sets employed in the original (Slavíček *et al.*, 2003) calculations. The extension of $V(R)$ to shorter separations was made in order to provide accurate values for regions of the repulsive wall accessed by high-energy beam scattering (collision energies of order 1–10 eV) and high-temperature (2000 K to 10^4 K) viscosity and thermal conductivity measurements. An analytic fit to the 47 values of $V(R)$ thereby obtained can be given through eqns (3.90) with A, all c_i, all $C_{2n'+1}$ set to zero, $m' = m = 1$, and $q = 8$ or, more explicitly, by

$$V_{\rm SM}(R) = (d_{-1}R^{-1} + d_0 + d_1 R)\mathrm{e}^{-\alpha R - \beta R^2} - \sum_{n=3}^{8} f_{2n}(\rho R)\,\frac{C_{2n}}{R^{2n}}\,, \tag{3.95}$$

in which d_1, d_0, d_{-1}, α, β, and ρ are treated as free parameters determined by nonlinear least-squares fitting. Accurate experimental (Kislyakov, 1999) and *ab initio* (Hättig and Hess, 1996) values of C_6, C_8, C_{10} were employed in the fitting, while C_{2n} $(n = 6,$ $7, 8)$ were derived from C_6, C_8, and C_{10} via eqn (3.10c). Because eqn (3.95) leads to unphysical values of $V_{\rm SM}(R)$ for separations less than 0.1Å, it was spliced (Patkowski *et al.*, 2005) to a simple quasi-Coulombic short-range form $V_{\rm SR}(R)$ given by

$$V_{\rm SR}(R) = d_{\rm SR}R^{-1}\mathrm{e}^{-\alpha_{\rm SR}R}\,,$$

with parameters $d_{\rm SR}$ and $\alpha_{\rm SR}$ determined by equating $V_{\rm SR}(R)$ and $V_{\rm SM}(R)$ and their (analytic) first derivatives at $R = 0.15$Å.

The results obtained by Patkowski *et al.* (2005) indicate that extrapolation to the CBS limit plays an important role in the determination of accurate dispersion-dominated interaction energies even when large orbital basis-sets have been utilized in the calculations. The contribution to the CCSD(T) interaction energy of the $\mathrm{Ar_2}$ dimer arising from extrapolation to the CBS limit in the vicinity of $R_{\rm e}$ is of the order of $-0.37\,\mathrm{cm^{-1}}$: its contribution to $V_{\rm SM}(R)$ for $R < R_{\rm e}$ is more substantial, and has a magnitude similar to that of the core and core–valence electron correlation correction for separations corresponding to the short-range repulsion region of the $\mathrm{Ar_2}$ PEF. Both corrections are important if an *ab initio* PEF possessing an accuracy of better than $\pm 1\,\mathrm{cm^{-1}}$ is desired. Patkowski *et al.* (2005) found that inclusion of these two corrections gives a depth $\mathfrak{D}_{\rm e} = 99.24\,\mathrm{cm^{-1}}$ ($99.27\,\mathrm{cm^{-1}}$ from the fitted analytic form), which differs from the depth $\mathfrak{D}_{\rm e}(\text{Aziz}) = 99.55\,\mathrm{cm^{-1}}$ of the empirical PEF by only 0.3%. However, because the fitted analytic form $V_{\rm SM}(R)$ for $\mathrm{Ar_2}$ incorporates larger (and more realistic) values of the C_6, C_8, and C_{10} dispersion coefficients, it is slightly deeper than the Aziz empirical PEF for separations $R > R_{\rm e}$. As this $\mathrm{Ar_2}$ PEF has no corrections for omitted triples and higher excitations or for relativistic effects, which should contribute at roughly the $1\,\mathrm{cm^{-1}}$ level, a few details remained to be clarified. For short separations, $V_{\rm SM}(R)$ becomes progressively more repulsive than the Aziz empirical PEF (Aziz, 1993) and is in rather good agreement with the short-range behavior of the $\mathrm{Ar_2}$ PEF recommended by Phelps *et al.* (2000) based upon an analysis of high-energy (10–100 eV) experimental Ar–Ar collision cross-section data.

Because the intrinsic uncertainty of the *ab initio* PEF obtained in Patkowski *et al.* (2005) almost certainly exceeds the $0.3\,\mathrm{cm^{-1}}$ level, Jäger *et al.* (2009, 2010) and, nearly simultaneously, Patkowski and Szalewicz (2010) undertook the recomputation of the $\mathrm{Ar_2}(X^1\Sigma_g^+)$ interaction energy as a function of Ar–Ar separation [for 38 separations (Jäger *et al.*, 2009; Jäger *et al.*, 2010) in the interval $1.8\text{Å} \le R \le 15\text{Å}$, and for 33 separations (Patkowski and Szalewicz, 2010) in the interval $2\text{Å} \le R \le 20\text{Å}$] using high-level basis-sets developed by Dunning and coworkers (Brenner and Peters, 1982; Dunning, 1989; Dunning *et al.*, 2001; Kendall *et al.*, 1992; Peterson and Dunning, 2002;

Wilson *et al.*, 1999; Woon and Dunning, 1993). In addition, Patkowski and Szalewicz (2010) introduced a number of new basis-set sequences for their Ar_2 calculations and provided a detailed discussion of the development and testing of all augmented and non-augmented sets of basis functions that were employed in their calculations. Suffice it to say here that in both studies, the basis-sets had to be capable of determining final interaction energy values accurate to at least $0.1\,\mathrm{cm}^{-1}$ for each separation R. This rather stringent requirement necessitated that calculations be carried out with high-X doubly augmented basis-sets, minimally at the CCSD(T) frozen-core level.

As Patkowski and Szalewicz discovered that none of the basis-set families tested by them performed uniformly well at all separations, they found it necessary to carry out full calculations at each of the 33 Ar–Ar separations using four separate basis-set sequences: their recommended value for the interaction energy for a given separation R was then determined by discarding the computed CBS-limit interaction energy that deviated most from the other three and assigning the arithmetic mean of the remaining three as $E_{\mathrm{int}}(R)$. At the same time, an uncertainty $\sigma(R)$ was assigned to each final interaction energy according to a requirement that all three CBS basis-set-sequence extrapolations, plus at least one of the largest-X unextrapolated results fall within $\pm\sigma$ of the assigned value.

In both sets of calculations, the full Ar_2 interaction energy was represented in terms of the CCSD(T) interaction energy at the frozen-core level, with full BSSE counterpoise corrections and extrapolated to the CBS limit, designated here as $E^{\mathrm{CCSD(T)/FC}}(R) \equiv E^{(0)}_{\mathrm{int}}(R)$, plus correction terms giving:

(i) the difference, $\Delta E^{\mathrm{FC}}_{\mathrm{int}}(R)$, between CCSD(T) calculations with full electron correlation and the frozen-core bases,

(ii) the difference, $\Delta E^{\mathrm{T-(T)}}_{\mathrm{int}}(R)$, between CCSD(T) and CCSDT calculations, both at the frozen-core level,

(iii) the difference, $\Delta E^{\mathrm{(Q)-(T)}}_{\mathrm{int}}(R)$, between CCSD(Q) and CCSD(T) calculations at the frozen-core level,

(iv) a relativistic correction, $\Delta E^{\mathrm{rel}}_{\mathrm{int}}(R)$, and

(v) the difference, $\Delta E^{\mathrm{Q-(Q)}}_{\mathrm{int}}(R)$, between CCSDQ and CCSD(Q) calculations (Patkowski and Szalewicz, 2010) at the frozen-core level.

Inclusion of at least the first four corrections should enable the Ar_2 PEF to be obtained with an accuracy of the order of $\pm 0.1\,\mathrm{cm}^{-1}$, particularly in the region of the minimum. The interaction energy can thus be expressed as

$$E_{\mathrm{int}}(R) = E^{(0)}_{\mathrm{int}}(R) + \Delta E^{\mathrm{FC}}_{\mathrm{int}}(R) + \Delta E^{\mathrm{T-(T)}}_{\mathrm{int}}(R) + \Delta E^{\mathrm{(Q)-(T)}}_{\mathrm{int}}(R) + \Delta E^{\mathrm{rel}}_{\mathrm{int}}(R) + \Delta E^{\mathrm{Q-(Q)}}_{\mathrm{int}}(R)\,. \tag{3.96}$$

All energies were obtained from counterpoise-corrected supermolecule calculations extrapolated, whenever possible, to the CBS limit.

In carrying out their corrections for the frozen-core approximation, Patkowski and Szalewicz (2010) found that the core correction (of $-1.701\,\mathrm{cm}^{-1}$, based upon calculations for the aug-cc-pV5Z{33221} basis-set) employed (Patkowski *et al.*, 2005; Slavíček *et al.*, 2003) for $R = 3.7500\,\mathrm{Å}$ is roughly twice as large as it should be due to irregular convergence associated with the aug-cc-pVXZ{33221} sequence of basis-sets. They also found that even though all-electron (AE) optimized basis sequences converged distinctly more slowly in CCSD(T)/AE interaction energy calculations than do FC-optimized basis-set sequences in corresponding CCSD(T)/FC calculations, their difference, which gives the correction to the frozen-core approximation, converges reasonably well, to give $\Delta E_{\rm int}^{\rm FC} = -0.817 \pm 0.024\,{\rm cm}^{-1}$, rather than $-1.701\,{\rm cm}^{-1}$, at $R = 3.7500$ Å.

Because high-level CC contributions to $E_{\rm int}(R)$ are progressively more difficult (and expensive) to obtain, corrections estimating the contributions arising from CCSDT and higher-level CC contributions must perforce be computed with significantly smaller basis-sets (e.g. using 294, 178, 76 contracted functions, respectively, for the $\Delta E_{\rm int}^{\rm T-(T)}$, $\Delta E_{\rm int}^{\rm (Q)-(T)}$, and $\Delta E_{\rm int}^{\rm Q-(Q)}$ corrections, vs. 767 contracted functions employed at the $E_{\rm int}^{(0)}$ level; see Patkowski and Szalewicz 2010). The first two higher-level CC contributions were each found to have magnitudes greater than $1\,{\rm cm}^{-1}$, and both were found to be difficult to bring to the CBS limit. Fortuitously, however, they have opposite signs, and thus largely cancel one another: for $R = 3.7500$ Å, for example, $\Delta E_{\rm int}^{\rm T-(T)} = (1.444 \pm 0.171)\,{\rm cm}^{-1}$ and $\Delta E_{\rm int}^{\rm (Q)} = (-1.945 \pm 0.239)\,{\rm cm}^{-1}$. Patkowski and Szalewicz (2010) were able to estimate the correction for the contribution from full quadruple excitations as $\Delta E_{\rm int}^{\rm Q-(Q)} \simeq (0.086 \pm 0.043)\,{\rm cm}^{-1}$, to give a final nonrelativistic Ar–Ar well-depth $\mathfrak{D}_{\rm e} = (98.676 \pm 0.304)\,{\rm cm}^{-1}$. Relativistic corrections to $E_{\rm int}^{(0)}$ were generally in excess of 0.1% at all Ar–Ar separations for which calculations were carried out. In particular, at $R_{\rm e}$ the relativistic correction was estimated to be $\Delta E_{\rm int}^{\rm rel}(R_{\rm e}) = (-0.615 \pm 0.092)\,{\rm cm}^{-1}$: this value is consistent with corrections $-0.7\,{\rm cm}^{-1}$ obtained by Faas *et al.* (2000) and $-0.62\,{\rm cm}^{-1}$ (also $-0.48\,{\rm cm}^{-1}$) obtained earlier by Patkowski and Szalewicz (2007). Thus, according to eqn (3.96), $R_{\rm e} = 3.7500$ Å and $V(R_{\rm e}) = -\mathfrak{D}_{\rm e} = -99.291 \pm \sigma(R_{\rm e})$, with uncertainty $\sigma(R_{\rm e}) = 0.318\,{\rm cm}^{-1}$ determined by adding the uncertainties associated with the individual contributions quadratically. The corresponding values for $R_{\rm e}$ and $\mathfrak{D}_{\rm e}$ obtained by Jäger *et al.* (2009, 2010) are $R_{\rm e} = 3.7620$ Å, $\mathfrak{D}_{\rm e} = 99.48\,{\rm cm}^{-1}$.

Comparison of the computed values of the component parts of $V_{\rm SM}(R) \equiv E_{\rm int}(R)$ in the format given in eqn (3.96) reported by Jäger *et al.* (2009, 2010) and by Patkowski and Szalewicz (2010) shows that there is generally very good agreement between these two independent calculations. The two sets of results are typically within a fraction of one percent for the CCSD(T) frozen-core interaction energies, $E_{\rm int}^{(0)}(R)$, and often agree within a few percent for most of the individual correction terms. The agreement between the final interaction energies, $V_{\rm SM}(R)$, is even better than that found between the CCSD(T) frozen-core values. Generally speaking, but not surprisingly, the worst agreement in all cases is obtained for values of R in the vicinity of σ. Indeed, the Jäger *et al.* (2009, 2010) PEF values generally fall within the uncertainty envelope computed by Patkowski and Szalewicz (2010) and differs from it on average by less than 0.25% except for separations exceeding 13 Å($25\,a_0$). By comparison, no other *ab initio* or empirical multi-property-fit Ar_2 PEF falls within this rather tight uncertainty envelope, although the empirical multi-property-fit PEF of Boyes (1994) gives the next closest behavior for $R \gtrsim 3$ Å.

On the one hand, Jäger *et al.* (2009, 2010) fitted their *ab initio* interaction energies to a version of eqns (3.90) in which all c_i, d_j, and $C_{2n'+1}$ coefficients are set to zero, as for their $Ne_2(X^1\Sigma_g^+)$ PEF (Hellmann *et al.*, 2008*a*), Hellmann *et al.* 2008*b*, so that $V_{SM}^{J}(R)$ for $Ar_2(X^1\Sigma_g^+)$ is represented by eqn (3.94b). On the other hand, Patkowski and Szalewicz (2010) fitted their *ab initio* interaction energies to a version of eqns (3.90) in which A, γ, all c_i, and all $C_{2n'+1}$ coefficients are set to zero, and $m' = 1$, $m'' = 3$, to give

$$V_{SM}^{PS}(R) = (d_{-1}R^{-1} + d_0 + d_1R + d_2R^2 + d_3R^3)e^{-\alpha R} - \sum_{n=3}^{8} f_{2n}(\rho R)\frac{C_{2n}}{R^{2n}}\,, \tag{3.97a}$$

with d_j ($j = -1, 0, 1, 2, 3$), α, ρ, C_6, and C_8 free fitting parameters, and with C_{10} set to the *ab initio* value of Hättig and Hess (1996), while C_{2n}, ($n = 6, 7, 8$) were taken from Patkowski *et al.* (2005). To make eqn (3.97a) compatible with the small-separation interaction energies obtained in Patkowski *et al.* (2005), it was spliced to the function

$$\widetilde{V}_{SM}^{PS} = (\widetilde{d}_{-1}R^{-1} + \widetilde{d}_0 + \widetilde{d}_1R + \widetilde{d}_2R^2)e^{-\widetilde{\alpha}R-\widetilde{\beta}R^2} \tag{3.97b}$$

for $R \leq R_s \equiv 1.3\,\text{Å}$, in which $\widetilde{d}_{-1} = 18^2E_ha_0$ expresses the Coulombic repulsion between a pair of argon nuclei. The parameters $\widetilde{d}_j$ ($j = 0, 1, 2$), $\widetilde{\alpha}$, and $\widetilde{\beta}$ were fitted to the small-separation *ab initio* results obtained by Patkowski *et al.* (2005) for separations R lying in the range $0.25 \leq R \leq 1.2\,\text{Å}$, subject to the constraints that the functions and their first derivatives match at $R = R_s$. eqn (3.97b) provides a reasonable fit (within $\pm 3.3\%$ on average) of the *ab initio* values reported Patkowski *et al.* (2005). The final fitted PEF lies well within the uncertainties in the *ab initio* data of Patkowski and Szalewicz (2010) for all separations less than 13 Å (25 a_0). Note that the fit-value $30.9833\times10^4\text{cm}^{-1}\text{Å}^6$ obtained for C_6 lies well within the $\pm1\%$ error bounds of the most recent recommended value, $C_6 = 31.0464\times10^4\,\text{cm}^{-1}\text{Å}^6$, for Ar (Kumar and Thakkar, 2010).

The computed *ab initio* Ar–Ar interaction energies obtained by Jäger *et al.* (2009, 2010) and by Patkowski and Szalewicz (2010) are (for the most part) indistinguishable. Jäger *et al.* (2009, 2010) employed the analytic form (3.97b) to fit their *ab initio* data, while Patkowski and Szalewicz (2010) fitted their values to the analytic form (3.97a). The parameters are shown in Table 3.8. The main differences between these two *ab initio* PEFs occurs in the vicinity of the equilibrium separation, R_e, for the Ar–Ar interaction: these somewhat subtle differences particularly affect the shape of the PEF in the region around R_e, and show up as small but meaningful changes in the DRMSD values determined by differences between predicted and experimentally observed spectroscopic properties and equilibrium and nonequilibrium bulk gas properties (Myatt *et al.*, 2018).

The original publications should be consulted for details of the fitting procedure and for values for the specific fit-parameters obtained by Jäger *et al.* (2009, 2010) and Patkowski and Szalewicz (2010) for their *ab initio* PEFs. Values for the MLR best-fit-parameters for the empirical $Ar_2(X^1\Sigma_g^+)$ PEF of Myatt *et al.* (2018), for MLR fits to the *ab initio* $Kr_2(X^1\Sigma_g^+)$ PEF, of Jäger *et al.* (2016), and to the *ab initio* $Xe_2(X^1\Sigma_g^+)$

Table 3.8 Parameters obtained from fitting of *ab initio* computed interaction energies for the $Ne_2(X^1\Sigma_g^+)$ and $Ar_2(X^1\Sigma_g^+)$ van der Waals dimers to analytic functional forms. All quantities are expressed in units of cm^{-1} for energy and Å for distance

	Ne_2	Ar_2		Ar_2
	a	b		c
$10^{-7}A$	2.800301559	3.18365779	$10^{-7}d_0$	12.76418964
a_1	−4.28654039586	−2.98337630	d_1	−225.0623273
$10^2 a_2$	−3.33818674327	−9.71208881	d_2	33.5895075
$10a_{-1}$	−5.34644860719	−2.75206827	d_{-1}	−278.8986103
a_{-2}	0.501774999491	−1.01489050	d_3	−1.794786725
ρ	4.92438731676	4.0251721	ρ	4.523717745
			α	2.935475675
$10^{-4}C_6$	3.062749244	30.77589571		30.98330807
$10^{-5}C_8$	1.146069050	22.7065238		20.44399988
$10^{-5}C_{10}$	5.493874011	170.7338480		189.8651533
$10^{-6}C_{12}$	3.374200223	130.8330994		200.8809383
$10^{-7}C_{14}$	2.655025451	102.1750024		109.2209845
$10^{-8}C_{16}$	2.676530594	81.32062525		383.3475918

a, Hellmann *et al.*, Jäger *et al.*, 2008*a*; b, 2009, 2010; c, Patkowski and Szalewicz 2010.

PEF of Hellmann *et al.* (2017) are given in Table 3.9. Similarly, the properties traditionally employed to characterize a PEF are given in Table 3.9.

The most recent (Myatt *et al.*, 2018) determination of the $Ar_2(^1\Sigma_g^+)$ PEF is based upon the MLR form for the PEF initially developed (Le Roy and Henderson, 2007; Le Roy *et al.*, 2011) for chemically bound diatomic molecules, then extended to less strongly bound $a^3\Sigma_u^+$ spin-aligned diatomic molecules (Dattani and Le Roy, 2011) and to the Ar–Xe rare gas dimer (Piticco *et al.*, 2010). The weakly-bound Ar_2 dimer provided an almost ideal case for the determination of a highly accurate MLR fit-potential due to the availability of an extensive set of spectroscopic data (Herman *et al.*, 1988) spanning most of the potential well, in addition to pressure and, especially, high-quality acoustic virial data. A set of seven MLR fit-parameters were determined (in addition to three predetermined parameters and literature values for the C_6, C_8, and C_{10} long-range dispersion coefficients for argon) based upon a fitting to 1634 primary data (1397 spectroscopic transition frequencies plus 178 pressure virial data and 59 acoustic virial data) and vetted against a set of 609 secondary (transport property) data, based upon a minimization of the DRMSD value obtained for the full set of 2243 data.

A comparison between the $\overline{dd}$ value 1.043 obtained for the full set of 2243 data using the empirical MLR (Myatt *et al.*, 2018) PEF and $\overline{dd} = 2.951$ for the Aziz (1993) HFD-

Table 3.9 Properties characterizing the $Ar_2(X^1\Sigma_g^+)$, $Kr_2(X^1\Sigma_g^+)$, and $Xe_2(X^1\Sigma_g^+)$ van der Waals dimers and their interaction energies

R_e /Å	σ /Å	$\mathfrak{D}_e$ /cm^{-1}	$\langle R\rangle_0$ /Å	$\mathfrak{D}_0$ /cm^{-1}	Ref.
		$Ar_2(X^1\Sigma_g^+)$			
3.7785		96.9858	3.8567	82.3798	a
3.783		97.6004			b
3.763	3.3562	99.3627			c
3.762	3.357	99.3474		84.64	d
3.7545		99.291			e
3.7644		99.5454		84.74	f
3.7662		99.494			g
		$Kr_2(X^1\Sigma_g^+)$			
R_e /Å	σ /Å	$\mathfrak{D}_e$ /cm^{-1}	$\langle R\rangle_0$ /Å	$\mathfrak{D}_0$ /cm^{-1}	Ref.
4.0592		132.76	4.1040	121.547	h
4.027		135.09		120.71	i
4.017		138.61			j
4.0135		140.06			k
4.0158		139.62			l
4.0110	3.5708	139.91	4.0537	128.42	m
		$Xe_2(X^1\Sigma_g^+)$			
R_e /Å	σ /Å	$\mathfrak{D}_e$ /cm^{-1}	$\langle R\rangle_0$ /Å	$\mathfrak{D}_0$ /cm^{-1}	Ref.
4.525		156.45			n
4.382		196.77			o
4.3780		194.593			p
4.3656	3.8910	196.557			q

a, Cybulski and Toczyłowski, 1999; b, Fernández and Koch, 1998; c, Laschuk *et al.*, 2003; d, Jäger *et al.*, 2009; e, Patkowski and Szalewicz, 2010; f, Boyes 1994; g, Myatt *et al.*, 2018; h, López Cacheiro *et al.*, 2004; i, Slavíček *et al.*, 2003; j, Nasrabad and Deiters, 2003; k, Waldrop *et al.*, 2015; l, Jäger et al., 2016; m, Dham *et al.*, 1989; n, Runeberg and Pyykkö, 1998; o, Hanni *et al.*, 2004; p, Hellmann *et al.*, 2017; q, Dham *et al.*, 1990.

ID PEF, $\overline{dd} = 1.124$ for the Boyes (1994) HFD-ID PEF shows not only that the MLR Ar_2 PEF is the best empirical Ar_2 PEF, but also its comparison with the values $\overline{dd} =$ 1.062 for the Jäger *et al.* (2009, 2010) and $\overline{dd} = 1.337$ for the Patkowski and Szalewicz (2010) Ar_2 PEFs also establishes the efficacy of the *ab initio* calculations of the Ar–Ar interaction energy. Indeed, both *ab initio* PEFs fare rather well in predicting the set of 2243 data: it is also clear from the DRMSD values $\overline{dd} = 2.442$, $\overline{dd} = 4.867$ predicted for the 59 $\beta_{2a}(T)$ data, $\overline{dd} = 1.180$, $\overline{dd} = 1.689$ predicted for the $B_2(T)$ data, and $\overline{dd} = 0.938$, $\overline{dd} = 1.015$ predicted for the 1397 spectroscopic transition frequencies by the Jäger *et al.* (2009, 2010), and Patkowski and Szalewicz (2010) PEFs, respectively, that the former *ab initio* PEF appears to be slightly superior to the latter, although not quite matching the best-fit empirical MLR PEF, for which $\overline{dd} = 2.360$ for the $\beta_{2a}(T)$ data set, $\overline{dd} = 1.162$ for the $B_2(T)$ data set, and $\overline{dd} = 0.922$ for the spectroscopic data set. As these three PEFs are indistinguishable on the scale of the full potential well, differences in the DRMSD values associated with the three specific data sets appear to be associated with relatively small differences in the shapes of the potential wells in the vicinity of the potential minimum. An illustration of the general statement on the TT03 PEFs given in Section 3.3 are the DRMSD values $\overline{dd} = 3.047$ for the spectroscopic data, 15.759 for the accurate set of pressure virial data, 114.10 for the acoustic virial data set, and 6.454 for the transport data employed in determining the empirical MLR PEF (Myatt *et al.*, 2018).

3.8.4 The $Kr_2(X^1\Sigma_g^+)$ interaction

Until relatively recently, it had not been possible to carry out *ab initio* calculations of the pair interaction energy for krypton with accuracies comparable to those obtained for helium, neon, and argon, or to that obtained by the best empirical krypton PEFs (Aziz and Slaman, 1986*b*; Dham *et al.*, 1989). Early *ab initio* calculations of the Kr–Kr interaction energy all utilized a supermolecule approach, either using MBPT calculations (Tao, 1999), CCSD(T) calculations with an effective core potential (Slavíček *et al.*, 2003), or CCSD(T) calculations using a frozen-core approximation (Haley and Cybulski, 2003; Nasrabad and Deiters, 2003). The earliest supermolecule calculation (Tao, 1999) was carried out at the FC-MP4 level with a [9s7p4d3f] basis coupled with {3321} mid-bond functions: the resultant PEF was characterized by a minimum $(R_e, -\mathfrak{D}_e) = (4.037\,\text{Å}, -135.4\,\text{cm}^{-1})$, lying $4.5\,\text{cm}^{-1}$ above the minimum $V_{\min} = -132.8\,\text{cm}^{-1}$ for the XC(fit) PEF (Dham *et al.*, 1989). Two of the three CCSD(T) calculations (Nasrabad and Deiters, 2003; Slavíček *et al.*, 2003) were based upon the frozen-core *aug*-cc-pV5Z{33221} basis-set combination, and obtained very similar results for the minimum, essentially $R_e = 4.06\,\text{Å}$, $V_{\min} = -\mathfrak{D}_e = -132.8\,\text{cm}^{-1}$, which lies about 5.1% above $V_{\min}$ for the XC(fit) PEF. By making core corrections using an electron core potential at the *aug*-cc-pVQZ{spdfg} level, plus a CBS extrapolation and relativistic corrections, Slavíček *et al.* (2003) were able to attain a depth $V_{\min} = -138.2\,\text{cm}^{-1}$, which lies only 1.2% above the empirical value. The third calculation (Nasrabad and Deiters, 2003) employed X = T, Q CCSD(T) results to extrapolate to the CBS limit, giving $R_e = 4.0170\text{Å}$, $V_{\min} = -138.61\,\text{cm}^{-1}$, within 0.2% and 1%, respectively, of the empirical values (Aziz *et al.*, 2004; Dham *et al.*, 1989). However, none of these studies included the full set of corrections that were made for the Ar–Ar interaction (Patkowski and Szalewicz, 2010) to give an additional contribution of $-1.847\,\text{cm}^{-1}$ to the final value for $V_{\min}(Ar_2)$. For this reason, it would therefore not be unreasonable to anticipate a modestly larger overall correction to the Kr_2 interaction energy (i.e. by perhaps as much as 2.0–$2.5\,\text{cm}^{-1}$) beyond the empirical value. Given that the *ab initio* PEF obtained by Nasrabad and Deiters (2003) also does not include corrections for the frozen-core approximation, the difference between CCSD(T) and CCSDT calculations, the omission of quadruple and higher excitation contributions, or relativistic effects, it thus seems likely that the empirical PEFs may indeed prove to be slightly too shallow. Note that the *ab initio* $Kr_2(X^1\Sigma_g^+)$ PEF of Nasrabad and Deiters (2003) was computed for use with the Axilrod–Teller triple–dipole three-body interaction (Axilrod and Teller, 1943) in order to carry out Monte Carlo simulations of the thermodynamic properties of bulk krypton.

Two high-level sets of *ab initio* calculations of the Kr–Kr interaction energy as a function of the internuclear separation R have been carried out using the CCSD(T) method at the frozen-core level plus corrections for core–core and core–valence correlations, CC calculations beyond the CCSD(T) level, and relativistic effects (Jäger *et al.*, 2016; Waldrop *et al.*, 2015). Waldrop *et al.* (2015) carried out calculations using standard basis-sets developed by Dunning and coworkers (Brenner and Peters, 1982; Dunning, 1989; Dunning *et al.*, 2001; Kendall *et al.*, 1992; Peterson and Dunning, 2002; Wilson *et al.*, 1999; Woon and Dunning, 1993), and included up to perturbative quadruple excitations, CCSDT(Q), within the CC scheme. Their final results included BSSE counterpoise corrections (Boys and Bernardi, 1970) and CBS limit extrapolation (Gdanitz, 2001; Lee, 2005). Interaction energies were obtained for 25 nuclear separations between 2.6 Å and 12.0 Å, and their results were fitted to a modified TT function using seven fit-coefficients, including values for effective C_6 and C_8 long-range dispersion energy coefficients. Their modified TT function was splined to a repulsive short-range PEF form at $R = 1.8$ Å in order to avoid the turnover that occurs at very short internuclear separations for a typical TT PEF. Their fitted *ab initio* PEF, which closely resembles the HFD-B2 empirical PEF determined by Dham *et al.* (1989), supports 16 bound vibrational levels.

For the determination of the $Kr_2(^1\Sigma_g^+)$ PEF, Jäger *et al.* (2016) chose first to develop a new sextuple-zeta basis-set along lines followed by Dunning and coworkers (Brenner and Peters, 1982; Dunning, 1989; Dunning *et al.*, 2001; Kendall *et al.*, 1992; Peterson and Dunning, 2002; Wilson *et al.*, 1999; Woon and Dunning, 1993), but extended to a larger number of basis functions with d, f, g polarization functions optimized using a $(30s19p14d)/[9s8p1d]$ basis-set. Their optimized basis-set (V6Z) for the krypton atom is characterized by $(28s20p15d5f4g3h2i)/[10s9p7d5f4g3h2i]$. In addition to full BSSE counterpoise corrections to the results obtained using this sextuple-zeta basis-set, the correlation component was extrapolated to the CBS limit by including corrections for CC methods beyond the CCSD(T) level, by accounting for core–core and core–valence electron correlations, and by carrying out extensive analysis of relativistic corrections, including a full calculation of the interaction between the spin of one electron with the magnetic moment of a second electron. Values for the Kr–Kr interaction energies, together with uncertainty intervals, were obtained for 36 nuclear separations between 2.2 Å and 15.0 Å, then fitted to the modified TT function (3.94b) using an eight-parameter fit. An additional two parameters were employed to spline a modified Coulombic repulsion term to the TT fit-function for nuclear separations less than $0.3R_e$. Their *ab initio* PEF corresponds to $\mathfrak{D}_e = 139.61\,\mathrm{cm}^{-1}$ and $R_e = 4.0158$ Å.

3.8.5 The $Xe_2(X^1\Sigma_g^+)$ and $Rn_2(X^1\Sigma_g^+)$ interactions

An accurate multi-property best-fit empirical HFD-B PEF for the $Xe_2(X^1\Sigma_g^+)$ interaction was first constructed by Aziz and Slaman (1986*a*). It was characterized by a minimum $-\mathfrak{D}_e = -196.2\,\text{cm}^{-1}$ at an equilibrium internuclear separation $R_e = 4.3627\,\text{Å}$. A slightly improved HFD-B PEF and a XC(fit)-type PEF were also constructed a few years later (Dham *et al.*, 1990). The revised HFD-B PEF is characterized by $(R_e, \mathfrak{D}_e) = (4.3581\,\text{Å}, 195.8\,\text{cm}^{-1})$, which agrees reasonably well with values $(4.37773\pm\ 0.0049\,\text{Å}, 196.1\pm1\,\text{cm}^{-1})$ obtained from high-resolution VUV laser spectroscopy measurements (Wüest *et al.*, 2004).

Although there had been considerable interest in the computation of *ab initio* interaction energies for a pair of Xe atoms, the main focus was initially upon obtaining accurate values for the minimum in the interaction. An initial detailed *ab initio* study (Runeberg and Pyykkö, 1998) of the long-range and equilibrium properties of the $Xe_2(X^1\Sigma_g^+)$ interaction based upon the CCSD(T) supermolecule procedure using an appropriately modified cc-pVQZ{33221} basis-set, including a relativistic effect core potential (RECP) and a core polarization potential (CPP), led to a minimum interaction characterized by ($R_e = 4.525\,\text{Å}$, $\mathfrak{D}_e = 156.45\,\text{cm}^{-1}$), and the long-range interaction energy could be represented by long-range disperson coefficients C_6, C_8 that were in good agreement (0.9% for C_6 and 8.5% for C_8) with the best available empirical fit values (Dham *et al.*, 1990).

Shortly after the Runeberg and Pyykkö study (1998), Slavíček *et al.* (2003) carried out a more extensive set of CCSD(T) supermolecule computations, using an *aug*-cc-pVQZ{3322} basis together with a RECP, of the interaction energies at 28 internuclear separations R lying between 2.6 Å and 20.0 Å to obtain a full *ab initio* $Xe_2(X^1\Sigma_g^+)$ PEF, with a minimum characterized by ($R_e = 4.420\,\text{Å}$, $\mathfrak{D}_e = 183.1\,\text{cm}^{-1}$), a minimum that is only 0.7% shallower and lying 1.3% further out than the minimum in the multi-property best-fit PEF (Dham *et al.*, 1990). At about the same time, Hanni *et al.* (2004) carried out similar *ab initio* computations of the Xe–Xe interaction energy for 14 internuclear separations lying between 3.8 Å and 8.0 Å using the CCSD(T) supermolecule method and a modified version of the basis-set employed by Slavíček *et al.* (2003), including a CPP and a {33221} set of bond functions. This set of 14 interaction energies was fitted to the functional form (3.94a) with the dispersion series restricted to the three leading terms and the individual damping functions replaced by an overall damping function having the form of eqn (3.6d). These calculations led to a slightly deeper well and a shorter equilibrium internuclear separation, namely, $R_e = 4.420\,\text{Å}$, $\mathfrak{D}_e = 196.8\,\text{cm}^{-1}$, lying only 0.15% deeper than, but shifted to a 1.4% shorter internuclear separation in comparison with the multi-property XC(fit) PEF (Dham *et al.*, 1990).

The most recent *ab initio* $Xe_2(X^1\Sigma_g^+)$ PEF, was obtained by Hellmann *et al.* (2017) based upon a set of all-electron basis functions constructed along the same lines that had been employed to develop a new V6Z basis-set (Jäger *et al.*, 2016) for krypton. A new 6VZ basis-set was employed for the xenon atom in order to ensure smoother extrapolations of the calculated Xe–Xe interaction energies to their CBS limits. All interaction energies were obtained using CC supermolecule computations with full BSSE counterpoise corrections and including sets of bond functions located at the internuclear midpoint in order to reduce basis-set incompleteness errors. Interaction energies were obtained for 39 internuclear separations R lying between 2.4 Å and 15.0 Å. Final interaction energies were expressed in terms of the CCSD(T) nonrelativistic interaction energies plus a correction for higher level (up to CCSDTQ) CC terms and a number of relativistic corrections.

The 39 Xe–Xe interaction energies were fitted to a modified TT function (Tang and Toennies, 1984) having the form of eqns (3.90), specifically,

$$V(R) = c_0 \exp\{a_1 R + a_2 R^2 + a_{-1} R^{-1} + a_{-2} R^{-2}\} - \sum_{n=3}^{8} f_{2n}(\rho R)\, \frac{C_{2n}}{R^{2n}}\,.$$

The parameters c_0, a_1, a_2, a_{-1}, a_{-2}, and ρ, together with the dispersion coefficients C_6, C_8, and C_{10}, were treated as independent fit-parameters, while the higher dispersion coefficients C_{12}, C_{14}, and C_{16} were obtained using the semi-empirical recursion relation (3.11) (Thakkar, 1988). Energy uncertainties were also provided in order to enable computational uncertainties to be determined for the various calculated properties of xenon and to enable more complete comparisons with experimental results. The well-depth $\mathfrak{D}_e$ and the equilibrium bond length for the $Xe_2(X^1\Sigma_g^+)$ dimer were obtained as $\mathfrak{D}_e = (194.59 \pm 1.75)\,\mathrm{cm}^{-1}$ and $R_e = (4.3780 \pm 0.0036)$ Å. In particular, the *ab initio* result for R_e is in excellent agreement with the value $R_e = (4.3773 \pm 0.0049)$ Å determined by Wüest *et al.* (2004) from the rotational constants for the $v = 0$ and 1 vibrational levels of the $^{129}Xe^{132}Xe$ dimer.

In general, the final interaction energy values for separation R obtained via the CCSD(T) supermolecular method are obtained sequentially as

$$V(R) = V_{\mathrm{CCSD(T)}}(R) + \Delta V_{\mathrm{post\text{–}CCSD(T)}}(R) + \Delta V_{\mathrm{rel}}(R)\,, \tag{3.98}$$

with $V_{\mathrm{CCSD(T)}}(R)$, $\Delta V_{\mathrm{post\text{–}CCSD(T)}}(R)$ the nonrelativistic CCSD(T) interaction energy and a correction for contributions to $V(R)$ associated with higher-order CC terms (typically up to CCSDTQ), while $\Delta V_{\mathrm{rel}}(R)$ arises from relativistic corrections at the CCSD(T) level. The main contribution, $\Delta V_{\mathrm{CCSD(T)}}(R)$, to $V(R)$ is also expressed as an additive set of terms, namely,

$$V_{\mathrm{CCSD(T)}}(R) = V_{\mathrm{SCF}}(R) + V_{\mathrm{corr}}^{\mathrm{FC}}(R) + \Delta V_{\mathrm{corr}}^{\mathrm{IFC}}(R) + \Delta V_{\mathrm{corr}}^{\mathrm{AE}}(R)\,. \tag{3.99}$$

The dominant contributions to $V_{\mathrm{CCSD(T)}}(R)$ are associated with the SCF interaction energy, $V_{\mathrm{SCF}}(R)$, and the frozen-core electron-correlation interaction energy, $V_{\mathrm{corr}}^{\mathrm{FC}}(R)$. The two correction terms are $\Delta V_{\mathrm{corr}}^{\mathrm{IFC}}(R)$, which represents additional electron-correlation contributions determined using an inner (i.e. smaller) frozen core to estimate core–core and core–valence correlation effects involving the original frozen core, and $\Delta V_{\mathrm{corr}}^{\mathrm{AE}}(R)$, which provides an estimate of the additional contribution to the correlation energy arising from the inclusion of all electrons. Although this all-electron additional contribution to the correlation energy is typically calculated using lower-level basis-sets, extrapolation to the CBS limit is nonetheless carried out.

Typically, post-CCSD(T) corrections $\Delta V_{\mathrm{post-CCSD(T)}}(R)$ are expressed in terms of sequential estimates of the differences between interaction energies determined at the full and perturbative triples levels, perturbative quadruples and full triples levels, and full and perturbative quadruples levels. Hellmann *et al.* (2017) did, however, find it necessary to include additional post-CCSD(T) corrections in their computation of $V(R)$ for $Xe_2(X^1\Sigma_g^+)$. Relativistic corrections, $\Delta V_{rel}(R)$, become increasingly relatively more important for heavier atoms due to the relativistic speeds associated with the inner core electrons. For the $Ar_2(^1\Sigma_g^+)$ interaction, for example, Jäger *et al.* (2009, 2010) found it sufficient to compute the relativistic contribution, $\Delta_{rel}V(R)$, at what is referred to as the Cowan–Griffen approximation (Cowan and Griffin, 1976) to the scalar relativistic energy, while for the $Kr_2(^1\Sigma_g^+)$ and $Xe_2(^1\Sigma_g^+)$ interaction energies Hellmann *et al.* (2017, Jäger *et al.* (2016) found it necessary to include additional scalar contributions using a method referred to as direct perturbation theory (Klopper, 1997; Kutzelnigg, 1989; Rutkowski, 1986) and various spin-orbit (or "vector") contributions, all evaluated at the CCSD(T) FC level.

The situation for radon is much less satisfactory than that for xenon. Apart from time-dependent HF and MP2 calculations (Hättig and Hess, 1996) of the frequency-dependent dipole polarizability of ground state Rn atoms, from which the value $C_6 = 1.957\times10^6\,\mathrm{cm^{-1}\AA^6}$ (including a relativistic correction of $-0.086 \times 10^6\,\mathrm{cm^{-1}\AA^6}$) was obtained for the leading long-range dispersion coefficient, only a preliminary study (Runeberg and Pyykkö, 1998) of the equilibrium separation and interaction energy for $Rn_2(X^1\Sigma_g^+)$ has been carried out, giving $R_e = 4.639\,\mathrm{\AA}$ and $\mathfrak{D}_e = 222.6\,\mathrm{cm^{-1}}$. Two empirical corrections, one to allow for the inclusion of g-orbital basis elements (estimated from MP2-level calculations) and the second based upon the difference between the *ab initio* and experimentally determined R_e, $\mathfrak{D}_e$ values for the $Xe_2(X^1\Sigma_g^+)$ dimer, led Runeberg and Pyykkö (1998) to recommend final values $R_e = 4.478\,\mathrm{\AA}$ and $\mathfrak{D}_e = 276.6\,\mathrm{cm^{-1}}$ for radon in its ground electronic state. In addition to these values for R_e and $\mathfrak{D}_e$, Runeberg and Pyykkö computed values for the two leading long-range dispersion coefficients, obtaining $C_6 = 2.027\times10^6\,\mathrm{cm^{-1}\AA^6}$ and $C_8 = 2.600\times10^7\,\mathrm{cm^{-1}\AA^8}$.

3.8.6 Noble gas atom hetero-atomic $X^1\Sigma^+$ pair interactions

Obtaining a reliable heterodiatom PEF from BP mixture measurement data is both in principle and in practice considerably more challenging than it is to determine a homodiatom PEF from BP measurement data for a pure substance. The essence of the problem is that a BP measurement in a binary mixture of two substances A and B has three components,

$$(\mathrm{BP})_{\mathrm{mix}} = x_{\mathrm{A}}^2(\mathrm{BP})_{\mathrm{AA}} + 2x_{\mathrm{A}}x_{\mathrm{B}}(\mathrm{BP})_{\mathrm{AB}} + x_{\mathrm{B}}^2(\mathrm{BP})_{\mathrm{BB}}\,,$$

so that the data needed for determination of a heterodiatom PEF requires the acquisition of three sets of BP measurements, followed by an extraction from the mixture data of the dependence, $(\mathrm{BP}_{\mathrm{AB}})$, of the BP upon the heterodiatom PEF. The resultant set of data obtained via such a procedure for fitting to a given PEF form will be considerably less accurate than each individual measurement of the BP for the set of binary mixtures and for each pure substance. The availability of crossed-beam differential (Beneventi *et al.*, 1986; Danielson and Keil, 1988; Keil *et al.*, 1988; Keil *et al.*, 1991) and integral (van den Biesen *et al.*, 1982) scattering cross-section data and of microwave spectral data (Grabow *et al.*, 1995; Jäger *et al.*, 1993; Xu *et al.*, 1995) associated with pure rotational state-to-state transitions in the ground vibrational state, in addition to the more traditional bulk gas mixture data, for many noble gas heterodimers has enabled the determination of a number of accurate empirical noble gas heterodimer PEFs (Aziz and van Dalen, 1983*a*; Barrow and Aziz, 1988; Barrow *et al.*, 1989; Barrow *et al.*, 1992; Beneventi *et al.*, 1986; Danielson and Keil, 1988; Grabow *et al.*, 1995; Keil *et al.*, 1988; Keil *et al.*, 1991; Jäger *et al.*, 1993; van den Biesen *et al.*, 1982; Xu *et al.*, 1995).

There have also been six studies of the interaction energies between unlike rare gas atoms based upon *ab initio* computations in which high-level quantum chemical methods have been employed to obtain the dependence of the interaction energy upon the internuclear separation R (Cybulski and Toczyłowski 1999; Haley and Cybulski 2003; Jäger and Bich 2017; López Cacheiro *et al.* 2004; Partridge *et al.* 2001). The earliest (Cybulski and Toczyłowski, 1999) utilized CCSD(T) calculations with a number of aug-cc-pVXZ (X = T, Q, 5) correlation-consistent basis-sets, both with and without sets of mid-bond function, to obtain the interacton energies of the He–Ne, He–Ar, and Ne–Ar van der Waals dimers at 13 internuclear separations R lying between 2.0 Å and 6.0 Å. Shortly thereafter, Haley and Cybulski (2003) extended this study based upon CCSD(T) calculations employing aug-cc-pVXZ{33221} (X = T, Q, 5) sets of basis and mid-bond functions to obtain interaction energies at 13 internuclear separations R lying between 2.5 Å and 7.5 Å for the corresponding He–Kr, Ne–Kr, and Ar–Kr van der Waals dimers. At about the same time, Cacheiro *et al.* (2004) carried out calculations using much the same CCSD(T) method and an aug-cc-pV6Z{33221} basis-set to obtain interaction energies for the He–Ne, He–Ar, and Ne–Ar heterodimers at 20–23 internuclear separations lying within the interval $2.0\,\text{Å} \lesssim R \lesssim 10.5\,\text{Å}$. The sets of R_e values obtained by Cybulski and Toczyłowski (1999) and by Cacheiro *et al.* (2004) typically differ by no more than 0.01%, while the two sets of $\mathfrak{D}_e$ values obtained using the Z = 6 basis-set (López Cacheiro *et al.*, 2004) are in close agreement with those obtained by Cybulski and Toczyłowski (1999) but lie, on average, approximately 0.07% deeper, which suggests that the latter calculations must give values close to the CBS limit. Both sets of computed interaction energies were fitted to the functional form (3.94a).

As the earlier *ab initio* calculations did not include extrapolation to the CBS limit, corrections for the frozen-core approximation, missing triple- and higher-level excitation contributions to the electron correlation energy, or relativistic effects, the resulting interaction energies will likely have overall inaccuracies of the order of a few cm^{-1} due to their absence. Indeed, all such corrections must be included before an accuracy exceeding $\pm 1\,\text{cm}^{-1}$ can be attained. Moreover, as the *ab initio* interaction energies obtained without making such corrections are already very close to the empirical values determined from fits to experiment by Keil *et al.* (1991) for the He–Ne, He–Ar, and He–Kr interactions, by Aziz and Van Dalen (1983*a*) for the Ar–Kr interaction, and by Barrow and Aziz (1988) for the Ne–Kr interaction, these observations suggest, as previously noted by Cybulski and Toczyłowski (1999), that new empirical determinations may be required: such determinations may, however, also require the acquisition of additional higher-quality experimental mixture data.

As part of an extended study of the binary interactions between He and other ground electronic state atoms from the first three rows of the periodic table, Partridge *et al.* (2001) computed interaction energies for the He–Ne, He–Ar, and He–Kr van der Waals dimers at 23, 25, and 26 internuclear separations R, respectively, based upon the frozen-core CCSD(T) method and atom-centered correlation-consistent polarized-valence aug-cc-pVQZ basis-sets developed by Dunning and co-workers (Brenner and Peters, 1982; Dunning, 1989; Dunning *et al.*, 2001; Kendall *et al.*, 1992; Peterson and Dunning, 2002; Wilson *et al.*, 1999; Woon and Dunning, 1993), with all energies corrected using full BSSE counterpoise corrections. The dependence of the binary interaction energy on the separation R between a He atom and atoms in the first three rows of the periodic table were computed by Partridge *et al.* (2001) for separations lying between 1.3 Å ($2.5a_0$) and 6.4 Å ($12a_0$) using high-level CCSD(T) calculations with aug-cc-pVQZ basis-sets (Brenner and Peters, 1982; Dunning, 1989; Dunning *et al.*, 2001; Kendall *et al.*, 1992; Peterson and Dunning, 2002; Wilson *et al.*, 1999; Woon and Dunning, 1993) plus a [33221] bond-function set. Their results for the He–Ne, He–Ar, and He–Kr interactions, obtained as part of a systematic study, agree closely with the interaction energies computed for these dimers by Cybulski and Toczyłowski (1999) (He–Ne, He–Ar) and by Haley and Cybulski (2003) (He–Kr), and by Cacheiro *et al.* (2004) (He–Ne, He–Ar).

By extrapolating the results of CCSD(T) calculations of the *ab initio* Ne–Ar interaction energies obtained by Cybulski and Toczyłowski (1999) from aug-cc-pVQZ{33221} and aug-cc-pV5Z{33221} and basis-sets to the CBS limit, Nasrabad *et al.* (2004) were able to construct a new *ab initio* NeAr($X^1\Sigma^+$) PEF that closely resembled Barrow and Aziz (1988) best-fit empirical NeAr PEF, with a minimum characterized by $R_e = 3.490$ Å, $\mathfrak{D}_e = 45.785\,\text{cm}^{-1}$, versus $R_e = 3.489$ Å, $\mathfrak{D}_e = 46.978\,\text{cm}^{-1}$ for the empirical PEF. They also computed Ar–Kr interaction energies at 14 internuclear separations in the interval $3.0\,\text{Å} \leq R \leq 10.0\text{Å}$ using the CCSD(T) supermolecule method and the Boys and Bernardi (1970) counterpoise-correction procedure to correct for the BSSE, followed by extrapolation to obtain CBS limit values for the interaction energies. Their final results were then fitted to a modified (Korona *et al.*, 1997) TT functional form to obtain an *ab initio* ArKr($X^1\Sigma^+$) PEF characterized by an equilibrium position $R_e = 3.9008$ Å and depth $\mathfrak{D}_e = 112.16\,\text{cm}^{-1}$, which may be compared with Aziz and Van Dalen's (1983*a*) best-fit empirical ArKr PEF, for which $R_e = 3.881$ Å, $\mathfrak{D}_e = 116.28\,\text{cm}^{-1}$, and Haley and Cybulski (2003) *ab initio* ArKr results, for which $R_e = 3.9226$ Å and $\mathfrak{D}_e = 111.96\,\text{cm}^{-1}$.

Finally, Jäger and Bich (2017) carried out a detailed *ab initio* study of the He–Kr interaction using the correlation-consistent basis-sets of Dunning and coworkers (Brenner and Peters, 1982; Dunning, 1989; Dunning *et al.*, 2001; Kendall *et al.*, 1992; Peterson and Dunning, 2002; Wilson *et al.*, 1999; Woon and Dunning, 1993) for the He atom and employing correlation-consistent basis-sets introduced by Wilson *et al.* (1999) and extended by Jäger *et al.* (2016) for the Kr atom. The interaction energy $V(R)$ between He and Kr was computed according to eqns (3.98) and (3.99), in which $V_{\text{CCSD(T)}}(R)$ includes the correction for the influence of core electrons on the correlation energy in addition to core–core and core–valence correlation effects beyond the FC approximation. Convergence to the interaction energy CBS limits were assisted by the introduction of sets of bond functions at the midpoint between the He and Kr nuclei. The corrections accounting for higher-order couple-cluster excitations and relativistic effects were found to increase the depth of the minimum in the He–Kr interaction energy by about $0.45\,\mathrm{cm}^{-1}$ (approximately 2.1%). The original reference should be consulted for additional details of the quantum chemical calculations.

Table 3.10 shows values for the position, R_e, and well-depth, $\mathfrak{D}_e$, characterizing the equilibrium interaction energy, as well as the average internuclear separation, $\langle R \rangle_{v=0}$, in the ground vibrational level, the dissociation energy, $\mathfrak{D}_0$, and the rotational constant, B_0, for the ground vibrational level, features that may be considered to characterize an individual dimer, while Table 3.11 shows the equilibrium position R_e and interaction depth $\mathfrak{D}_e$ characterizing empirical PEFs obtained for ground state heterodimers of Xe with the other noble gas atoms. Parameters characterizing best-fit empirical PEFs have been included as final rows for each rare-gas heterodimer appearing in Table 3.12 for comparison with the characteristics associated with the *ab initio* results.

An alternative procedure for the bulk gas properties could be to compute the pure gas contributions to a mixture BP using accurate *ab initio* (or empirical) PEFs and to subtract them from the mixture values in order to obtain the contributions to the mixture property arising from the heterodiatom interaction. Sets of BP data obtained in this manner could then be employed, together with available spectroscopic data, initial values for the C_6, C_8, C_{10} long-range dispersion coefficients, the characteristic depth $\mathfrak{D}_e$ and position R_e of the equilibrium minimum of the PEF obtained via accurate (Tang and Toennies, 1986) combining rules, to extract the relevant heterodiatom PEFs.

3.9 Interactions Involving Open-Shell Atoms

3.9.1 Alkali atom–atom interaction energies

Alkali atoms are the natural multi-electron equivalents of the hydrogen atom. The interaction between a pair of alkali atoms, like that between two H atoms, takes the form of a singlet ground electronic term, $X^1\Sigma_g^+$, and a close-lying triplet first excited electronic term, $a^3\Sigma_u^+$: these two terms are well separated from the next-nearest excited electronic term. Not surprisingly, the $B^1\Sigma_u^+$ next-lowest excited electronic term for H_2, located $91{,}700\,\mathrm{cm}^{-1}$ above the ground electronic term, is much higher in energy than the $A^1\Sigma_u^+$ terms of $^7Li_2$ (at $14{,}068.3\,\mathrm{cm}^{-1}$), $^{23}Na_2$ (at $14{,}680.6\,\mathrm{cm}^{-1}$), or $^{39}K_2$ (at $11{,}681.9\,\mathrm{cm}^{-1}$) (Huber and Herzberg, 1979). The composition of an alkali metal vapor is very different from that of H_2 because the equilibrium constant for the dimerization

$$A_2 \rightleftharpoons A + A,$$

is very different from that for one of the alkali metals, and greatly favors $H_2(g)$ over $H(g)$ except for very high temperatures, while alkali vapors are composed predominantly of alkali atoms at all temperatures.

Due to the open-shell nature of alkali atoms, it has proven challenging to perform *ab initio* calculations of alkali dimer interaction energies as functions of the interatomic separation R with spectroscopic accuracy, that is, giving rise to equilibrium depths

Table 3.10 Properties characterizing $X^1\Sigma^+$ *ab initio* electronic ground states of heteronuclear He, Ne, Ar, and Kr noble gas dimers

R_e /Å	σ /Å	$\mathfrak{D}_e$ /cm^{-1}	$\langle R \rangle_0$ /Å	$\mathfrak{D}_0$ /cm^{-1}	B_0 /cm^{-1}	Ref.
			HeNe($X^1\Sigma^+$)			
3.0282		14.610	3.9791	2.570	0.3599	a
3.03	2.70	14.60				b
3.028		14.621	3.9776	2.5812		c
3.0355		14.52				d
			HeAr($X^1\Sigma^+$)			
3.4925		20.6635	4.0754	7.0144	0.2957	e
3.49	3.31	20.653				f
3.4920		20.6842	4.0745	7.0318	0.2957	g
3.4780		20.61				h
			HeKr($X^1\Sigma^+$)			
3.711	3.32	20.543				i
3.9074		20.740	4.2542	7.8063	0.2564	j
3.6822		21.8355				k
3.6912		20.471	4.248	7.651	0.2574	l
			NeAr($X^1\Sigma^+$)			
3.4934		45.183	3.6137	32.471	0.0965	m
3.493		45.196	3.645	32.482	0.0965	n
3.490		45.785				o
3.4889	3.1140	46.9778				p
			NeKr($X^1\Sigma^+$)			
3.6742		47.0383	3.8056	36.1233		q
3.621	3.250	49.75				r
			ArKr($X^1\Sigma^+$)			
3.9226		111.96	3.9839	98.764	0.0394	s
3.9008	3.4868	112.16				t
3.8810	3.468	116.28	3.9428	102.74	0.0402	u

a, Cybulski and Toczyłowski, 1999; b, Partridge *et al.*, 2001; c, López Cacheiro *et al.*, 2004; d, Keil *et al.*, 1991; e, Cybulski and Toczyłowski, 1999; f, Partridge *et al.*, 2001; g, López Cacheiro *et al.*, 2004; h, Keil *et al.*, 1991; i, Partridge *et al.*, 2001; j, Haley and Cybulski, 2003; k, Jäger and Bich, 2017; l, Keil *et al.*, 1991; m, Cybulski and Toczyłowski, 1999; n, López Cacheiro *et al.*, 2004; o, Nasrabad *et al.*, 2004; p, Barrow and Aziz, 1988; q, Haley and Cybulski, 2003; r, Barrow *et al.*, 1989; Barrow et al., 1992; s, Haley and Cybulski, 2003; t, Nasrabad *et al.*, 2004; u, Aziz and van Dalen, 1983*a*.

Table 3.11 Parameters characterizing best-fit empirical PEFs for the $X^1\Sigma^+$ electronic ground states of the He–Xe, Ne–Xe, Ar–Xe, and Kr–Xe heterodimers

Dimer	R_e /Å	σ /Å	$\mathfrak{D}_e$ /cm^{-1}	Source
He–Xe	3.997	3.565	25.252	Danielson and Keil 1988
	3.978	3.558	19.527	Keil *et al.* 1991
Ne–Xe	3.861	3.465	51.575	Keil *et al.* 1988
Ar–Xe	4.0668	3.645	131.11	Aziz and van Dalen 1983*b*
	4.0958		129.81	Piticco *et al.* 2010
Kr–Xe	4.174	3.741	162.28	Aziz and van Dalen 1983*b*

Table 3.12 Fit-parameters for the unlike pair interaction energy $V_{SM}(R)$ for the He–Ne, He–Ar, Ne–Ar, He–Kr, Ne–Kr, and Ar–Kr pairs. All quantities are expressed in terms of cm^{-1} units for energy and Å units for distance

Dimer	He–Ne[a]	He–Ar[a]	Ne–Ar[a]
$10^{-6}d_0$	5.35360473	5.05463011	16.56183766
α	3.90644137	3.08121911	3.30608091
10β	0.11076982	1.67126507	1.32021668
ρ	4.90092220	3.07549099	3.07326174
$10^{-4}C_6$	1.47584567	4.53424327	9.16001775
$10^{-5}C_8$	0.42753920	2.23383848	5.30193147
$10^{-5}C_{10}$	1.61646709	14.35007919	39.91816836
$10^{-6}C_{12}$	0.81753317	1.23307906	40.20158629
$10^{-6}C_{14}$	2.29881591	58.9108203	225.0993662
$10^{-8}C_{16}$	0.45195110	19.65602934	88.02687284
Dimer	**He–Kr[b]**	**Ne–Kr[b]**	**Ar–Kr[b]**
$10^{-6}d_0$	5.68729	17.0092	17.7574
α	2.93996	3.03441	2.43735
10β	1.35024	1.20275	1.52536
ρ	3.30957	4.25982	3.87626
$10^{-4}C_6$	6.62669	13.7659	45.8167
$10^{-5}C_8$	3.20214	6.91515	29.0897
$10^{-5}C_{10}$	20.1306	45.1935	242.2671
$10^{-6}C_{12}$	76.9280	39.507	262.008
$10^{-6}C_{14}$	79.1435	192.0218	1549.90
$10^{-8}C_{16}$	25.8423	65.18171	619.556

[a] López Cacheiro *et al.* 2004; [b] Haley and Cybulski 2003.

$\mathfrak{D}_e$ and separations R_e accurate to within the experimental uncertainties typically associated with the values obtained from the most accurate available spectroscopic data. As this is especially the case for the ground electronic $X^1\Sigma_g^+$ terms of alkali dimers, this section focuses only upon the lowest excited electronic term, that is, the $a^3\Sigma_u^+$ electronic term for homo-atomic electron-spin-aligned alkali dimers. Very reliable PEFs for the relevant ground electronic terms may be obtained from fittings of the type discussed in Section 3.5, in which the data set to be fitted includes either accurate *ab initio* calculations of the interaction energies at short range (i.e. $R \ll R_e$) or atomic beam scattering data in order to provide a better representation of the repulsive wall than can be obtained from the fitting of spectroscopic data alone.

The need to have highly accurate interatomic interaction energies in order to understand the formation and properties of Bose–Einstein condensates of alkali metal atoms at ultralow temperatures has served to hasten the development of appropriate *ab initio* methods for dealing with open-shell atom–atom interactions (Hutson and Soldán, 2007). There are two methods in particular that are used for such calculations. One is a single-reference-determinant, size-extensive, all-electron method based upon a quadratic configuration interaction that includes all contributions from single and double excitations plus perturbative triple excitations [referred to by the acronym QCISD(T,full) (Pople *et al.*, 1987)]. The other is a single-reference-determinant restricted open-shell version (Knowles *et al.*, 1993; Knowles *et al.*, 2000) of the CCSD(T) CC supermolecule method [with the acronym RCCSD(T)].

The interaction energy for the $a^3\Sigma_u^+$ electronic term of Li_2 deriving from a pair of spin-polarized $Li(2\,^2S_{\frac{1}{2}})$ atoms has been computed at 122 equidistant separations R covering the range $2.00\text{Å} \leq R \leq 32.00\text{Å}$ via QCISD(T, full) calculations (Halls *et al.*, 2001*a*) based upon a high-level cc-pV5Z basis-set. It has also been obtained from high-level RCCSD(T) calculations (Halls *et al.*, 2001*c*) carried out for 56 separations between 2Å and 20Å as part of a more extended study of three-body nonadditive interactions in homo-atomic alkali trimers formed from spin-polarized ground electronic ($n\,^2S_{\frac{1}{2}}$, n = 3, 4, 5, 6) alkali atoms. The values of $\mathfrak{D}_e$ and R_e obtained for the $Li_2(a^3\Sigma_u^+)$term in these two studies agree well with one another and with the experimental values (Le Roy *et al.*, 2011) (see Table 3.13). The $\mathfrak{D}_e$ and R_e values obtained for $Na_2(a^3\Sigma_u^+)$ from RCCSD(T) calculations (Simoni *et al.*, 2009) and from spectroscopic data for the sodium dimer (Ivanov *et al.*, 2003) are in reasonable agreement (differing by $0.704\,\text{cm}^{-1}$ and $0.028\,\text{Å}$, respectively). Knoop *et al.* 2011 improved the sodium ground-state potentials by using a coupled-channels calculation. Note that even though the *ab initio* interaction energies for the $a^3\Sigma_u^+$ electronic terms of the Li_2 and Na_2 dimers formed from ground electronic term atoms differ from the corresponding experimental values by relatively small, though spectroscopically significant, amounts, they are still sufficiently accurate to provide computed values of second virial coefficients and transport properties of atomic Li and Na vapors that fall within the experimental uncertainties for these macroscopic properties.

Initial *ab initio* calculations (Quéméner *et al.*, 2005; Soldán *et al.*, 2003) of $V_{SM}(R)$ for the K_2 dimer were carried out at 42 interatomic separations within the interval $2.1\,\text{Å} \leq R \leq 14.0\,\text{Å}$; the long-range behavior of $V_{SM}(R)$ was extrapolated to the form of the (undamped) dispersion series given by fixing the C_6 and C_8 dispersion coefficients at the values obtained by Derevianko and coworkers Derevianko *et al.* (1999) and Porsev and Derevianko (2003), while allowing C_{10} to vary within the uncertainties in the *ab initio* value obtained by Porsev and Derevianko (2003). The extrapolation procedure gave (Quéméner *et al.*, 2005; Soldán *et al.*, 2003) $C_{10} = 7.6502 \times 10^{12}\,\text{cm}^{-1}\text{Å}^{10}$, which was well within the uncertainty attributed to the *ab initio* value. As shown in Table 3.13, the values for R_e and $\mathfrak{D}_e$ for the $K_2(a^3\Sigma_u^+)$ dimer were found, as were those for the $Li_2(a^3\Sigma_u^+)$ and $Na_2(a^3\Sigma_u^+)$ dimers, to be in good agreement with experiment.

It is clear from Table 3.13 that the level of agreement between the computed *ab initio* and experimentally determined values for the locations, R_e, and depths, $\mathfrak{D}_e$, characterizing the interaction energies for the $Rb_2(a^3\Sigma_u^+)$ and $Cs_2(a^3\Sigma_u^+)$ dimer ground triplet terms is not as satisfactory as that obtained for the corresponding Li_2, Na_2, and K_2 dimers. Even when a rather high-level RCCSD(T) calculation (Soldán, 2010), in which the nine $4s, 4p, 5s$ electrons of Rb are all treated as valence electrons (with the remaining 28 electrons treated as core electrons described in terms of relativistic small-core potentials), the computed and experimentally determined values of $\mathfrak{D}_e$ differ approximately by $11.86\,\text{cm}^{-1}$, while the corresponding R_e values differ by 0.033 Å: these values may be compared with differences of approximately $20.0\,\text{cm}^{-1}$ and 0.16 Å obtained from an earlier computation (Quéméner *et al.*, 2005; Soldán *et al.*, 2003) using a significantly smaller basis-set (70 vs. 154 primitive Gaussian functions per Rb atom). The rather large differences of nearly $33.5\,\text{cm}^{-1}$ and 0.35 Å between computed

Table 3.13 Characteristics of $a^3\Sigma_u^+$ first excited electronic term homo-atomic alkali dimers and their interaction energies. The upper entries give the best *ab initio* results, and the lower entries the best empirical (experimentally derived) results, where applicable. All quantities are expressed in terms of cm^{-1} units for energy and Å units for distance

	$^7Li_2$	$^{23}Na_2$	$^{39}K_2$	$^{85}Rb_2$	$^{133}Cs_2$
R_e	4.1686[a]	5.194[b]	5.786[c]	6.102[d]	6.581[c]
	4.17005[e]	5.16609[f]	5.7725[g]	6.06897[h]	6.2354[i]
$\mathfrak{D}_e$	334.145[a]	172.946[b]	252.567[c]	229.6[d]	246.786[c]
	333.758[e]	173.6496[f]	252.74[g]	241.4529[h]	279.349[i]
$\mathfrak{D}_0$	301.989[a]				
	301.842[e]	161.895[f]			
ω_0	65.400[a]			14[d]	
		26.14[f]			

[a] Halls *et al.*, 2001; [b] Simoni *et al.*, 2009; [c] Quéméner *et al.*, 2005; Soldán *et al.*, 2003; [d] Soldán, 2010; [e] Le Roy *et al.*, 2011; [f] Ivanov *et al.*, 2003; [g] Zhao *et al.*, 1996; [h] Beser *et al.*, 2009; [i] Xie *et al.*, 2009.

(Quéméner *et al.*, 2005; Soldán *et al.*, 2003) and experimental (Xie *et al.*, 2009) values of $\mathfrak{D}_e$, R_e, respectively, for the $Cs_2(a^3\Sigma_u^+)$ dimer show that considerable progress has yet to be made in the *ab initio* computation of the interaction energies for such dimers.

Although empirical multiparameter fits of spectroscopic transition frequencies to specified functional forms for the interatomic interaction energy are available for both the $X^1\Sigma^+$ and $a^3\Sigma^+$ electronic terms of many of the mixed alkali dimer species (Pashov *et al.*, 2007), high-quality *ab initio* calculations of such dimer interaction energies have been carried out for the $a^3\Sigma^+$ term for only a relatively small number of these dimers (Korek *et al.*, 2007; Soldán and Špirko, 2007). However, Geum *et al.* (2001) carried out a detailed *ab initio* study of the dependence on internuclear separation of the interaction energies associated with both the $X^1\Sigma^+$ and $a^3\Sigma^+$ electronic terms for the analogous alkali halides (Ak $\equiv$ Li, Na, K, Rb, Cs). Their MRCI-level calculations included all double and triple excitations, and extend from well inside the short-range repulsive region for the $a^3\Sigma^+$ term, typically $1.5\text{Å} < R < 5.25\text{Å}$, to the near-asymptotic region, $R \simeq 16\text{Å}$. Moreover, all electron correlations were included for the Li–H dimer, while valence–core correlation was represented by augmenting the full valence configurations with single core–excitations. Reasonable agreement (i.e. within $\pm 2\%$) was obtained with empirical PEF parameters characterizing the minimum in the $X^1\Sigma^+$ ground electronic terms (with generally better agreement for R_e than for $\mathfrak{D}_e$).

3.9.2 Interactions between closed-shell and open-shell atoms

Thus far we have focused on interactions between noble gas atoms, for which the closed-shell ground electronic terms are energetically well separated from their lowest excited electronic terms, or upon open-shell atoms with a single unpaired ns valence electron $(n = 1, 2, \cdots, 6)$. Due to the closed-shell nature of noble gas atoms, the attractive component of the interaction between a pair of ground term atoms is entirely determined by dispersion, thereby giving rise to a weakly bound ground $X^1\Sigma^+_{(g)}$ molecular ground electronic term that is energetically well separated from the next-nearest molecular term. We have also seen that because the energy associated with interatomic electron exchange between a pair of $(ns)^1$ open-shell atoms has a sizeable attractive component associated with interatomic electron exchange, the (chemically bound) ground $X^1\Sigma^+_{(g)}$ molecular electronic term is paired with an energetically relatively low-lying and weakly bound first excited molecular electronic $a^3\Sigma^+_{(u)}$ term that corresponds to the same asymptotic separated $^2S_{\frac{1}{2}}$-atom limit. This excited electronic term governs the interaction between pairs of $^2S_{\frac{1}{2}}$ spin-polarized atoms in their ground electronic terms, and hence, is characterized by a larger value of its associated characteristic equilibrium separation, R_e.

The interaction between pairs of more complex open-shell atoms typically gives rise to several diatomic molecular electronic terms associated with the same asymptotic separated-atom limit. For example, two ground-term $N(^4S)$ atoms give rise to four diatomic molecular electronic terms, including the $X^1\Sigma_g^+$ ground term, that correlate with the $N(^4S)$–$N(^4S)$ separated-atom limit. The *ab initio* calculation of interaction energies between pairs of open-shell multi-electron atoms is very much more challenging than the calculation of the interaction energy between a pair of closed-shell atoms.

Nonetheless, such calculations have been carried out successfully for alkali atom pairs and for homo-atomic pairs of several first-row atoms as well as for some hetero-atomic pairs, especially those involving the atmospherically important C, N, and O atoms (Levin *et al.*, 1990; Partridge *et al.*, 1986; Stallcop *et al.*, 2000*b*).

Because high-level *ab initio* calculations of the interaction energies between atoms from the second and higher rows in the periodic table remain extremely challenging, Partridge *et al.* (2001) proposed a means for constructing homo-atomic PEFs for atoms in the first three rows of the periodic table based upon the success of atom–atom combining rules (Olson and Smith, 1972; Tang and Toennies, 2003, 1986) and the relative ease with which accurate He–B interaction energies can be obtained via CCSD(T) supermolecule calculations using high-level basis-sets. However, rather than employ combining rules to obtain hetero-atomic PEFs, they proposed that the parameters characterizing the B–B interaction be obtained from the parameters for the He–He and He–B interaction energies via combining rules. We adopt their terminology, namely, the "aufbau method," for this procedure.

The aufbau method for obtaining $V(R)$ for the B–B interaction energy, given that $V_{\rm SM}(R)$ is known for both the A–A and A–B interactions, may be summarized by the following three-step procedure.

(1) Carry out a multiparameter fit of a given functional representation $V_{\rm AB}(R)$ of the A–B interaction energy to the set of *ab initio* values obtained for $V_{\rm SM}^{\rm AB}(R)$: if accurate values are available for one or more of the dispersion coefficients, they may be utilized either to reduce the number of fit-parameters or to constrain the parameter space.

(2) Determine $C_6^{\rm BB}$, $C_8^{\rm BB}$, and $C_{10}^{\rm BB}$ using combining rules and predetermined values of C_6, C_8, and C_{10} for the He–He and He–B interactions. Values for the higher $C_{2n}^{\rm BB}$ ($n = 6, 7, 8$) dispersion coefficients may then also be generated by employing expressions like those in eqns (3.8c) and (3.15).

(3) Employ combining rules to generate Born–Mayer $A_{\rm BB}$ and $\alpha_{\rm BB}$ parameters for the repulsive component of $V_{\rm BB}(R)$ and the scale factor $\rho_{\rm BB}$ for the dispersion-damping functions.

3.9.2.1 The process

The combination rules of Section 3.2 deal with the determination of the hetero-atomic AB interaction energy $V_{\rm AB}(R)$ from a knowledge of the homo-atomic interaction energies $V_{\rm AA}(R)$ and $V_{\rm BB}(R)$ when all three interactions are dispersion dominated and each interaction can be accurately represented by a PEF having the form of a Born–Mayer short-range repulsion term plus a damped-dispersion series long-range attraction term. The interaction between an arbitrary pair of $^1S_0$ ground electronic term atoms, typified by noble gas atoms, to form a $X^1\Sigma_{(g)}^+$ ground electronic term dimer is very accurately represented by this form of interaction, as is the high-spin $a^3\Sigma_{(u)}^+$ first excited electronic term dimer formed by a pair of ground electronic term $^2S_{\frac{1}{2}}$ atoms, typified by H and alkali atoms. Combination rules (Olson and Smith, 1972; Tang and Toennies, 1986) have been employed successfully to generate rare gas hetero-atomic interaction energies (Tang and Toennies, 1986), noble gas hydride interaction energies

(Tang and Toennies, 1991) and, more recently, the interaction energies between Ca and noble gas atoms (Yang *et al.*, 2009). For the ten mixed rare gas pairs the average absolute error, defined as $100|(R_{\rm min}^{\rm calc} - R_{\rm min}^{\rm expt})/R_{\rm min}^{\rm expt}|$, in the values of $R_{\rm min}$ generated from the combining rule was 0.59%; similarly, combining rule values obtained for $\mathfrak{D}_e$ had average absolute errors of 1.6% for the He–Ar, He–Kr, He–Xe pairs, 5.2% for the Ne–He, Ne–Ar pairs, and 0.22% for the remaining mixed rare gas pairs.

The aufbau method effectively inverts the normal usage of these combining rules in the sense that it determines homo-atomic $V_{\rm BB}(R)$ interaction energies from a specific *ab initio* homo-atomic interaction energy $V_{\rm AA}(R)$ together with *ab initio* hetero-atomic interaction energies $V_{\rm AB}(R)$ between atom A and atoms B. This process requires accurate values of the relevant atomic polarizabilities for atoms A and B, plus all relevant potential parameters characterizing the interactions $V_{\rm AA}(R)$ and $V_{\rm AB}(R)$ as input. Moreover, it requires that the long-range dispersion coefficients $C_6^{\rm AB}$, $C_8^{\rm AB}$, and $C_{10}^{\rm AB}$ (or an equivalent set of multipole-multipole dispersion coefficients) and the static dipole, quadrupole, and octupole polarizabilities for both atoms A and B be available.

This aufbau procedure thus requires an inversion of the $C_6^{\rm AB}$ and $C^{\rm AB}(1,2)$ combining rules (3.17) to give

$$C_6^{\rm BB} = \frac{C_6^{\rm AB} C_6^{\rm AA} \alpha_1^2({\rm B})}{2\alpha_1({\rm A})\alpha_1({\rm B})C_6^{\rm AA} - \alpha_1^2({\rm A})C_6^{\rm AB}} \tag{3.100a}$$

for the leading long-range dispersion coefficient, and

$$C_8^{\rm BB} = \frac{2\alpha_1({\rm B})\alpha_2({\rm A})C_6^{\rm AA}C^{\rm AB}(1,2)}{\alpha_1({\rm A})[15\alpha_2({\rm B})C_6^{\rm AA} - \alpha_1({\rm A})C^{\rm AB}(1,2)]} \tag{3.100b}$$

for the next-most important dispersion coefficient for the B–B interaction. If a value for $C^{\rm AB}(1,2)$ is not readily available, it will be necessary to obtain it from the known values of $C_6^{\rm AA}$, $C_8^{\rm AA}$, and $C_6^{\rm BB}$ [obtained via eqn (3.100a)] via

$$C^{\rm AB}(1,2) = C_8^{\rm AB} - C^{\rm AB}(2,1)\,,$$

in which the quadrupole–dipole dispersion coefficient $C^{\rm AB}(2,1)$ is given in terms of $C_6^{\rm AA}$, $C_8^{\rm AA}$, and $C_6^{\rm BB}$ as

$$C^{\rm AB}(2,1) = \frac{5}{2}\,\frac{\alpha_1({\rm B})\alpha_2({\rm A})C_6^{\rm AA}C_8^{\rm AA}C_6^{\rm BB}}{5C_6^{\rm BB}[4\alpha_1({\rm A})\alpha_2({\rm A})C_6^{\rm AA} - 6\alpha_1^2({\rm A})C_8^{\rm AA}] + 2\alpha_1^2({\rm B})C_6^{\rm AA}C_8^{\rm AA}}\,.$$

Determination of $C_{10}^{\rm BB}$ is more involved, as we see from Section 3.3.3 that $C_{10}^{\rm ij}$ has three multi-polar contributions, two of which are equal for homo-atomic interactions, giving

$$C_{10}^{\rm BB} = 2C^{\rm BB}(1,3) + C^{\rm BB}(2,2) \tag{3.101a}$$

for the B–B interaction. The quadrupole–quadrupole dispersion contribution $C^{\rm BB}(22)$ is given in terms of $C_6^{\rm BB}$ and $C_8^{\rm BB}$ by

$$C^{\rm BB}(2,2) = \frac{35}{3}\,\frac{\alpha_2({\rm B})C_8^{\rm BB}C_6^{\rm BB}}{10\alpha_1({\rm B})\alpha_2({\rm B})C_6^{\rm BB} - \alpha_1^2({\rm B})C_8^{\rm BB}}\,, \tag{3.101b}$$

with all components already known or previously calculated at this stage of the determination. The dipole–octupole dispersion coefficient $C^{\rm BB}(1,3)$, given by

$$C^{\rm BB}(1,3) = \frac{28\alpha_1({\rm B})\alpha_3({\rm B})C_6^{\rm BB}C_6^{\rm AA}C^{\rm AB}(1,3)}{C_6^{\rm BB}[28\alpha_1({\rm B})\alpha_3({\rm B})C_6^{\rm AA} - 3\alpha_1^2({\rm A})C^{\rm AB}(1,3)] + 3\alpha_1^2({\rm B}C_6^{\rm AA}C^{\rm AB}(1,3)}\,, \tag{3.101c}$$

is more complicated, however, as it involves $C^{\rm AB}(1,3)$, which may or may not be available from *ab initio* computations. In the event that only $C_{10}^{\rm AB}$ is available from *ab initio* calculations, it is possible to use a procedure analogous to that discussed for $C^{\rm AB}(1,2)$. To wit, we employ eqn (3.16c) for $C_{10}^{\rm AB}$ to obtain

$$C^{\rm AB}(1,3) = C_{10}^{\rm AB} - C^{\rm AB}(3,1) - C^{\rm AB}(2,2)\,, \tag{3.102a}$$

with $C^{\rm AB}(2,2)$ given by eqn (3.3)–(3.13d) and $C^{\rm AB}(3,1)$ is given explicitly by

$$C^{\rm AB}(3,1) = \frac{28\alpha_3({\rm A})\alpha_1({\rm B})C_6^{\rm AA}C_6^{\rm BB}C^{\rm AA}(1,3)}{C_6^{\rm BB}[28\alpha_1({\rm B})\alpha_3({\rm A})C_6^{\rm AA} - 3\alpha_1^2({\rm A})C^{\rm AA}(1,3)] + 2\alpha_1^2({\rm B})C_6^{\rm AA}C^{\rm AA}(1,3)} \tag{3.102b}$$

and $C^{\rm AA}(1,3)$ obtained from

$$C^{\rm AA}(1,3) = \tfrac{1}{2}[C_{10}^{\rm AA} - C^{\rm AA}(2,2)]$$

and $C^{\rm AA}(2,2)$ obtained by interchanging B and A in eqn (3.101b).

The Born–Mayer parameters $\alpha_{\rm BB}$ and $A_{\rm BB}$ for $V_{\rm rep}^{\rm BB}(R)$ are obtained via the Smith combining rules (Olson and Smith, 1972) as

$$\alpha_{\rm BB} = \frac{\alpha_{\rm AB}\alpha_{\rm AA}}{2\alpha_{\rm AA} - \alpha_{\rm AB}} \tag{3.103a}$$

and

$$\ln A_{\rm BB} = \frac{2\alpha_{\rm AA}\ln[\alpha_{\rm AB}A_{\rm AB}] - \alpha_{\rm AB}\ln[\alpha_{\rm AA}A_{\rm AA}]}{2\alpha_{\rm AA} - \alpha_{\rm AB}} - \ln\alpha_{\rm BB}\,, \tag{3.103b}$$

while the scaling factor for the long-range damping function is given as

$$\rho_{\rm BB} = \frac{\rho_{\rm AB}\rho_{\rm AA}}{2\rho_{\rm AA} - \rho_{\rm AB}}\,. \tag{3.103c}$$

3.9.2.2 Dispersion-dominated interactions and the aufbau procedure

It is clear that *ab initio* methods cannot currently provide interaction potential energies that are of spectroscopic accuracy (i.e. accurate to ± 0.005 cm^{-1}) for atomic interactions that are not dispersion dominated. It thus makes little sense to attempt to generate *ab initio* PEFs that do not attain accuracies comparable to those typically obtained with direct potential fits to myriad very precise spectroscopic transition frequencies associated with chemically bound diatomic species (see, for example, Section 3.5). It is not uncommon, however, for pairs of open-shell atoms to form both low-spin, often chemically bound, and high-spin, often dispersion-dominated, diatomic molecular electronic terms, all of which may play significant roles in atomic scattering. Although dispersion-dominated interactions are especially important for the scattering of spin-polarized atoms, they normally support rather few (sometimes no) bound vibrational states due to the relative weakness of attractive dispersion interactions, so that there are few, if any, relevant spectrocopic data to which direct potential fits can be made.

Collisions between two ground electronic term atoms provide a useful illustration of the need for a knowledge of the interaction energies for excited electronic molecular terms in order to carry out reliable calculations of the transport properties of atomic gases. The classic case is that of H atoms: two electronic terms arise from the interaction of a pair of $H(^1S)$ atoms, namely, the chemically bound ground electronic singlet $H_2(X^1\Sigma_g^+)$ term, which has 15 bound vibrational levels, and the very shallow lowest electronic triplet $H_2(a^3\Sigma_u^+)$ term that, because it does not support any bound vibrational levels, is inaccessible to spectroscopy. Fortunately, however, the interaction energy for this lowest triplet term of the H_2 molecule was determined long ago via *ab initio* methods by Kołos and Wolniewicz (1974), as it is particularly important for the scattering of spin-polarized H atoms (H↑). Moreover, as we have seen in Section 3.3, the $H_2(a^3\Sigma_u^+)$ open-shell dispersion-dominated atom–atom interaction has also served as the prototypical interaction that enabled the development of several useful models, such as the HFD and TT models of Section 3.3.1 and the XC models of Section 3.3.4, describing the interaction energy between atom pairs.

It is worth considering a slightly more complicated, yet similarly important, open-shell atom–atom interaction between two ^4S electronic ground-term N atoms. Angular momentum coupling within the Russell–Saunders coupling scheme dictates that the interaction between a pair of S-term atoms determines the potential energy of an electronic Σ term of the resultant diatomic molecule. Coupling of the two $S = \frac{3}{2}$ total electronic spin-angular momenta associated with the pair of ^4S term atoms leads to four separate electronic Σ terms being available for the N_2 diatomic molecule formed from ground term N atoms, one for each of the four values 3, 2, 1, 0, of the resultant total (coupled) spin-angular momentum. Details of the scattering between a pair of ground electronic term N atoms will thus largely be determined by the repulsive wall of one of these four molecular terms, identified by molecular spectroscopists as the ground $X^1\Sigma_g^+$ molecular term, plus the $A^3\Sigma_u^+$, $A'^5\Sigma_g^+$, and $^7\Sigma_u^+$ excited terms, in order of increasing energy from the bottom of the ground term PEF. The $^7\Sigma_u^+$ interaction, which is clearly dispersion dominated, has never been detected via molecular spectroscopy: consequently, it does not even appear in Huber and Herzberg's (1979) classic compendium.

Not only is collisional scattering between pairs of spin-polarized N atoms (N↑) determined by the potential energy of the $N_2(^7\Sigma_u^+)$ dimer, but due to its high (sevenfold) electronic spin degeneracy, this dispersion-dominated interaction also plays an important role in the scattering of pairs of nonpolarized $N(^4S)$ atoms, which are normally the atoms obtained via thermal dissociation of N_2. More specifically, the $^7\Sigma_u^+$ PEF contributes $\frac{7}{16}$ of the scattering intensity versus $\frac{1}{16}$ for the ground-term $^1\Sigma_g^+$ PEF. Partridge *et al.* (1986) careful *ab initio* study of the interaction of two $N(^4S)$ atoms to give $N_2(^7\Sigma_u^+)$ using a large basis-set ($7s6p5d3f$ Slater-type orbitals) at the

SDCI level. They noted, in particular, that correlation of the $2s$ electrons made an important contribution to the overall interaction energy and that BSSE counterpoise corrections were of the order of 20–25% of $\mathfrak{D}_e$ for the ${}^7\Sigma_u^+$ term. Moreover, they were able to represent their *ab initio* results quite well using the HFD model form (3.6) with $A = 41.014\,\mathrm{cm}^{-1}$, $b = 0.9165\,\text{Å}^{-1}$, $C_6 = 24.12\,\mathrm{cm}^{-1}\text{Å}^6$, $C_8 = 475.5\,\mathrm{cm}^{-1}\text{Å}^8$, and $C_{10} = 12{,}247.9\,\mathrm{cm}^{-1}\text{Å}^{10}$, giving rise to a minimum in the PEF that is characterized by $R_e = 3.979\text{Å}$, $\mathfrak{D}_e = 21 \pm 1\,\mathrm{cm}^{-1}$.

The set of combining rules upon which the aufbau procedure (Partridge *et al.*, 2001) is based have been thoroughly tested for dispersion-dominated interactions like the hetero-atomic noble gas dimers (Tang and Toennies, 2003) , the noble gas hydrides (Tang and Toennies, 1991), and the calcium dimer and calcium rare gas complexes (Yang *et al.*, 2009). Stallcop *et al.* (1992) high-level MRCI calculations of the ground $X^3\Sigma^-$ and first excited ${}^5\Sigma^-$ electronic term interaction energies for the N–H radical in order to compute the temperature dependence of NH transport cross-sections. They compared the reduced *ab initio* PEF, that is, of $U(x) \equiv V_{\mathrm{SM}}(x)/\mathfrak{D}_e$ versus $x \equiv R/R_{\min}$, for $\mathrm{NH}({}^5\Sigma^-)$ with those obtained for $\mathrm{H}_2(a^3\Sigma_u^+)$ (Kolos and Wolniewicz, 1974) and $\mathrm{N}_2({}^7\Sigma_u^+)$ (Partridge *et al.*, 1986) and showed that the three interactions are conformal. Of course, it is perhaps not surprising that they should be so, as these three interactions have a common physical origin, specifically, repulsion due to electron charge overlap combined with attraction due to dispersion interactions. Of greater import, however, is the comparison between the *ab initio* $\mathrm{NH}({}^5\Sigma^-)$ interaction energies and the $\mathrm{NH}({}^5\Sigma^-)$ HFD model PEF obtained via combining rules employing the HFD parameters determined from fits to the *ab initio* $\mathrm{N}_2({}^7\Sigma_u^+)$ and $\mathrm{H}_2(a^3\Sigma_g^+)$ data. The level of agreement between the PEF determined from combining rules and the *ab initio* results is remarkable. (Stallcop *et al.*, 1992) provide a more extensive discussion of the reasoning for the level of agreement obtained by the combination rule result in this case.

Although *ab initio* quantum structure calculations are not currently competitive with spectroscopic direct potential fits, especially for chemically bonded diatomic species, such calculations can be particularly useful for the determination of the (relatively small) energies associated with the attactive wells of dispersion-dominated interactions for which, in many cases, there is an insufficient body of spectroscopic data to allow a direct potential fit to be carried out successfully. Accurate determination of such dispersion-dominated interactions is not only typically required for the calculation of scattering cross-sections for spin-polarized species, but also may contribute significantly to scattering cross-sections for nonpolarized species, as has already been mentioned for scattering between $\mathrm{N}({}^4S)$ atoms via the $\mathrm{N}_2({}^7\Sigma_u^+)$ PEF. The *ab initio* calculations of Partridge *et al.* (2001) used either the CCSD(T) approach (Pople *et al.*, 1987) for closed-shell systems or the RCCSD(T) approach (Knowles *et al.*, 1993, 2000) for open-shell systems, with only valence-electron correlation for interactions between He and atoms X of the main group elements, but including outer-core-electron correlations in addition for the interactions between He and alkali or alkaline earth atoms. A few special cases were also considered, notably the He–Zn, Cd, Ge interactions, for which the outer d-shell electrons were correlated, and the He–Ga interaction, for which $3s$, $3p$, and $3d$ electrons were correlated. Generally speaking, high-level atom-centered aug-cc-pVQZ basis-sets, further augmented with a [33211] centered set of bond functions, were employed for all calculations.

The BSSE counterpoise-corrected interaction energies

$$V^{\text{He-X}}(R) \equiv E^{\text{He-X}}(R) - E_{\text{BSSE}} - E^{\text{He}} - E^{\text{X}} \tag{3.104}$$

of He with 24 ground electronic term atoms X (excluding transition elements) from the first three rows of the periodic table, plus Zn and Cd, were determined for between 23 and 34 interatomic separations R lying in the interval 0.9261Å ($1.75\,a_0$) to 7.9377Å ($15\,a_0$), plus four to five additional separations between 8.4668Å ($16\,a_0$) and 12.7002Å ($24\,a_0$) for the He–Li, He–Na, and He–K dimers. Interactions between He($^1S_0$) and alkali ($^2S_{\frac{1}{2}}$) and alkaline earth ($^1S_0$) atoms give rise to $^2\Sigma^+$ and $^1\Sigma^+$ electronic term van der Waals molecules, respectively, while interactions between He($^1S_0$) and various $2S+1$ spin multiplet np open-shell atoms give rise to $^{2S+1}\Sigma$ term van der Waals molecules in all cases, plus $^{2S+1}\Pi$ terms for 1, 2, 4, or 5 valence electrons in the np shell of the X atom. The terms Σ and Π correspond to the values M_L =0 and $M_l = \pm 1$, respectively, of the ground electronic P term of atom X. For $(np)^1$ and $(np)^2$ open-shell configurations, the $^{2S+1}\Pi$ diatomic molecular term lies lower in energy relative to the separated stationary atoms (SSA) limit than does the $^{2S+1}\Sigma$ term, while for $(np)^4$ and $(np)^5$ open-shell configurations, the energy ordering of these two diatomic molecular terms is reversed.

In a number of cases accurate values for the long-range dispersion coefficients are not available, and the computed X–He interaction energies do not generally extend to sufficiently large separations R to allow an accurate value of the C_6^{HeX} dispersion coefficient to be extracted from the long-range results. However, if C_8^{HeX} and C_{10}^{HeX} are approximated by (Partridge *et al.*, 2001; Staemmler and Jaquet, 1985; Starkschall and Gordon, 1972)

$$C_8^{\text{HeX}} = \frac{3C_6^{\text{HeX}}}{2}\left[\frac{\langle R_{\text{He}}^4\rangle}{\langle R_{\text{He}}^2\rangle} + \frac{\langle R_{\text{X}}^4\rangle}{\langle R_{\text{X}}^2\rangle}\right] \tag{3.105a}$$

and

$$C_{10}^{\text{HeX}} = C_6^{\text{HeX}}\left[2\frac{\langle R_{\text{He}}^6\rangle}{\langle R_{\text{He}}^2\rangle} + 2\frac{\langle R_{\text{X}}^6\rangle}{\langle R_{\text{X}}^2\rangle} + \frac{21}{5}\frac{\langle R_{\text{He}}^4\rangle\,\langle R_{\text{X}}^4\rangle}{\langle R_{\text{He}}^2\rangle\,\langle R_{\text{X}}^2\rangle}\right], \tag{3.105b}$$

in terms of C_6^{HeX} and tabulated values (Desclaux, 1973) of the atomic radial expectation values $\langle R_\alpha^n\rangle$ (α = He, X, n = 2, 4, 6), and the C_{2n}^{HeX} (n = 6, 7, 8) are obtained via the semi-empirical recursion relation (3.11), then the full set of $V_{\text{SM}}^{\text{HeX}}(R)$ *ab initio* data of Partridge *et al.* (2001) can be fitted to the TT model form (3.10) to obtain appropriate values for A^{HeX}, α^{HeX}, and C_6^{HeX}. Two such sets of fit-coefficients will be obtained for the He–X interactions in which X is a P ground-term atom. Values for the TT potential parameters for He–X dimers between He($^1\text{S}_0$) and S ground-term X atoms are given in Table 3.14.

Values of R_e and $\mathfrak{D}_e$ obtained for the Ne–He, Ar–He, and Kr–He dimers are in good agreement with values extracted from atomic beam scattering experiments (Keil *et al.*, 1991) and with other high-level *ab initio* results (Haley and Cybulski 2003; López Cacheiro *et al.* 2004).

Table 3.14 A, α, C_6 values for He–X dimers of He($^1S_0$) atoms with S ground-term atoms (all in amu)

X	A	α	C_6
He	41.96	2.523	1.461
Be	0.6821	2.058	18.24
N	0.4965	3.264	5.70
Ne	98.02	2.496	3.029
Mg	0.7722	1.717	19.78
P	0.5997	3.168	14.69
Ar	124.3	2.153	9.538
Ca	0.9038	1.459	35.34
Zn	0.6926	2.144	16.37
As	0.6190	3.170	17.16
Kr	118.9	2.025	13.40
Cd	0.7214	2.223	22.31
Xe	95.90	1.853	19.54

* Values for the interaction between He($^1S_0$) and $^1S_0$ noble gas atoms are from Tang and Toennies (2003), while those for the interaction between He($^1S_0$) and other ground-term atoms are from Partridge *et al.* (2001).

Partridge *et al.* (2001) point out that a comparison between the Cl–He results appearing and earlier high-level *ab initio* results (Burcl *et al.*, 1998; Naumkin and McCourt, 1998) for the Cl–He $^2\Sigma^+$ and $^2\Pi$ interaction energies nicely illustrates the utility of employing bond functions in the computation of dispersion-dominated interaction energies. Values of $\mathfrak{D}_e$ and R_{min} for the HeCl($^2\Sigma^+$) and HeCl($^2\Pi$) dimers agree well with results obtained in Burcl *et al.* (1998) using the CCSD(T) method together with the same aug-cc-pVTZ basis-set plus bond functions: note, however, that the values for $\mathfrak{D}_e$ are, respectively, 11.6% ($^2\Sigma^+$) and 12.7% ($^2\Pi$) larger than the values of $\mathfrak{D}_e$ obtained by Naumkin and McCourt (1998) using an aug-cc-pVQT basis-set with the CCSD(T) method, but without the addition of bond functions, while the equilibrium separations are, respectively, 1.5% ($^2\Sigma^+$) and 1.3% ($^2\Pi$) shorter than the corresponding R_e values.

The results of the present study show that the aufbau method, which allows the determination of the effective potential energies for molecular interactions from the interaction energies of simpler systems, gives a way to investigate the interactions of larger molecules with more complex symmetry (Stallcop *et al.*, 2000*a*).

4 Comparison Between Theory and Experiment

Although Chapter 2 alluded to the rapid convergence of the Chapman–Cowling approximation scheme, it was not addressed quantitatively. The question of the rate of convergence can be answered quite straightforwardly in principle, provided that the binary interaction potential is precisely known, the transport coefficients have been measured with sufficient precision and accuracy, and reliable calculation procedures are available. However, until recently it has proven rather difficult to satisfy all three of these requirements simultaneously. One particular problem has been that the elimination of systematic errors from transport coefficient measurements is generally rather difficult to achieve, so that disagreements between calculated and experimentally measured values thus cannot readily be attributed to the degree of approximation of the distribution function. However, as virtually all systematic experimental errors have now been eliminated from viscosity measurements, so that experimental viscosity values with uncertainties as small as $\pm\,0.04\%$ are now generally available, while thermal conductivity and binary diffusion measurements having accuracies of the order of $\pm 0.1\%$ are also possible in the ambient temperature regime ($273\,\mathrm{K} \lesssim T \lesssim 325\,\mathrm{K}$) (Assael *et al.*, 1991; Nieuwoudt and Shankland, 1991) sufficiently accurate experimental data has now become available for such comparisons. Moreover, as reliable calculational routines are now readily available for spherical interaction potentials, we can now make accurate evaluations of the contributions to the transport coefficients arising from higher Chapman–Cowling approximations to the nonequilibrium distribution function. Thermal conductivity and binary diffusion measurements for temperatures outside the ambient regime are still subject to uncertainties of the order of $\pm\,2\%$ or greater, while thermal diffusion factor measurements are prone to even larger uncertainties.

Prior to any discussion of the nature of the contributions to the various transport coefficients arising from higher-order Chapman–Cowling approximations, it is useful, however, to consider briefly how well a number of interrelationships between pairs of transport coefficients obtained at the level of the first Chapman–Cowling approximation are satisfied by experimentally determined values. The Eucken factor

$$\mathrm{f}_{\mathrm{Eu}} \equiv \frac{\lambda m}{\eta c_V}$$

relating the shear viscosity and thermal conductivity coefficients of a dilute monatomic gas provides one such interrelation. For the first Chapman–Cowling approximations

Transport Properties and Potential Energy Models for Monatomic Gases. Hui Li and Frederick R.W. McCourt, Oxford University Press.
 DOI: 10.1093/oso/9780198888253.003.0004

$[\eta]_1$ and $[\lambda]_1$ of eqns (2.70) and (2.75), f_{Eu} has been found to have the value $\frac{5}{2}$ for monatomic gases. Experimental values of f_{Eu} for the noble gases obey this relation rather well, thus implying that the inclusion of higher–degree Laguerre polynomial contributions to the nonequilibrium distribution function actually change the transport coefficients relatively little (typically by less than 0.5%). For mixtures, there is also a relation between the thermal diffusion coefficient and the temperature derivative of the binary diffusion coefficient (Mason *et al.*, 1966): it has, however, been of only rather limited use, as thermal diffusion measurements are seldom sufficiently accurate (i.e. having uncertainties smaller than ±1%) and accurate diffusion coefficient measurements typically have experimental uncertainties of the order of ±0.5% or more and are often not available over a temperature domain sufficiently extensive to allow determination of accurate temperature derivatives.

Higher-order corrections to the first Chapman–Cowling approximations for the transport coefficients of rare gases have been shown to be typically well below 1% for the viscosity and no more than 1.5% for the thermal conductivity for reduced temperatures $T^* \equiv k_B T/\mathfrak{D}_e$ below 50 for moderately realistic two-parameter interaction models (characterized by well–depth $\mathfrak{D}_e$ and position R_e), such as the Lennard–Jones (12-6) or Buckingham (exp-6) PEFs (Ferziger and Kaper, 1972; Hirschfelder *et al.*, 1964; Maitland *et al.*, 1980). For λ, η, and the self-diffusion coefficient D_{AA}, deviations between experimental values and first-approximation calculations are often smaller than 0.2% for reduced temperatures T^* below 2, rising to about 1% for reduced temperatures of the order of 50. The diffusion of particles with unequal masses is, however, not as well represented by its first-approximation expression, and deviations of up to ± 3% may occur. Even larger deviations may occur for thermal diffusion. It is somewhat surprising that the first Chapman–Cowling approximation, which is exact for a gas of Maxwell molecules, gives such good agreement for real gases, which are often far from interacting via the simple Maxwell R^{-4} power law, especially in the case of thermal diffusion, which vanishes identically for gases of Maxwell molecules (Chapman and Cowling, 1970).

4.1 Comparison between Theory and Experiment

This chapter focuses on atomic gases for which the interaction potential energy is isotropic, that is, the potential energy is a function $V(R)$ that depends solely upon the separation R between a pair of colliding atoms. Callaway and Bauer (1965), Reid and Dalgarno (1969), Reid (1973), and Aquilanti and Grossi (1980) all showed that the interaction energy between a non-S-state open-shell atom and a closed-shell atom is anisotropic, and this section is thus restricted to comparisons between *ab initio* computational and experimental results for dilute gases of S-state atoms. Moreover, as it is not possible to measure transport properties of gases of open-shell S-state atoms, transport propery comparisons are further restricted to dilute gases of closed-shell atoms and their (binary) mixtures, as exemplified by the noble gases He, Ne, Ar, Kr, Xe, and Rn.

Developments in *ab initio* quantum chemical methodology, coupled with advancements in computational hardware and software during the final two decades of the twentieth century has enabled the computation of highly accurate interaction energies between a pair of multi-electron atoms using so-called supermolecule methods, in which the interaction energy between a pair of atoms is determined as a function of their separation R by a sequence of quantum mechanical computations determining the differences between the total energies computed for the atom pair with the two nuclei separated by a fixed distance R and the sum of the energies of the separated atoms. These computational procedures and results obtained using them to compute *ab initio* potential energy functions (PEFs) for a number of atomic systems have been described in greater detail in Chapter 3 (see, in particular, Sections 3.7 and 3.8).

Apart from the interactions between a helium atom and another rare gas atom, post-Born–Oppenheimer corrections to the interaction energy are found to make negligible contributions to the thermophysical properties. As a consequence, the interaction energy for all isotopologues of a given dimer may be represented by a single PEF characterized by a minimum $V(R_\mathrm{e})$ at a common (equilibrium) internuclear separation R_e. If the infinite separation of the atom pair is employed to define the zero of the interaction energy, then the minimum of the PEF may be expressed as $V(R_\mathrm{e}) = -\mathfrak{D}_\mathrm{e}$, with $\mathfrak{D}_\mathrm{e}$ referred to as the depth of the PEF. The energy eigenstates of a pair of atoms may then be obtained from the solutions of the Schrödinger equation that governs the motion of the atom pair. Typically, this Schrödinger equation may be separated into two decoupled Schrödinger equations, one governing the trivial and uninteresting center-of-mass (CM) translational motion of the atom pair, the other governing their relative motion (vibrational, rotational, translational). The full set of energy eigenvalues governing the relative motion of an atom pair can be split into two subsets, one consisting of a set of discrete energy eigenvalues representing the bound-state energies $E_{\mathrm{v}J}$ characterized by pairs of (integer) quantum numbers (v, J) associated with the ro-vibrational internal energies of the pair of nuclei, while the other set consists of a continuum of scattering states corresponding to (positive) values associated with the relative translational energy of the two nuclei.

At this point, it suffices to say that the ability to obtain a set of high-level *ab initio* interaction energies for a pair of multi-electron atoms as a function of their separation R, especially when coupled with the ability to compute equilibrium and nonequilibrium properties with uncertainties equaling or exceeding experimental precision, provides a promising means for obtaining highly accurate interaction PEFs. For the interaction energy between a pair of atoms having relatively few electrons, such as the four-electron He–He pair, the ten-electron He–Ne pair, or even the twenty-electron He–Ar or Ne–Ne pair, it may well be that, once a sufficiently extensive set of interaction energies has been computed and fitted to a realistic functional form, the calculated temperature dependence of the transport properties will fall within the relatively tight experimental uncertainties currently available. This would imply that the computed PEF can then be utilized to calculate values of equilibrium and nonequilibrium properties at high temperatures with an accuracy that far exceeds experimental accuracies. For larger atoms, the computed *ab initio* potential energies could, if needed, be fitted to a realistic (multi-parameter) functional form to provide an initial function that could then be further fitted iteratively to the most accurate experimental virial and transport data, thereby providing a final (semi-empirical) PEF for the interaction.

A majority of the bound states correspond to the negative eigenvalues, and are referred to as truly bound states, while the eigenstates corresponding to the positive discrete eigenvalues are referred to as quasi-bound states. It has not been possible to observe spectroscopic transitions between the energy eigenstates of rare gas homodimers because they do not have electric dipole moments. Moreover, were any such transitions to occur, they would still be extremely difficult to detect, because they would lie in the far microwave region of the electromagnetic spectrum. It has been possible to utilize vacuum ultraviolet (VUV) transitions from the ground electronic state into one or more relatively strongly bonded excited electronic states, thereby resulting in a series of R-branch and P-branch transitions from which the bound-state energies of the ground state dimer could be deduced by employing difference methods. Data obtained in this manner normally serves primarily as a preliminary check on the efficacy of a proposed PEF.

In molecular spectroscopy it is traditional to employ *term values* $G \equiv E/(hc)$, with E the (nontranslational) energy (h and c are, respectively, the Planck constant and the speed of light) to represent energies associated with the internal energy of a molecule, so that spectroscopic transitions are associated with differences between the term values representing the upper and lower (spectroscopic) states of the molecule. Term value differences may thus be obtained from differences between pairs of spectroscopic transitions that have a common upper (or lower) term. They are thus *derived* quantities and, as such, are subject to much larger experimental uncertainties (as large as several percent) than are typical spectroscopic transition frequencies themselves (uncertainties often smaller than 0.005%). Term differences for pure vibrational transitions are designated by $\Delta G_{\mathrm{v}+\frac{1}{2}}$ for vibrational transitions from v+1 to v. A simple harmonic oscillator (SHO) corresponds to all $\Delta G_{\mathrm{v}+\frac{1}{2}}$ values being identical, while an anharmonic oscillator will have a series of $\Delta G_{\mathrm{v}+\frac{1}{2}}$ values, decreasing from a largest value that corresponds to the difference $\Delta G_{\frac{1}{2}} \equiv G_{\frac{3}{2}} - G_{\frac{1}{2}}$ between the ground vibrational level and the first excited vibrational level of a molecule.

An extensive literature is associated with the experimental determination of equilibrium (equation-of-state and speed-of-sound) virials and nonequilibrium (i.e. transport) phenomena in the rare gases and their mixtures. Prior to the advent of modern high-powered digital computers and the development of quantum chemical procedures that enabled accurate *ab initio* determinations of the interaction energies between multi-electron chemical species, the temperature dependence of equilibrium phenomena, such as the second virial coefficient (Mason and Spurling, 1969), and nonequilibrium phenomena, such as the shear viscosity and thermal conductivity in particular, were used for the determination of optimal fits to sets of parameters associated with various empirical functions $V(R)$ employed to model the interaction potential energy between a pair of atoms as a function of their separation R. Moreover, from about 1970, it became possible to determine shear viscosity coefficients with experimental uncertainties of the order of $\pm 0.1\%$ or better and thermal conductivity coefficients with uncertainties of the order $\pm 0.3\%$ or better (Assael *et al.*, 1991; Nieuwoudt and Shankland, 1991), especially in the ambient temperature range (roughly speaking, 290–350 K). Further, the determination of equation of state data has also improved sufficiently to enable the extraction of second virial coefficients with accuracies of order $\pm 0.5\%$ or better, and it has become possible to acquire speed of sound data from which second acoustic virial coefficients (McCourt, 2003; Moldover *et al.*, 2001; Trusler, 1991) $\beta_a(T)$ may be determined with accuracies of order $\pm 0.1\%$ or better. The acquisition of data of such accuracy has enabled fitted potential function parameters to be determined (in principle) with high reliability. Nevertheless, there remains one caveat, namely, that it is necessary to have eliminated all systematic errors in the measurements because, ultimately, an empirical PEF that has been obtained from fitting, even to extensive sets of experimental data, is only as good as the data employed for its determination.

An important class of equilibrium gas properties is provided by equation of state measurements in which the molar compressibility $P\overline{V}/(N_0 k_B T)$ is expressed as a power series in the molar density $\overline{\rho}$ as (Mason and Spurling, 1969; McCourt, 2003)

$$\frac{P\overline{V}}{N_0 k_B T} = 1 + B_2(T)\overline{\rho} + B_3(T)\overline{\rho}^2 + \cdots , \tag{4.1}$$

with P the (fluid) pressure, $\overline{V}$ the molar volume, $\overline{\rho}$ its reciprocal, and the coefficients $B_n(T)$, $n = 2, 3, \cdots$, the various density virial coefficients which are, as indicated, functions only of the equilibrium temperature T. As the second density virial coefficient $B_2(T)$ depends only upon the two-particle interatomic PEF $V(R)$, comparisons can thus meaningfully be made between its calculated and measured values. It is, however, still rather difficult to obtain highly accurate experimental values for $B_2(T)$, especially at nonambient temperatures. Typical experimental uncertainties for values of $B_2(T)$ obtained prior to about 1980 are $\pm 0.6\%$ to $\pm 1.5\%$, although more recent experimental values may only have uncertainties as small as $\pm 0.3\%$ or so.

It has also become possible to make high-accuracy speed of sound measurements due to significant improvements in acoustic resonators. Speed of sound data can be represented by a virial equation (McCourt, 2003; Moldover *et al.*, 2001; Trusler, 1991)

$$u_{\mathrm{ad}}^2(T, \overline{\rho}) = \frac{RT\gamma^\circ}{\overline{M}} \left[1 + \beta_{2\mathrm{a}}(T)\overline{\rho} + \beta_{3\mathrm{a}}(T)\overline{\rho}^2 + \cdots \right],$$

with u_{ad} the adiabatic speed of sound, $\gamma^\circ \equiv C_P/C_V$ the ideal gas value for the heat capacity ratio (which equals $\frac{5}{3}$ for a closed-shell noble gas), $(N_0 k_B T\gamma^\circ/\overline{M})^{\frac{1}{2}} = u_{\mathrm{ad}}^\circ$ (the ideal gas adiabatic speed of sound), and $\beta_{n\mathrm{a}}(T)$, $n = 2, 3, \cdots$, the acoustic virial coefficients. The second acoustic virial coefficient, $\beta_{2\mathrm{a}}(T)$, depends only upon the two-particle PEF, and is related to the second density virial coefficient $B_2(T)$ and its first and second temperature derivatives by (McCourt, 2003; Moldover *et al.*, 2001; Trusler, 1991)

$$\beta_{2\mathrm{a}}(T) = 2B_2(T) + 2(\gamma^\circ - 1)T\,\frac{\mathrm{d}B_2}{\mathrm{d}T} + \frac{(\gamma^\circ - 1)^2}{\gamma^\circ}\,T^2\,\frac{\mathrm{d}^2 B_2}{\mathrm{d}^2 T^2}\,. \tag{4.2}$$

Values of $\beta_{2\mathrm{a}}(T)$ can readily be extracted from acoustic speed of sound data with experimental uncertainties at least as small as $\pm 0.1\%$ (Moldover *et al.*, 2001).

As it is now possible to compute zero-density values of the viscosity and thermal conductivity with accuracies of the order of 0.05% or better, it is important not only to eliminate, so far as is possible, sources of experimental error, but also to consider any residual density dependence that the measured data may contain. This is a relatively straightforward task when the measurements have been carried out isothermally as a function of the mass density ρ, so that extrapolation to $\rho = 0$ can be made. However, as many of the earlier measurements were carried out only for a mass density that corresponds to a pressure of the order of 1 bar, it is vital ot use a different means of correcting for the initial density dependence. In such cases, Bich and Vogel (1996) have employed Rainwater–Friend moderately dense gas corrections (Bich and Vogel, 1991; Friend and Rainwater, 1984; Rainwater, 1984; Rainwater and Friend, 1987) for the initial density dependence of these transport properties: the viscosity, for example, may be approximated as

$$\eta(T,\rho) \simeq \eta_0(T)[1 + B_{2,\eta}(T)\rho]\,, \tag{4.3}$$

in which $B_{2,\eta}(T)$, referred to as the second viscosity virial coefficient, is given by

$$B_{2,\eta}(T) = \frac{\overline{M}\eta_1(T)}{\eta_0(T)}\,, \tag{4.4}$$

with $\overline{M}$ the molar mass, and $\eta_0(T)$, $\eta_1(T)$ the leading terms in a series expansion of $\eta(T,\rho)$ as a power series in the density. The zero-density viscosity is then expressed as (Bich and Vogel, 1996)

$$\eta_0(T) = \frac{\eta(T,\rho)}{1 + \dfrac{N_0\sigma^3 B^*_{2,\eta}(T^*)\rho}{\overline{M}}}\,, \tag{4.5a}$$

in which $T^* \equiv k_{\rm B}T/\varepsilon$ is the reduced temperature, $B^*_{2,\eta}(T^*)$ is a reduced second viscosity virial coefficient defined by

$$B^*_{2,\eta}(T^*) \equiv \frac{B_{2,\eta}(T)}{N_0\sigma^3}\,, \tag{4.5b}$$

while σ, ε are length and energy scale factors. Values of σ and ε have been determined for the noble gases by Rainwater and Friend (1984).

Agreement between calculated and experimental values of a thermophysical property may be illustrated at three levels. The first and most rudimentary level is, of course, provided by a plot showing the temperature dependence of the calculated transport property together with the available experimental data values. Such a plot illustrates the general level of agreement between calculated and experimentally determined behaviors, and provides a visual measure of the overall level of agreement among the various experimental measurements over the full range of temperatures for which measurements have been made. A more quantitative visual comparison of the level of agreement between the calculated and experimental values of a thermophysical property may also be provided by a relative difference plot in which the percentage difference $\delta y(i) \equiv \{[y_{\rm expt}(i) - y_{\rm calc}(i)]/y_{\rm calc}(i)\} \times 100\%$ between the experimental value $y_{\rm expt}(i)$ and the calculated value $y_{\rm calc}(i)$ is displayed as a function of temperature. Finally, a quantitative statistical measure of the level of agreement between a particular experimental data set (identified by a label α) and the calculated behavior may be provided by the dimensionless root-mean-square deviation (DRMSD) value (Myatt *et al.*, 2018)

$$\overline{dd}_{\alpha}(y) \equiv \left[\frac{1}{N_{\text{dat}}(\alpha)} \sum_{i=1}^{N_{\text{dat}}(\alpha)} \left(\frac{y^{\alpha}_{\text{calc}}(i) - y^{\alpha}_{\text{expt}}(i)}{u^{\alpha}_{i}} \right)^{2} \right]^{\frac{1}{2}}, \tag{4.6}$$

in which $N_{\text{dat}}(\alpha)$ is the number of values in the data set under consideration and u^{α}_{i} is the experimental uncertainty for the datum $y^{\alpha}_{\text{expt}}(i)$. An overall DRMSD value, $\overline{dd}_{\text{tot}}$, for the thermophysical property y may also be determined from the individual DRMSD values $\overline{dd}_{\alpha}(y)$ via

$$\overline{dd}_{\text{tot}} \equiv \overline{dd}(y) = \left[\frac{1}{N_{\text{dat}}} \sum_{\alpha} N_{\text{dat}}(\alpha) \overline{dd}^{2}_{\alpha}(y) \right]^{\frac{1}{2}}. \tag{4.7}$$

A $\overline{dd}(y)$ value of 1 implies that the differences between the calculated and experimental results for property y are found, on average, to correspond to the experimental uncertainties, while $\overline{dd}(y)$ values smaller than 1 correspond to calculated values that, on average, differ from the corresponding experimental values by amounts that are smaller than the stated experimental uncertainties. Moreover, the smaller the $\overline{dd}(y)$ value, the better will be the level of agreement between the average calculated and experimental values for y. Naturally, values of $\overline{dd}(y)$ greater than 1 indicate that the calculated behavior falls, on average, either systematically above or below the observed experimental behavior: in this case, the greater the value obtained for $\overline{dd}(y)$, the larger the discrepancies between calculated and experimental values obtained for y.

4.1.1 The He–He interaction

Although helium, in addition to being the simplest of the noble gases, is the second most abundant element in the galaxy, it is a relatively rare, but incredibly important, element on the Earth. Its rarity in the atmosphere is due to the attainment by a small fraction of the helium in thermal equilibrium in the atmosphere of velocities that exceed the Earth's escape velocity, which then enables those atoms to depart the Earth into interplanetary space. Hurly and Moldover (2000) emphasized the importance of helium to metrology, in particular, and pointed out the role played by low-density helium in thermometry standardization and the calibration, for example, of acoustic resonators employed in the measurement of the speed of sound in various gases and of vibrating wire and acoustic viscometers.

The interaction (potential) energy between a pair of ground electronic state helium atoms has been computed *ab initio* by Hellmann *et al.* (2007) using an extensive high-level basis-set for a set of separations R ranging from a_0 ($a_0 \equiv 1$ bohr $\simeq$ 52.91772 pm) to 8 a_0: in addition to the usual non-relativistic Born–Oppenheimer contribution, $V_{\text{BO}}(R)$, the final potential energy includes the adiabatic correction (also known as the Born–Oppenheimer breakdown correction), $V_{\text{ad}}(R)$, together with the Casimir–Polder (1948) retardation correction, and the Cowan–Griffin (1976) relativistic correction. Przybytek *et al.* (2010) performed an even more extensive *ab initio* computation of the He–He interaction energy (including statistical uncertainties) as a function of interatomic separation (see also Cencek *et al.* (2012)). As there are four distinct types of contribution to the He–He interaction energy, it is convenient to write $V(R)$ as

$$V(R) = V_{\rm BO}(R) + V_{\rm ad}(R) + V_{\rm rel}(R) + V_{\rm QED}(R)\,, \tag{4.8}$$

in terms of the order of their contribution to the total interaction energy. The two most important contributions are the predominant Born–Oppenheimer energy and the adiabatic correction associated with the coupling of the electronic and nuclear motions. The third term in this expression contains all relativistic corrections proportional to the square of the fine-structure constant (Mohr *et al.*, 2008) α, while the fourth term contains all quantum electrodynamic (QED) corrections proportional to the cube of the fine-structure constant. The resulting *ab initio* PEF should be valid for He–He separations up to approximately $100a_0$. For calculations involving the very weakly bound He–He ground vibrational level, higher-order Casimir–Polder retardation corrections not already included in $V_{\rm rel}(R)$ and $V_{\rm QED}(R)$ may be added to $V(R)$ to give the fully corrected retardation He–He PEF. Neither the ^{3}He–^{3}He nor the ^{3}He–^{4}He interaction supports a bound state, while Przybytek *et al.* (2010) showed how the ^{4}He–^{4}He interaction supports a single bound (dimer) state with an energy $D_0 = (1.13 \pm 0.02) \times 10^{-3}\,\mathrm{cm}^{-1}$ below the dissociation limit, with a corresponding average nuclear separation $\langle R\rangle \simeq (47.1 \pm 0.5)$Å. A more complete discussion of the *ab initio* He–He interaction can be found in Section 3.8.1. Cencek *et al.* (2012) provide in-depth discussions of the various contributions and for detailed comparisons with earlier computational results. In particular, note that the interaction energies obtained by Hellmann *et al.* (2007) generally agree well with the more extensive later results (Przybytek *et al.*, 2010), as seen from the detailed comparison made between the two sets of *ab initio* values in Table 13 Cencek *et al.* (2012).

Although the dimer of the ^{4}He isotope can form a bound state, it is so weakly bound that for all practical purposes, helium can be considered not to exist as a bound diatomic species. This means, of course, that there will be no spectroscopy data for helium, so that experimental second pressure and acoustic virial coefficient data must provide the primary equilibrium property test for any newly proposed He–He PEF. Importantly, note that, as pointed out explicitly by Cencek *et al.* (2012), values of the density second virial coefficient $B_2(T)$ computed from their *ab initio* PEF (Przybytek *et al.*, 2010) are in very good agreement with previous theoretical values (Bich *et al.*, 2007; Hurly and Mehl, 2007; Hurly and Moldover, 2000; Mehl, 2009). Agreement with a set of 13 recommended low-temperature-fit values for $B_2(T)$ lying between 2.0 K and 40.0 K given for ^{4}He in the Dymond *et al.* (2002*a*) tabulation is characterized by a DRMSD value $\overline{dd} \simeq 0.1$ (due primarily to the rather large experimental uncertainties associated with much of the earlier data). Similarly, comparison with a set of 21 experimental $B_2(T)$ values obtained (Dymond *et al.*, 2002*a*) at temperatures of 2.00–300.00 K for which corresponding *ab initio* values have been published gives a DRMSD value $\overline{dd} \simeq 1.02$. Comparisons (Bich *et al.*, 2007; Cencek *et al.*, 2012) with more recent (and considerably more accurate) results (Gaiser and Fellmuth, 2009; McLinden and Lösch-Will, 2007) for $B_2(T)$ gave excellent agreement between the calculated values and experiment. A similar comparison (Cencek *et al.*, 2012) between calculated and experimental values of the second acoustic virial coefficient (McCourt, 2003; Moldover *et al.*, 2001; Trusler, 1991) $\beta_a(T)$ gave very good agreement between experimental results obtained by Pitre *et al.* (2006) over the temperature ranges $7\,\text{K} \leq T \leq 25\,\text{K}$ and $77.0\,\text{K} \leq T \leq 235.0\,\text{K}$: excellent agreement was also obtained between the highly accurate experimental value $\beta_a = (22.2201 \pm 0.0024)\ \text{cm}^3\text{mol}^{-1}$ determined by Gavioso *et al.* (2011) at temperature $T = 273.16\,\text{K}$ and the calculated value (Cencek *et al.*, 2012) $\beta_a = (22.2202 \pm 0.0012)\ \text{cm}^3\text{mol}^{-1}$.

Cencek *et al.* (2012) showed that previous transport property calculations carried out by Hurly and Mehl (2007) and by Bich *et al.* (2007) agree with calculations using the most recent *ab initio* He–He PEF (Przybytek *et al.*, 2010) to within 0.1% at the lowest temperatures and to within 0.01% for temperatures above 20 K. They also established that the statistical uncertainties in the interaction energies obtained by Przybytek *et al.* (2010) only affected the calculated transport properties by amounts ranging from approximately 0.05% at the lowest temperatures to as little as 0.002% at temperatures in excess of 50 K. Moreover, as collisions between pairs of helium atoms must be treated quantum mechanically in terms of scattering phase shifts, the relevant omega integrals entering the expressions for the viscosity and thermal conductivity coefficients in Sections 2.4.1 and 2.4.2 must be evaluated using the quantum mechanical expressions of Section 2.5.4. Finally, because the nuclear spin-$\frac{1}{2}$ ^{3}He isotope is a fermion, while the nuclear spin-0 ^{4}He isotope is a boson, the quantum statistical phase-shift sums must be carried out according to the appropriate Fermi–Dirac or Bose–Einstein statistical recipe of eqns (2.182).

Figure 4.1 compares the temperature dependences of the ^{3}He and ^{4}He shear viscosity and thermal conductivity coefficients, η and λ, computed (Bich *et al.*, 2007) in the Chapman–Cowling fifth approximation using the Hellmann *et al.* (2007) *ab initio* PEF, with a number of representative experimental values of η and λ for low-temperature ($0\,\text{K} < T \leq 35\,\text{K}$) helium. The fourth- and fifth-approximation results were found to differ by less than $\pm\,0.01\%$, essentially in agreement with the calculations carried out by Hurly and Moldover (2000) and updated by Hurly and Mehl (2007) for temperatures in the range 10–10,000 K. In particular, the most accurate value obtained experimentally for ^{4}He at room temperature ($T = 298.15\,\text{K}$), $\eta = (19.842 \pm 0.007)\ \mu\text{Pa s}$,(Berg, 2005; Berg, 2006) differs by only 0.08% from the calculated value (Bich *et al.*, 2007) $19.826\ \mu\text{Pa s}$.

Since *ab initio* values for the viscosity and thermal conductivity of helium can be obtained with uncertainties that are at least an order of magnitude smaller than the best corresponding experimental uncertainties, it is now accepted practice, as initially recommended by Hurly and Moldover (2000), to employ *ab initio* results for helium as the standard for the calibration of instruments that rely upon the use of thermophysical properties of gases. In this context, Berg and Burton (2013) proposed that the *ab initio* value $\eta(^4\text{He}) = 19.8253\ \mu\text{Pa s}$ obtained by Cencek *et al.* (2012) for gaseous ^{4}He be accepted as the fixed standard for the calibration of gas viscometers.

The ^{4}He thermal conductivity data shown in Table 4.1 correspond with the 11 values spanning the temperature range 4.0–20.0 K and represented by black squares (■) in the right-hand panel of Fig. 4.1, and serve to illustrate the information conveyed by a DRMSD value, defined in eqn (4.6), as a quantitative representation of

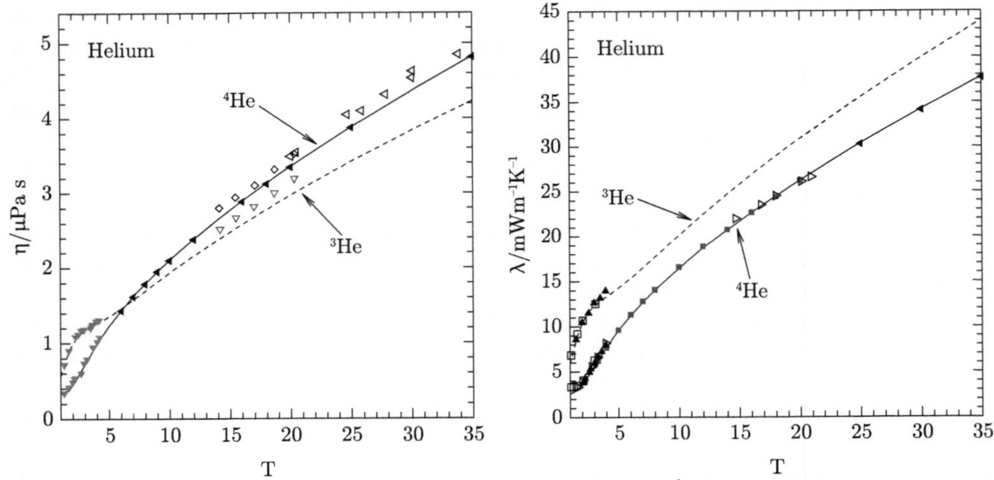

Fig. 4.1 Comparison between computed[a] and experimental[b] values of the viscosity and thermal conductivity of helium at low temperatures.

[a] Bich *et al.*, 2007.

[b] Viscosity data: ◀, Coremans *et al.*, 1958; ▼ (pink), Becker *et al.*, 1954; ◇, ^{4}He), ▽(blue), ^{3}He), Becker and Misenta 1955. Thermal conductivity data: ▶, Ubbink and de Haas 1943; □, Fokkens *et al.*, 1964; △, Kerrisk and Keller, 1969; ■ (blue), Acton and Kellner 1977.

the level of agreement between the calculated and experimental behaviors of a dilute gas transport coefficient. The relevant data set and the results of its various manipulations can be found in this table. The $\pm 1\%$ experimental uncertainties u in the thermal conductivity measurements under consideration are given in column 3, while the differences $\delta\lambda \equiv \lambda_{\text{expt}} - \lambda_{\text{calc}}$ are given in column 5 of Table 4.1. As seen by comparing the entries in columns 3 and 5, the calculated values of λ generally satisfy the condition $\lambda_{\text{expt}} - u < \lambda_{\text{calc}} < \lambda_{\text{expt}} + u$. This same impression is also adequately conveyed by the single DRMSD value $\overline{dd} = 0.8020$ obtained via eqn (4.6) and the values contained in column 6 of this table. A similar comparison may also be achieved graphically from a plot of the percentage values $\delta_\lambda \equiv 100\,\delta\lambda/\lambda_{\text{calc}}$ given in column 7 of this table versus temperature: such a plot is referred to as a difference plot. Note that a quantitative measure of the quality of a difference plot may also be conveyed by the rms value 0.8118 obtained using the data that appears in column 7 of this table.

Figure 4.2 illustrates the temperature dependences of the shear viscosity coefficient η and the thermal conductivity coefficient λ for ^{3}He and ^{4}He. The curves represent the calculated values obtained by Bich *et al.* (2007) in the Chapman–Cowling fifth approximation from the Hellmann *et al.* (2007) *ab initio* PEF. For comparison, 87 experimental values of the shear viscosity of ^{4}He determined over the temperature range $40\,\text{K} < T \leq 700\,\text{K}$ are also shown in the left-hand panel of this figure; similarly, a selection of 14 experimental values of the thermal conductivity of ^{4}He obtained at temperatures in the range $80\,\text{K} < T \leq 800\,\text{K}$ are shown in the right-hand

Table 4.1 Analysis of a sample set of thermal conductivity data* for ^{4}He

T /K	$\lambda_{\text{expt}}{}^a$	u^a	$\lambda_{\text{calc}}{}^b$	$\delta\lambda$	$\delta\lambda/u$	$100\delta\lambda/\lambda_{\text{calc}}$ %
4.0	7.679	0.077	7.6619	0.0171	0.2221	0.22
5.0	9.621	0.096	9.5375	0.0835	0.8698	0.88
6.0	11.32	0.11	11.179	0.1410	1.2818	1.26
7.0	12.81	0.13	12.645	0.1650	1.2692	1.30
8.0	14.68	0.14	13.987	0.0930	0.6643	0.66
10.0	16.52	0.17	16.423	0.1470	0.8647	0.90
12.0	18.88	0.19	18.637	0.2430	1.2789	1.30
14.0	20.75	0.21	20.695	0.0550	0.2619	0.27
16.0	22.68	0.23	22.635	0.0450	0.1957	0.20
18.0	24.42	0.24	24.478	−0.0580	−0.2417	−0.24
20.0	26.76	0.26	26.242	−0.0420	−0.1615	−0.16

* All λ values have units $\text{mWm}^{-1}\text{K}^{-1}$.
[a] Acton and Kellner, 1977
[b] Bich *et al.*, 2007, their Table 3.

panel, together with 21 values of λ selected from the Bich *et al.* (1990) experimental correlation. The considerably fewer independent experimental values for the thermal conductivity coefficient is a reflection of the difficulty in obtaining accurate values for this thermophysical property. Note also that no experimental values for either $\eta(^3\text{He})$ or $\lambda(^3\text{He})$ appear to have been obtained (i.e. published in the open literature) for temperatures exceeding 50 K.

Bich *et al.* (2007) provided a set of figures in which the differences between experimental and calculated values of the shear viscosity and thermal conductivity coefficients are illustrated graphically via difference plots: it is clear from these plots that the most accurate measurements of both the shear viscosity and the thermal conductivity coefficients are those obtained near room temperature. Moreover, it is clear that the more removed the temperature at which measurements are to be made is from an ambient temperature, the more difficult it is to carry out highly accurate thermophysical property measurements. Measurements, especially those made at higher temperatures, often differ from the corresponding calculated values by amounts in the order of a few percent rather than by the fractions of 1% that are typical for measurements in the ambient temperature range.

The calculated values of Bich *et al.* (2007) are in good agreement (within, say, ±0.5%) with the absolute measurements obtained using capillary viscometers (Clarke and Smith, 1969; Kestin and Nagashima, 1964; Kao and Kobayashi, 1967) and a rotating-disk viscometer (Evers *et al.*, 2002), with better agreement (within ±0.2%) between calculated and experimental values, as (for example) the measurements of Ross and co-workers (1964, 1969) and of Evers *et al.* (1959) obtained for temperatures closer to room temperature (typically $293\,\text{K} < T < 305\,\text{K}$). The oscillating disk viscometer

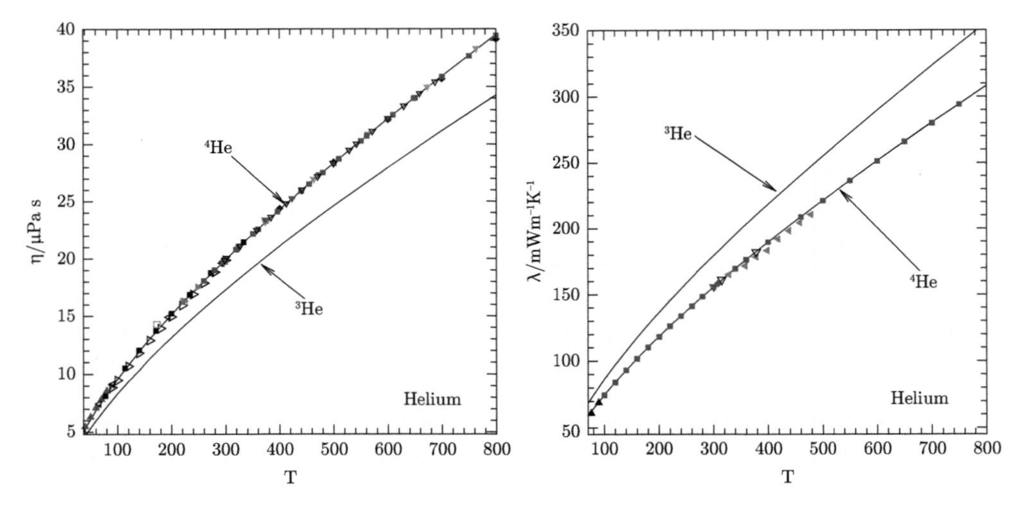

Fig. 4.2 Comparison between computed[a] and experimental[b] values of the viscosity and thermal conductivity of helium at intermediate temperatures.

[a] Bich *et al.*, 2007.

[b] Viscosity data: ▷, Johnston and Grill, 1942; ▲ (blue), Coremans *et al.*, 1958; ◁, Rietveld *et al.*, 1959; ► (pink), Flynn *et al.*, 1963; ▼, Kestin and Nagashima, 1964; ▼ (green), Kestin *et al.*, 1972*a*; ▼ (red), Kestin and Wakeham, 1983; □ (pink), Gracki *et al.*, 1969; ■, Clarke and Smith, 1969; ♦ (grey), Dawe and Smith, 1970; ■ (red), ▽, E. Vogel, private communication, 2018; ■ (blue), Bich *et al.*, 1990. Thermal conductivity data: ▲, Ubbink and de Haas, 1943; ◁ (pink), Haarman, 1973; ▼ (grey), Kestin *et al.*, 1980; ■ (blue), Assael *et al.*, 1981; ▽, Bich *et al.*, 1990.

measurements of Kestin and co-workers (Kestin and Leidenfrost, 1959; Kestin *et al.*, 1972*a*, 1972*e*), which are normally regarded as absolute measurements,[1] also fall into this category. The closest agreement (better than $\pm 0.1\%$) between calculated and experimental results was obtained for the (reanalyzed) relative viscosity data of the Vogel group (Strehlow, 1987; Vogel, 1984) and the datum of Berg (2005, 2006). Truely relative viscosity measurements, which depend critically upon the specific value of the absolute viscosity employed in the calibration of the viscometer, typically did not fare as well, due in large measure to incorrect viscometer calibrations.

A plot showing the temperature dependence of the percentage difference $\delta_\eta(T)$, defined as $\delta_\eta(T) \equiv 100\% \, [\eta_{\rm expt}(T) - \eta_{\rm calc}(T)]/\eta_{\rm calc}(T)$, between the experimentally determined and calculated values of $\eta(T)$ is presented in Fig. 4.3 for temperatures ranging from about 100 K to 1600 K. Six sets of independent experimental data have been illustrated in this way: the data sets of Johnston and Grilly (1942) and Gough

[1] As noted by Bich *et al.* (2007), these measurements should not, strictly speaking, be so designated, as they depend (albeit implicitly) upon an absolute measurement by Beardon (1939) of the room temperature shear viscosity of air.

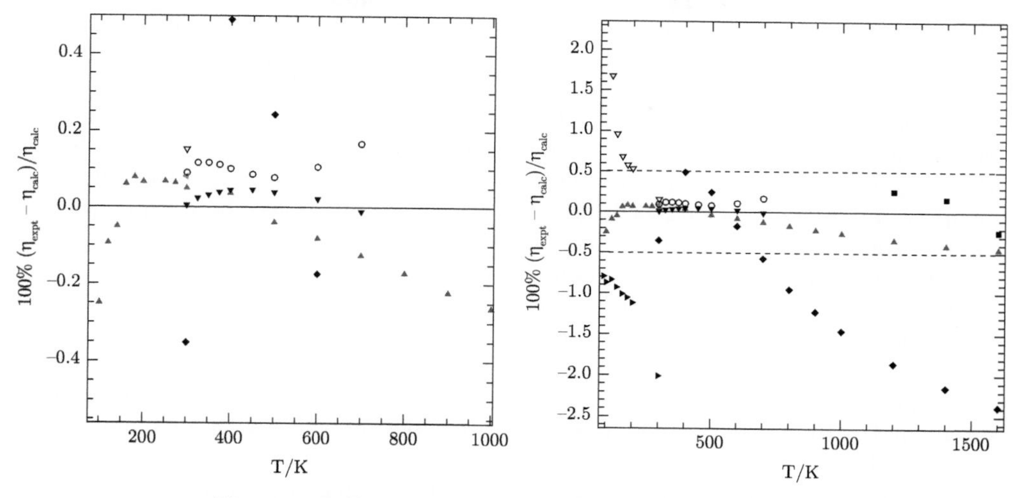

Fig. 4.3 Difference plot for the shear viscosity of ^{4}He.

►, Johnston and Grilly, 1942; ♦, Dawe and Smith, 1970; ■, Guevara *et al.*, 1969; ▽, Gough *et al.*, 1976; Berg, 2005; ◄ Pink, Berg, 2006; ○, Vogel, 1984; ▲ (red), Bich *et al.*, 1990; ▼, reanalyzed results courtesy of E. Vogel, private communication (2018).

et al. (1976) span roughly the same temperature range (100–300 K), a subset of data from Guevara *et al.* (1969) covers the temperature range 1200 K to 1600 K, and the data set of Dawe and Smith (1970) spans the intervening temperature range. A measurement by Berg (2005, 2006) at $T = 298.15\,\text{K}$, together with a set of smoothed data determined from the fit function provided by Vogel (1984, 1987) for his experimental helium data, but with the original experimental reference value $\eta_{298.15\text{K}}$ replaced by the calculated *ab initio* value of (Bich, Hellmann, and Vogel, 2007), have also been included in this figure, as they demonstrate the highest accuracy (better than $\pm\,0.1\%$) that can be achieved. Finally, a number of representative values from the Bich *et al.* (1990) experimental correlation are included. The pair of dashed horizontal lines demarcate percentage differences between theory and experiment that are smaller than $\pm\,0.5\%$.

A difference plot, such as Fig. 4.3, serves two main functions: it provides a qualitative visual measure of the level of agreement, typically between calculated and experimentally determined zero-density-limit values of a transport coefficient, while at the same time providing a comparison between the theoretical (i.e. calculated) and experimental temperature dependence of that transport coefficient. A set of experimentally determined values that steadily increases or decreases relative to the behavior associated with the *ab initio* PEF likely indicates that there are (perhaps as yet unidentified) uncompensated systematic experimental errors. It is clear, for example, from the right-hand panel of Fig. 4.3 that both for temperatures significantly above or below the ambient temperature range, the experimental trends for η are distinctly different from the temperature dependence exhibited by the calculated values. It is clear from this figure that experimentally determined and calculated values of η are in excellent agreement for temperatures lying between 280 K and 700 K, but show systematic differences for temperatures that lie either above or below this range. Figure 4.3 also show that, even though values of $\eta(T)$ predicted by the experimental correlation of Bich *et al.* (1990) deviate systematically from the corresponding calculated values as the temperature increases, the percentage difference does not exceed -0.50% until the temperature is in excess of 1600 K. Note also that the differences between correlation values (Bich *et al.*, 1990) and calculated values (Bich *et al.*, 2007) of $\eta(T)$ do not exceed approximately -0.35% for temperatures lying below 100 K.

Qualitatively, much the same can be said about the level of agreement between experimental values of λ and the values calculated by Bich *et al.* (2007) from the *ab initio* He–He PEF (Hellmann *et al.*, 2007). The level of agreement between calculated and experimental values of λ is best at ambient temperatures, with differences that are often smaller than $\pm 0.3\%$. The differences are typically worse both for temperatures significantly above or below the ambient temperature range. As has been seen for the viscosity coefficient, values of λ derived from the theory-based correlation of Bich *et al.* (1990) compare on the whole very well with the corresponding values computed (Bich *et al.*, 2007) from the *ab initio* PEF. The percentage differences for the correlation-predicted values of $\lambda(T)$ do show a systematic trend with increasing temperature, decreasing from a minimal value of approximately -0.01% at ambient temperatures to -0.47% at $T = 1600\,\mathrm{K}$, and to -0.64% by $T = 4000\,\mathrm{K}$: for temperatures below the ambient temperature range, however, the difference remains smaller than $\pm 0.40\%$ for temperatures below 100 K.

As measurement of the isotopic diffusion coefficient, $D_{\mathrm{A'A}}(T)$, or the thermal diffusion factor, $\alpha(T)$, for binary mixtures of ^{4}He and ^{3}He is very challenging, experimental values are typically accompanied by fairly large experimental uncertainties. Moreover, as the central values for $D_{\mathrm{A'A}}(T)$ also do not show a clear functional temperature dependence, it seems best to display the differences between values of $D_{\mathrm{A'A}}$ calculated from an *ab initio* PEF and experiment in terms of a plot of the relative percentage difference, $100(D_{\mathrm{expt}} - D_{\mathrm{calc}})/D_{\mathrm{calc}}$, as a function of temperature, as by Hurly and Moldover (2000). Neither Cencek *et al.* (2012) nor Bich *et al.* (2007) carried out calculations of the mutual diffusion coefficient $D_{\mathrm{A'A}}(T)$ or of the thermal diffusion factor $\alpha_{\mathrm{T}}(T)$. It is clear, however, from comparisons between calculated values of the viscosity and thermal conductivity coefficients for ^{4}He obtained (Bich *et al.*, 2007; Hurly and Mehl, 2007; Hurly and Moldover, 2000; Mehl, 2009) using He–He *ab initio* PEFs that the results of such calculations would essentially be indistinguishable from the comparisons given for these transport properties by Hurly and Moldover (2000).

4.1.2 The Ne–Ne interaction

The interaction energy between a pair of ground electronic state neon atoms has been computed *ab initio* by Hellmann *et al.* (2008*a*) for 32 neon internuclear separations R lying in the interval $1.400\,\text{Å} \leq R \leq 8.000\,\text{Å}$. All calculations were carried out using the supermolecule approach within the frozen-core approximation: extensive high-level basis-sets were employed, full counterpoise corrections were applied, and extrapolation to the complete basis-set (CBS) limit was carried out for each separation. A more in-depth discussion of *ab initio* computations of the Ne–Ne interaction appears in Section 3.8.2.

Unlike helium, for which the dominant naturally occurring isotope ^{4}He is readily available with a purity of 99.999% or better, naturally occurring neon is a mixture of three Ne isotopes, namely, ^{20}Ne (90.48%), ^{21}Ne (0.27%), and ^{22}Ne (9.25%). Naturally occurring neon in principle thus involves six types of pair interactions, each of which corresponds to a different reduced mass, m_r. As the atomic mass of a neon atom is of the order of 20 amu, the reduced masses for Ne–Ne binary collisions will be of the order of 10 amu: in order to be assured that the required level of accuracy is obtained for the transport properties of gases of atoms giving such relatively low reduced masses it is important, especially for temperatures below ambient temperatures, that they be evaluated quantum mechanically. Further, because ^{20}Ne and ^{22}Ne have nuclear spin $I = 0$, while ^{21}Ne has nuclear spin $I = \frac{3}{2}$, evaluation of the effective (collision) cross sections entering into the expressions that determine the transport coefficients in terms of quantum phase shifts δ_ℓ must be carried out for pure ^{20}Ne and ^{22}Ne using Bose–Einstein statistics, for pure ^{21}Ne using Fermi–Dirac statistics, and for unlike-isotope interactions using Boltzmann statistics.

Because the adiabatic correction makes a negligible contribution to the interaction energy between a pair of neon atoms, the interaction energy for all Ne_2 isotopologues[2] may therefore be represented by a single PEF with a minimum characterized by depth $\mathfrak{D}_e = 29.297\,\mathrm{cm}^{-1}$ and internuclear separation $R_e = 3.0895\,\text{Å}$. Upon employing the Le Roy (2017*b*) LEVEL code to determine the energies E_{vJ} of the bound eigenstates for the homonuclear $^{20}Ne_2$ dimer, Hellmann *et al.* (2008*a*) found ten truly bound plus two quasi-bound eigenstates. The ground state energy was determined to be $E_{00} = -16.7462\,\mathrm{cm}^{-1}$ (equivalently, the dissociation energy $D_0 = 16.7462\,\mathrm{cm}^{-1}$), while the remaining truly bound (excited) states corresponded to ($\mathrm{v} = 0,\ J = \{2, 4, \cdots, 10\}$), ($\mathrm{v} = 1,\ J = \{0, 2, 4\}$), and ($\mathrm{v} = 2,\ J = 0$), with E_{20} lying only $0.0130\,\mathrm{cm}^{-1}$ below the dissociation limit. Similarly, they found 24 bound eigenstates for the heteronuclear ^{20}Ne^{22}Ne isotopologue, with the ground ro-vibrational state having energy E_{00} $-16.9913\,\mathrm{cm}^{-1}$, thereby giving it a zero-point energy of $12.3016\,\mathrm{cm}^{-1}$. This isotopologue has 17 truly bound excited states, with ($\mathrm{v} = 0,\ \{J = 0, 1, 2, \cdots, 10\}$), ($\mathrm{v} = 1,\ \{J = 0, 1, \cdots, 5\}$), and ($\mathrm{v} = 2,\ J = 0$), with E_{20} lying $0.0307\,\mathrm{cm}^{-1}$ below the dissociation limit, plus six quasi-bound excited states, with ($\mathrm{v} = 0,\ J = \{11, 12, 13, 14\}$) and ($\mathrm{v} = 1,\ J = \{6, 7\}$).

Although it is not possible to observe spectroscopic transitions between the bound states of a Ne_2 isotopologue using microwave and infrared spectroscopy due to the absence of a permanent electric dipole moment in the ground electronic state, it is possible to employ VUV transitions from the bound energy levels of the ground electronic state of a neon isotopologue into sets of ro-vibrational states of one or more excited electronic states of the isotopologue and then to utilize the resultant sequences of (resolved) P- and R-branch transitions to obtain values for some or all of the energies E_{vJ} of the truly bound states of the particular isotopologue by a process referred to as a

[2] Isotopologues are molecules that differ only in their isotopic compositions, while *isotopomers* have the same isotopic compositions but differ due to the specific molecular sites at which an isotopic substitution has been made.

combination difference procedure. As the natural abundances of the neon isotopes are 90.48% ^{20}Ne, 9.25% ^{22}Ne, and 0.27% ^{21}Ne, the equilibrium dimer population of neon will therefore consist of 81.87% $^{20}Ne_2$, 16.74% $^{22}Ne_2$, 0.86% $^{20}Ne^{22}Ne$, respectively, with the largest fraction of the population containing a ^{21}Ne atom being 0.21% for the $^{20}Ne^{21}Ne$ isotopologue.

Wüest and Merkt (2003) employed VUV excitations to optically accessible excited electronic neon dimer states of the two most abundant neon isotopologues, and were able to determine the energies (relative to the ground-state energy) of eight of the nine truly bound rovibrational excited states of the $^{20}Ne_2$ isotopologue (with an average uncertainty of 0.6%) and to determine the relative energies of four of the 17 truly bound states of the $^{22}Ne_2$ isotopologue (with an average uncertainty of 4.4%) by using combination differences between members of respective P- and R-branch transition sequences. The energies determined by Hellmann *et al.* (2008*a*) from the *ab initio* Ne–Ne PEF for ro-vibrational excited states of the $^{20}Ne_2$ isotopologue agree on average within 0.34% with the experimental values obtained for the ($\mathrm{v} = 0$, $J = \{2,4,6,8\}$) states, and within 0.19% with the experimental values obtained for the ($\mathrm{v} = 1$, $J = \{0,2,4\}$) states. Similarly, the calculated energies obtained for the $^{20}Ne^{22}Ne$ isotopologue agree within $\pm$1.36% on average with the experimental values obtained for the ($\mathrm{v} = 0$, $J = \{2,3,4,5\}$) states.

Because naturally occurring neon is a ternary mixture of the three stable isotopes ^{20}Ne, ^{21}Ne, and ^{22}Ne, comparable calculated values of η and λ for neon should, in principle, be obtained using ternary mixture expressions for them. However, as even second Chapman–Cowling approximation expressions for η and λ have yet to be developed for ternary and higher multi-component mixtures, other appropriate means for calculating the transport properties of neon must be introduced. One logical approach to the treatment of naturally occurring neon would be, for example, given that the ^{21}Ne isotope makes up only 0.27% of naturally occurring neon, to treat it as a binary mixture consisting of 90.48% ^{20}Ne and 9.52% ^{22}Ne. Another logical approach would be to treat naturally occurring neon as a pure gas consisting of atoms with an atomic mass equal to 20.1797 amu, which is its relative weighted atomic mass. Yet a third logical approach would be to attempt to treat naturally occurring neon as a ternary mixture, starting from the first Chapman–Cowling approximation expressions for η_{mix} and λ_{mix}.

Bich *et al.* (2008) considered two approaches to compute the viscosity and thermal conductivity coefficients for naturally occurring neon using their *ab initio* (Hellmann *et al.*, 2008*a*) PEF. An important observation in treating neon as a ternary mixture is to note that the first Chapman–Cowling approximation expressions for $[\eta_{\mathrm{mix}}]_1$ and $[\lambda_{\mathrm{mix}}]_1$ are given (Ern and Giovangigli, 1994; Maitland *et al.*, 1980) as ratios of pairs of determinants, in which the determinant in the numerator in each case is given in terms of the determinant of the corresponding denominator augmented by a row and a column with elements consisting of the three mole fractions, plus 0 as a common fourth element. That every element of the denominator determinants is given in terms of appropriate first Chapman–Cowling approximations, $[\eta_{\mathrm{AB}}]_1$ and $[\lambda_{\mathrm{AB}}]_1$, respectively, for the six isotopic neon pair interactions suggests the replacement of all first Chapman–Cowling approximations to η_{AB} and λ_{AB} by corresponding higher Chapman–Cowling approximation expressions, as given in eqns (2.74d) and (2.79d). Expressions of this type using fifth Chapman–Cowling approximation expressions were evaluated by Bich *et al.* (2008) using the explicit expressions and computer codes provided by Viehland *et al.* (1995), with the omega integrals computed quantum mechanically. The resulting expression should, however, perhaps be termed a pseudo-fifth-order approximation. Naturally occurring neon was also approximated as a pure gas consisting of neon atoms having an atomic mass of 20.1797 amu, with η and λ also evaluated using fifth Chapman–Cowling approximation expressions, but in this case with the omega integrals determined via classical trajectory calculations. Classical calculations were used here in part to enable the establishment of the range of temperatures for which quantum mechanical calculations are necessary in order to obtain results suitable for meaningful comparison with the most accurate experimental values obtained for η and λ.

Comparisons between the results obtained from quantum scattering and classical trajectory calculations showed (Bich *et al.*, 2008), for example, that classical trajectory results obtained using the first Chapman–Cowling approximation are not reliable for temperatures below about 100 K, at which they lie approximately 1% below the pseudo-fifth-order quantum mechanical results, and that for temperatures above 200 K, the classical results level off at approximately 0.8% below the corresponding pseudo-fifth-order values. Similarly, first approximation expressions evaluated using quantum mechanical scattering fall progressively below the pseudo-fifth-order quantum mechanical results, effectively joining the classical first approximation results for temperatures above 600 K. Finally, the fifth Chapman–Cowling approximation classical trajectory results for neon, while also unreliable for temperatures below about 100 K, become increasingly reliable as the gas temperature increases, with the difference between the classical result and the pseudo-fifth-order quantum result decreasing until approximately 600 K, at which point the quantal and classical results (essentially) agree.

Figure 4.4. shows a gross comparison between the experimentally determined and calculated values of η and λ for naturally occurring neon, and it clearly demonstrates that theory and experiment are in overall excellent agreement for both η and λ. In addition, an examination of the relative percentage deviations in the ambient temperature range indeed shows that most of the experimental data obtained for temperatures between 290 K and 305 K fall within $\pm 0.2\%$ of the values calculated (Bich *et al.*, 2008) using the *ab initio* (Hellmann *et al.*, 2008*a*) PEF. The absolute measurement carried out by Kestin and Nagashima (1964) at four temperatures (223.15 K, 248.15 K, 298.15 K, 373.15 K) using a capillary viscometer all lie within $\pm 0.3\%$ of the corresponding calculated values, with the value $\eta_0 = (31.370 \pm 0.010)\ \mu\text{Pa s}$ obtained at temperature $T = 298.15$ K lying only 0.01% above the calculated value of 31.728 μPa s. Better than $\pm 0.1\%$ agreement between theory and experiment was also attained with the absolute measurement of η_0 by Evers *et al.* (2002) at $T = 348$ K using a rotating cylinder viscometer. Further, a recalibration of the Vogel quartz oscillating disk viscometer based upon the computed value (Bich *et al.*, 2007) for the viscosity of ^{4}He at 298.15 K led to revised neon viscosity values lying typically within the $\pm 0.1\%$ uncertainties assigned to the corresponding calculated values. Note also that a similar

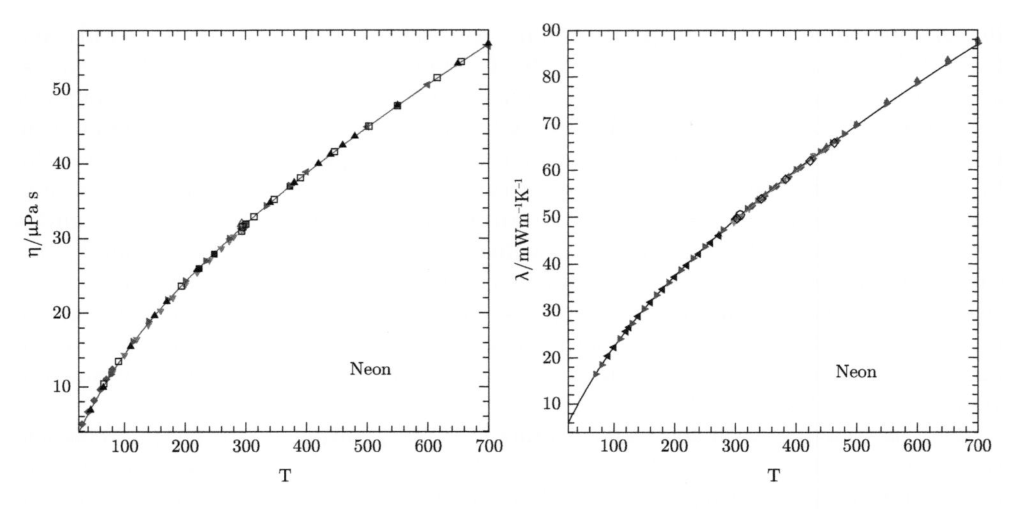

Fig. 4.4 Comparison between computed[a] and experimental values[b] of the viscosity and thermal conductivity of neon.

[a] Bich *et al.*, 2008

[b] Viscosity data: • (green), Berg and Burton, 2013; ▲, Bich *et al.*, 1990; ▼ (pink), Johnston and Grilly, 1942; ♦ (red), Coremans *et al.*, 1958; □, Rietveld *et al.*, 1959; ■ (grey), Flynn *et al.*, 1963; ▷, Clarke and Smith, 1969; ◀ (red), Dawe and Smith, 1970; △ (blue), Kestin and Leidenfrost, 1959; ■ (grey), Vogel, 1984. Thermal conductivity data: ▶ (blue), Bich *et al.*, 1990; ♦, Haarman, 1973; ◇ (grey), Kestin *et al.*, 1980; ○, Assael *et al.*, 1981; □ (pink), Millat *et al.*, 1988; ▼ (green), Sengers *et al.*, 1964; ◀ (grey), Nesterov and Sudnik, 1976; ▲ (red), Ziebland, 1981; ◇, Hemminger, 1987.

re-evaluation of relative data of Coremans *et al.* (1958) obtained using an oscillating disk viscometer, in which the theoretical value (Bich *et al.*, 2007) for ^{4}He at 20 K was employed as a new calibration value, led to a significant improvement in the level of agreement between the calculated and experimental results for low-temperature neon. Finally, even though Bich *et al.*, (2008) noted significant and systematic deviations between experiment and theory for temperatures above 500 K, it is clear that the most accurate experimental viscosity data obtained for neon at temperatures below 500 K still enable a clear discrimination to be made between the various proposed (Aziz and Slaman, 1989*a*; Cybulski and Toczyłowski, 1999; Dham *et al.*, 1989; Wüest and Merkt, 2003) PEF for neon.

The most accurate thermal conductivity data appear to be those obtained using a technique termed the transient hot-wire method, which is more or less restricted to ambient temperatures. Data obtained by Kestin *et al.* (1980) and by Assael *et al.* (1981) at room temperature deviate from the quantum mechanical values calculated (2008) from the Hellmann *et al.* (2008*a*) *ab initio* PEF by less than ±0.2%, in each case, lying within the reported experimental uncertainties, while the data obtained by Haarman (1973) and by Hemminger (1987) in the temperature range 328–468 K lie below the corresponding calculated values by less than 0.5%, and the smoothed experimental values of λ reported by Nesterov and Sudnik (1976) over the temperature range 90–273.15 K fall below the correponding calculated (2008) values by about 0.1% at 90 K, increasing to about 0.7% below the calculated values at 273.15 K.

Bich *et al.* (2007) also provided comparisons with other *ab initio* (Cybulski and Toczyłowski, 1999) and empirical (Aziz and Slaman, 1989*b*; Wüest and Merkt, 2003) Ne–Ne PEFs in their difference plots for both η and λ. The shear viscosity data for the ambient temperature range clearly allow a distinction to be made between the Hellmann *et al.* (2008*a*) *ab initio* PEF and its three (Aziz and Slaman, 1989*a*; Cybulski and Toczyłowski, 1999; Dham *et al.*, 1989; Wüest and Merkt, 2003) closest competitors. The same is true, although to a lesser extent, for the lower temperature and for the high temperature (i.e. 500–2000 K) viscosity data. The thermal conductivity data for the temperature range 100–500 K clearly indicates, in agreement with the viscosity data, that the empirical PEF proposed by Wüest and Merkt (2003) on the basis of an iterative fit to their spectrscopic data does not fare well in predicting transport data: this outcome is consistent with the fact that spectroscopic data, while very sensitive to the depth and shape of an attractive region of a PEF, are relatively insensitive to the repulsive region of such a function. Typically, thermal conductivity measurements made at temperatures above 400 K tend to have much larger experimental uncertainties (of the order of ±1% or more) and fractional difference plots often show temperature trends that differ from that obtained from an *ab initio* PEF (or of a well-behaved empirical PEF). The end result is that such data cannot be used for meaningful comparisons with thermal conductivity values calculated from *ab initio* PEFs or even to distinguish among different PEFs.

4.1.3 The Ar–Ar interaction

The interaction energy between two ground electronic state argon atoms has been computed *ab initio* for 38 internuclear separations R spanning the interval $1.8\,\text{Å} \leq R \leq 15\,\text{Å}$ by Jäger *et al.* (2009, 2010) and for 33 values of R spanning the interval $2\,\text{Å} \leq R \leq 20\,\text{Å}$ by Patkowski and Szalewicz (2010). In each case, the set of *ab initio* energies has been fitted to an analytical function of the Tang–Toennies (Tang and Toennies, 1984) (TT) type. The computations of Jäger *et al.* (2009, 2010) used supermolecule coupled-cluster (SMCC) methods at the CCSDT(Q) level, employing high-quality basis-sets supplemented with bond functions, and included corrections for correlations involving core electrons and for relativistic effects (determined in the Cowan–Griffin (Cowan and Griffin, 1976) approximation). The Patkowski–Szalewicz computations were also based upon SMCC methods: the largest component of the Ar–Ar interaction energy was computed using the frozen-core CCSD(T) approximation plus corrections associated with differences arising from the inclusion of correlations involving core electrons, from higher excitations included in the next three levels of frozen-core CC computations, and due to relativistic effects. Moreover, by developing additional, more extensive basis-sets, and typically extrapolating their results to CBS limits, they were able to provide uncertainty estimates for the *ab initio* Ar–Ar interaction energies.

Figure 4.5 compares five accurate PEFs proposed for the Ar–Ar interaction. Two PEFs are represented by nonlinear least-squares fits of sets of *ab initio* Ar–Ar interaction energies generated by high-level SMCC calculations to modified TT (MTT)

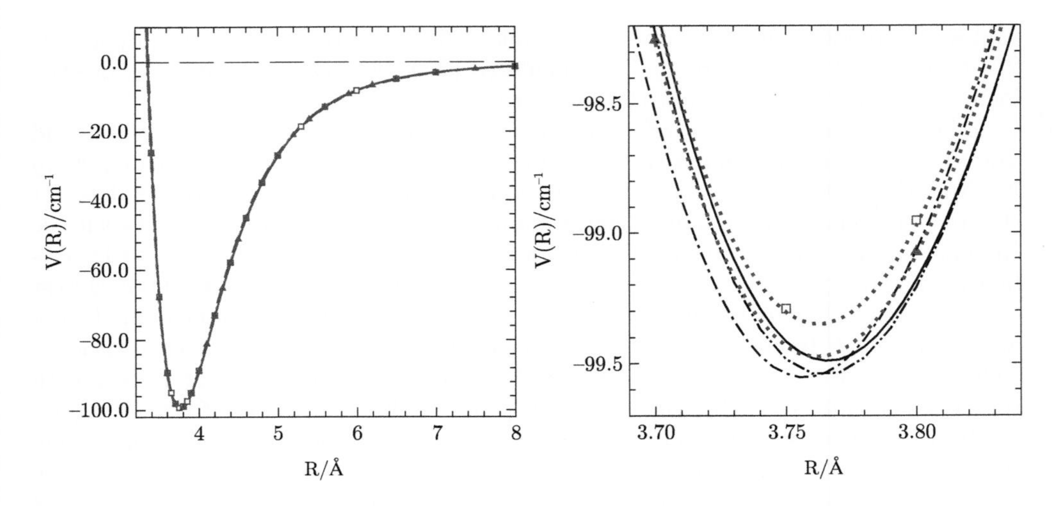

Fig. 4.5 Comparison between the MTT1 Ar–Ar PEF (dotted curve plus ▲ *ab initio* values), the MTT2 PEF (dotted curve plus □ *ab initio* values), and the HFDID1 (dash-dot curve), HFDID2 (dash-doubledot curve) and the MLR (solid curve) empirical PEFs for the Ar–Ar interaction. After Myatt *et al.* (2018).

functional forms: these PEFs will be referred to as the MTT1 Jäger *et al.* (2009, 2010) and MTT2 (Patkowski and Szalewicz, 2010) PEFs. Two empirical PEFs (Aziz, 1993; Boyes, 1994) that used the individually damped Hartree–Fock plus damped-dispersion (HFD) function form, referred to as the HFDID1 (Aziz, 1993) and HFDID2 (Boyes, 1994) PEFs, and a more recent (Myatt *et al.*, 2018) empirical PEF having the Morse/long-range (Le Roy and Pashkov, 2017; Le Roy *et al.*, 2011) (MLR) form, all obtained from fits to extensive sets of experimental data, are also shown in this figure.[3]

Both HFDID PEFs employed four free fit-parameters: initial values for the HFDID1 parameters were determined firstly by requiring that the attractive well of the PEF be consistent with the Ar_2 dimer ro-vibration band spectrum (Herman *et al.*, 1988), then further adjusted to bring the repulsive wall of the PEF into agreement with high-energy beam scattering data and the viscosity data of Vogel (1984). Because the HFDID1 Ar_2 PEF was found to predict second acoustic virial coefficient values that seriously disagreed with experiment (Ewing *et al.*, 1989; Ewing and Goodwin, 1992; Moldover *et al.*, 1999), a new HFDID2 PEF for argon was redetermined (Boyes, 1994) by optimization of the same four fit-parameters using a multi-property non-linear least-squares regression analysis based on (re-analyzed) spectroscopic data of Herman, LaRocque, and Stoicheff (1988), the viscosity data of Vogel (1984), the high-energy beam data employed in Aziz (1993), plus the then-available acoustic second virial coefficient data. The empirical MLR PEF was determined (Myatt *et al.*, 2018)

[3]The various HFD forms and the MLR form are discussed in greater detail in Section 3.3.1 and Section 3.5, respectively.

by employing the automated nonlinear least-squares simultaneous fitting procedure **dPotFit** (Le Roy, 2017*a*). An initial fit was made to a set of 1634 experimental fit-data consisting of 1397 spectroscopic transition frequencies of the Ar_2 dimer, 178 second (pressure) virial coefficient values, and 59 second acoustic virial coefficient values using **dPotFit**. A final optimization was carried out via a fit cycle that consisted of a systematic (manual) augmentation of the primary fit-data set of 1634 experimental values by an energy value at 2.8Å (in the repulsive region), followed by a refit of the augmented primary data set to the MLR form and a determination of the DRMSD value for a set of 609 (secondary) transport property data. The final MLR PEF corresponded to a minimum in the DRMSD value for the secondary experimental data set and a minimal change to the DRMSD value for the original primary experimental data set.

The internal energy associated with an argon dimer can be characterized by a vibrational quantum number v and a rotational (end-over-end) angular momentum quantum number J, and is thus designated E_{vJ}. Table 4.2 illustrates the values obtained for the set of term differences $\Delta G_{v+\frac{1}{2}}$, defined via $hc\Delta G_{vJ} \equiv E_{v+1,0} - E_{v,0}$, for the Ar_2 dimer. Because the only available spectroscopic data for the Ar_2 dimer necessarily involves excited electronic states in addition to its ground electronic state, information about term-value differences for the Ar_2 ground electronic state must be determined indirectly from combination differences between various experimental directly observed transitions, the spectral data serve more as an initial filter for the ground-state Ar_2 PEF. This observation has been demonstrated explicitly by Myatt *et al.* (2018), who found that their "best-fit" MLR empirical PEF for the argon ground electronic state based upon fitting to the 1397 available spectroscopic transitions alone led to DRMSD ($\overline{dd}$) values of 8.273 and 43.661, respectively, for the 178 pressure and 59 acoustic second virial coefficient values that were, together with the spectroscopic data, considered to form their primary set of fit-data. Moreover, the rough equivalence (all of order 1) of the DRMSD values obtained for the set of 1397 spectroscopic data using the MLR, MTT1, MTT2, HFDID1, and HFDID2 PEFs is also confirmed by the results shown in Table 4.2.

Table 4.2 Term differences* $\Delta G_{v+\frac{1}{2}}$ for the Ar_2 dimer

	Experiment			AI PEF	Empirical PEFs	
v	CD[a]	HLS[b]	u^b	MTT1[c]	HFDID1[d]	HFDID2[e]
0	25.74	25.69	0.01	25.71	25.68	25.68
1	20.41	20.58	0.02	20.52	20.56	20.59
2	15.60	15.58	0.02	15.52	15.58	15.57
3	10.91	10.91	0.03	10.88	10.92	10.87
4	6.78	6.84	0.07	6.82	6.83	6.82

* All term differences are given in cm^{-1} units.
[a] Colbourn and Douglas, 1976; [b] Herman *et al.*, 1988; [c] Jäger *et al.*, 2009;
[d] Aziz, 1993; [e] Patkowski and Szalewicz, 2010

The sensitivity of thermophysical properties to details of a PEF is typified by the temperature dependence of the equation of state (pressure) or the acoustic second virial coefficient, $B_2(T)$ or $\beta_a(T)$, as illustrated by the comparisons shown in Fig. 4.6. These equilibrium thermophysical properties have been found to provide good measures of the quality of the pair interaction energy. The right-hand panel of Fig. 4.6 contains a plot showing the temperature dependence of the difference between values of the second pressure virial coefficient $B_2(T)$ computed (Vogel, 2010) using the Jäger *et al.* (2009, 2010) *ab initio* Ar_2 PEF and a set of 98 "experimental" values (Fig. 4.6) for $B_2(T)$ obtained in the temperature range $90\,K < T < 450\,K$. These values have been determined in the usual manner (Myatt *et al.*, 2018), either by re-analysis (data from Estrada-Alexanders and Trusler 1995; Gilgen *et al.* 1994; Michels *et al.* 1949; Michels *et al.* 1958) or first analysis (data from Cencek *et al.* 2013; Klimeck *et al.* 1998) of the original experimental equation of state data, and have much tighter statistical uncertainties (as much as a factor 10 smaller) than those that had typically been obtained previously. In addition to showing a difference plot for the 98 experimentally determined results, differences between second pressure virial coefficient values obtained using the Patkowski–Szalewicz (2010) *ab initio* PEF, and the Jäger *et al.* (2009, 2010) PEF are shown by the dash-dotted curve. Similarly, differences between results computed using the HFDID1, HFDID2, and MLR empirical PEFs are shown, respectively, by the dotted, dashed, and dash-doubledotted curves. The left-hand panel of Fig. 4.6 shows a similar difference plot for 59 values of the second acoustic virial coefficient $\beta_a(T)$ determined from speed of sound data (Fig. 4.6) for argon.

DRMSD values may also, in principle, be quite informative. Their usefulness, however, does rely heavily upon the availability of realistic experimental uncertainties, as overly conservative uncertainties will typically result in unrealistically small DRMSD values that could then be interpreted inappropriately to indicate excellent agreement between calculated and experimental values of a physical property. Of course, just the opposite can occur for overly optimistic uncertainties. DRMSD values do provide good representations of the levels of agreement between theory and experiment for realistic uncertainties. Table 4.3 was constructed using the data that went into the preparation of Fig. 4.6 and provides an illustration of the role that can be played by DRMSD values. The tight, but statistically meaningful, experimental uncertainties in the data employed in the construction of Table 4.3 ensure that differences among the DRMSD (or $\overline{dd}$) values obtained for the five Ar–Ar PEFs reflect the global sensitivities of the temperature dependences of the pressure and acoustic second virial coefficients to features of the PEFs under consideration.

The level of sensitivity exhibited by these two (equilibrium) thermophysical properties may also be seen from Fig. 4.5, in which the left-hand panel shows all five PEFs and the two sets of *ab initio* points (Jäger *et al.*, 2010; Patkowski and Szalewicz, 2010) on an energy scale that is appropriate for a display of the attractive well of the interaction between a pair of ground electronic state argon atoms. Although it is clear that on this scale it is essentially impossible to draw any distinctions between the PEFs, it is also evident from Table 4.3 that $B_2(T)$ and, especially, $\beta_a(T)$ are sensitive to features of the PEFs that are not apparent on this scale. Differences between the five PEFs are, however, apparent on the energy scale employed in the right-hand panel of

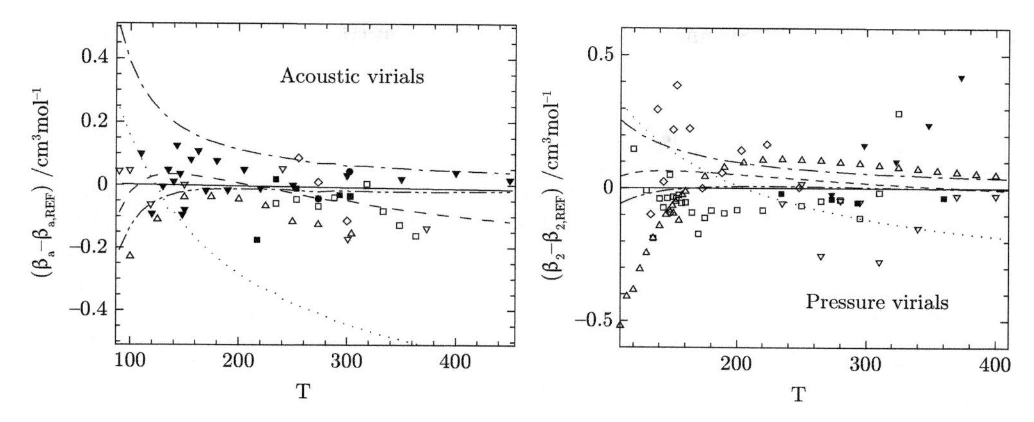

Fig. 4.6 Comparison between computed[a] and experimental temperature dependences of the pressure[b] and acoustic[c] virial coefficients of argon. The dotted, dashed, dash-doubledotted, and dash-dotted curves represent the differences between virial coefficient values computed using the MTT1 *ab initio* PEF and, respectively, the HFDID1, HFDID2, MLR empirical PEFs, and the MTT2 *ab initio* PEF.

[a] Vogel, 2010
[b] ▼, Michels *et al.*, 1949; ◇, Michels *et al.*, 1958; □, Gilgen *et al.*, 1994; △, Estrada-Alexanders and Trusler, 1995; ▽, Klimeck *et al.*, 1998; ■, Cencek *et al.*, 2013.
[c] △, Ewing *et al.*, 1989; ◇, Ewing and Goodwin, 1992; ■, Moldover *et al.*, 1999; •, Moldover and Trusler, 1988; ▽, Ewing and Trusler, 1992*b*; ▼, Estrada-Alexanders and Trusler, 1995; □, Benedetto *et al.*, 2004.

Fig. 4.5. In particular, the HFDID1 and MTT2 PEFs are clearly differentiated from the other three PEFs, with the minimum in the HFDID1 PEF clearly shifted inward (and slightly downward), while the minimum in the MTT2 PEF is shifted significantly upward relative to the other PEFs. The MLR empirical PEF shares characteristics with three other PEFs: specifically, it almost coincides with the inner wall of the attractive well of the MTT2 *ab initio* PEF, has a minimum that is similar in magnitude to that of the MTT1 *ab initio* PEF at a separation that is close to that for the minimum of the HFDID2 PEF, and has an outer wall of the attractive well that effectively coincides with that of the HFDID2 PEF. These subtle differences are reflected in the sets of DRMSD values in Table 4.3.

The differences between the "all-data" DRMSD values for $B_2(T)$ and $\beta_a(T)$ for the MTT1, MLR, and HFDID2 are likely closely related to the greater precision and the considerably smaller experimental uncertainties associated with the acoustic second virial coefficient. It is likely that the significantly larger differences between the overall DRMSD values obtained from comparison between calculated and measured values of $\beta_a(T)$ for the MTT2 *ab initio* and the HFDID1 empirical PEFs indicate inadequacies in the region of the attractive wells of these PEFs. These observations serve to emphasize the importance of being able to carry out very precise measurements of these thermophysical properties as well as being able to obtain tight experimental uncertainties.

Table 4.3 DRMSD analysis of a selection of pressure and acoustic virial data

$B_2(T)$ Reference	# Data	$\overline{dd}$ values MTT1	MTT2	MLR	HFDID1	HFDID2
a	7	0.946	0.828	0.941	1.376	0.925
b	12	0.676	0.417	0.667	0.688	0.594
c	26	1.176	1.448	1.231	2.189	1.804
d	34	1.330	1.947	1.214	2.564	1.541
e	13	1.079	1.584	1.113	1.616	1.257
f	6	0.290	0.556	0.308	0.229	0.376
All data	98	1.123	1.514	1.024	2.019	1.415
$\beta_a(T)$						
g	9	3.108	5.242	2.529	14.468	1.539
h	9	1.973	5.208	1.463	5.228	1.916
i	4	4.686	5.896	5.232	34.402	6.015
j	8	1.216	3.372	1.477	3.694	1.006
d	21	1.405	3.296	1.656	7.222	1.756
k	2	4.297	10.343	3.749	40.118	4.672
l	6	2.940	6.374	2.568	20.164	2.694
All data	59	2.442	4.867	2.360	15.253	2.467

[a] Michels *et al.*, 1949; [b] Michels *et al.*, 1958; [c] Gilgen *et al.*, 1994; [d] Estrada-Alexanders and Trusler, 1995; [e] Klimeck *et al.*, 1998; [f] Cencek *et al.*, 2013; [g] Ewing et al., 1989; [h] Benedetto *et al.*, 2004; [i] Ewing and Goodwin, 1992; [j] Ewing and Trusler, 1992*b*; [k] Moldover and Trusler, 1988; [l] Moldover *et al.*, 1999.

Because argon at temperature $T = 298.15\,\text{K}$ (often referred to simply as "room temperature") has served as a traditional calibration standard for viscometers employed to determine fluid viscosities relative to an accepted standard substance, a major effort was exerted by experimental groups to determine an accurate absolute experimental value for the room temperature viscosity of argon. The results of this effort are summarized in Table 4.4. Both Berg and Moldover (2012) and Berg and Burton (2013) made concerted efforts to reduce the experimental uncertainty in the proposed calibration value for the viscosity of argon at temperature 298.15 K. Table 4.4 shows that, even though the two proposed values differ by only 0.06%, there is no overlap between the two very tight uncertainty intervals that were determined. The most recent absolute measurement of $\eta_{\text{Ar}}(T = 298.15\,\text{K})$ by Lin *et al.* (2014) using a two-capillary viscometer lies approximately midway between these two values and has an uncertainty interval that encompasses both values. The Berg and Burton (2013) value does, however, lie very close to the calculated value obtained from the *ab initio* PEF(s).

Table 4.4 Argon viscosity determinations at the standard temperature $T = 298.15\,\mathrm{K}$

Year	Reference	Experimental Value		Difference	
		η_{expt} $/\mu\mathrm{Pa\,s}$	$\pm u$ $/\mu\mathrm{Pa\,s}$	$\delta\eta \equiv$ $\eta_{\mathrm{expt}} - \eta_{\mathrm{calc}}$	$\delta\eta/\eta_{\mathrm{calc}}$ $/\%$
2014	a	22.562	0.018	0.010	0.044
2013	b	22.5539	0.0050[†]	0.0019	0.008
2012	c	22.5666	0.0060	0.014	0.065
2007	d	22.570	0.019	0.018	0.080
2006	e	22.578	0.019	0.026	0.115
2005	f	22.582	0.008	0.030	0.133

[†]Obtained by multiplying the Berg and Burton (2013) experimental value by $u(\eta_{\mathrm{Ar}}/\eta_{\mathrm{He}}) = 0.00030$ given in their Table 1.
[a] Lin *et al.*, 2014; [b] Berg and Burton, 2013; [c] Berg and Moldover, 2012;
[d] May *et al.*, 2007; [e] Clifford *et al.*, 1975; [f] Berg, 2005; Berg, 2006.

Unfortunately, it has now also been established (Song *et al.*, 2016; Vogel, 2010) that the Kestin *et al.* (1972*e*) (KRW) values that served as noble gas viscosity standards for more than thirty years were affected by a systematic error that led to incorrect determinations of the gas sample temperatures, with the consequence that their recommended viscosity values differ from the corresponding values computed from high-level *ab initio* PEFs by amounts that vary from about 0.2% at room temperature to as much as 0.8% at their highest measured temperatures. As the all-quartz oscillating disk viscometer employed by Vogel (1972, 2000) and the vibrating wire viscometer used by Wilhelm and Vogel (1984) to determine the viscosity of argon had been calibrated using the KRW recommended viscosity for argon at $T = 298.15\,\mathrm{K}$, their data were reanalyzed by employing the zero-density value $\eta_{\mathrm{Ar}}(T{=}298.15\mathrm{K}) = 22.552\,\mu\mathrm{Pa\,s}$ computed from their *ab initio* Ar–Ar PEF as the calibration viscosity. Argon viscosities obtained from reanalysis of the vibrating disk viscometer data were found to lie typically less than 0.1% above while those obtained from reanalysis of the vibrating wire viscometer data were found to lie typically about 0.1% below the corresponding computed values (Vogel, 2010).

Figure 4.7 illustrates the overall agreement between the calculated and experimental behaviors of the shear viscosity (left panel) and the thermal conductivity (right panel) in the temperature range $120\,\mathrm{K} \leq T \leq 750\,\mathrm{K}$. The curves represent values of η and λ obtained from classical calculations (Vogel, 2010) based upon the *ab initio* PEF of Jäger *et al.* (2009, 2010), with the (zero-density) transport coefficients evaluated using the fifth Chapman–Cowling approximation. Only a representative selection of zero-density limiting values of the viscosity are shown. The most recent results obtained by Vogel in 2010 used an all-quartz oscillating disk viscometer (Song *et al.*, 2016; Vogel, 2010) designed specifically to enable very accurate viscosity measurements to be made over an extensive temperature range, typically $290\,\mathrm{K} \leq T \leq 700\,\mathrm{K}$, not only for the noble gases but also for a number of common molecular gases. It was calibrated using the value $\eta(T{=}298.15\,\mathrm{K}) = 22.552\,\mu\mathrm{Pa}$ s calculated (Vogel, 2010) using

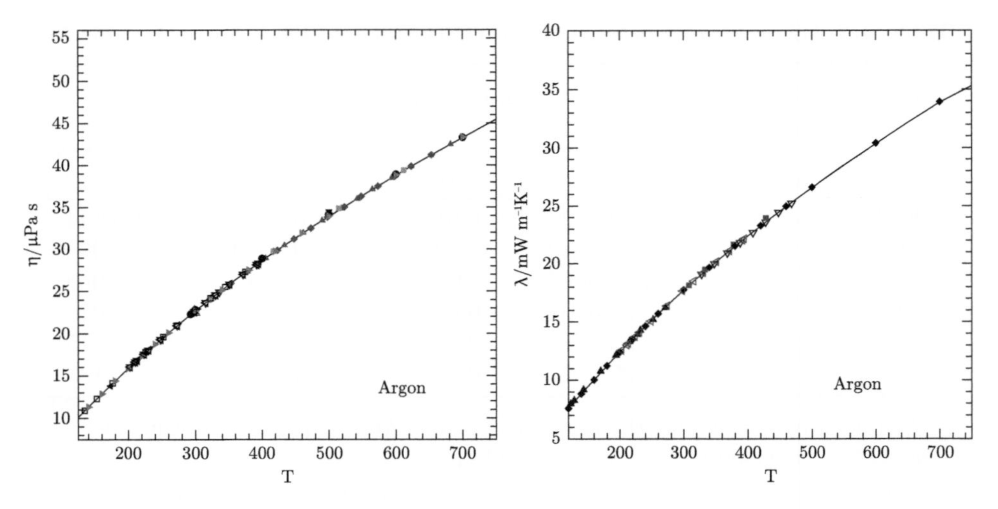

Fig. 4.7 Calculated[a] and experimental[b] temperature dependences of the shear viscosity, η (left panel), and thermal conductivity, λ (right panel), coefficients of gaseous argon in the temperature range $120\,\text{K} \leq T \leq 750\,\text{K}$.

[a] Vogel, 2010

[b] Viscosity data: ▶, Kestin *et al.*, 1966; □, Clarke and Smith, 1968; ◀, Gracki *et al.*, 1969; •, Dawe and Smith, 1970; ■ (green), Clifford *et al.*, 1975; ▼, Vargaftik and Vasilevskaya, 1984; ▼ (magenta), Berg and Burton, 2013; ▶ (magenta), Bich *et al.*, 1990; ▽, May *et al.*, 2006; ▲ (blue), Vogel, 2010; ◁, May *et al.*, 2007; △, Berg and Moldover, 2012; ♦ (violet), Lin *et al.*, 2014. Thermal conductivity data: ▽, Haarman, 1973; ▲ (grey), Shashkov *et al.*, 1978; ■ (blue), Millat *et al.*, 1987; ♦, Bich *et al.*, 1990; ◁ (red), May *et al.*, 2007.

the MTT1 *ab initio* Ar–Ar PEF. Song *et al.* (2016) showed that values of $\eta_{\text{A}}(T)$ and $\lambda_{\text{A}}(T)$ calculated using the MTT2 *ab initio* Ar–Ar PEF are indistinguishable from the curves shown in Fig. 2.11, as they typically differ from them by less than 0.1% over the entire temperature range shown. Indeed, the MTT1, MTT2 *ab initio*, and the MLR empirical Ar–Ar PEFs are very similar, as can be seen from the DRMSD values $\overline{dd} =$ 1.062, 1.337, 1.043, respectively, obtained from comparisons between computed and experimental values of a set of 2243 spectroscopic, equilibrium, and nonequilibrium thermophysical property data (Myatt *et al.*, 2018).

In order to avoid an overcrowding of experimental viscosity points in the left panel of Fig. 4.7, only the Series 2 viscosity results reported by Vogel (2010) have been included: the Series 1 and Series 3 data sets (a total of 34 values) are consistent with the Series 2 values. Finally, note that the argon viscosity values obtained by May *et al.* (2007) with a viscometer calibrated using the value $\eta_{\text{He}}(T{=}298.15\text{K})$ computed from the He–He *ab initio* PEF of Przybytek *et al.* (2010) and Cencek *et al.* (2012) are also consistent with the Vogel data, as they too lie within 0.1% of corresponding values computed (2010) from the MTT1 and MTT2 Ar–Ar *ab initio* PEFs.

It is clear from Fig. 5 of Ref. Vogel 2010 that the percentage differences between experimental values of the viscosity and corresponding values $\eta_{\text{AI}}(T)$ computed using an *ab initio* Ar–Ar PEF (Jäger *et al.*, 2010; Patkowski and Szalewicz, 2010) typically vary systematically from about +1% to −1% for temperatures above 400 K, while corresponding differences between viscosity values calculated from high-quality best-fit empirical PEFs (Aziz, 1993; Boyes, 1994; Myatt *et al.*, 2018) exhibit temperature behaviors consistent with (i.e. approximately parallel to) that of the MTT1 *ab initio* PEF. A more detailed comparison between much of the published experimental argon viscosity data and the corresponding calculated values can be found in Vogel (2010).

Figure 4.8 shows a comparison between argon viscosity data obtained for temperatures above 700 K and the computed (Vogel, 2010) temperature dependence of $\eta_{\text{Ar}}(T)$ for the temperature range 750 K $\leq T \leq$ 4000 K. Although a number of experimental groups have carried out viscosity measurements in the temperature range 180 K $\leq T \leq$ 750 K, only a much smaller number of groups have made such measurements in the temperature range 1000 K $\leq T \leq$ 2500 K, and essentially no reliable experimental measurements have been carried out for temperatures higher than 2500 K. Because a number of studies (Bich *et al.*, 1990; Kestin *et al.*, 1984; Ziebland, 1981) have employed "extended corresponding-states" (Kestin *et al.*, 1972*d*) correlations (see also Section 2.10.6) based upon experimental viscosity measurements to obtain consistent sets of values of $\eta_{\text{Ar}}(T)$ and/or $\lambda_{\text{Ar}}(T)$ over an extensive range of temperatures, including temperatures above 2500 K, a set of representative values obtained via these correlations have also been included in this figure.

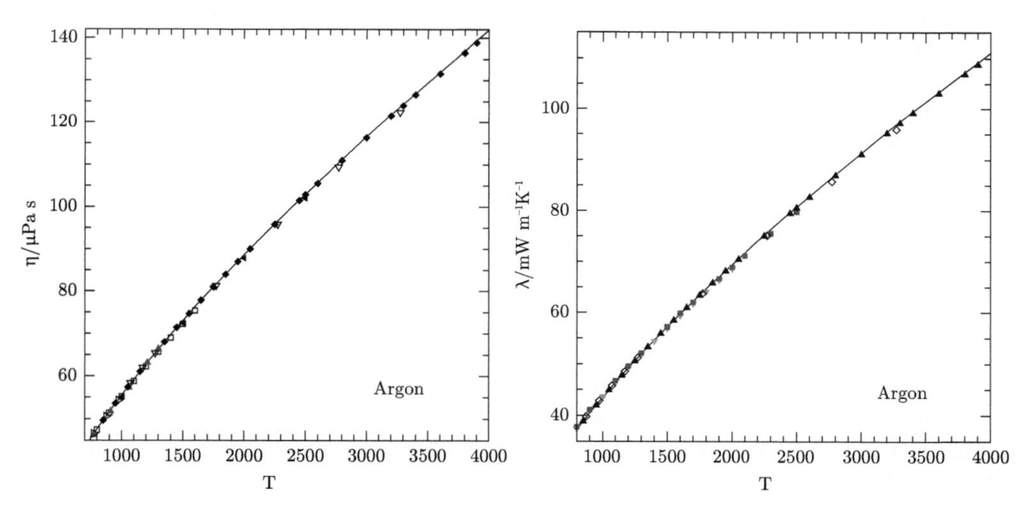

Fig. 4.8 Calculated[a] and experimental[b] temperature dependences of the shear viscosity, η (left panel), and thermal conductivity, λ (right panel), coefficients of gaseous argon in the temperature range 750 K $\leq T \leq$ 4000 K.

[a] Vogel, 2010

[b] Viscosity data: □, Dawe and Smith, 1970; ▷, Hellemans *et al.*, 1974; ▲ (blue), Clifford *et al.*, 1975;◀, Vargaftik and Vasilevskaya, 1984; ▽, Kestin *et al.*, 1984; ♦, Bich *et al.*, 1990. Thermal conductivty data: ■ (blue), Springer and Wingeier, 1973; ▼ (green), Ziebland, 1981; ◊, Kestin *et al.*, 1984; ▲ (grey), Bich *et al.*, 1990.

Transient hot-wire (THW) measurements have been found to be capable of providing relatively precise (uncertainties of the order of $\pm 0.3\%$) values of the thermal conductivity of a fluid at densities sufficiently below the critical density, as opposed to the more traditional coaxial cylinder/hot-wire or parallel plate geometry steady-state measurements (uncertainties of the order of $\pm 1\%$ to $\pm 3\%$) (Assael *et al.*, 1991; Le Neindre *et al.*, 1991). Note, however, that experimental uncertainties associated with THW measurements degrade from about $\pm 0.3\%$ at higher densities/pressures to about $\pm 2\%$ at lower pressures (1 bar or less), thereby complicating the determination of accurate experimental zero-density limiting values for the thermal conductivity via the THW method. Moreover, steady-state hot-wire measurements (as seen for argon, e.g. Springer and Wingeier 2000) typically have minimal experimental uncertainties (of the order of $\pm 2\%$) for pressures in the vicinity of 1 bar.

It appears that, apart from the measurements of Haarman (1973), Johns *et al.* (1986), and Li *et al.* (1994) in the ambient temperature range lying between about 270 K and 470 K, which have uncertainties of the order of $\pm 0.5\%$ or smaller and exhibit temperature dependencies that are approximately parallel to that obtained from calculations with the *ab initio* PEF(s), most of the published thermal conductivity data, especially for temperatures above or below the ambient temperature range, typically have associated uncertainties of the order of $\pm 1\%$ or more. A more detailed discussion of experimental thermal conductivity measurements and comparisons with calculations based upon the MTT1 Ar–Ar PEF can be found in Vogel (2010). Given the magnitudes of the experimental uncertainties inherent in the experimental determination of the thermal conductivity, it would likely be prudent to rely more on thermal conductivity values computed from the MTT1, MTT2, or MLR PEFs for temperatures that lie outside the ambient temperature range.

Figure 4.9 compares calculated and experimental behaviors of the "self-diffusion" coefficient,[4] $D_{\mathrm{AA}}(T)$, at pressure $p = 1.01325\,\mathrm{bar}$ (left panel) and the reduced isotopic thermal diffusion factor, $\alpha_0(T)$, (right panel) for argon. The computed values for $D_{\mathrm{AA}}(T)$ have been obtained at the level of the second Chapman–Cowling approximation, while values for $\alpha_0(T)$ have been obtained from the lowest-order nonvanishing Chapman–Cowling approximation to $\alpha_{\mathrm{T}}(T)$ given in eqn (2.77a). The second-order Chapman–Cowling approximation expression for the self-diffusion coefficient should suffice at this time, as experimental uncertainties for all diffusion coefficient measurements are typically of the order of ±2.5% to ±1.75%, while differences between results computed using the *ab initio* MTT (Jäger *et al.*, 2009, 2010; Patkowski and Szalewicz, 2010) or the empirical HFDID (Aziz, 1993; Boyes, 1994) and MLR (Myatt *et al.*, 2018) PEFs typically lie between ±0.02% and ±0.15%. Unfortunately, the situation is even more dramatic when thermal diffusion measurements are considered, as the experimental uncertainties for isotopic thermal diffusion results may even be as high as ±5% to ±10%. Note also that error bars have not been given for the correlation values of Kestin *et al.* (1973) as their assigned error bars of ±1.0% are approximately the same size as the symbols on the scale employed in this figure.

[4]Note that the values for $D_{\mathrm{AA}}(T)$ of Myatt *et al.* 2018 (listed in Table 8 and illustrated in Fig. A5) are actually for $100D_{\mathrm{AA}}(T)$ rather than for $D_{\mathrm{AA}}(T)$.

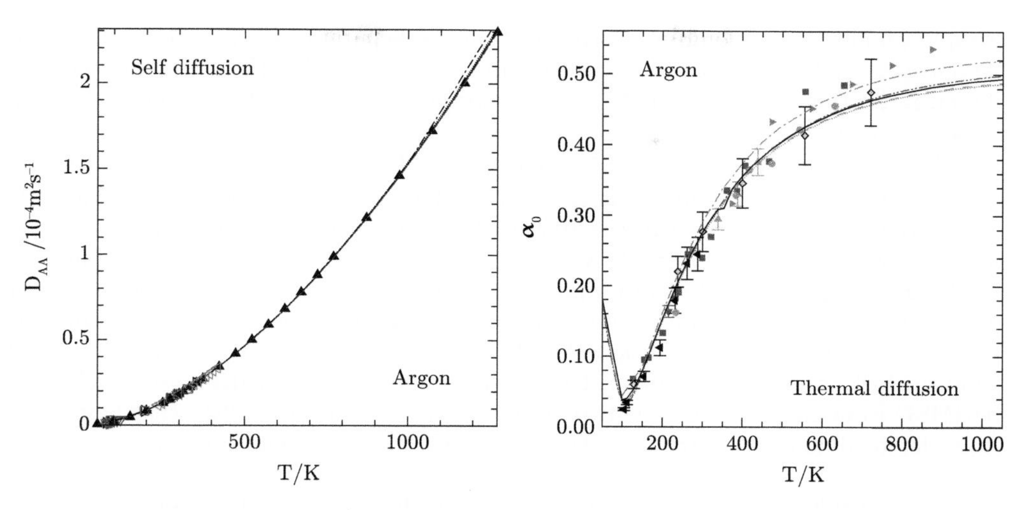

Fig. 4.9 Calculated[a] and experimental[b] temperature dependences of the self-diffusion coefficient, $D_{AA}(T)$, (left panel) and isotopic reduced thermal diffusion factor, α_0, (right panel) for argon in the temperature range $90\,\mathrm{K} \leq T \leq 1200\,\mathrm{K}$: the solid, dash-dotted, dotted, dashed, and dash-doubledotted curves represent calculations for the MTT1, MTT2, HFDID1, HFDID2, and MLR PEFs, respectively.

[a] Myatt *et al.*, 2018

[b] □ (red), Hutchinson, 1949; ▼ (blue), De Paz *et al.*, 1967; ▷, Winn, 1950; △ (pink), Bich *et al.*, 1990; □, Vugts *et al.*, 1969; ▲, Rutherford, 1973. Thermal diffusion data: ▼ (red), Taylor, 1975; ▶ (pink), Taylor and Weissman, 1973; ■ (blue), Paul *et al.*, 1963; ◀, Stevens and De Vries, 1968; • (green), Moran and Watson, 1958; ♦ (yellow), Stier, 1942.

4.1.4 The Kr–Kr interaction

The interaction energies between two ground electronic state krypton atoms separated by fixed distances have been computed *ab initio* for 25 internuclear separations (2.20Å $\leq R \leq$ 15.00Å) by Waldrop *et al.* (2015) via four-step CC calculations at the CCSD(T) level using high-level basis-sets, and including mid-bond hydrogenic functions. Counterpoise-corrected CBS limits for the interaction energies at the frozen-core CCSD(T) level were supplemented by corrections for core–core and core–valence electron correlations, CC excitations up to the CCSDT(Q) level and, finally, by a correction term accounting for relativistic effects. Similar calculations were also carried out by Jäger *et al.* (2016) using a high-quality sextuple-zeta level basis-set developed explicitly to improve the CBS convergence of the correlation energy, including corrections for CC contributions beyond the CCSD(T) level, for core–core and core–valence electron correlations, and for contributions associated with relativistic effects. The set of 25 interaction energies computed by Waldrop *et al.* (2015) were fitted to a MTT2 functional form using seven fit-parameters, while the 36 interaction energies computed by Jäger *et al.* (2016) were fitted to a MTT1 functional form using eight fit-parameters. Both the MTT1 and MTT2 analytic functional forms were spliced to simple (modified) Coulomb repulsive functions for separations typically smaller than about 1.8Å. Although a more detailed discussion of the Kr–Kr interaction can be found in Section 3.8.4, more detailed descriptions of the computations appear in the original papers.

Tanaka *et al.* (1973) were able to extract a set of nine vibrational spacings $\Delta G_{v+\frac{1}{2}}$ for the vibrational quantum numbers v = 0 to 8 for the $^{84}Kr_2$ isotopologue using VUV absorption spectra obtained in the 1050Å–1260Å wavelength range, while LaRocque *et al.* (1986) were able to obtain values for the first two vibrational level spacings, $\Delta G_{\frac{1}{2}}$ and $\Delta G_{\frac{3}{2}}$, for the $^{86}Kr_2$ isotopologue. Table 4.5 provides a comparison between the experimentally determined vibrational term-value differences and those computed using the MTT1, MTT2 *ab initio* PEFs (Waldrop *et al.*, 2015) and the HFD-B2, XC-3 empirical PEFs (Aziz and Slaman, 1989*a*; Dham *et al.*, 1989). As demonstrated by the DRMSD values obtained for the two *ab initio* and two empirical PEFs that have been considered, all four generate DRMSD values for the $^{84}Kr_2$ isotopologue that are smaller

Table 4.5 Term differences* $\Delta G_{v+\frac{1}{2}}$ for the $^{84}Kr_2$ and $^{86}Kr_2$ dimers

			$^{84}Kr_2$		
v	Experiment[a]	MTT1[b]	MTT2[c]	HFD-B2[d]	XC-3[e]
0	21.56 ± 0.54	21.455	21.466	21.41	21.42
1	19.09 ± 0.57	19.285	19.307	19.30	19.26
2	16.76 ± 0.60	17.133	17.166	17.20	17.11
3	14.76 ± 0.75	15.010	15.053	15.11	15.00
4	12.23 ± 0.51	12.930	12.980	13.02	12.94
5	10.49 ± 0.50	10.911	10.963	10.97	10.92
6	8.92 ± 0.44	8.976	9.026	9.01	8.98
7	6.92 ± 0.63	7.154	7.199	7.17	7.17
8	5.54 ± 0.30	5.481	5.517	5.49	5.50
$\overline{dd}(^{84}Kr_2)$:		0.6181	0.6765	0.7053	0.6515
			$^{86}Kr_2$		
v	Experiment[f]	MTT1[b]	MTT2[c]	HFD-B2[d]	XC-3[e]
0	21.175 ± 0.010	21.229	21.240	21.167	21.180
1	19.093 ± 0.020	19.109	19.131	19.073	19.071
$\overline{dd}(^{86}Kr_2)$:		3.8601	4.7885	0.9055	0.8544

* All term differences are given in cm^{-1} units.

[a] Tanaka *et al.*, 1973; [b] Jäger *et al.*, 2016; [c] Waldrop *et al.*, 2015; [d] Aziz and Slaman, 1989*a*; [e] Dham *et al.*, 1989; [f] LaRocque *et al.*, 1986.

than unity, but largely because of the large ($\overline{u} = \pm 4.73\%$) associated experimental uncertainties. Table 4.5 also shows that although the MTT1 *ab initio* PEF gives rise to the smallest DRMSD value for the nine term-value differences obtained by Tanaka *et al.* (1973) for the $^{84}Kr_2$ isotopologue, it also gives the largest DRMSD value for the more precise results of LaRocque *et al.* (1986) for the $^{86}Kr_2$ isotopologue.

A comparison between second virial coefficient values extracted by means of power series fits to the mass-density dependence of isothermal equation of state data sets and the temperature dependence of $B_2(T)$ computed using the Kr–Kr *ab initio* PEF (Jäger *et al.*, 2016) is illustrated in Fig. 4.10 for the temperature range $190\,\mathrm{K} \lesssim T \lesssim 650\,\mathrm{K}$. Second virial coefficient values for krypton determined prior to the year 2000 have been summarized by Dymond *et al.* (2002*a*) for the new series of Landolt–Börnstein tables. Figure 4.10 shows only a subset of error bars for the experimentally extracted values of $B_2(T)$ in order to avoid overcrowding. Values of $B_2(T)$ determined for temperatures lying below about 170 K, unlike values for $B_2(T)$ obtained above 200 K, deviate strongly from the computed temperature dependence, which leads to a suspicion that these values may be compromised by adsorption onto surfaces of the apparatus, as has been suggested by Jäger *et al.* (2016). As values of $B_2(T)$ computed from the *ab initio* PEF of Waldrop *et al.* (2015) or from the best empirical (Aziz and Slaman, 1989*a*; Dham *et al.*, 1989) PEFs are indistinguishable from the theoretical (solid) curve of Fig. 4.10, it is clear that the experimental uncertainties for the $B_2(T)$ values are too large to enable differentiation to be made among the four *ab initio* and empirical PEFs considered in Table 4.5.

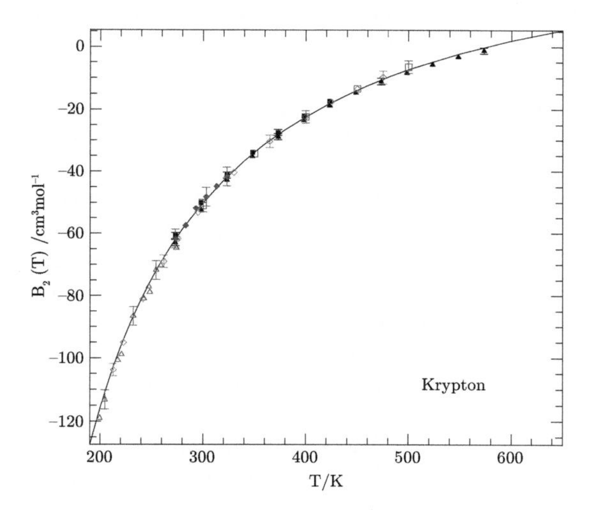

Fig. 4.10 Calculated[a] and experimental[b] temperature dependences of the second virial coefficient, $B_2(T)$, of krypton in the temperature range $190\,\mathrm{K} \leq T \leq 650\,\mathrm{K}$.

[a] Jäger *et al.*, 2016.

[b] Solid curve: Sharipov and Benites, 2017; ▲, Beattie *et al.*, 1970; ■, Trappeniers *et al.*, 1966; ♦ (red), Santafe *et al.*, 1976; □ (blue), Pérez *et al.*, 1980; ◇ (pink), Schmiedel *et al.*, 1980; △ (blue), Pollard, 1971.

The most accurate experimental determinations of the viscosity of krypton have been carried out at room temperature. In particular, researchers at the U.S. National Institute of Standards and Technology (NIST) have obtained viscosity values of 25.3062 μPa s (Berg and Moldover, 2012) and 25.2813 μPa s (Berg and Burton, 2013) at $T = 298.15$ K, each of which has uncertainty $\pm 0.0080\,\mu$Pa s. These two values differ from the value $25.279 \pm 0.020\,\mu$Pa s computed by Jäger *et al.* (2016) from their *ab initio* Kr–Kr PEF by +0.01%, and +0.009%, respectively, and from the value $25.275 \pm 0.025\,\mu$Pa s computed by Waldrop *et al.* (2015) from their *ab initio* PEF by 0.025% and 0.12%, respectively. More generally, viscosity measurements performed with viscometers operating in an absolute measurement mode (Trappeniers *et al.*, 1965; van den Berg, 1979; Evers *et al.*, 2002) often give values of η that lie within $\pm$ 0.3% of values computed from the *ab initio* Kr–Kr PEF over the temperature range 273–350 K. Many of the viscosity values η_{REF} employed prior to 2010 for the calibration of oscillating disk, (Kestin and Leidenfrost, 1959; Kestin *et al.*, 1972*c*: 5837; Vogel, 1984) vibrating wire, (Wilhelm and Vogel, 2000) or capillary (Clarke and Smith, 1968; Dawe and Smith, 1970; Maitland and Smith, 1974; Gough *et al.* 1976) viscometers operating in a relative mode have since been superseded. A re-evaluation of the Rostock data obtained with a quartz-glass oscillating disk viscometer (Vogel, 1984; Wilhelm and Vogel, 2000) using the $T = 298.15$ K value of η computed from the Ar–Ar *ab initio* PEF of Jäger *et al.* (2009, 2010) to calibrate the viscometer resulted in an experimental temperature dependence of the krypton viscosity data (Vogel, 2016) that is fully consistent with that shown by the viscosity computed from either of the two *ab initio* Kr–Kr PEFs Waldrop *et al.*, (2015) over the temperature range 295 K–690 K. Moreover, a similar re-evaluation Jäger *et al.*, (2016) of the other relative-mode viscosity data resulted in all experimental krypton data showing deviations no larger than $\pm$ 0.5% from the temperature dependence predicted by the *ab initio* PEF, with much of the data actually lying within $\pm$ 0.3% of the computed values. Plots of the percentage differences between viscosity values computed using the empirical Kr–Kr PEFs from Aziz and Slaman (1989*a*) and Dham *et al.* (1989) and the *ab initio* PEFs from Waldrop *et al.* (2015) and Jäger *et al.* (2016) show much the same trend with temperature, with differences of approximately +0.5% for temperatures between 100 K and 300 K, then rising steadily with increasing temperature to about 3.5% by $T = 2000$ K. The available viscosity data are thus sufficiently accurate to be able to discriminate between the "best" empirical PEFs and the *ab initio* PEFs, but are still not precise enough to discriminate between the *ab initio* PEFs from Waldrop *et al.* (2015) and Jäger *et al.* (2016).

Figure 4.11 illustrates the overall agreement between the calculated and experimental behaviors of the viscosity η (left panel) and thermal conductivity λ (right panel) of krypton in the temperature range $90\,\mathrm{K} \leq T \leq 800\,\mathrm{K}$. The solid curves represent values of η and λ obtained from classical calculations (Waldrop *et al.*, 2015) in which the (zero-density) transport coefficients have been evaluated using Chapman–Cowling fifth approximation expressions. Detailed comparisons between the computed temperature dependence of η (respectively, λ) obtained from the two *ab initio* PEFs and a number of higher-precision experimental results have been carried out (Waldrop *et al.*, 2015) using percentage differences $\delta_\eta \equiv 100(\eta_{\mathrm{expt}} - \eta_{\mathrm{calc}})/\eta_{\mathrm{calc}}$ (respectively, δ_λ) as

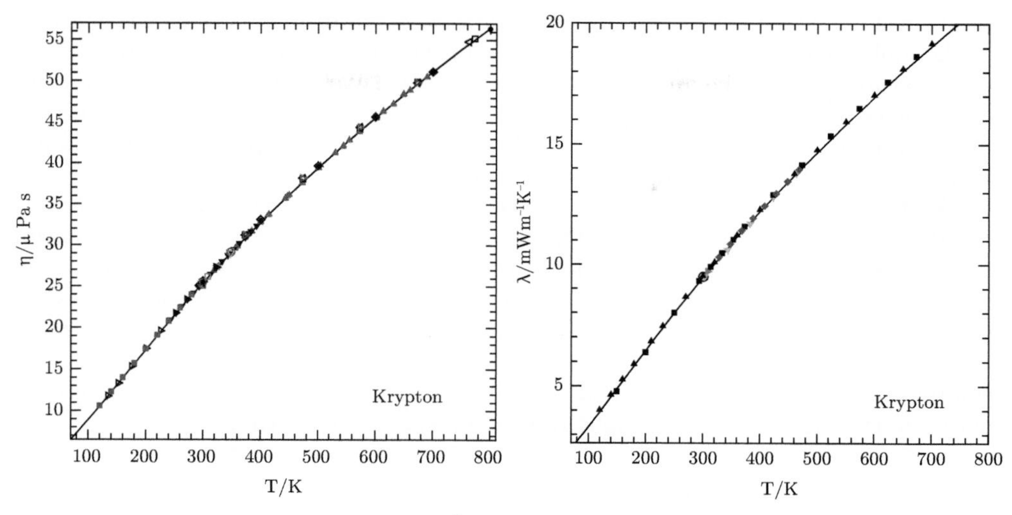

Fig. 4.11 Calculated[a] and experimental[b] temperature dependences of the shear viscosity, η (left panel), and thermal conductivity, λ (right panel), coefficients of gaseous krypton in the temperature range $80\,\mathrm{K} \leq T \leq 800\,\mathrm{K}$.

[a] Waldrop et al., 2015

[b] □, Trappeniers *et al.*, 1965; ►, Clarke and Smith, 1968; ◇, Dawe and Smith, 1970; □ (green), Kalelkar and Kestin, 1970; ◁ (orange), Kestin *et al.*, 1972*a*; ▷ (red), Kestin *et al.*, 1972*e*; ■ (red), Gough *et al.*, 1976; ▽ (green), van den Berg, 1979; ■ (magenta), Kestin *et al.*, 1980; ○ (purple), Wilhelm and Vogel, 2000; ◄ (blue), Berg and Moldover, 2012; ■ (blue), Berg and Burton, 2013; ▼ (blue), Lin *et al.*, 2016; ▲ (blue), Vogel, 2016; Maitland and Smith, 1974; ○, Kestin *et al.*, 1980; ► (magenta), Assael *et al.*, 1981; ▼ (green), Hemminger, 1987; ■, Kestin *et al.*, 1984; ▲, Bich *et al.*, 1990; ♦, Haarman, 1973; •, Snel *et al.*, 1979.

the metrics with which to determine the level of agreement between computed and experimental values. As the two sets of computed values of η and λ typically differ on average by approximately 0.08%, they will be indistinguishable on the scale of Fig. 4.11. Note, however, that although it is currently not difficult to determine η with an accuracy of the order of $\pm 0.5\%$ or better, it is actually quite difficult, except for temperatures in the vicinity of room temperature, to determine λ with an accuracy that exceeds $\pm 0.5\%$. For temperatures that lie either above or below the ambient temperature range, it is often difficult to determine λ with an accuracy that is smaller than $\pm 1\%$.

The right-hand panel of Fig. 4.11 illustrates the overall agreement between calculated and experimental temperature dependence of the thermal conductivity of krypton. THW measurements are widely considered to provide the most accurate and reliable thermal conductivity values, with uncertainties of order $\pm 0.2\%$ to $\pm 0.5\%$, in the ambient temperature range ($273.15\,\mathrm{K} \leq T \leq 323.15\,\mathrm{K}$). The uncertainties associated with THW measurements increase to the order of $\pm 1\%$ or more for temperatures either below or above the ambient temperature range. Similarly, thermal conductivity values obtained via hot-wire or concentric cylinder measurements under steady-state conditions are less reliable (typical uncertainties of order $\pm 1\%$) than THW values in the ambient temperature range and have uncertainties of order $\pm 2\%$ to $\pm 3\%$ for temperatures outside the ambient temperature range. We know that the temperature dependence of the resulting thermal conductivity data will not be useful for discriminating among the competing Kr–Kr PEFs. More detailed comparisons can be found in Waldrop *et al.*, (2015) and Jäger *et al.*, (2016).

As it is even more difficult to obtain accurate data for the self-diffusion coefficient, it will be clear that current self-diffusion data will not be useful for discriminating between competing Kr–Kr PEFs. Moreover, given that the existing data typically lie within $\pm 2\%$ of the values computed from the *ab initio* PEFs and that the values computed from the two *ab initio* PEFs agree within $\pm 0.06\%$ over the temperature range 100–2000 K, it will also be clear that self-diffusion coefficient values obtained from computations with the *ab initio* PEFs will be more reliable than either experimentally determined values or values computed using an empirical Kr–Kr PEF.

4.1.5 The Xe–Xe interaction

The interaction energies between a pair of ground electronic state xenon atoms separated by fixed distances have been computed *ab initio* for 39 internuclear separations ($2.40\text{Å} \leq R \leq 15.00\text{Å}$) by Hellmann *et al.* (2017) using counterpoise-corrected supermolecule CC calculations up to the CCSDTQ level. Calculations were carried out using a number of correlation-consistent nonrelativistic Gaussian basis-sets of up to sextuple-zeta quality consisting of basis elements having exponents that have been optimized via minimization of the Xe atom self-consistent field (SCF) energy. The atomic basis-sets were typically supplemented by an appropriate set of mid-bond functions in order to reduce basis-set incompleteness errors. As the SCF contribution to the supermolecule energy converges much more rapidly to the CBS limit than does the correlation energy contribution, the total interaction energy in the CBS limit could be approximated in terms of the sum of the extrapolated correlation contribution and the SCF contribution evaluated using the largest basis-set. The Xe–Xe interaction energy obtained by Hellmann *et al.* is represented as the sum of a nonrelativistic interaction energy $V_{\mathrm{CCSD(T)}}$ computed at the CCSD(T) level, its relativistic correction, and a nonrelativistic estimation of post-CCSD(T) corrections up to the CCSDTQ level.

The nonrelativistic interaction energy consists of four terms, namely, the SCF interaction energy, the frozen-core correlation contribution, a correction to account for core–core and core–valence correlation effects associated with the inclusion of $4s$, $4p$, and $4d$ electrons in the correlation treatment, and a final (and relatively smaller) correction term for the correlation contribution arising from the inclusion of all electrons in the correlation treatment. Relativistic corrections were carried out at the CCSD(T) level using second-order direct perturbation theory (Klopper, 1997; Kutzelnigg *et al.*, 1995) to compute the scalar relativistic correction within the frozen-core approximation and its correction for core–core and core–valence correlation effects. In addition, corrections to account for relativistic effects associated with spin-(own orbit) and spin-(other orbit) contributions were also made. Five additional post-CCSD(T) corrections were included, representing, respectively, the differences between energies computed at the CCSDT and CCSD(T) levels, the CCSDT(Q) and CCSDT levels, and the CCSDTQ and CCSDT(Q) levels within the frozen-core approximation, plus corrections for the inclusion of $4d$ electrons in the calculations of the correlation energy. This hierarchy of post-CCSD(T) corrections was illustrated by Hellmann *et al.* (2017) with a set of values obtained for a separation $R = 4.4$Å (the minimum occurs at $R_{\mathrm{e}} = 4.3780 \pm 0.0036$Å) showing that the CCSD(T) energy represents 98.90%, while the relativistic and post-CCSD(T) corrections represent 4.27% and −3.16% of the total energy, respectively. As the two corrections have opposite signs, the result is a net contribution of only 1.10% to the total attractive energy at a separation of 4.4 Å. The original paper (Hellmann *et al.*, 2017) provides a more detailed description of the computations. In addition, a more detailed discussion of the Xe–Xe interaction appears in Section 3.8.5.

In order to enable the evaluation of uncertainties for physical properties computed from their *ab initio* PEF, Hellmann *et al.* (2017) employed the uncertainty estimates $u(R)$ for the values of $V_{\mathrm{tot}}(R)$ obtained at each separation R to construct sets of values for a pair of perturbing PEFs, and have made available MTT1 fit-parameters for them. The MTT1 PEF fitted to the set of *ab initio* Xe–Xe interaction energies is characterized by the equilibrium distance $R_{\mathrm{e}} = (4.3780 \pm 0.0036)$Å and depth $\mathfrak{D}_{\mathrm{e}} = (194.60 \pm 1.79)\,\mathrm{cm}^{-1}$, with the uncertainties determined from the perturbed PEFs. This value for R_{e} is only 0.09% shorter than the *ab initio* value $R_{\mathrm{e}} = 4.3818$Å obtained by Hanni *et al.* (2004) and is in excellent agreement with the experimental value $R_{\mathrm{e}} = (4.3773 \pm 0.0049)$Å determined by Wüest *et al.* (2004) from values of the $\mathrm{v} = 0$ and $\mathrm{v} = 1$ rotational constants of the $^{129}\mathrm{Xe}^{132}\mathrm{Xe}$ isotopologue. Despite spacings $\Delta G_{\mathrm{v}+\frac{1}{2}}$ between adjacent pairs of vibrational levels being derived quantities, they may nonetheless provide a fairly rigorous test of the quality of a PEF. Although Freeman *et al.* (1974) were able to obtain a set of $\Delta G_{\mathrm{v}+\frac{1}{2}}$ values associated with the first ten vibrational levels of $\mathrm{Xe_2}$ dimers, they were unable to resolve individual contributions made by each of the 45 isotopologues formed by the nine stable xenon isotopes. These experimental vibrational spacings should thus be treated as averages over the set of naturally occurring isotopologues of xenon. Average vibrational spacings $\langle \Delta G_{\mathrm{v}+\frac{1}{2}} \rangle$ determined from the corresponding vibrational spacings computed for each member of the 45 $\mathrm{Xe_2}$ isotopologues according to the weighted average

$$\langle \Delta G_{\mathrm{v}+\frac{1}{2}} \rangle = \sum_i x_i^2 \Delta G_{\mathrm{v}+\frac{1}{2};i,i} + 2 \sum_{i<j} x_i x_j \Delta G_{\mathrm{v}+\frac{1}{2};i,j}\,,$$

in which x_i is the mole fraction of the stable xenon isotope i, agreed within the mutual uncertainties with the experimental results obtained by Freeman *et al.* (1974) for $\mathrm{v} = 1$ to $\mathrm{v} = 9$. While the computed value (Hellmann *et al.*, 2017) $\langle \Delta G_{\frac{1}{2}} \rangle = (19.31 \pm 0.08)\,\mathrm{cm}^{-1}$ disagreed significantly with the experimental value $\langle \Delta G_{\frac{1}{2}} \rangle = (19.90 \pm 0.3)\,\mathrm{cm}^{-1}$ of Freeman *et al.* (1974), it was, however, in excellent agreement with the experimental value $\langle \Delta G_{\frac{1}{2}} \rangle = (19.3485 \pm 0.0005)\,\mathrm{cm}^{-1}$ obtained by Wüest *et al.* (2004) based upon measurement of $\Delta G_{\mathrm{v}+\frac{1}{2}}$ for the $^{129}\mathrm{Xe}^{132}\mathrm{Xe}$ and $^{131}\mathrm{Xe}^{136}\mathrm{Xe}$ isotopologues. Values of the vibrational spacings for $\mathrm{v} = 0$ to $\mathrm{v} = 9$ obtained from the MTT1 $\mathrm{Xe_2}$ PEF and averaged over the set of $\mathrm{Xe_2}$ isotopologues were also consistent with values obtained in the same manner using the two best-fit empirical PEFs (Aziz and Slaman, 1986*a*) for the Xe–Xe interaction.

Experimental values for the xenon equation of state second (density) virial coefficient $B_2(T)$ published prior to 1995 have been summarized by Dymond *et al.* (2002*a*) for the New Series Landolt–Börnstein tables. The most recent data (Hurly *et al.*, 1997) were obtained using a Burnett apparatus designed to determine the equation of state, $\overline{\rho}(P,T)$, for the molar mass density of a fluid up to pressures of the order of 20 MPa over the temperature range $273\,\mathrm{K} \leq T \leq 473\,\mathrm{K}$. The precision with which equation of state data could be acquired with this Burnett apparatus was established indirectly from a comparison between measured values of the helium pressure and corresponding pressures calculated from the virial equation of state using second and third virial coefficient values computed from an *ab initio* He–He PEF (Aziz *et al.*, 1995): almost all of the measured helium pressures obtained using the Burnett apparatus were within $\pm\,0.01\%$ of the corresponding computed values. A comparison between 13 experimentally determined values of $B_2(T)$ ($210.00\,\mathrm{K} \leq T \leq 400.00\,\mathrm{K}$) for xenon obtained from the function

$$B_2(T) = 57.96870[1 - 3.86084(\mathrm{e}^{181.7638/T} - 1)]$$

fitted to the $B_2(T)$ values extracted from four isotherms at temperatures 258.15 K, 273.15 K, 288.15 K, and 303.15 K and the corresponding values obtained from the MTT1 *ab initio* PEF (Hellmann *et al.*, 2017) resulted in anrms percentage deviation of 0.11%, with a corresponding DRMSD ($\overline{dd}$) value of 0.835 when uncertainties of $\pm\,0.1\%$ are assigned to the $B_2(T)$ values given by the fit-function.

Values of $B_2(T)$ for xenon determined from experimental equation of state data (Beattie *et al.*, 1951; Hahn *et al.*, 1974; Michels *et al.*, 1954; Rentschler and Schram, 1977; Schramm *et al.*, 1977; Whalley *et al.*, 1955) in the temperature range $250\,\mathrm{K} \lesssim T \lesssim 650\,\mathrm{K}$ typically differ by less than $\pm\,2\%$ from the corresponding values obtained using the MTT1 PEF (Hellmann *et al.*, 2017) or either of the two best-fit empirical PEFs (Aziz and Slaman, 1986*a*; Dham *et al.*, 1990), and hence, show much the same temperature trend as that obtained theoretically for the same temperature range. However, the behavior of experimentally determined values of $B_2(T)$, and reflected by the correlation of Dymond *et al.* (2002*a*), for temperatures outside this range differs strongly and systematically from the theoretically predicted dependence.

Figure 4.12 illustrates the overall agreement between the calculated and experimental behaviors of the viscosity η (left panel) and thermal conductivity λ (right panel) of xenon in the temperature range $150\,\mathrm{K} \leq T \leq 800\,\mathrm{K}$. The solid curves represent the zero-density behaviors of η and λ obtained via classical Chapman–Cowling fifth approximation calculations (Hellmann *et al.*, 2017) using the MTT1 *ab initio* Xe–Xe PEF, while the symbols represent experimental values. However, as already noted, more detailed comparisons between the temperature dependences of η and λ predicted by a given PEF and observed experimentally requires the application of appropriate metrics, such as the percentage difference δ_λ for thermal conductivity (introduced in Table 2.3) and its equivalent, δ_η, for viscosity. Hellmann *et al.* (2017) used these metrics to examine the level of agreement between the behavior predicted by their MTT1 *ab initio* PEF and experiment.

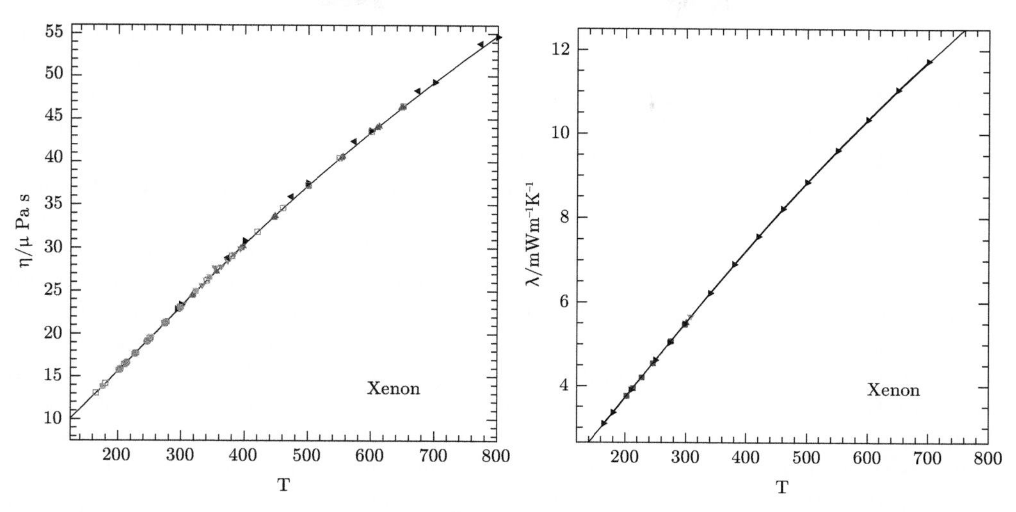

Fig. 4.12 Calculated[a] and experimental[b] temperature dependences of the shear viscosity, η (left panel), and thermal conductivity, λ (right panel), coefficients of gaseous xenon in the temperature range $80\,\mathrm{K} \leq T \leq 800\,\mathrm{K}$.

[a] Hellmann *et al.*, 2017.

[b] ■ (green), Clarke and Smith, 1968; ►, Dawe and Smith, 1970; ◄, Kestin *et al.*, 1972*e*; □ (magenta), Bich *et al.*, 1990; • (orange), May *et al.*, 2007; ▼, Berg and Moldover, 2012; ♦ (blue), Berg and Burton, 2013; ▼ (magenta), Lin *et al.*, 2016; ▲ (blue), Vogel, 2016; ▲ (red), Kestin *et al.*, 1980; ▼ (green), Assael *et al.*, 1981; ■ (blue), May *et al.*, 2007; ►, Bich *et al.*, 1990.

The most accurate experimental determinations of the viscosity and thermal conductivity of xenon have been carried out in the vicinity of room temperature ($T = 298.15\,\mathrm{K}$). In particular, researchers at NIST have reported very accurate viscosity values, $\eta = 23.0183\,\mu\mathrm{Pa\,s}$ (Berg and Moldover, 2012) and $\eta = 23.0162\,\mu\mathrm{Pa\,s}$ (Berg and Burton, 2013), each of which has an uncertainty $u = \pm\,0.0072\,\mu\mathrm{Pa\,s}$, for temperature $T = 298.15\,\mathrm{K}$: these values differ by -0.045% and -0.056%, respectively, from the value $\eta = 23.029 \pm 0.27\,\mu\mathrm{Pa\,s}$ computed by Hellmann *et al.* (2017) for the MTT1 Xe–Xe PEF. Moreover, viscosity values computed using the empirical best-fit HFD-B PEF (Aziz and Slaman, 1986*a*), the XC-3 (Dham *et al.*, 1990) PEF, or the earlier *ab initio* PEF of Hanni *et al.* (2004), give values of $\eta(298.15\,\mathrm{K})$ that differ from $\eta_{\mathrm{MTT1}}(298.15\,\mathrm{K})$ by $+0.26\%$, $+0.34\%$, and -0.67%, respectively (Hellmann *et al.*, 2017). The xenon viscosity measurements of May *et al.* (2007) and of Lin *et al.* (2016) together span the temperature range $200\,\mathrm{K} \leq T \leq 400\,\mathrm{K}$, and agree with the viscosity values computed from the MTT1 PEF to better than $\pm\,0.1\,\%$. Similarly, the viscosity values reported by Vogel (2016) for the temperature range $300\,\mathrm{K} \leq T \leq 650\,\mathrm{K}$ typically agree with the theoretical viscosity values to better than $\pm\,0.3\%$. In comparison, the temperature behavior of the viscosity values computed from the HDF-B best-fit empirical Xe–Xe PEF (Aziz and Slaman, 1986*a*) approximately mirrors that of the MTT1 PEF, but gives δ_η values that lie roughly 0.35% above those obtained for the MTT1 PEF, while the XC-3 empirical PEF (Dham *et al.*, 1990) typically gives δ_η values that exceed those obtained using the MTT1 PEF by at least 0.5%, and values of δ_η obtained from the earlier *ab initio* PEF of Hanni *et al.* (2004) are generally more than 0.5% smaller than corresponding values obtained using the MTT1 PEF. Finally, note that many of the early viscosity measurements either differ from the computed MTT1 values by amounts well in excess of $\pm 0.2\%$, exhibit temperature dependences that depart significantly from that shown by the MTT1 PEF, or, in some cases, both.

Apart from two thermal conductivity measurements carried out at or near room temperature by Kestin *et al.* (1980) and Assael *et al.* (1981) using the transient hot-wire technique, with reported experimental uncertainties of only $\pm 0.3\%$ and $\pm 0.2\%$, respectively, all other thermal conductivity measurements typically have experimental uncertainties of $\pm 1.0\%$ or larger. Moreover, because the traditional transport coefficients η and λ are strongly dominated by their first Chapman–Cowling approximations, with all higher approximations contributing only rather small changes to the first approximation values, and because the first approximation, $[\lambda]_1$, to the thermal conductivity is given in terms of $[\eta]_1$, via eqns (2.70b, 2.75a) and eqn (2.133), by

$$[\lambda]_1(T) = \frac{15k_\mathrm{B}}{4m}\,[\eta]_1\,,$$

thermal conductivity data should, in general, be most useful in confirming the conclusions obtained from the viscosity data. Of the four PEFs (two *ab initio* PEFs and two best-fit empirical PEFs), only the MTT1 *ab initio* PEF was found by Hellmann *et al.* (2017) to give values lying within the relatively tight error bounds of both measurements.

On the scale employed in the right-hand panel of Fig. 4.12, the level of agreement between the experimental thermal conductivity data and the calculated temperature dependence of λ appears to be quite satisfactory. However, even a cursory examination of the behavior of the percentage differences δ_λ between the measured and calculated values of λ (Figure 3, Hellmann *et al.*, 2017) demonstrates that such data (except for the accurate data obtained near room temperature) are not sufficiently accurate to aid in differentiating between various proposed PEFs. The same can be said of self-diffusion data, which are inherently difficult to obtain, let alone with high accuracy.

4.2 Correlation Concept

It is known from statistical thermodynamics (see, for example, Section 7.9 in Kestin and Dorfman, 1971) that, when the interaction energy $V(R)$ between constituent particles of a substance separated by a distance R can be represented in terms of a function

$$V(R) = \varepsilon f(R/\sigma)$$

characterized by a pair of scale parameters, ε and σ, whose values differ from system to system, equilibrium thermodynamic quantities, such as the Helmholtz energy A, the second virial coefficient B, or the pressure P, can be expressed in reduced forms $A^*(T^*, \overline{V}^*)$, $B^*(T^*)$, or $P^* = T^*/\overline{V}^*$ as universal functions of the reduced temperature

$T^* \equiv k_B T/\varepsilon$ and the reduced molar volume $\overline{V}^* \equiv \overline{V}/(N_0\sigma^3)$. In most cases, the scale parameters will be similar, but not necessarily equal, to the depth $\mathfrak{D}_e$ and position of the nonasymptotic-zero, σ, of the binary interaction between the constituents of a particular system. This same concept can be extended to more general functional forms $V(R) = \varepsilon f(R/\sigma, \{\alpha_i\})$ that may include, for example, a set of dimensionless parameters having forms $C^*_{2n} \equiv C_{2n}/(\mathfrak{D}_e\sigma^{2n})$, that refines the shape of the attractive well of the PEF, provided that the set of PEFs generated thereby can be brought into congruence using only the two scaling factors, ε and σ.

Kestin *et al.* (1972*b*) and Naja *et al.* (1983) proposed an extended law of corresponding states that aims to unify both equilibrium and nonequilibrium (transport) properties of gases, effectively by employing the corresponding states concept to define characteristic quantities that enable construction of correlations of gas transport properties. A relevant correlation for the viscosity η may be obtained from eqn (2.74d), namely,

$$[\eta]_n = f_\eta^{[n]}[\eta]_1 = f_\eta^{[n]} \frac{k_B T}{\overline{c}_r \mathfrak{S}(20)},$$

giving the nth Chapman–Cowling approximation to η. Should $[\eta]_n$ be considered to provide a good approximation to the experimentally measured viscosity $\eta(T)$, the ratio

$$\frac{\mathfrak{S}(20)}{f_\eta^{[n]}(T)} = \frac{k_B T}{\overline{c}_r \eta(T)}$$

may then be considered to represent the temperature dependence of a functional $\mathfrak{S}(20)/f_\eta^{[n]}$ of $V(R)$. Division of this expression by the value $\mathfrak{S}_{rs}(20) = 4\pi\sigma^2/5$ for rigid spheres of diameter σ then gives

$$\frac{\mathfrak{S}^\star(20)}{f_\eta^{[n]}(T)} = \frac{5 k_B T}{4\overline{c}_r \pi\sigma^2 \eta(T)},$$

with $\mathfrak{S}^\star(20)$ defined as the ratio

$$\mathfrak{S}^\star(20) \equiv \frac{\mathfrak{S}(20)}{\mathfrak{S}_{rs}(20)} \tag{4.9a}$$

of the value of the effective cross-section $\mathfrak{S}(20)$ for a gas whose constituent atoms/molecules interact according to a realistic PEF to that of a gas of rigid spheres of diameter σ. A further substitution of $\overline{c}_r \equiv [8k_B T/(\pi m_r)]^{\frac{1}{2}}$ then gives (Bich *et al.*, 1990; Hirschfelder *et al.*, 1964; Kestin *et al.*, 1972*b*; Najafi *et al.*, 1983)

$$\Omega_{22}(T^*; \{\alpha_i\}) \equiv \frac{\mathfrak{S}^\star(20)}{f_\eta^{[n]}(T)} = \frac{5}{16}\left(\frac{m k_B T}{\pi}\right)^{\frac{1}{2}} \frac{1}{\sigma^2 \eta(T)}, \tag{4.9b}$$

with $T^* \equiv k_B T/\mathfrak{D}_e$ the reduced temperature. This functional of $V(R)$ is traditionally referred to as a viscosity correlation.

Figure 4.13 illustrates the form taken by the viscosity correlation $\Omega_{22}(T^*; \{\alpha_i\})$. In particular, the left-hand panel shows the behavior obtained by employing a set of

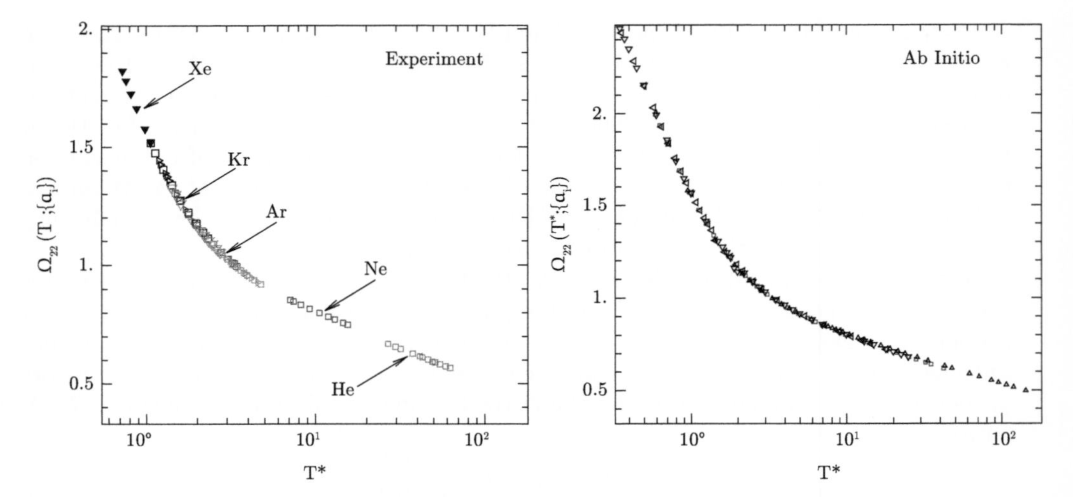

Fig. 4.13 Universal viscosity correlation for the noble gases. The left panel shows the correlation obtained using experimentally determined viscosities: □ (magenta, blue), Vogel, private communication, 2018; □ (green), Vogel, 2010; □ (black, red), Vogel, 2016; ▷ (red, black), Lin *et al.*, 2014; ▽ (green), ▼, May *et al.*, 2007, while the right panel gives the correlation obtained from viscosity values computed from the *ab initio* PEFs for Ne[a], Ar[b], Kr[c], and Xe[d].

[a]Hellmann *et al.*, 2008*a*; [b]Jäger *et al.*, 2009; [c] Waldrop *et al.*, 2015; Jäger *et al.*, 2016; [d]Hellmann *et al.*, 2017.

relatively recent highly accurate viscosity measurements for neon, (Hellmann *et al.*, 2007) argon (Vogel, 2010), krypton (Vogel, 2016), and xenon (Vogel, 2016) in conjunction with eqn (4.9b). Appropriate values for the reduced temperature have been obtained via the relation $T^* \equiv k_B T/\mathfrak{D}_e$, with $\mathfrak{D}_e$ the depth of the relevant *ab initio* pure gas PEF. Corresponding rigid-sphere diameters σ_{rs} have been equated with the separations σ associated with the finite-R zeroes, $V(\sigma) = 0$, of the *ab initio* pure gas PEFs. The solid curve shown has been generated from eqn (4.9b) using argon viscosities (Vogel, 2010) for $0.58 \leq T^* \leq 9.80$ and neon viscosities (Bich *et al.*, 2008) for $10.4 \leq T^* \leq 240$ calculated from the respective (Bich *et al.*, 2008; Jäger *et al.*, 2009, 2010; Patkowski and Szalewicz, 2010) Ar–Ar and Ne–Ne *ab initio* PEFs using fifth Chapmann–Cowling approximation expressions. The viscosity correlation displayed in the right-hand panel is a fully *ab initio* correlation based upon sets of viscosity values (Hellmann *et al.*, 2008; Vogel *et al.*, 2010; Jäger *et al.*, 2016; Hellmann *et al.*, 2017) that have been obtained from PEFs fitted to high-quality *ab initio* quantum chemical computations for the Ne–Ne (Hellmann *et al.*, 2008*a*), Ar–Ar (Jäger *et al.*, 2009, 2010), Kr–Kr (Jäger *et al.*, 2016), and Xe–Xe (Hellmann *et al.*, 2017) interaction energies.

Because the law of corresponding states is primarily based upon thermodynamic reasoning, it may be considered to be an essentially classical mechanical concept. This reasoning is supported by the comparison given in the left-hand panel of Fig. 4.13

between the curve generated from the *ab initio* PEFs and sets of experimental viscosity values for xenon, krypton, argon, and neon. Very good agreement is obtained between theory and experiment for xenon, krypton, argon, and to a lesser extent, neon. No helium data have been employed in obtaining the correlation shown in Fig. 4.13 as the essentially quantum mechanical nature of He effectively precludes its use in the construction of such a correlation unless a number of (nontrivial) quantum corrections are made. An illustration of the level of deviation from the classical correlation that can occur for a quantum system is provided in the left-hand panel of Fig. 4.13 by a group of values of $\Omega_{22}(T^*;\{\alpha_i\})$ generated from a set of very accurate helium viscosity data (Hellmann *et al.*, 2007) lying in the reduced temperature interval $T^* \simeq 15$ and $T^* \simeq 75$. Note that these experimental values of η are in excellent agreement ($\pm\,0.1\%$) with values obtained from fully quantum mechanical scattering calculations based upon an *ab initio* He–He PEF (Hellmann *et al.*, 2007). As pointed out by Najafi *et al.* (1983), these deviations for helium do not result from a failure of two-parameter congruence for the He–He interaction, but arise because He–He collisions are governed by the laws of quantum, rather than classical, mechanics. For $T^* \lesssim 20$, this problem may be overcome in principle by the introduction of the atomic mass via the reduced de Broglie wavelength $\Lambda^* \equiv h/(\sqrt{m\varepsilon})$. More complex corrections, whose natures are not yet well understood, would be required (Kestin *et al.*, 1972*b*; Najafi *et al.*, 1983) for high values of T^*.

Upon following the same line of reasoning employed to obtain the viscosity correlation $\Omega_{22}(T^*,\{\alpha_i\}) \equiv \mathfrak{S}^\star(20)/f_\eta^{[n]}$ via eqn (2.70a), a thermal conductivity correlation $\mathfrak{S}^\star(11)/f_\lambda^{[n]}$ may be obtained from eqn (2.75a) as

$$\frac{\mathfrak{S}^\star(11)}{f_\lambda^{[n]}} = \frac{75k_\mathrm{B}}{64}\left(\frac{k_\mathrm{B}T}{\pi m}\right)^{\frac{1}{2}}\frac{1}{\sigma^2\lambda(T)}\,.$$

In principle, this should enable an independent correlation to be obtained from thermal conductivity measurements. However, the generalized cross-sections $\mathfrak{S}(20)$ and $\mathfrak{S}(11)$ for atomic interactions are proportional to one another and to the omega integral $\Omega^{(2,2)}(T)$ via

$$\mathfrak{S}(11) = \tfrac{2}{3}\mathfrak{S}(20) = \frac{16}{15\bar{c}_\mathrm{r}}\Omega^{(2,2)}(T)$$

and, as a consequence, $\mathfrak{S}^\star(11)$ and $\mathfrak{S}^\star(20)$ are equal, given by

$$\mathfrak{S}^\star(11) = \mathfrak{S}^\star(20) = \frac{\Omega^{(2,2)}(T)}{\Omega^{(2,2)}_\mathrm{hs}(T)} = \frac{2\Omega^{(2,2)}(T)}{\pi\sigma^2}\,.$$

In practice, the natural logarithm $\ln\,\Omega_{22}(T^*;\{\alpha_i\})$ of the functional $\Omega_{22}(T^*;\{\alpha_i\})$ should be fitted to a power series in $\ln T^*$ in order for it to be employed to determine viscosities from the expression

$$\eta(T^*;\{\alpha_i\}) = \frac{5}{16}\left(\frac{mk_\mathrm{B}T}{\pi}\right)^{\frac{1}{2}}\frac{1}{\sigma^2\Omega_{22}(T^*;\{\alpha_i\})}\,. \tag{4.10a}$$

Because $\mathfrak{S}^{\star}(11) = \mathfrak{S}^{\star}(20)$, the thermal conductivity $\lambda(T^*, \{\alpha_i\})$ may also be obtained in terms of the viscosity correlation $\Omega_{22}(T^*, \{\alpha_i\})$ as

$$\lambda(T^*; \{\alpha_i\}) = \frac{75k_{\rm B}}{64}\left(\frac{k_{\rm B}T}{\pi m}\right)^{\frac{1}{2}} \frac{1}{\sigma^2\Omega_{22}(T^*; \{\alpha_i\})} \frac{f_{\lambda}^{[n]}(T)}{f_{\eta}^{[n]}(T)}. \tag{4.10b}$$

Because both $\eta(T^*; \{\alpha_i\})$ and $\lambda(T^*; \{\alpha_i\})$ can be obtained from $\Omega_{22}(T^*; \{\alpha_i\})$, sets of experimental thermal conductivity measurements will likely serve more as consistency checks on the viscosity measurements (or, in principle, vice versa), rather than provide an independent correlation. However, as it is also well known that it is extremely difficult to reduce the experimental error in a typical thermal conductivity measurement significantly below 1% (apart from the room temperature regime, for which it may perhaps be reduced to $\pm 0.2\,\%$) while viscosity measurements can be obtained with experimental uncertainties as small as $\pm 0.03\,\%$ (Berg and Burton, 2013), in the ambient temperature regime, and of the order of $\pm 0.2\%$ for temperatures outside the ambient regime, it would likely be more productive to employ the viscosity correlation to calculate values of the thermal conductivity, especially for temperatures either significantly above or below the room temperature regime, rather than to attempt to determine them experimentally.

Table 4.6 compares values of the viscosities and thermal conductivities of room temperature ($T = 298.15\,$K) noble gases obtained by direct measurement with room temperature viscosities derived from two correlations and obtained by calculations from *ab initio* PEFs. The comparison set includes very accurate experimental (uncertainties of the order of $\pm 0.02\%$) room temperature viscosities (Berg and Burton, 2013) obtained for each of the five noble gases using a single-capillary viscometer (May *et al.*, 2006, 2007) calibrated to the room temperature viscosity of ^{4}He calculated (Cencek *et al.*, 2012) from an *ab initio* He–He PEF (Przybytek *et al.*, 2010), room temperature viscosities calculated from *ab initio* PEFs for the Ne–Ne (Hellmann *et al.*, 2008*a*), Ar–Ar (Jäger *et al.*, 2009, 2010), Kr–Kr (Jäger *et al.*, 2016), and Xe–Xe (Hellmann *et al.*, 2017) interactions, and values obtained from two correlations, one (Kestin *et al.*, 1984) (labeled KK) based solely upon experimental results, the other (Bich *et al.*, 1990) (labeled BMV) based upon a set of near-conformal empirical PEFs of the HFD-B type (Aziz and Slaman, 1986*a*, 1989*b*; Aziz *et al.*, 1987). The theoretical value of the room temperature viscosity of ^{4}He has been employed to calibrate the viscometer as it can be calculated with an absolute uncertainty ($\pm 0.001\%$) that is more than ten times smaller than the corresponding experimental uncertainty.

The top portion of Table 4.6 contains a comparison between a set of very accurate experimental values obtained for the room temperature viscosities of the five noble gases, a set of values calculated from *ab initio* PEFs, and sets of values obtained from two correlations: the metric employed for these comparisons is the percentage difference $\delta_{\rm X} \equiv 100(\eta_{\rm exp} - \eta_{\rm X})/\eta_{\rm X}$, with X = AI, BMV, KK. As there are no thermal conductivity measurements for any of the rare gases at $T = 298.15\,$K, and as the room temperature values of η calculated from the *ab initio* PEFs are all in excellent agreement with the recommended experimental values for η, values of the thermal conductivity calculated using the *ab initio* PEFs have been employed in the bottom portion of Table 2.8 for comparison with thermal conductivity values generated from the viscosity correlations

Table 4.6 Comparisons among recommended experimental,[a] calculated,[b] and correlation values[c] of the viscosity* and thermal conductivity* of room temperature ($T = 298.15\,\text{K}$) noble gases

	η_{exp}	$\delta\eta_{\text{AI}}$	η_{AI}	η_{BMV}	$\delta\eta_{\text{BMV}}$	η_{KK}	$\delta\eta_{\text{KK}}$
He	19.8253†	−0.003	19.826	19.84	−0.071	19.73	0.483
Ne	31.6990	−0.091	31.728	31.79	−0.286	31.60	0.313
Ar	22.5539	0.008	22.552	22.59	−0.160	22.39	0.732
Kr	25.2813	0.009	25.279	25.39	−0.428	25.02	1.044
Xe	23.0162	−0.056	23.029	23.09	−0.320	22.62	1.751
			λ_{AI}	λ_{BMV}	$\delta\lambda_{\text{BMV}}$	λ_{KK}	$\delta\lambda_{\text{KK}}$
He			155.01	155.0	0.006	154.23	0.506
Ne			49.203	49.30	−0.197	49.00	0.414
Ar			17.617	17.65	−0.187	17.49	0.726
Kr			9.4082	9.449	−0.438	9.32	0.947
Xe			5.4691	5.483	−0.254	5.37	1.845

* Units: $\eta/\mu\text{Pa s}$, $\lambda/\text{mWm}^{-1}\,\text{s}^{-1}$, $\delta/\%$.

† Fixed at the $T = 298.15\,\text{K}$ *ab initio* value obtained in Cencek *et al.* (2012).

[a] Berg and Burton, 2013.

[b] Bich *et al.*, 2007; Bich *et al.*, 2008; Vogel *et al.*, 2010; Jäger *et al.*, 2016; Hellmann *et al.*, 2017.

[c] Bich *et al.*, 1990; Kestin *et al.*, 1984.

of Bich *et al.* (1990) and Kestin *et al.* (1984). The table shows that the BMV correlation consistently gives closer agreement with the calculated thermal conductivity values than does the KK correlation.

Table 4.7 illustrates the level of agreement obtained over an extensive range of temperatures (roughly, temperatures of 90–900 K) between the theory-based BMV noble gas correlation and the zero-density values of η and λ calculated from high-level *ab initio* PEFs for the five noble gases. The DRMSD values $\overline{dd}_\eta$ and $\overline{dd}_\lambda$, based upon approximately 30 representative temperatures spanning each temperature range, indicate that the BMV correlation can reasonably be relied upon to give viscosity and thermal conductivity values that lie within $\pm\,0.2\%$, respectively, $\pm\,0.5\%$, $\pm\,0.3\%$ of the calculated *ab initio* values of η and λ for Ne and Ar, Kr, Xe.

Given that the interaction between a pair of He atoms is well characterized (Cencek *et al.*, 2012; Przybytek *et al.*, 2010) and the transport properties of both ^{4}He and ^{3}He can be calculated with accuracies of the order of 0.005% or better, there seems to be little need for any future helium correlations. This would be especially the case were fit-functions giving the calculated temperature dependence of the various transport coefficients of helium (accurate to $\pm\,0.02\%$ or so) for ^{4}He, ^{3}He, and perhaps an equimolar ^{3}He–^{4}He mixture, available. Moreover, given the accuracy with which

Table 4.7 Comparisons among recommended experimental,[a] calculated,[a] and correlation values[b] of the viscosity* and thermal conductivity* of room temperature ($T = 298.15\,\mathrm{K}$) noble gases

	$\pm u_\eta{}^\dagger$	$\overline{dd}_\eta$	δ_η	$\pm u_\lambda{}^\dagger$	$\overline{dd}_\lambda$	δ_λ	range
Ne	0.20	0.796	0.160	0.20	0.715	0.143	$90 \le T \le 900$
Ar	0.20	0.751	0.205	0.20	0.774	0.154	$90 \le T \le 900$
Kr	0.50	0.872	0.438	0.50	0.866	0.435	$120 \le T \le 900$
Xe	0.30	0.879	0.264	0.30	0.884	0.266	$170 \le T \le 900$

* RMS percentage differences given by $\delta_X \equiv 100(X_{\mathrm{BMV}} - X_{\mathrm{AI}})/X_{\mathrm{AI}}$, $X = \eta, \lambda$.
† Percentage uncertainty associated with the BMV correlation value.
[a] Bich *et al.*, 2007; Bich *et al.*, 2008; Vogel, 2010; Jäger *et al.*, 2016; Hellmann *et al.*, 2017.
[b] Bich *et al.*, 1990.

noble gas interaction energies may now be computed, there does not seem to be much to recommend further attempts to construct additional experimentally based correlations for the noble gases, again provided that appropriate fit-functions be made available.

4.3 Binary Mixtures of Noble Gases

It is well known that highly accurate equation of state measurements of compressibility isotherms enable the extraction of a set of virial coefficients from the experimental data. For a pure nonideal gas the set of second virial coefficient data, $B_2(T)$, may then in principle be inverted to obtain the interaction PEF, $V(R)$, for the gas (Maitland and Smith, 1972). Similarly, it is also possible to employ equation of state measurements of compressibility isotherms for binary gas mixtures to extract a set of mixture virial coefficients. For a binary mixture of gases A and B, the mixture second virial coefficient, $B_{2,\mathrm{mix}}(T)$, thereby obtained is given in terms of the two pure gas second virial coefficients $B_{2,\mathrm{A}}(T)$, $B_{2,\mathrm{B}}(T)$ and the interaction second virial coefficient $B_{2,\mathrm{AB}}(T)$ by

$$B_{2,\mathrm{mix}}(T, x_\mathrm{A}, x_\mathrm{B}) = x_\mathrm{A}^2 B_{2,\mathrm{A}}(T) + 2x_\mathrm{A}x_\mathrm{B}B_{2,\mathrm{AB}}(T) + x_\mathrm{B}^2 B_{2,\mathrm{B}}(T)\,, \tag{4.11}$$

with x_A, x_B the mole fractions of A and B in the binary mixture. Thus, should reliable values of $B_{2,\mathrm{A}}(T)$ and $B_{2,\mathrm{B}}(T)$ be available either from accurate calculations based upon high-level *ab initio* homo-atomic PEFs or from accurate experimental pure gas measurements, then equally accurate equation of state binary mixture measurements may directly serve as important sources of data for the extraction of the interaction energies between pairs of unlike atoms (or molecules). The essential issue for equation of state measurements is thus the ability to carry out very accurate measurements of the relevant experimental data. A similar statement should also apply to speed of sound measurements and the extraction of acoustic second virial coefficients, $\beta_{\mathrm{a,mix}}(T)$.

The situation is somewhat more complex for transport properties. To begin with, as shown via the relevant expressions in Sections 2.4.3 and 2.4.4, not only do the traditional transport properties, namely the viscosity and the thermal conductivity, of a binary gas mixture vary with mixture composition in the same fashion as the mixture second (pressure) virial coefficient of eqn (4.11) but also, because the first Chapman–Cowling expression is not exact, the composition dependence of $\eta_{\mathrm{mix}}(T, x_{\mathrm{A}})$, or of $\lambda_{\mathrm{mix}}(T, x_{\mathrm{A}})$, is further complicated by contributions from higher Chapman–Cowling approximations. This additional embedding of the composition dependence makes disentanglement of the experimental data without a significant loss of accuracy essentially impossible. In this context, recall that the rather complicated expressions for the interaction viscosity $\eta_{\mathrm{AB}}(T)$ and the interaction thermal conductivity $\lambda_{\mathrm{AB}}(T)$ given by Hirschfelder *et al* (1964), for example, have been obtained from first Chapman–Cowling approximations to $\eta_{\mathrm{mix}}(T, x_{\mathrm{A}})$ and $\lambda_{\mathrm{mix}}(T, x_{\mathrm{A}})$. Thus, these expressions should not be used for the analysis of precision transport coefficient data obtained for binary mixtures. Although it is true that the Chapman–Cowling approximations for these transport coefficients tend to converge fairly rapidly, with the second approximation typically contributing between 0.5% and 2% to a transport coefficient, and fifth and higher approximations being accurate to perhaps 0.02% to 0.1%, the employment of a first-order expression to extract values of $\eta_{\mathrm{AB}}(T)$ or, especially, of $\lambda_{\mathrm{AB}}(T)$ from binary mixture data reduces the accuracy of the extracted results by about one order of magnitude.

In addition to the increased complexity of the expressions obtained for the binary mixture viscosity and thermal conductivity coefficients arising from the energy (heat) and momentum fluxes associated with nonequilibrium temperature and flow velocity gradients, the existence of a concentration gradient in a nonequilibrium binary mixture gives rise to a mass flux and a corresponding transport property, namely, diffusion. In addition, however, as mass and heat fluxes are both vector quantities that can be coupled by collisions, a pair of cross-effects, specifically thermal diffusion (i.e. diffusion driven by a temperature gradient) and its reciprocal effect, referred to as the diffusiothermal, or Soret, effect (i.e. heat conduction driven by a concentration gradient) arise. As the first Chapman–Cowling approximation for the binary diffusion coefficient, $D_{\mathrm{AB}}(T)$, is independent of the component concentrations, it might be thought that its measurement should provide a simpler route to the experimental determination of the interaction PEF between pairs of unlike atoms/molecules. Indeed, this would be true were it not for two problems: (i) it is rather difficult to determine diffusion coefficients with accuracies better than $\pm 1\%$, and (ii) higher Chapman–Cowling approximation corrections are typically larger for the diffusion coefficient than for either the viscosity or thermal conductivity coefficients. Similarly, because thermal diffusion vanishes identically in the traditional first Chapman–Cowling approximation, it might be expected that thermal diffusion data could therefore provide a good source of data for the experimental determination of interaction PEFs. However, it is, unfortunately, quite difficult to obtain thermal diffusion coefficients $D_{\mathrm{T}}(T)$ or, equivalently, thermal diffusion factors $\alpha_{\mathrm{T}}(T)$ with accuracies that exceed ± 5–10%. It thus appears that, as apart from the first Chapman–Cowling approximation for the binary diffusion coefficient, all relevant experimentally accessible properties depend upon the unlike-atom interaction and upon both like-atom interactions, so that progress toward truly accurate unlike-interaction PEFs may best be achieved through combined theoretical–experimental efforts, in which accurate calculations of mixture properties obtained using a high-level *ab initio* PEF for the unlike-atom interaction are used in tandem with similar already-vetted high-level like-atom *ab initio* PEFs for direct comparison with high-accuracy experimental mixture transport data.

Shortly after Woon (1994) established that CCSD(T)-level supermolecule calculations could generate reliable interaction energies between pairs of homonuclear rare gas atoms, Cybulski and Toczyłowski (1999) carried out CCSD(T) computations of the interaction energies for the He–He, Ne–Ne, Ar–Ar, He–Ne, He–Ar, and Ne–Ar closed-shell atom pairs at 13 internuclear separations. They then fitted each set of interaction energies thereby obtained to MTT (1984) PEF so that their results could be employed to calculate thermophysical properties, such as second virial coefficients and transport properties. This *ab initio* computational study of rare gas interaction energies was extended by Haley and Cybulski (2003) to include the He–Kr, Ne–Kr, Ar–Kr, and Kr–Kr interactions. Note in particular that these studies found that, although the CCSD(T) approach was rather successful in providing accurate predictions of $\mathfrak{D}_e$ and R_e values for the rare gas complexes when sufficiently large basis-sets were employed, its greatest limitation occurred for short internuclear separations. Similar CCSD(T)-based calculations were also carried out by Partridge *et al.* (2001) for the He–Ne, He–Ar, and He–Kr pairs (for 23–36 separations between 1.32Å and 6.35Å) as part of a more extended study of the interaction between He and 18 other ground state atoms. An additional set of *ab initio* computations (López Cacheiro *et al.*, 2004) of the He–Ne, He–Ar, and Ne–Ar interaction energies (at 21 to 23 internuclear separations between 2.1Å and 10.6Å and fitted to the same MTT functional form) resulted in a set of PEFs that are nearly indistinguishable from those obtained earlier by Cybulski and Toczyłowski (1999). Finally, CCSD(T) computations of the He–Kr interaction energy for 34 internuclear separations between 1.3Å and 9.0Å employing high-level correlation-consistent basis-sets for both He and Kr, together with a large set of mid-bond functions, has also been carried out by Jäger and Bich (2017) in order to obtain fully *ab initio* calculations of the thermophysical properties of krypton–helium gas mixtures based upon *ab initio* PEFs obtained for the He–He, Kr–Kr, and He–Kr interactions from CCSD(T) computations using similarly large, high-level, consistent sets of basis functions.

Table 4.8 lists the depths, $\mathfrak{D}_e$, and locations, R_e, associated with the *ab initio* interaction energies for all but the four mixed interactions involving Xe computed prior to 2010 (Cybulski and Toczyłowski, 1999; Haley and Cybulski, 2003; López Cacheiro *et al.*, 2004; Partridge *et al.*, 2001). A comparison between values of the locations of the minima, R_e, thus obtained for the interaction energies of the He–Ne, He–Ar, and Ne–Ar pairs shows that they differ by no more than 0.01%, while the corresponding values, $\mathfrak{D}_e$, obtained for the minima typically differ by no more than 0.13%, so that thermophysical properties calculated from each pair of PEFs will essentially be indistinguishable. The He–Ne, He–Ar, and He–Kr *ab initio* PEFs of Partridge *et al.* (2001) are similar to those obtained by Cybulski and coworkers (Cybulski and Toczyłowski, 1999; Haley and Cybulski, 2003).

Based upon comparisons between experimentally determined ro-vibrational frequencies and values calculated from the *ab initio* PEFs, Cybulski and Toczyłowski (1999) and Haley and Cybulski (2003) concluded that their *ab initio* PEFs were slightly too shallow. For the He–Kr interaction, this is further confirmed by the comparison

Table 4.8 The minimum $\mathfrak{D}_e$ in the interaction energy between a pair of unlike noble gas atoms and the corresponding equilibrium separation R_e between the atom pair

System	$\mathfrak{D}_e/\mathrm{cm}^{-1}$			$R_e/\text{Å}$		
	a	b	c	d	e	f
He–Ne	14.640	14.951	14.6207	3.0283	3.029	3.0285
He–Ar	20.663	20.648	20.6842	3.4925	3.494	3.492
He–Kr	20.7404	20.551	–	3.7094	3.711	–
Ne–Ar	45.1833	–	45.1959	3.4934	–	3.4931
Ne–Kr	47.6283	–	–	3.6742	–	–
Ar–Kr	98.764	–	–	3.9226	–	–

a, Cybulski and Toczyłowski, 1999; Haley and Cybulski, 2003; b, Partridge *et al.*, 2001; c, López Cacheiro *et al.*, 2004; d, Cybulski and Toczyłowski, 1999; Haley and Cybulski, 2003; e, López Cacheiro *et al.*, 2004; f, Partridge *et al.*, 2001.

in panel (a) of Fig. 4.14 between their PEF and the more recent *ab initio* PEF of Jäger and Bich (2017). The solid, dotted, and dashed curves represent, respectively, the He–Kr PEFs (Cybulski *et al.* 2001; Haley and Cybulski, 2003; Jager and Bich, 2017). Note that the PEF from Partridge *et al.* (2001) has a minimum that is 0.91% shallower and corresponds to an equilibrium separation R_e that is about 0.04% larger than R_e for the PEF from Haley and Cybulski (2003), while the He–Kr PEF (Jäger and Bich, 2017), with $\mathfrak{D}_e = 21.8355\,\mathrm{cm}^{-1}$ and $R_e = 3.68220\text{Å}$, is 5.28% deeper and has an equilibrium separation that is 0.73% shorter than the PEF from Haley and Cybulski (2003) listed in Table 4.8. Panel (b) of Fig. 4.14 illustrates the temperature dependence of the interaction second virial coefficient, $B_{2,\mathrm{AB}}(T)$, for the He–Kr interaction. The solid curve gives $B_{2,\mathrm{AB}}(T)$ obtained using the *ab initio* PEF from Jäger and Bich (2017), while the dotted curve gives $B_{2,\mathrm{AB}}(T)$ obtained (Song *et al.*, 2011) for the He–Kr PEF (Haley and Cybulski, 2003). Both calculations include the classical contribution plus the first three quantum corrections to the second virial coefficient. Also included in panel (b) for comparison are three sets of experimental values for $B_{2,\mathrm{AB}}(T)$: for clarity, only two of the experimental uncertainties for the Brewer data have been included in this figure. Detailed comparisons between the experimental values shown in Fig. 4.14 and values computed using the *ab initio* He–Kr PEFs show (Jäger and Bich, 2017; Song *et al.*, 2011) that values of $B_{2,\mathrm{AB}}(T)$ computed at the corresponding temperatures using the Jäger–Bich *ab initio* PEF lie, except for $T = 273.16\,\mathrm{K}$, within the experimental uncertainties, while values of $B_{2,\mathrm{AB}}(T)$ computed using the Haley–Cybulski *ab initio* PEF typically lie well outside the experimental uncertainties. This clearly indicates that the Jäger–Bich PEF provides the better representation of the He–Kr interaction.

Song and coworkers (Song *et al.*, 2010, 2011) made a systematic study of thermophysical properties, including the interaction second virial coefficient, $B_{2,\mathrm{AB}}(T)$, the mixture viscosity, $\eta_{\mathrm{mix}}(x_{\mathrm{A}};T)$, the mixture steady-state thermal conductivity,

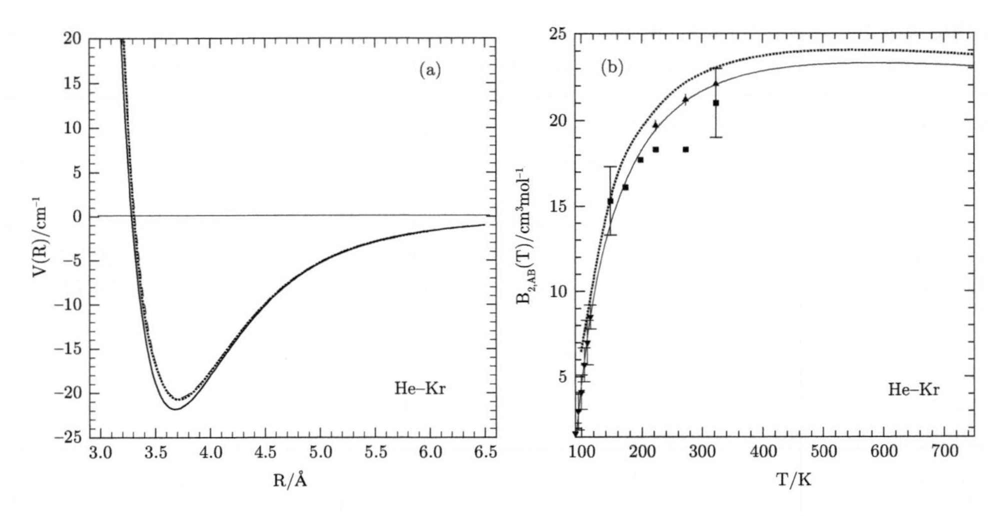

Fig. 4.14 The He–Kr interaction.

Panel (a), *ab initio* PEFs: solid line, Jäger and Bich, 2017; dashed line: Haley and Cybulski, 2003; dotted line: Partridge *et al.*, 2001. Panel (b), interaction second (pressure) virial coefficient: solid line, Jäger and Bich, 2017; dashed line, Song *et al.*, 2011; ▲, Dillard *et al.*, 1978; ■, Dymond *et al.*, 2002*b*; ▼, Kate and Robinson, 1973.

$\lambda_{\text{mix}}(x_{\text{A}};T)$, the binary diffusion coefficient, $D_{\text{AB}}(x_{\text{A}};T)$, and the thermal diffusion factor, $\alpha_{\text{T}}(x_{\text{A}};T)$. For their initial study (Song *et al.*, 2010), they utilized empirical PEFs obtained by Aziz (1993) Janzen and Aziz (1997) for pure helium and argon, *ab initio* PEFs determined by Cybulski and Toczyłowski (1999) for pure neon and for the He–Ne, He–Ar, and Ne–Ar interactions, the *ab initio* PEFs of Nasrabad and Deiters (2003) and Nasrabad *et al.* (2004) for pure krypton and the Ar–Kr interaction, plus PEFs for pure xenon and the Ar–Xe interaction based upon a Najafi *et al.* (1983) correlation. In their follow-up study, they employed more recent *ab initio* PEFs determined by the Rostock group (Hellmann *et al.*, 2007, 2008*a*; Jäger *et al.*, 2009, 2010) for pure helium, neon, and argon, plus the Nasrabad and Deiters (2003) *ab initio* PEF for pure krypton, together with the *ab initio* He–Ne, He–Ar, Ne–Ar PEFs of López Cacheiro *et al.* (2004), and *ab initio* He–Kr, Ne–Kr, and Ar–Kr PEFs obtained by Haley and Cybulski (2003). The initial study focused on the prediction of $\eta_{\text{mix}}(x_{\text{A}};T)$ and $\lambda_{\text{mix}}(x_{\text{A}};T)$ for the He–Ne, He–Ar, Ne–Ar, Ar–Kr, and Ar–Xe binary mixtures, while the follow-up study examined the temperature dependences of the interaction second virial coefficient $B_{2,\text{AB}}(T)$, the binary diffusion coefficient $D_{\text{AB}}(0.5;T)$, and the thermal diffusion factor $\alpha_{\text{T}}(0.5;T)$ for equimolar binary mixtures of He, Ne, Ar, and Kr.

Sharipov and Benites have carried out detailed and thorough studies of the transport properties of low density helium–argon (Sharipov and Benites, 2015) and helium–neon (Sharipov and Benites, 2017) binary mixtures based upon *ab initio* PEFs obtained both for pure He (Cencek *et al.*, 2012; Hellmann *et al.*, 2007; Przybytek *et al.*, 2010), pure Ne (Hellmann *et al.*, 2008*a*), and pure argon (Jäger et al., 2009, 2010) as well as for the He–Ar (Cybulski and Toczyłowski, 1999) and He–Ne (Haley and Cybulski, 2003) unlike interactions. The He–Ar study used classical mechanical scattering to evaluate the omega integrals needed for the calculation of fifth- and higher-order Chapman–Cowling approximations to the transport coefficients, while the He–Ne study employed quantum mechanical scattering to evaluate the omega integrals. Jäger and Bich (2017) performed a similarly complete study of the transport properties of He–Kr binary mixtures using quantum mechanical scattering to evaluate the relevant omega integrals. Figure 4.15 illustrates the dependences of the mixture viscosity $\eta_{\mathrm{mix}}(x_{\mathrm{A}};T)$ and thermal conductivity $\lambda_{\mathrm{mix}}(x_{\mathrm{A}};T)$ on the mole fraction x_{He} of helium for He–Ne and He–Kr binary mixtures for temperatures $T = 300\,\mathrm{K}$ (solid lines) and $T = 400\,\mathrm{K}$ (dashed lines).

As the pure gas contributions to $\eta_{\mathrm{mix}}(x_{\mathrm{A}};T)$ and $\lambda_{\mathrm{mix}}(x_{\mathrm{A}};T)$ have been determined (Jäger and Bich, 2017; Sharipov and Benites, 2015, 2017) using He–He (Cencek *et al.*, 2012; Hellmann *et al.*, 2007; Przybytek *et al.*, 2010), Ne–Ne (Hellmann *et al.*, 2008*a*), Ar–Ar (Jäger *et al.*, 2009, 2010; Patkowski and Szalewicz, 2010), and Kr–Kr (Jäger *et al.*, 2016) *ab inito* PEFs that give excellent agreement with the most accurate experimental values obtained for the pure gases (see Sections 4.1.1, 4.1.2, and 4.1.4), contributions to the overall differences between computed and experimental values arising from the pure gas interactions will be minimized, thereby enabling more meaningful vettings of proposed *ab initio* PEFs for the unlike He–Kr, He–Ar, and He–Ne interactions. Also included in Fig. 4.15 are mixture viscosity and thermal conductivity values computed for He–Ar and He–Ne mixtures (mole fractions $x_{\mathrm{He}} = 0.25$, 0.50,

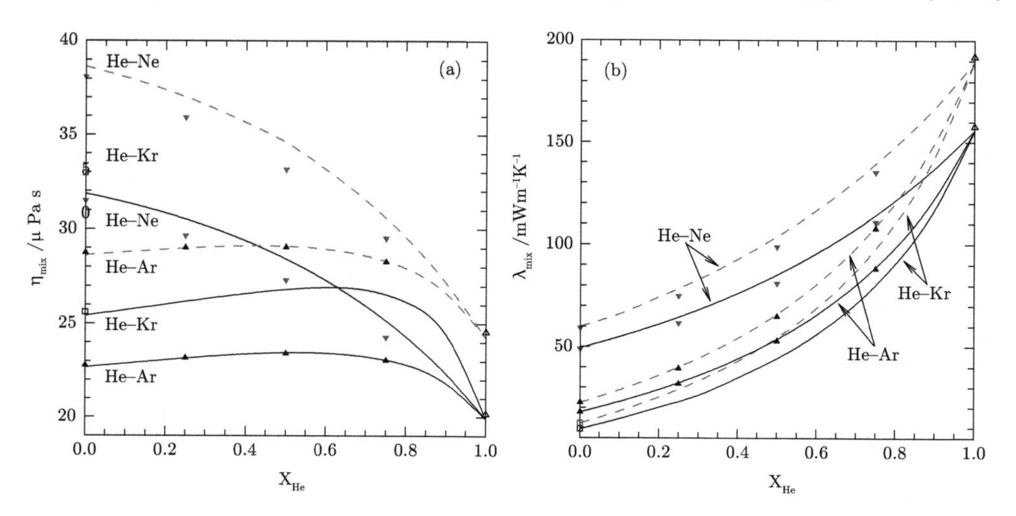

Fig. 4.15 Mole-fraction dependences of $\eta_{\mathrm{mix}}(x_{\mathrm{He}};T)$ and $\lambda_{\mathrm{mix}}(x_{\mathrm{He}};T)$ for He–Ne, He–Ar, and He–Kr binary mixtures for temperatures $T = 300\,\mathrm{K}$ and 400 K. Solid lines, $T = 300\,\mathrm{K}$; dashed lines, $T = 400\,\mathrm{K}$; He–Kr curves, Jäger and Bich 2017; He–Ar curves, Sharipov and Benites 2015; He–Ne curves, Sharipov and Benites 2017; the symbols ▲, ▼, and □ represent calculated values reported by Song *et al.*, 2010.

0.75) by Song *et al.* (2010) using the same (He–Ar) or very similar (He–Ne) PEFs, but employing slightly different He–He (Aziz, 1993; Janzen and Aziz, 1997), Ar–Ar (Aziz, 1993; Boyes, 1994), and Ne–Ne (Cybulski and Toczyłowski, 1999) PEFs. Note that the three pure gas PEFs employed by Song *et al.* (2010), while considered the best available prior to 2010, do give considerably poorer agreement with the 298.15 K ultra-precise (experimental uncertainties of order $\pm 0.03\%$) viscosity measurements of Berg and Burton (2013) than do the corresponding *ab initio* pure gas PEFs employed by Sharipov and Benites (2015, 2017). Values for the pure gas viscosities and thermal conductivities computed by Song *et al.* (2010) using these PEFs have also been included in Fig. 4.15 at $x_{\mathrm{He}} = 0$ (Ar–Ar, ▲; Ne–Ne, ▼) or $x_{\mathrm{He}} = 1$ (He–He, △). Finally, as Song *et al.* (2010) also computed values of $\eta_{\mathrm{Kr}}(T)$ and $\lambda_{\mathrm{Kr}}(T)$ using the *ab initio* Kr–Kr PEF of Nasrabad and Deiters (2003), these values have also been included (as open squares, □) at $x_{\mathrm{He}} = 0$ for comparison with the pure krypton values obtained by Jäger and Bich (2016) using their (Jäger *et al.*, 2016) *ab initio* Kr–Kr PEF.

It is often convenient to employ a percentage difference δ_{X}, defined as

$$\delta_{\mathrm{X}} \equiv 100\% \, \frac{\mathrm{X}_i(y) - \mathrm{X}(y)}{\mathrm{X}(y)} \,, \tag{4.12}$$

to express the level of agreement between a specified value $\mathrm{X}_i(y)$ of an attribute $\mathrm{X}(y)$ and a continuous curve that represents $\mathrm{X}(y)$. Normally, $\mathrm{X}_i(y)$ represents the outcome of a specific experimental measurement of attribute X and $\mathrm{X}(y)$ represents an interpolation curve to a large set of experimental data. This same format can be employed to express the percentage differences δ_η and δ_λ between calculated values of $\eta_{\mathrm{mix}}(x_{\mathrm{A}}; T)$ and $\lambda_{\mathrm{mix}}(x_{\mathrm{A}}; T)$ obtained for He–Ar and He–Ne mixtures by Song *et al.* (2010) for mole fractions $x_{\mathrm{He}} = 0.25$, 0.50, and 0.75 relative to those obtained by Sharipov and Benites (2015, 2017). Table 4.9 shows the results of such a comparison.

It is clear both from Fig. 4.15 and from Table 4.9 that the differences between the two sets of calculated values are much smaller for the He–Ar mixtures than for the He–Ne mixtures. As transport property calculations for binary He–Ar mixtures have employed both quantum (Song *et al.*, 2010) and classical (Sharipov and Benites, 2015) collision dynamics, the comparisons in Table 4.9 have been made for tempera-

Table 4.9 Comparison between calculated viscosities and thermal conductivities of He, Ne, Ar, and He–Ar, He–Ne binary mixtures

	He–Ar				He–Ne			
	$T = 300\,\mathrm{K}$		$T = 400\,\mathrm{K}$		$T = 300\,\mathrm{K}$		$T = 400\,\mathrm{K}$	
x_{He}	$\delta_\eta/\%$	$\delta_\lambda/\%$	$\delta_\eta/\%$	$\delta_\lambda/\%$	$\delta_\eta/\%$	$\delta_\lambda/\%$	$\delta_\eta/\%$	$\delta_\lambda/\%$
0.0	0.45	0.45	0.49	0.49	−1.20	−1.17	−1.44	−1.42
0.25	0.10	−1.52	0.13	−1.57	−2.85	−4.65	−2.98	−4.70
0.50	−0.15	−1.61	−0.14	−1.70	−3.91	−4.98	−4.11	−4.94
0.75	−0.03	−0.63	−0.07	−0.73	−3.48	−2.94	−3.46	−2.90
1.0	1.32	1.32	1.27	1.28	1.32	1.32	1.27	1.28

tures $T = 300\,\text{K}$ and $400\,\text{K}$ so that differences due to the use of classical and quantum collision dynamics for He–Ar collisions represent contributions to the mixture transport coefficients that are typically smaller than 0.1%. Because the same He–Ar PEF has been employed for both sets of mixture calculations, differences for Ar–He mixtures must be due to differences between the pure gas PEFs and the levels of Chapman–Cowling approximations employed for the two sets of calculations. Values of $\eta(T)$ and $\lambda(T)$ for pure Ar and He were computed by Song *et al.* using fifth Chapman–Cowling approximation expressions and by Sharipov and Benites using twelfth Chapman–Cowling approximation expressions: all pure gas properties should thus be converged to six figures. However, Song *et al.* (2010) employed an *ad hoc* modification of the first Chapman–Cowling approximation expression to compute the contributions to $\eta_{\text{mix}}(x_{\text{A}};T)$ and $\lambda_{\text{mix}}(x_{\text{A}};T)$, while Sharipov and Benites (2015) employed full twelfth Chapman–Cowling approximation expressions for the mixture properties. Differences between the two sets of δ_η and δ_λ values listed in Table 4.9 are likely largely due to differences between the computed pure gas values (as shown by the values of δ_η and δ_λ for $x_{\text{He}} = 0.0$, 1.0) in addition to any errors introduced by the employment of Song *et al.* (2010) of a modified first Chapman–Cowling approximation.

Although it is clear from Fig. 4.15 and Table 4.9 that there are significant differences between the two sets (Song *et al.*, 2010; Sharipov and Benites, 2017) of calculated values of $\eta_{\text{mix}}(x_{\text{A}};T)$ and $\lambda_{\text{mix}}(x_{\text{A}};T)$ for He–Ne mixtures, it is very difficult to separate contributions associated with the use of different PEFs for the He–He (Cencek *et al.*, 2012; Janzen and Aziz, 1997; Przybytek *et al.*, 2010), Ne–Ne (Cybulski and Toczyłowski, 1999; Hellmann *et al.*, 2008*a*), and He–Ne (Cybulski and Toczyłowski, 1999; López Cacheiro *et al.*, 2004) interactions from contributions arising from the employment (Song *et al.*, 2010) of the modified first Chapman–Cowling approximation expressions for the viscosity and thermal conductivity of a binary mixture without additional information. That the values of δ_λ are significantly larger than the values of δ_η for the binary mixtures is consistent with the finding of Jäger and Bich (2017) that for He–Kr mixtures at temperature $T = 300\,\text{K}$, the percentage changes to λ_{mix} in passing from first to second Chapman–Cowling approximation calculations were significantly larger than the corresponding percentage changes to η_{mix}. For example, for mixtures having mole fractions $x_{\text{He}} = 0.2$, 0.4, 0.6, and 0.8, increases of 1.97%, 2.38%, 2.24%, and 1.74%, respectively, were obtained for the mixture thermal conductivities, while the corresponding increases to the mixture viscosities were only 0.18%, 0.31%, 0.42%, and 0.49%, respectively. These contributions to the transport properties of He–Kr mixtures at $T = 300\,\text{K}$ may also be compared with contributions of 0.048% and 0.076% to the viscosity and thermal conductivity, respectively, of pure krypton, and of 0.59% and 0.91% to the viscosity and thermal conductivity, respectively, of pure helium. It is also clear from results obtained for equimolar He–Ar (Sharipov and Benites, 2015) and He–Ne (Sharipov and Benites, 2017) mixtures that fifth Chapman–Cowling approximation calculations of $\eta_{\text{mix}}(x_{\text{A}};T)$, $\lambda_{\text{mix}}(x_{\text{A}};T)$, and $D_{\text{AB}}(x_{\text{A}};T)$ provide values for these transport properties that have accuracies that exceed 0.1% for temperatures below 2000 K. Similar detailed calculations of Jäger and Bich (2017) for He, Kr, and He–Kr mixtures show that values for the mixture viscosity, thermal conductivity, and binary diffusion coefficients computed for temperature $T = 300\,\text{K}$ at the fifth

Chapman–Cowling level of approximation have accuracies that exceed 0.05%. These results also show that the total changes in the viscosity and thermal conductivity coefficients from the first to the sixth Chapman–Cowling approximation are, for $x_{\mathrm{He}} = 0.6$ for example, 0.47% for η_{mix} versus 2.71% for λ_{mix}, with both values converged to six figures.

Relatively few systematic experimental studies have been carried out for the viscosity or thermal conductivity of noble gas binary mixtures. Some of the experimental studies have been carried out over an extensive temperature range (0–700°C) and for a several different mixture compositions. Similarly, a more limited number of systematic computational studies have covered a very large temperature range (from just above liquifaction temperature to several thousand Kelvin) and at a series of compositions. Occasionally, however, there are mismatches between the sets of temperatures and compositions selected for these studies, thereby making it necessary to resort to interpolation schemes in order to make comparisons between the two sets of results.

It is unlikely that it will be possible to extract interaction viscosity values that have percentage uncertainties significantly smaller than $\pm 1\%$ from experimental measurements of mixture viscosities, even though it is possible to measure gas viscosities with an accuracy of the order of $\pm 0.1\%$. However, as it is possible to compute binary mixture viscosities for a specified set of three PEFs with accuracies exceeding $\pm 0.1\%$, high-accuracy experimental results could be employed to test a specific hetero-atomic PEF by comparing the experimental results with computed values obtained using a pair of already-tested (accurate) homo-atomic PEFs together with the hetero-atomic PEF. Such a combined study would obviate the need to employ interpolation schemes that introduce additional uncertainties into the calculated results. Table 4.10 illustrates the feasibility of such a procedure, which utilizes the results of high-level mixture viscosity computations for He–Ar (Sharipov and Benites, 2015), He–Ne (Sharipov and Benites, 2017), and He–Kr (Jäger and Bich, 2017) mixtures using in each case pure-gas AI PEFs (that have previously been tested and found to compare very favorably with the best experimental measurements) together with proposed *ab initio* unlike-atom PEFs for comparison with mixture viscosity measurements that have accuracies between $\pm 0.2\%$ and $\pm 0.5\%$. As many of the *ab initio* values for the mixture viscosities, designated $\eta(\mathrm{AI})$, have been obtained from the computational studies via interpolations from high-precision values, the results displayed in Table 4.10 serve more as an indication of the type of agreement that *may* be achieved rather than as test of the qualities of the *ab initio* hetero-atomic PEFs employed in the computational studies. Jäger and Bich (2017) provide a more detailed analysis of the percentage differences between experimental values of $\eta_{\mathrm{mix}}(x_{\mathrm{He}};T)$ and those computed using a high-level *ab initio* He–Kr PEF, and also show explicitly the differences between values calculated using an earlier (Haley and Cybulski, 2003) *ab initio* He–Kr PEF and values obtained from their *ab initio* PEF. They also show the effect of employing classical evaluations, rather than fully quantum evaluations, of the effective cross-sections (equivalently, omega integrals) for both the He–He and He–Kr interactions.

Sharipov and Strapasson (2012*b*) introduced and applied an extension of the direct simulation Monte Carlo (DSMC) method (Bird, 1994) incorporating arbitrary interatomic PEFs to the computation of $\eta_{\mathrm{mix}}(x_{\mathrm{A}};T)$ and $\lambda_{\mathrm{mix}}(x_{\mathrm{A}};T)$ for helium–argon

Table 4.10 Comparison between calculated and experimentally measured viscosities of He–Ar, He–Ne, and He–Kr binary mixtures

He–Ar ($T = 298.15\,\mathrm{K}$)				He–Kr ($T = 298.15\,\mathrm{K}$)			
x_{He}	$\eta(\mathrm{AI})^*$	η^{m}	$\delta_\eta/\%$	x_{He}	$\eta(\mathrm{AI})^*$	η^{m}	$\delta_\eta/\%$
0.0	22.552	22.632	0.35	0.0	25.277	25.374	0.38
0.2184	22.949	23.074	0.54	0.2165	25.923	25.993	0.27
0.4300	23.252	23.418	0.71	0.4376	26.538	26.599	0.23
0.6218	23.156	23.460	1.31	0.5953	26.785	26.832	0.18
0.7506	22.953	23.122	0.74	0.7041	26.637	26.677	0.17
0.8953	21.744	21.885	0.65	0.8721	24.907	24.969	0.25
1.0	19.825	19.632	0.19	1.0	19.825	19.863	0.19
He–Ne ($T = 298.15\,\mathrm{K}$)				He–Ne ($T = 373.15\,\mathrm{K}$)			
x_{He}	$\eta(\mathrm{AI})^*$	η^{n}	$\delta_\eta/\%$	x_{He}	$\eta(\mathrm{AI})^*$	η^{n}	$\delta_\eta/\%$
0.0	31.727	31.777	0.16	0.0	36.887	37.073	0.50
0.1988	30.707	30.743	0.12	0.1988	35.704	35.877	0.48
0.3699	29.478	29.549	0.24	0.3699	34.289	34.464	0.57
0.5934	27.249	27.312	0.23	0.5934	31.707	31.830	0.39
0.7663	24.776	24.580	−0.75	0.7663	28.867	28.952	0.29
1.0	19.825	19.859	0.17	1.0	23.133	23.182	0.21

* *Ab initio* values based upon results for $\eta_{\mathrm{mix}}(x_{\mathrm{He}}; T)$ reported in Jäger and Bich, 2017 (He–Kr), Sharipov and Benites, 2015 (He–Ar), and Sharipov and Benites, 2017 (He–Ne).
[m] 1970; [n] Kestin *et al.*, 1972*c*.

mixtures with mole fractions $x_{\mathrm{He}} = 0.25, 0.50, 0.75$ at temperature $T = 300\,\mathrm{K}$ using a consistent set (Cybulski and Toczyłowski, 1999) of *ab initio* He–He, Ar–Ar, and He–Ar PEFs (Sharipov and Strapasson, 2012*a*). Although values of the viscosity and thermal conductivity coefficients obtained from the DSMC calculations for pure He and Ar were within 0.05% and 0.2%, respectively, of values (Bich *et al.*, 2007; Vogel, 2010) obtained using high-level kinetic theory Chapman–Cowling approximations, the DSMC values obtained for the mixture viscosities were typically between 0.3% and 0.5% (mixture thermal conductivities typically between 0.9% and 1.2%) higher than values obtained (Sharipov and Benites, 2015) using very similar pure gas PEFs (Bich *et al.*, 2007; Song *et al.*, 2016; Vogel, 2010) together with the same He–Ar PEF for calculations of $\eta_{\mathrm{mix}}(x_{\mathrm{He}}; 300\,\mathrm{K})$ and $\lambda_{\mathrm{mix}}(x_{\mathrm{He}}; 300\,\mathrm{K})$ using high-level kinetic theory expressions.

As noted for pure gases, obtaining a very accurate value for the thermal conductivity coefficient typically presents a much greater challenge to the experimentalist than does the determination of a very accurate value of the viscosity coefficient. It is particularly difficult to obtain accurate values of the thermal conductivity for temperatures that lie outside a relatively narrow range of temperatures around 300 K. Although experimental thermal conductivities have been obtained for a number of noble gas mixtures, high-level calculations have only been carried out for binary mixtures of

He and three other noble gases, specifically, for He–Ne (Sharipov and Benites, 2017), He–Ar (Sharipov and Benites, 2015), and He–Kr (Jäger and Bich, 2017) mixtures. A comparison of the level of overall agreement between binary mixture values computed using *ab initio* unlike-atom PEFs in conjunction with already-tested accurate *ab initio* like-atom interaction PEFs and sets of experimental mixture thermal conductivity values obtained at temperatures in the ambient temperature range (roughly 273–318 K) appearing in Table 4.11 illustrates the levels of agreement that are currently attainable.

The best RMSD values for δ_λ appearing in Table 4.11 for each of the three different binary helium mixtures is comparable with the respective RMSD values (0.35% for He–Ne, 0.76% for He–Ar, 0.34% for He–Kr) for δ_η obtained from the results appearing in Table 4.10. However, while it is now possible to obtain δ_η RMSD values considerably smaller than the values appearing in Table 4.10, the RMSD values $\delta_\lambda = 0.36\%$ for He–Ne and He–Ar mixtures (Assael *et al.*, 1981) likely represent the best experimental accuracy that can be achieved at the present time for thermal conductivity measurements, even for the ambient temperature range. The accuracy of both experimental viscosity and thermal conductivity measurements decreases significantly for temperatures external to the ambient temperature range. For nonambient temperatures, both mixture viscosities and mixture thermal conductivities computed from sets of reliable homo- and hetero-atomic PEFs will likely provide considerably more reliable values than will experiment.

Table 4.11 Comparison between calculated and experimentally measured thermal conductivities of some He–Ne, He–Ar, and He–Kr binary mixtures

Mixture	T /K	Source	# Values	δ_λ /%		
				High*	Low*	RMSD
He–Ne	296.8	a	9	−1.46	−0.74	1.04
	302.5	b	6	4.76	0.30	2.16
	308.15	c	6	−0.49	0.07	0.36
He–Ar	296.8	a	11	0.80	0.11	0.53
	300.65	d	6	1.81	−0.07	1.01
	302.15	b	6	−1.72	0.06	1.97
	308.15	e	5	−5.75	−1.42	3.73
	323.15	e	6	−5.47	0.28	3.39
He–Kr	302.15	b	8	−2.24	0.12	1.37
	308.15	e	6	−5.32	−1.17	4.15
	308.15	c	6	−0.49	0.07	0.36
	363.15	e	5	−5.90	−0.82	3.89

* "High" ("Low") refers to the largest (smallest) magnitude of δ_λ for the data set.
a, Van Dael and Cauwenbergh, 1968; b, Mason and Von Ubisch, 1960; c, Assael *et al.*, 1981; d, Clifford *et al.*, 1979; e, Gambhir and Saxena, 1966.

As mixtures are subject not only to temperature and pressure gradients, but also to concentration gradients, they give rise to (mutual) diffusion and to a pair of cross-effect phenomena referred to as thermal diffusion and the Soret (diffusio-thermal) effect. The dependences of the binary diffusion coefficient $D_{\rm AB}(x_{\rm He};T)$ and the thermal diffusion factor $\alpha_{\rm T}(x_{\rm He};T)$ upon the mole fraction $x_{\rm He}$ are displayed in Fig. 4.16 for He–Ne, He–Ar, and He–Kr mixtures at temperatures $T = 150\,{\rm K}$, $300\,{\rm K}$, and $500\,{\rm K}$. Although comparing Figs. 4.15 and 4.16 clearly shows that $D_{\rm AB}(x_{\rm A};T)$ has a much weaker dependence upon mixture composition than do $\eta_{\rm mix}(x_{\rm A};T)$, $\lambda_{\rm mix}(x_{\rm A};T)$, and $\alpha_{\rm T}(x_{\rm A};T)$, it is also clear that $D_{\rm AB}(x_{\rm A};T)$ is not independent of mixture composition, as predicted by its first Chapman–Cowling approximation expression. For example, Jäger and Bich (2017) found that for diffusion in He–Kr binary mixtures at $T = 300\,{\rm K}$, the difference between the sixth and first Chapman–Cowling approximations for the diffusion coefficient varies between 3.8% to 1.5% as $x_{\rm He}$ varies from 0.1 to 0.8. They also found that although higher Chapman–Cowling corrections were rather small for helium-rich mixtures, they were relatively large for krypton-rich mixtures, with corrections of 3.1%, 0.55%, 0.15%, 0.046%, and 0.005% associated with the first through sixth Chapman–Cowling approximations—a behavior that is typical of binary mixtures consisting of components having large mass differences.

The left-hand panel of Fig. 4.16 includes experimental values of $D_{\rm AB}(x_{\rm He};300\,{\rm K})$ obtained by Staker and Dunlop (1976) ($x_{\rm He} = 0.9$) and by Arora *et al.* (1979) ($x_{\rm He} \rightarrow 1.0$) for each He–Ne, He–Ar, and He–Kr binary mixture, while the right-hand panel includes experimental values of $\alpha_{\rm T}(x_{\rm He};300\,{\rm K})$ obtained by Trengove *et al.* (1981) for mixtures with mole fractions $x_{\rm He} = 0.2, 0.8$.

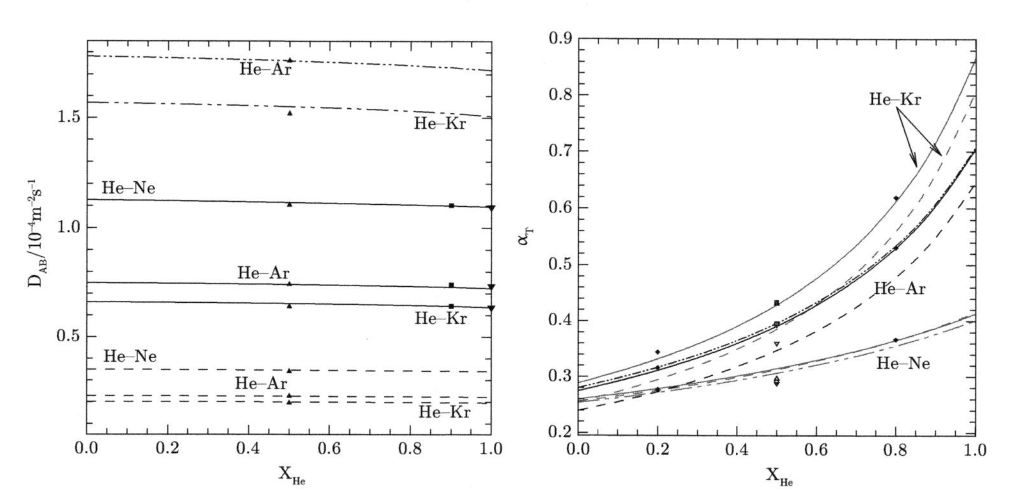

Fig. 4.16 Mole-fraction dependence of the binary diffusion coefficient $D_{\rm AB}(x_{\rm He};T)$ and the thermal diffusion factor $\alpha_{\rm T}(x_{\rm He};T)$ for He–Ne[a], He–Ar[b], and He–Kr[c] mixtures.

[a]Sharipov and Benites, 2017; [b]Sharipov and Benites, 2015; [c]Jäger and Bich, 2017.
Dashed lines, $T = 150\,{\rm K}$, solid lines, $T = 300\,{\rm K}$; dash-doubledotted lines, $T = 500\,{\rm K}$; ▲, △, ▽, □, Song *et al.*, 2011; ■, Staker and Dunlop, 1976; ▼, Arora *et al.*, 1979; ◆ ($T = 300\,{\rm K}$), Trengove *et al.*, 1981.

Values of $D_{AB}(x_{He};T)$ and $\alpha_T(x_{He};T)$ computed by Song *et al.* (2011) for equimolar mixtures of He with Ne, Ar, and Kr are included in Fig. 4.16 for comparison with the results obtained in the more extensive calculations (Jäger and Bich, 2017; Sharipov and Benites, 2015, 2017) carried out using much the same sets of *ab initio* homo- and heter-atom PEFs *ab initio* pure gas and hetero-atom PEFs. However, as Song *et al.* (2011) employed pseudo-first-order expressions for $D_{AB}(0.5;T)$ and $\alpha_T(0.5;T)$ coupled with classical collision dynamics for evaluation of the relevant effective cross-sections (in terms of omega-integrals) determining the transport coefficients for these He–Rg binary mixtures at all temperatures, while two of the later (Jäger and Bich, 2017; Sharipov and Benites, 2017) calculations employed fully quantum mechanical evaluations coupled minimally with fourth Chapman–Cowling approximations for these transport coefficients, significant differences can be expected between results obtained at these two levels of calculation, especially for low temperatures. Note that although Sharipov and Benites (2015) did utilize classical evaluations of the omega integrals for their He–Ar calculations, they did use higher Chapman–Cowling approximation expressions for the transport coefficients.

A quantitative comparison between the sets of results is given in Table 4.12 for four representative temperatures, specifically, $T = 150\,\text{K}$, at which quantum mechanical effects will definately play a significant role, $T = 300\,\text{K}$, at which quantum mechanical effects can make significant contributions, especially for He–Ne mixtures, $T = 500\,\text{K}$, at which quantum mechanical effects may still play a minor role and, finally, $T = 1000\,\text{K}$, at which quantum mechanical effects should be relatively unimportant for these mixtures.

Table 4.12 Comparison between Song *et al.* (2011) and high-level Chapman–Cowling calculations of $D_{AB}(x_{He};T)$ and $\alpha_T(x_{He};T)$ for equimolar He–Ne, He–Ar, and He–Kr binary mixtures

Mixture	T /K	$D_{AB}(0.5;T)$			$\alpha_T(0.5;T)$		
		Song	AI*	δ_D /%	Song	AI*	δ_α /%
He–Ne	150	0.3405	0.34568	−1.498	0.2881	0.31276	−7.885
	300	1.1044	1.11426	−0.885	0.2967	0.32458	−5.683
	500	2.6254	2.64338	−0.680	0.2914	0.30652	−4.933
	1000	8.5608	8.60612	−0.526	0.2769	0.28934	−4.286
He–Ar	150	0.2274	0.2261	0.619	0.3600	0.3474	3.633
	300	0.7425	0.7414	0.154	0.3941	0.3911	0.764
	500	1.7630	1.7621	0.057	0.3957	0.3963	−0.156
	1000	5.7168	5.7113	0.096	0.3826	0.3846	−0.520
He–Kr	150	0.1962	0.1999	−1.871	0.3933	0.3856	1.995
	300	0.6405	0.6537	−2.007	0.4313	0.4290	0.527
	500	1.5197	1.5514	−2.043	0.4399	0.4329	0.238
	1000	4.9223	5.0232	−2.009	0.4808	0.4191	0.406

* *Ab initio* (AI) results: Sharipov and Benites, 2017 (He–Ne), Sharipov and Benites, 2015 (He–Ar), Jäger and Bich, 2017 (He–Kr).

As both sets of computations (Sharipov and Benites, 2017; Song *et al.*, 2011) of $D_{\mathrm{AB}}(x_{\mathrm{He}};T)$ and $\alpha_{\mathrm{T}}(x_{\mathrm{He}};T)$ for He–Ne mixtures employ either the same PEFs (He–Ne, Ne–Ne) or effectively equivalent PEFs (He–He), differences between the He–Ne mixture results displayed in Table 4.12 must be due to a combination of the employment of classical versus quantum collision dynamics and pseudo-first-order versus minimally fourth Chapman–Cowling approximation mixture expressions, respectively, in Song *et al.* (2011) and Sharipov and Benites (2017). The values of δ_D and δ_α for $T = 150\,\mathrm{K}$ and $T = 1000\,\mathrm{K}$ indicate both that the use of classical dynamics to describe He–Ne collisions leads to a significant underestimation of both D_{AB} and α_{T} at low temperatures and that Chapman–Cowling approximations beyond the lowest-order expressions are necessary for accurate determination of values for these thermophysical properties. Similarly, because the same He–He, Ar–Ar, and effectively equivalent He–Ar, PEFs have been employed (together with classical collision dynamics to evaluate the omega integrals) in Song *et al.* (2011) and Sharipov and Benites (2015), an examination of the trends in δ_D and δ_α for He–Ar mixtures should provide some guidance on the need for higher Chapman–Cowling approximation expressions for the transport coefficients. It is more difficult to draw any conclusions about possible sources of the differences between δ_D and δ_λ obtained for the He–Kr equimolar mixtures since, in addition to the use of classical versus quantum dynamics to describe He–He and He–Kr collisions, there are differences between the *ab initio* He–Kr PEFs and the Kr–Kr PEFs.

5

From *Ab Initio* Calculations to Spectroscopic and Thermophysical Properties

Chapter 1–4 presented, the theoretical background related to *ab initio* calculations, potential energy functions (PEF), virial coefficients and transport properties in detail. This chapter shows step by step how the spectroscopic and thermophysical properties of a specific molecular system can be obtained theoretically.

The main process for obtaining and applying interaction PEFs based upon high precision *ab initio* methodology is to employ *ab initio* calculations at discrete interatomic separations r to obtain highly accurate PEFs firstly, and then to fit the *ab initio* values to a reliable PEF model to obtain state-of-the-art PEFs. Finally, the high-fidelity PEFs are used to predict ro-vibrational energy levels and absorption spectra of noble gas dimers, then to employ expressions from quantum statistical mechanics and the kinetic theory of gases to obtain the thermophysical properties of low-density atomic gases. Noble gas dimers are taken as examples to dissect each step in detail.

5.1 *Ab Initio* Calculations

The growing maturity of *ab initio* methods has benefited from the rapid development of computer computing power in recent years. When using the Hartree–Fock (HF) method to obtain interaction energy curves, the nonrelativistic and single electron approximation are considered first, leading to errors associated with relativistic effects and electron correlation. To correct these crude PEFs, post-HF methods must be taken into account. The coupled-cluster (CC) theory is considered to be one of the most perfect and widely used post-HF methods, especially when higher-order excitations are involved: such calculated values are rather close to the full configuration interaction (CI) energies. This section shows how to use the CC method to calculate pointwise PEFs and transition dipole moments.

The CC singles, doubles, and contracted triples (CCSD(T)) method is regarded as a "gold standard" because of its efficiency and accuracy. Thus, in practical calculations, the interaction potential energy is normally obtained from a CCSD(T) level calculation with frozen-core (FC) electron, and a series of correction energies, should it be necessary.

Transport Properties and Potential Energy Models for Monatomic Gases. Hui Li and Frederick R.W. McCourt, Oxford University Press.
 DOI: 10.1093/oso/9780198888253.003.0005

5.1.1 *Ab initio* calculations of pointwise potential energy

For a dimer, the unused basis functions of one unit may augment the basis-set of the other unit and lower the energy artificially, that is, bring the basis-set superposition error (BSSE) (Simon *et al.*, 1996). The counterpoise (CP) correction developed by Boys and Bernardi (1970) is often used to minimize the BSSE as

$$V_{\mathrm{CP}}(R) = V_{\mathrm{AB}}(R) - V_{\mathrm{A}}(R) - V_{\mathrm{B}}(R). \tag{5.1}$$

The CP correction is required for each term of the total interaction energy .

Regarding the calculation value at the CCSD(T) level with FC electrons as the most basic precision required value, the total interaction energy of a rare gas dimer is determined as (Hellmann *et al.*, 2017)

$$V_{\mathrm{total}} = V_{\mathrm{CCSD(T)}} + V_{\mathrm{post-CCSD(T)}}, \tag{5.2}$$

in which $V_{\mathrm{CCSD(T)}}$ and $V_{\mathrm{post-CCSD(T)}}$ are the contributions at the CCSD(T) level and from higher-order CC correlation effects, respectively.

The contributions at the CCSD(T) level can be represented effectively by the sum of three terms, namely,

$$V_{\mathrm{CCSD(T)}} = V_{\mathrm{CCSD(T)/FC}} + V_{\mathrm{CCSD(T)/core}} + V_{\mathrm{CCSD(T)/rel}}, \tag{5.3}$$

in which $V_{\mathrm{CCSD(T)/FC}}$, $V_{\mathrm{CCSD(T)/core}}$, $V_{\mathrm{CCSD(T)/rel}}$ are the contributions from valence electrons, inner nuclear electrons, and relativistic effects at the CCSD(T) level, respectively.

A calculation of the energy within the FC approximation at the CCSD(T) level $V_{\mathrm{CCSD(T)/FC}}$ accounts for up to 90% of the total interaction energy (Table 5.1), so that contributions from other components may be considered as corrections to $V_{\mathrm{CCSD(T)/FC}}$. For a simple qualitative case, computations terminating at $V_{\mathrm{CCSD(T)/FC}}$ suffice, but at state-of-the-art levels, additional corrections must be included.

$V_{\mathrm{CCSD(T)/FC}}$ is defined as the sum of the SCF interaction energy and the electronic correlation energy, that is,

$$V_{\mathrm{CCSD(T)/FC}} = V_{\mathrm{SCF}} + V_{\mathrm{corr}}. \tag{5.4}$$

Table 5.1 The individual energies as percentages (%) for the total interaction energy of homonuclear rare gas dimers

	$V_{\mathrm{CCSD(T)/FC}}$	$V_{\mathrm{CCSD(T)/core}}$	$V_{\mathrm{CCSD(T)/rel}}$	$V_{\mathrm{post-CCSD(T)}}$	Reference
Ne–Ne	98.208	−0.161	0.207	1.749	*a*
Ar–Ar	98.343	0.749	0.529	0.379	*b*
Kr–Kr	95.273	2.754	2.323	−0.351	*c*
Xe–Xe	91.271	7.624	4.270	−3.165	*d*

a, Hellmann *et al.*, 2008*a*; *b*, Jäger *et al.*, 2009; *c*, Jäger *et al.*, 2016; *d*, Hellmann *et al.*, 2017.

The self-consistent field (SCF) energy converges rapidly with completion of the basis-set, as shown in Table 5.2, as the convergence level is of order 0.01 cm^{-1}. Thus, it is appropriate to regard calculated values at the taV6Z level as final results. Although we need not do that generally, the SCF interaction energies can also be extrapolated to the complete basis-set (CBS) limit if necessary, as

$$V_{\mathrm{SCF}}^{\mathrm{CBS}} = V_{\mathrm{SCF}}^{X} + \frac{V_{\mathrm{SCF}}^{X} - V_{\mathrm{SCF}}^{X-1}}{\frac{X\exp(9\sqrt{X}-\sqrt{X-1})}{X+1} - 1}, \tag{5.5}$$

in which X is a cardinal number (Karton and Martin, 2005).

The correlation energy V_{corr} corrects the error arising from the single electron approximation at the CCSD(T) level using the FC approximation. Note that this component of the energy does not converge as well as the SCF interaction the interaction energy (Table 5.3), which means that the CBS limit extrapolation must be carried out as (Hellmann *et al.*, 2017)

$$V_{\mathrm{corr}}^{\mathrm{CBS}} = \frac{X^3 V_{\mathrm{corr}}^{X} - (X-1)^3 V_{\mathrm{corr}}^{X-1}}{X^3 - (X-1)^3}. \tag{5.6}$$

Table 5.2 The V_{SCF} energy of a rare gas dimer near the equilibrium internuclear separation with different basis-sets taVXZ(or daVXZ)

	R / Å	X = 4	X = 5	X = 6	Reference
Xe–He	4.0	20.223	20.231	20.230	Liu *et al.*, 2022
Xe–Ne	3.8	73.259	73.305	73.306	Liu *et al.*, 2022
Xe–Ar	4.0	188.251	188.099	188.049	Liu *et al.*, 2022
Xe–Kr	4.2	175.593	175.589	175.555	Hu *et al.*, 2022
Xe–Xe	4.4	204.817	204.810	204.727	Hellmann *et al.*, 2017

*All energies are in cm^{-1}.

Table 5.3 The V_{corr} energy of a rare gas dimer near the equilibrium internuclear separation with different basis-sets taVXZ(or daVXZ)

	R / Å	X=4	X=5	X=6	Reference
Xe–He	4.0	−39.535	−39.680	−39.745	Liu *et al.*, 2022
Xe–Ne	3.8	−44.250	−44.412	−44.570	Liu *et al.*, 2022
Xe–Ar	4.0	−114.366	−115.552	−116.143	Liu *et al.*, 2022
Xe–Kr	4.2	−324.123	−325.003	−325.561	Hu *et al.*, 2022
Xe–Xe	4.4	−379.203	−380.187	−380.998	Hellmann *et al.*, 2017

*All energies are in cm^{-1}.

Similarly, other correlation contributions to the total interaction energy described later can also be extrapolated to the CBS limit in this way in order to fulfill the accuracy requirement.

It has been shown previously that $V_{\mathrm{CCSD(T)/core}}$ is one of the most important correction energies, as it accounts for about 5% of the total interaction energy near the equilibrium internuclear separation. If higher precision is required for a potential curve, the FC approximation should be rejected, and additional inner nuclear electrons should be included in the calculation rather than only valence electrons. The number of inner electrons to be included can be selected according to the desired accuracy and acceptable time consumption from part of the inner electrons (IFC) to all electrons (AE).

Note that for a heteronuclear dimer, the energy of the electrons included in the calculation of the two atoms should be close to each other to prevent the computation process from crashing. For example, Hellman *et al.* (2017) included the 4s, 4p, and 4d electrons in their work on the Xe–Xe dimer: however, when dealing with the Xe–Ne and Xe–Ar dimers, the 3d, 4s, 4p, and 4d electrons of Xe should all be considered, as the energies of the 1s electron of Ne and the 2s and 2p electrons of Ar are near the energy of the 3d electrons of Xe.

The $V_{\mathrm{CCSD(T)/core}}$ is obtained from the difference between the interaction energies with the inner core treated as IFC (or AE) and the full FC approximations, that is,

$$V_{\mathrm{CCSD(T)/core}} = V_{\mathrm{CCSD(T)/IFC(or\ AE)}} - V_{\mathrm{CCSD(T)/FC}}. \tag{5.7}$$

Note that $V_{\mathrm{CCSD(T)/IFC}}$ can only be subtracted using same correlation-consistent basis-sets of $V_{\mathrm{CCSD(T)/FC}}$. The $V_{\mathrm{CCSD(T)/rel}}$ term corrects the error arising from the non-relativistic approximation at the CCSD(T) level: this affects the total interaction energy only slightly less than $V_{\mathrm{CCSD(T)/core}}$.

The total Hamiltonian of a system can be decoupled into spin-free and spin-orbit (SO) parts as (Boyd, 2019)

$$H = H^{\mathrm{SR}} + H^{\mathrm{SO}}, \tag{5.8}$$

where the H^{SR} is also known as scalar relativistic (SR) Hamiltonian.

Multiple methods can be utilized to determine the SR correction, such as the rough first-order with degenerate perturbation theory and the more rigorous second-order direct perturbation theory methods (Klopper, 1997; Ottschofski and Kutzelnigg, 1995). For a lighter dimer, the SR correction $V_{\mathrm{CCSD(T)/SR}}$ is much larger than the SO relativistic correction $V_{\mathrm{CCSD(T)/SO}}$, as shown in Table 5.4. Thus, it is sufficient to achieve high accuracy by including the SR correction. In general, calculations with the FC approximation are economical and accurate: if necessary, inclusion of additional inner nuclear electrons in such calculations is a good way to further improve the accuracy.

For heavier dimers, contributions from the SO coupling cannot be neglected anymore, which means that more complete Hamiltonians must be taken into account. Interaction energies can be calculated with four-component Dirac–Coulomb (DC) and Dirac–Coulomb–Gaunt (DCG) Hamiltonians to determine the contributions of the two types of SO coupling, respectively, within the molecular mean-field approach. That is,

Table 5.4 Individual energies of V_{rel} of rare gas dimers near the equilibrium internuclear separation

	R / Å	SR	spin-(own)-orbit	spin-(other)-orbit	Reference
He–He	3.0	0.001	—	—	*a*
Ne–Ne	3.1	−0.060	—	—	*b*
Ar–Ar	3.8	−0.524	—	—	*c*
Kr–Kr	4.0	−2.838	−0.275	−0.129	*d*
Xe–Xe	4.4	−6.217	−1.737	−0.343	*e*

a, Hellmann *et al.*, 2007; *b,* Hellmann *et al.*, 2008*a*; *c,* Jäger *et al.*, 2009; *d,* Jäger *et al.*, 2016; *e,* Hellmann *et al.*, 2017.
*All energies are in cm^{-1}, — indicates that a value is unavailable.

the energies differences between $V_{\mathrm{CCSD(T)/DC}}$ and $V_{\mathrm{CCSD(T)/SR}}$ give corrections to the spin-(own)-orbit effects, and the differences between $V_{\mathrm{CCSD(T)/DCG}}$ and $V_{\mathrm{CCSD(T)/DC}}$ represent corrections to the spin-(other)-orbit effects. These corrections may be calculated at the FC level or at the more time-consuming IFC and AE levels.

For the CC method, the time consumption increases exponentially with the order of excitation. Above the CCSD(T) level, there are the CC singles, doubles and triples (CCSDT) level, the CC singles, doubles, triples and quadruples (CCSDTQ) level, and so on, to the full configuration interaction. Note that at the CC singles, doubles, contracted triples [CCSD(T)] level, which replaces the triple excitations by the product of the single and double excitation operators in iterative calculations, is an economical but still accurate approximation to the CCSDT level, as is the CC singles, doubles, triples and contracted quadruples [CCSDT(Q)] level to the CCSDTQ level.

The interaction energy difference between the CCSD(T) and CCSDT levels is determined as

$$V_{\mathrm{T-(T)}} = V_{\mathrm{CCSDT}} - V_{\mathrm{CCSD(T)}}, \tag{5.9}$$

in which the V_{CCSDT} and $V_{\mathrm{CCSD(T)}}$ are the interaction energies at the CCSDT level and CCSD(T) level, respectively: note that these two energies should be calculated using the same basis-sets and approximations. Similar methods are used to calculate the energies differences between other higher-order CC levels. If limited by computing conditions in reality, segmented computation using different basis-sets will still be a good choice to implement the calculation of this component, such as aug-cc-pVQZ for $V_{\mathrm{T-(T)}}$ and aug-cc-pVTZ for $V_{\mathrm{(Q)-T}}$, and then add them as $V_{\mathrm{(Q)-(T)}}$. Previous work showed that the contribution becomes smaller at higher CC levels (see Table 5.5).

As demonstrated in Table 5.1, the more massive the system, the more corrections should not be neglected when performing state-of-the-art potential energy surface. The contributions and trend of change between these corrections with the interatomic distances and the system are shown in Fig. 5.1.

Table 5.5 Individual energies $V_{\text{post-CCSD(T)}}$ of rare gas dimers near the equilibrium internuclear separation

	R / Å	T-(T)	(Q)-T	Q-(Q)	Reference
Ne–Ne	3.1	−0.449	−0.063	—	Hellmann *et al.*, 2008*a*
Ar–Ar	3.8	1.036	−1.411	—	Jäger *et al.*, 2009
Kr–Kr	4.0	2.442	−2.099	0.147	Jäger *et al.*, 2016
Xe–Xe	4.4	4.761	−3.426	−0.375	Hellmann *et al.*, 2017

*All energies are in cm^{-1}, — indicates that a value is unavailable.

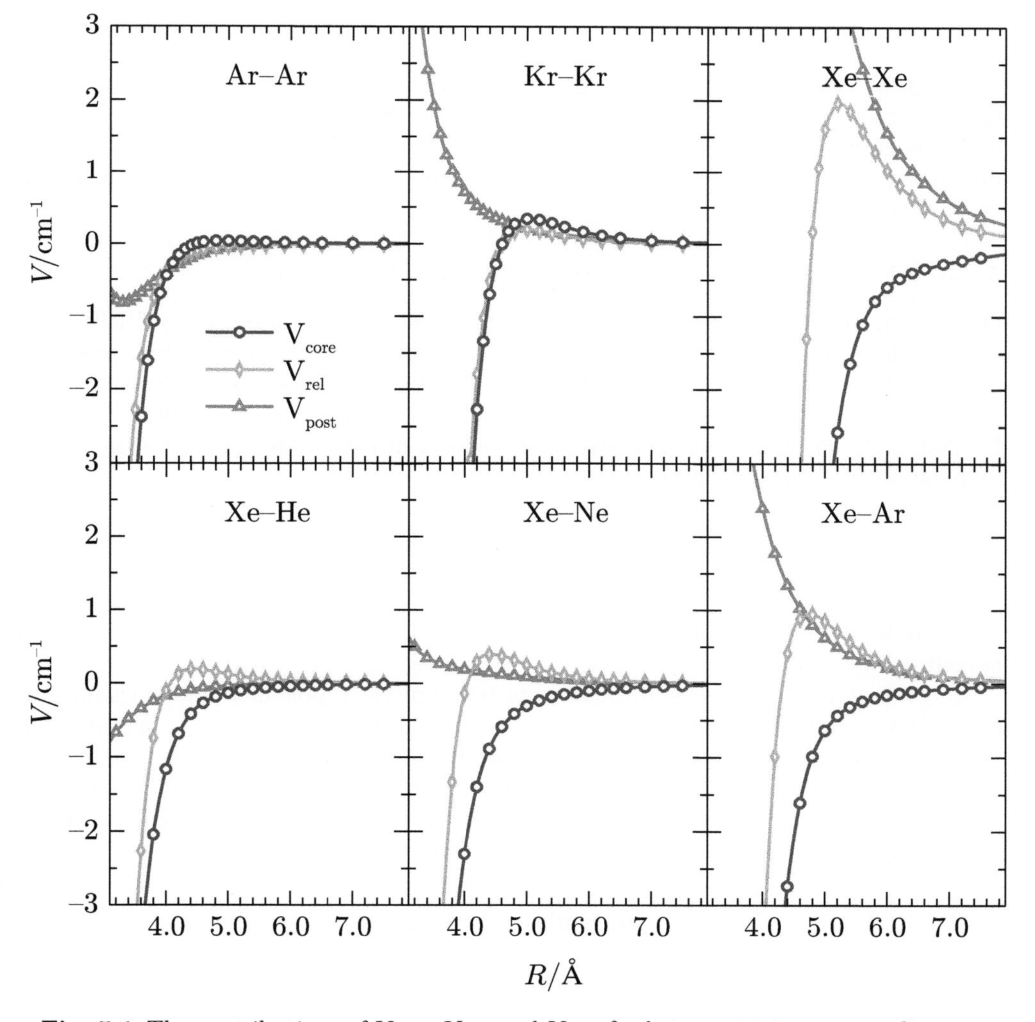

Fig. 5.1 The contributions of V_{core}, V_{rel}, and V_{post} for hetero-atomic rare gas dimers.

5.1.2 Estimates of the combined uncertainty

Each energy term has a corresponding uncertainty, and the combined uncertainty is estimated as the square root of the sum of squares of the uncertainties of the individual components (Hellmann *et al.*, 2017). The uncertainty for the HF interaction energy is

$$\delta_{\mathrm{SCF}} = |V_{\mathrm{SCF}}^{X} - V_{\mathrm{SCF}}^{X-1}|, \tag{5.10}$$

while the uncertainty δ for other components of the total interaction energy, for which basis-sets are extrapolations to the CBS limit, is represented by

$$\delta = \frac{|V^{\mathrm{CBS}} - V^{X}|}{2}. \tag{5.11}$$

To propagate the uncertainties for *ab initio* data into uncertainties for the calculated spectroscopic and thermophysical properties, the uncertainty of the total interaction energy estimates are added or subtracted at each separation to obtain the two perturbed potential curves $V_{U_1} = V^{total} + \delta^{total}$ and $V_{U_2} = V^{total} - \delta^{total}$ (Hellmann *et al.*, 2017).

5.1.3 *Ab initio* calculations of transition dipole moment

One of the most important applications of computational high-quality PEFs is to predict the vibrational spectra. In a vibrational spectrum, transition line positions indicate the energy gaps between ro-vibrational levels, and the transition line strengths are of great value for research in molecular thermodynamics and kinetics (Zhai and Li, 2022). As the transition line strength $S_{J'J''}$ is directly determined by the dipole moment μ_z as (Bernath, 2017)

$$S_{J'J''} = \sum_{M'} \sum_{M''} |\langle \psi_{v'J'M'} | \mu_z | \psi_{v''J''M''} \rangle|^2, \tag{5.12}$$

we need to obtain the dipole moment values in order to construct the absorption transition spectrum.

In general, there are two equivalent ways to define the dipole moment on theoretical grounds: the finite field method and expectation value method. The finite field method is to set a uniform electrostatic field of magnitude F_z in the z direction, so that the electrostatic potential is given by $V_0 - zF_z$. In this case, the dipole moment $\hat{\mu}_z$ is considered to be the derivative with respect to the external field (Zhai and Li, 2022). The energy of the molecular in the field is determined as (Stone, 1996)

$$\begin{aligned} U_{es} &= qV_0 - \sum_a e_a a_z F_z \\ &= qV_0 - \hat{\mu}_z F_z, \end{aligned} \tag{5.13}$$

in which q is the total charge.

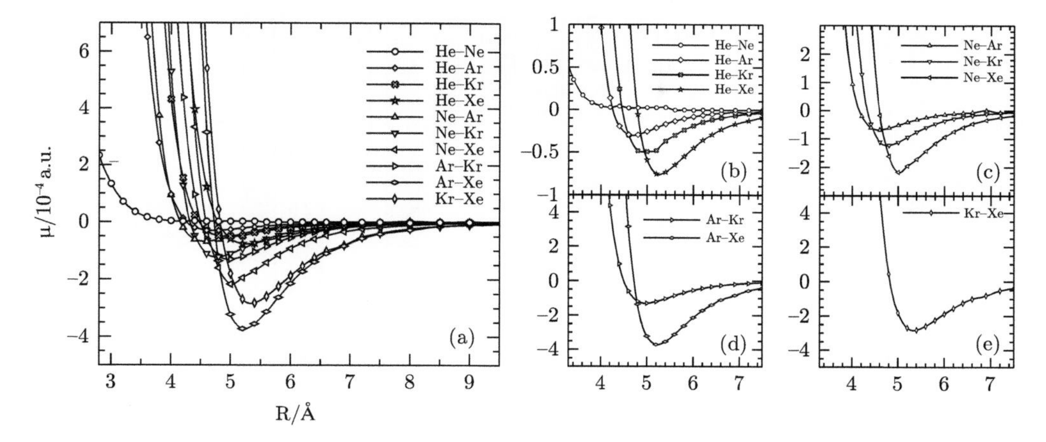

Fig. 5.2 Dipole moment curves for the heteronuclear dimers. Small panels: separate curves from (a). The positive direction points from left to right in an $A{-}X$ dimer.

It is obvious that $\hat{\mu}_z$ is an operator. Thus, we may also obtain its dipole moment in state $|n\rangle$ by calculating the expectation value (Stone, 1996)

$$\begin{aligned} \mu_z &= \langle n|\hat{\mu}_z|n\rangle \\ &= \int \rho_n(\mathbf{r}) r_z d^3\mathbf{r}, \end{aligned} \tag{5.14}$$

in which $\rho_n(\mathbf{r})$ is the molecular charge density.

To construct spectroscopically accurate dipole moment functions (DMFs), some important details on selecting the basis-sets must be noted, as interaction potential energies and dipole moments are also affected significantly by the BSSE. The CP corrected curves converge much faster than the uncorrected curves and eliminate artificial the unphysical behavior. Furthermore, mid-bond functions (bf) can also help the computational DMFs converge faster and become more reliable (Zhai and Li, 2022).

Figure 5.2 presents all DMFs of heteronuclear dimers. According to the rule of fixing the smaller atom in the dimer, four small panels (b), (c), (d), and (e) are abstracted from (a). The small panels show that, as the dimer increases in size, the position of the minimum increases, and its value becomes smaller as the larger atom is more negative. Some apparent confusions are that the largest heteronuclear dimer does not correspond to the smallest minimum value (see Fig. 5.3) and the minima of the DMFs and PECs do not occur at the same internuclear separation (see Fig. 5.4).

5.2 Fitting of Analytic Potential Energy Functions

Pointwise total potential energies are obtained from the *ab initio* calculations. As these energies alone are insufficient to develop a continuous and practical PEF, a model describing the interaction energy behavior reliably must be selected. These

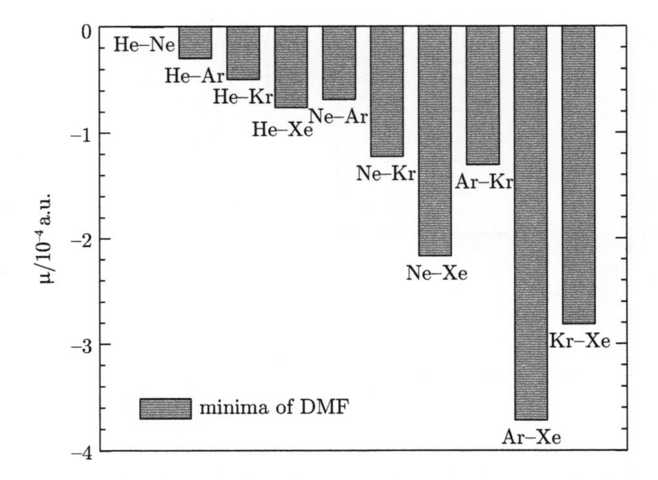

Fig. 5.3 The minima of the dipole moment curves shown in Fig. 5.2.

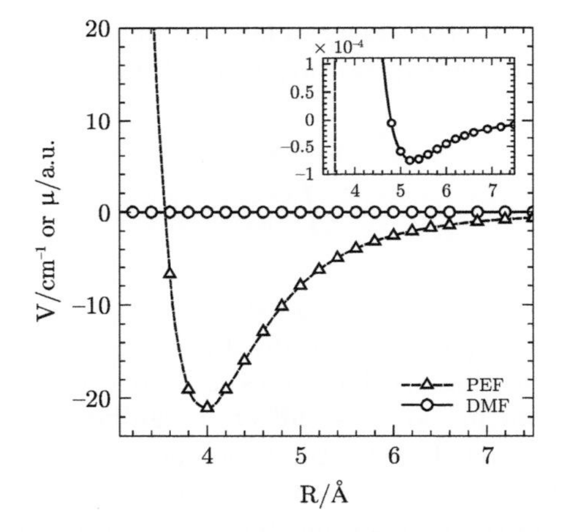

Fig. 5.4 The PEF and DMF of the Xe–He dimer. Small panel: zoom into the minimum regime of DMF.

interaction energies are then fitted to the model by an interpolation method. This section applies the widely used Morse/long-range function (Le Roy and Henderson, 2007) (detailed expressions can be found in Section 3.5) as an example. The quality of a multi-parameter fit is characterized by the value of the dimensionless root-mean-square deviation (DRMSD, or $\overline{dd}$).

The betaFIT program is a tool for fitting developed by Le Roy and Pashkov (2017) and widely employed to fit pointwise potentials to selected analytic functions (not only the MLR form). The fitting results in a set of parameters that describes the specific

Table 5.6 Parameters defining the recommended MLR potentials for ground-state ($X\,^1\Sigma_g^+$) Rg–Rg dimers

	parameters	He–He	Ne–Ne	Ar–Ar	Kr–Kr	Xe–Xe
fitted	$\mathfrak{D}_e$ /cm^{-1}	7.63534	29.3822	99.49	139.58207	194.51696
	r_e /Å	2.96754	3.0896	3.766	4.01606	4.37815
	β_0	0.62494	0.2034	0.04743	0.057092	−0.0547
	β_1	0.24	−0.162	0.0877	−0.0792	−0.0742
	β_2	2.55808	−0.03	0.112	−0.0809	−0.18022
	β_3	1.9218	0.1	0.57	0.12876	0.169
	β_4	−0.04	0.308	0.33	0.111	0.076
predetermined	C_6 cm^{-16}	7.04511×10^3	3.063×10^4	3.105×10^5	6.250×10^5	1.392×10^6
	C_8 cm^{-18}	1.90606×10^4	1.146×10^5	2.219×10^6	5.085×10^6	1.384×10^7
	C_{10} cm^{-110}	6.9442×10^4	5.494×10^5	1.899×10^7	5.446×10^7	1.994×10^8
	r_{ref} /Å	3.22	2.7	5.3	4.0	4.5
	ρ	0.88	1.1	1.1	0.9	0.9
	p	5	6	6	6	6
	q	1	3	3	3	3
data source	references	*a*	*b*	*c*	*d*	*e*
quality-of-fit	$\overline{dd}$	1.58	0.23	1.04	0.15	0.07

	parameter	He–Ne	He–Ar	He–Kr	He–Xe	Ne–Ar
fitted	$\mathfrak{D}_e$ /cm^{-1}	14.630	20.706	21.8466	21.097749	45.1964
	r_e /Å	3.028137	3.491612	3.6821	3.966172	3.4924735
	β_0	0.25269	0.683	0.0331	−0.082649	−0.166376
	β_1	0.058	−0.034	0.13221	0.1069	0.18707
	β_2	0.1477	0.26	−0.03	0.0347	−0.087
	β_3		−1.6	0.2137	0.22513	0.16
	β_4		3.9	0.15		−0.245
predetermined	C_6 cm^{-16}	1.4758457×10^4	4.5342434×10^4	6.4936783×10^4	9.41707×10^4	9.1600179×10^4
	C_8 cm^{-18}	4.2753922×10^4	2.2338386×10^5	3.4544808×10^5	6.8289×10^5	5.3019317×10^5
	C_{10} cm^{-110}	1.6164672×10^5	1.4350080×10^6	2.6420650×10^6	5.38×10^6	3.9918170×10^6
	r_{ref} /Å	3.1	2.4	4.3	4.1	3.9
	ρ	1.1	0.8	1.1	1.1	1.5
	p	5	5	5	5	5
	q	3	3	2	3	3
data source	references	*f*	*f*	*g*	*h*	*f*
quality-of-fit	$\overline{dd}$	0.05	0.15	0.38	1.51	0.01

	parameter	Ne–Kr	Ne–Xe	Ar–Kr	Ar–Xe	Kr–Xe
fitted	$\mathfrak{D}_e$ /cm^{-1}	49.677861	51.069600	116.414000	130.523000	165.895000
	r_e /Å	3.649039	3.887003	3.894061	4.092120	4.194350
	β_0	−0.059682	−0.11093	−0.173533	−0.2862	0.19664
	β_1	0.07499	0.0946	0.0159	0.1471	0.11471
	β_2	0.0311	0.04	−0.14065	−0.055	0.153
	β_3	0.0845	0.20482	0.19011	0.41731	0.0473
	β_4					
predetermined	C_6 cm^{-16}	1.31569×10^5	1.91137×10^5	4.3919×10^5	6.48207×10^5	9.24839×10^5
	C_8 cm^{-18}	7.061×10^5	1.3837×10^6	2.9921×10^6	5.507×10^6	9.6681×10^6
	C_{10} cm^{-110}	1.098×10^7	1.568×10^7	5.392×10^7	7.42×10^7	1.207×10^8
	r_{ref} /Å	3.9	4.0	4.0	4.2	4.3
	ρ	1.1	1.1	1.1	1.2	0.8
	p	5	5	6	5	5
	q	3	3	3	3	3
data source	references	*i*	*h*	*i*	*h*	*i*
quality-of-fit	$\overline{dd}$	1.04	1.10	0.25	1.25	0.80

a, Przybytek *et al.*, 2010; Cencek et al., 2012; *b*, Hellmann *et al.*, 2008*a*; *c*, Myatt *et al.*, 2018; *d*, Jäger and Bich, 2017; *e*, Hellmann *et al.*, 2017; *f*, López Cacheiro *et al.*, 2004; *g*, Jäger *et al.*, 2016; *h*, Liu *et al.*, 2022; *i*, Hu *et al.*, 2022.

PEF of a system. Table 5.6 gives the parameters defining the recommended MLR potentials for ground-state ($X\,^1\Sigma_g^+$) Rg–Rg dimers. These fitted MLR PEFs can be seen to be of high quality, as determined by their $\overline{dd}$ values.

5.3 Spectroscopic Properties

The calculation of ro-vibrational energy levels can be carried out using the program LEVEL (Le Roy, 2017) with the parameters given in Table 5.6. A series of spectroscopic parameters can be extracted from computation, for example, zero-point energy (ZPE), number and energy of vibrational and rotational energy levels, the ground state inertial rotational constant B_0 and the dissociation energy. Table 5.7 presents some of the spectral parameters of all rare gas dimers and Fig. 5.5 shows the microwave spectra of the Xe–He dimer.

Table 5.7 shows that the number of vibrational energy levels increases with dimer augmentation, while Figure 5.5 indicates that as the vibrational quantum number v increases, the number of rotational levels decreases, but only gradually.

Although we often think of ro-vibrational levels at "sublevels" of pure vibrational levels, in computational physics the rotational kinetic energy is often treated as part of an effective interatomic potential, as in eqn (5.15),

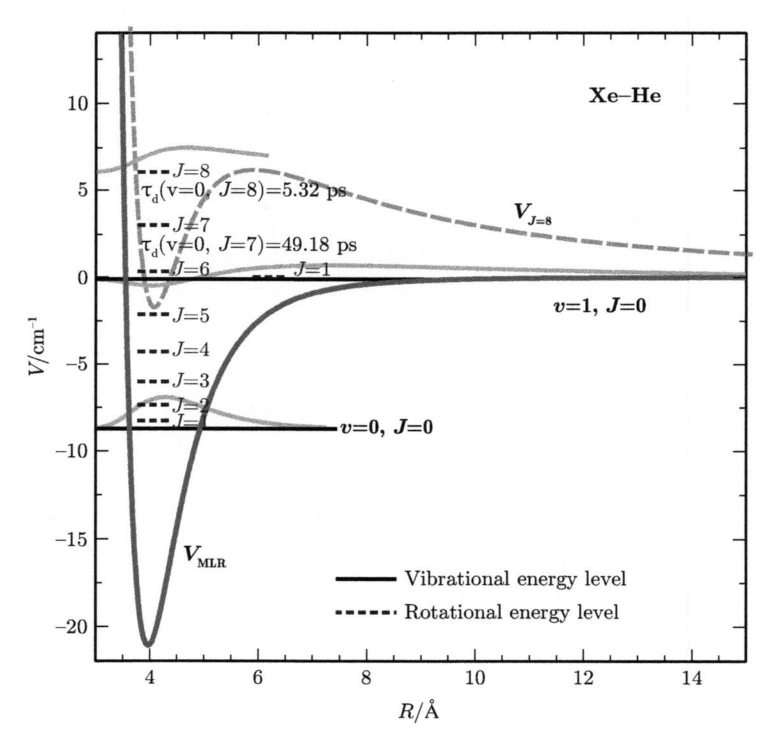

Fig. 5.5 MLR potential energy curve, energy levels, wave functions, and other spectral information for the Xe–He dimer. Red line: MLR potential curve, gray solid lines, wave functions of vibrational energy levels; grey dashed lines, effective radial potential. This image is from (Liu *et al.*, 2022), with permission of the publisher.

Table 5.7 Spectral parameters for ground-state ($X\,^1\Sigma_g^+$) Rg–Rg dimers

	ZPE	NVEL*	B_0	D_0
He–He	−7.6343	1	0.0431321449	−0.0010
He–Ne	−12.0449	1	0.3602044862	−2.5851
He–Ar	−13.6817	1	0.2958546479	−7.0243
He–Kr	−13.3771	2	0.2618967781	−8.4695
He–Xe	−12.3948	2	0.2257348374	−8.7029
Ne–Ne	−12.5838	3	0.1559808027	−16.7984
Ne–Ar	−12.7143	4	0.0965243461	−32.4821
Ne–Kr	−11.7606	5	0.0739756506	−37.9173
Ne–Xe	−10.9328	6	0.0610449360	−40.1368
Ar–Ar	−14.7949	8	0.0575079567	−84.6951
Ar–Kr	−13.4272	11	0.0400174930	−102.9868
Ar–Xe	−12.7977	13	0.0321159466	−117.7253
Kr–Kr	−11.5337	15	0.0244636990	−128.0484
Kr–Xe	−11.0270	19	0.0184743738	−154.8680
Xe–Xe	−10.0865	25	0.0132886609	−184.4305

*NVEL: number of vibrational energy levels. All energies are in cm^{-1}.

$$V_J(R) = V_{J=0}(R) + \frac{h}{8\pi^2\mu c}\frac{J(J+1)}{R^2}, \tag{5.15}$$

in which h is the Planck constant, μ is the reduced mass, and c is the speed of light in vacuum. In Fig. 5.5, the effective interatomic potential for $J = 8$ is plotted together with the original MLR potential curve, as well as the vibrational wave functions and ro-vibrational levels. The rotational effective potential, or kinetic energy, creates a centrifugal barrier (CB) on the potential energy curve (Piela, 2020). It is easy to obtain the CB for $J = 8$ for Xe–He located at $R = 5.90$, with $V_{J=8} = 6.1769$ cm^{-1}. It is just a little higher than that obtained for the $v = 0, J = 8$ resonance level (6.0509 cm^{-1}). This level has a short lifetime of $\tau_\text{d} = 5.32$ ps, which should, however, still be observed in high-accuracy experimental measurements.

A similar analysis can be carried out for all different Js for obtain an illustration of the ro-vibrational energy level structure of these systems. In Fig. 5.6, we label the lowest points (LP) and the CB for all nine (effective) potential curves. Together with the repulsive wall, they define the semi-classical allowed vibrational region. The mini-panel of Fig. 5.6 shows the depth of the region (Δ_J). Despite the increase in both LP and CB as J increase, the depth Δ_J decreases, which means that the valley becomes shallower. As such a system is highly anharmonic, it is arbitrary to take the harmonic approximation at the LP. The second-order derivatives of V_J at its LP decreases as J increases, which is favorable to having additional vibrational levels for a specific V_J.

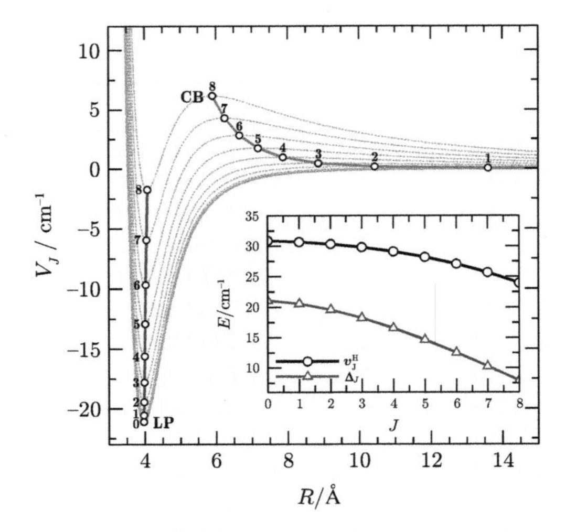

Fig. 5.6 The nine effective potential curves of the Xe–He dimer marked with the lowest points (LP) and the centrifugal barriers (CB). The inset panel presents the depth of the region Δ_J and the corresponding harmonic frequencies $\tilde{\nu}_J^{\mathrm{H}}$. This image is from (Liu *et al.*, 2022), with permission of the publisher.

A plot of the corresponding harmonic frequencies $\tilde{\nu}_J^{\mathrm{H}}$ for different V_J is shown in the mini-panel of Fig. 5.6. It is clear that $\tilde{\nu}_J^{\mathrm{H}}$ decrease much more slowly than the depth Δ_J, which indicates that there will be fewer vibrational levels for large J. Similar phenomena can also be found for the other rare gas dimers.

Note that only the bound and quasi-bound states are considered in this example, while continuous states are excluded. As rare gas dimers are bonded by weak van der Waals forces, in more rigorous work, continuum states should also be considered, especially in predicting the sensitive transition spectra.

The absorption spectrum, which can be predicted from the *ab initio* DMFs, implies the electronic transition information of a system. The line intensity S' (cm/molecule) is determined by (Bernath, 2017)

$$S' = \frac{2\pi^2\omega(e^{-E''/k_{\mathrm{B}}T} - e^{-E'/k_{\mathrm{B}}T})}{3\epsilon_0 hcQ(T)} S_{J'J''}, \tag{5.16}$$

in which ω is transition wavenumber, E'' and E' are the lower and upper state energies and $S_{J'J''}$ is the line strength defined in eqn (5.12). An example of the absorption spectrum is shown in Figs. 5.7 and 5.8.

5.4 Thermophysical Properties

The collision integrals $\Omega^{(l,s)}$ are derived from the PEF, $\varphi(r)$, and the angle of deflection in a collision, χ, and used to calculate the thermophysical properties. Applying the PEF information in Table 5.6, the second virial coefficient $B_{\mathrm{mix}}(T)$, the second

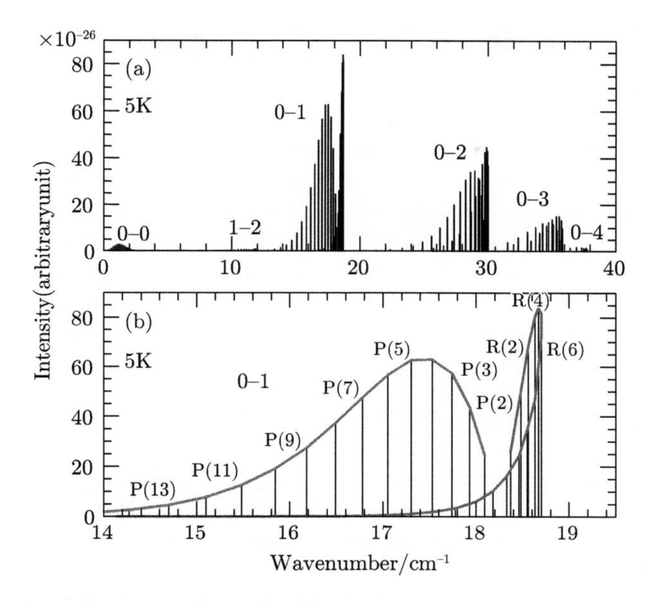

Fig. 5.7 Absorption spectrum of the Ke–Ne dimer at 5 K. This image is from (Hu *et al.*, 2022), with permission of the publisher.

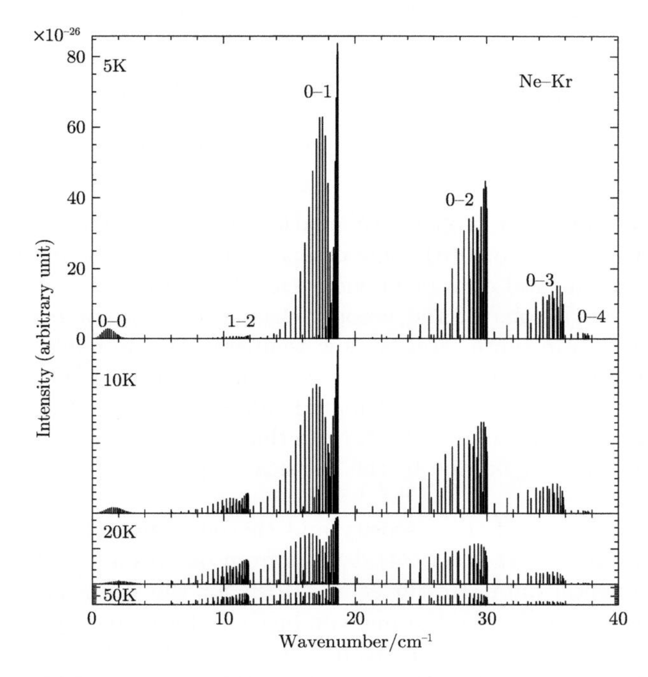

Fig. 5.8 Absorption spectrum of the Ke–Ne dimer at 5 K, 10 K, 20 K, and 50 K. This image is from (Hu *et al.*, 2022), with permission of the publisher.

acoustic virial coefficient $\beta_{12}(T)$, and the shear viscosity and thermal conductivity coefficients $\eta_{\text{mix}}(T)$ and $\lambda_{\text{mix}}(T)$, as well as the binary diffusion coefficient $D_{12}(T)$ and the thermal diffusion factor α_{T} of the corresponding dilute gas can be determined. Detailed theoretical background appears in Chapter 2 and Chapter 3.

Note that the properties of pure and mixed gases are calculated differently. In an experiment, the second virial coefficient of a mixture $B_{mix} = x_1^2 B_1 + 2x_1x_2B_{12} + x_2^2 B_2$ can only be measured with the same degree of accuracy as that for the two pure gases (B_1 and B_2), so that values of B_{12} extracted from experimental measurements will be considerably less accurate (by as much as an order of magnitude) than values for pure gases. Thus, the best source for B_{12} is therefore values calculated from accurate *ab initio* PEFs. The cross (or interaction) second virial coefficient of a heteronuclear dimer is described as (Jäger and Bich, 2017)

$$B_{12}(T) = B_{12}^{\text{cl}}(T) + \lambda B_{12}^{\text{qm},1}(T) + \lambda^2 B_{12}^{\text{qm},2}(T) + \lambda^3 B_{12}^{\text{qm},3}(T), \tag{5.17}$$

in which $\lambda = \hbar^2/24\mu_{12}$, $\mu_{12} = m_1m_2/(m_1+m_2)$. $B_{12}^{\text{cl}}(T)$, $B_{12}^{\text{qm},1}(T)$, $B_{12}^{\text{qm},2}(T)$, and $B_{12}^{\text{qm},3}(T)$, which formulae are the same as for the homonuclear dimer case considered by Bich *et al.* (2008), and depends on the derivative of the PEFs.

The interaction second acoustic virial coefficient is given as a function of the cross second pressure virial coefficient $B_{12}(T)$, the first temperature derivatives $B'_{12}(T)$, and the second temperature derivatives $B''_{12}(T)$ (Ewing and Trusler, 1992*a*; Vogel *et al.*, 2008) The computation of the second acoustic virial coefficient of the mixture β_{mix} follows the same arguments given for the computation of $\beta_{12}(T)$ (Ewing and Trusler, 1992*a*),

$$\begin{aligned}\beta_{mix}(T) = 2[B_{mix}(T) + (\gamma^\circ - 1)TB'_{mix}(T) \\ + \frac{(\gamma^\circ - 1)^2}{2\gamma^\circ}T^2B''_{mix}(T)].\end{aligned} \tag{5.18}$$

The procedure for transport property calculations appears in Chapter 2 and repeated here. Figure 5.9 to Figure 5.10(d) show comparisons between computed values from Table 5.6 and experimental data or previous calculated values.

The absolute deviations of the second virial coefficient of equimolar mixtures of xenon and argon are quite large at low temperatures, but decrease to less than 1.40 cm^{3}mol^{-1} whether compared with the experimental data of Schramm *et al.* (1977) and Rentschler *et al.* (1977) or the calculated values of Kestin *et al.* (1984) when the temperature exceeds room temperature. Note that the large errors shown at low temperatures in Fig. 5.9 may be due to the classical approach employed here to obtain collision integrals.

The calculated values of the viscosity and thermal conductivity of dilute Xe–He mixtures given by Kestin *et al.* (1984) deviate from our results by less than ±1.07%, and their experimental data (Kestin *et al.*, 1978) for the viscosity deviate from our calculated values by 1.53%. The thermal conductivity measured by Haire *et al.* (2007) is in agreement with our results to within ±5.38% at 291.2 K and −1.95% at 302.2 K. The maximum and minimum deviations from their work are 6.08% at 793.2 K and 0.16% at 291.2 K, respectively. The trend of viscosity with respect to the mole fraction

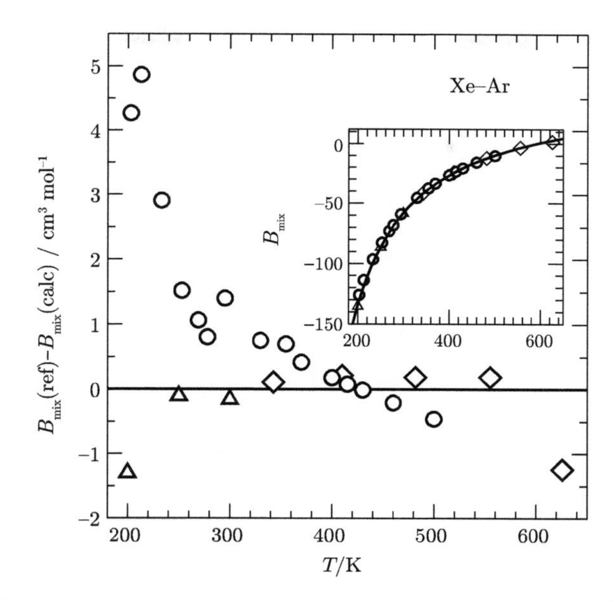

Fig. 5.9 Temperature dependence of the absolute deviations of experimental and theoretical values for the second virial coefficient of dilute equimolar mixtures of xenon and argon. **Inset panel:** Temperature dependence of the absolute values of the second virial coefficient. Experimental data: circle, Schramm *et al.*, 1977 ; diamond, Rentschler *et al.*, 1977. Black solid line, this work. This image is from (Liu *et al.*, 2022), with permission of the publisher.

x_{Xe} of Xe is nonmonotonic, whereas the thermal conductivity decreases rapidly as x_{Xe} increases.

The diffusion coefficient is only slightly affected by the mixture composition, but depends strongly upon temperature. Figure 5.10(c) shows the differences between the temperature dependence of the present calculated results for the diffusion coefficient in equimolar binary mixtures and both experimental and other calculated values. This figure shows that the deviations between the experimental values (Malinauskas, 1965; Taylor and Cain, 1983) and our calculated values for the binary diffusion coefficient for equimolar mixtures of Xe and Ar are widely distributed (between ±0.24% and ±5.40%), while the differences between our calculated values and those of previous calculations (Kestin *et al.*, 1984; Malinauskas, 1965) are more densely distributed (between ±2.78%).

Deviations between our calculated values for the thermal diffusion factor and experiment are relatively large: values of α_T obtained by Hurly *et al.* (1991) differ by as much as 9.24%, while those obtained by Kestin *et al.* (1984) differ by as much as −11.10%. The inset panel in Fig. 5.10(d) shows that the corresponding absolute deviations lie between 0.0003 and 0.0376.

Figure 5.11 shows the calculated values for the interaction second virial coefficient, $B_{12}(T)$, and the interaction second acoustic virial coefficient, $\beta_{12}(T)$, plus the mixture shear viscosity, η_{mix}, the mixture thermal conductivity, λ_{mix}, the binary diffusion

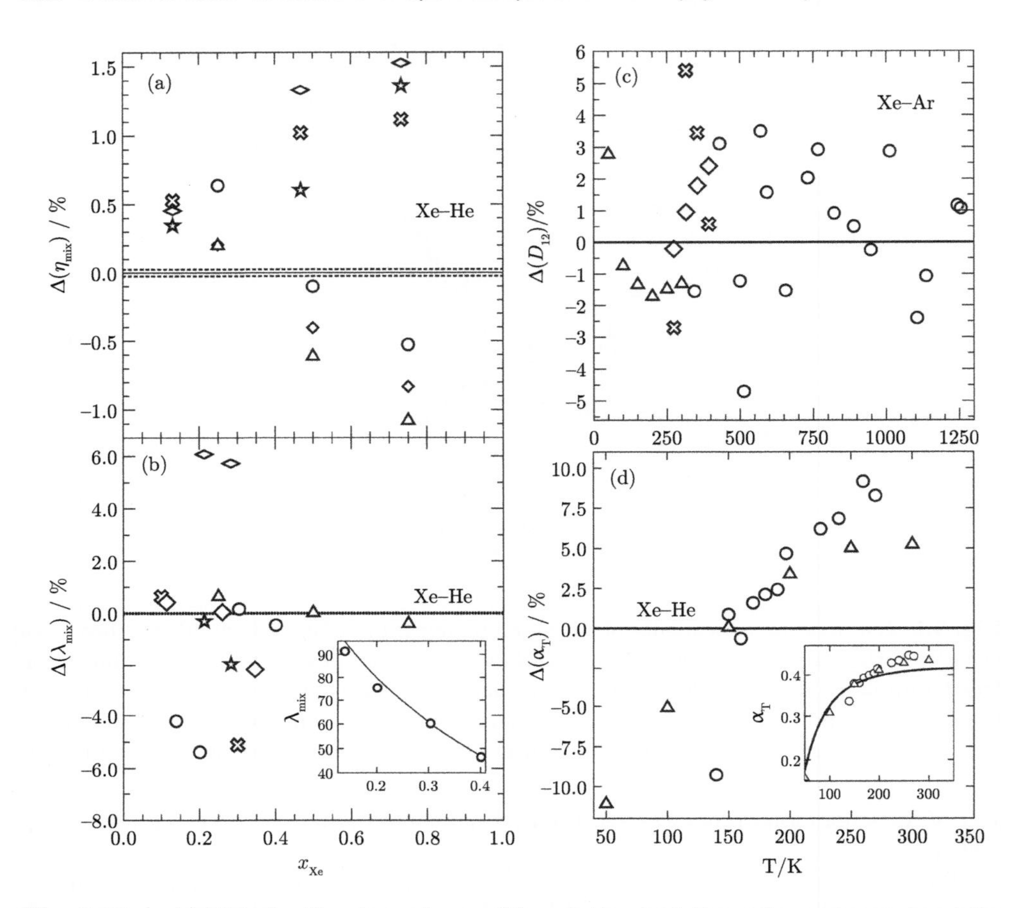

Fig. 5.10 (a, b) Mole fraction dependence of the relative deviations of experimental and theoretical values for the viscosity and thermal conductivity of dilute Xe–He gases. Black solid line, this work; dotted lines, uncertainty of this work. **Upper panel:** Viscosity. Experimental data: cross, 373.15 K; star, 568.15 K; horizontal lozenge, 778.15 K, Kestin *et al.* Calculated values: circle, 150 K; triangle, 200 K; diamond, 250 K, Kestin *et al.* **Lower panel:** Thermal conductivity. Experimental data: circle, 291.2 K; star, 302.2 K; diamond, 311.2 K; horizontal lozenge, 793.2 K; cross, 1500 K, Haire *et al.* Calculated values: triangle, 200 K, Kestin *et al.* **Inset panel:** Mole fraction dependence of the absolute values of the thermal conductivity.
(c) Temperature dependence of the relative deviations of experimental and theoretical values for the binary diffusion coefficient of dilute equimolar mixtures of xenon and argon. Experimental data: circle, Taylor *et al.*; cross, Malinauskas calculated values; diamond; Malinauskas triangle, Kestin *et al.* Black solid line, this work.
(d) Temperature dependence of the relative deviations of experimental and theoretical values for the thermal diffusion factor of dilute equimolar mixtures of xenon and helium. Experimental data: circle, Hurly *et al.* Calculated values: triangle, Kestin *et al.* Black solid line, this work. Inset panel: Temperature dependence of the absolute values of the thermal diffusion factor. This image is from (Liu *et al.*, 2022), with permission of the publisher.

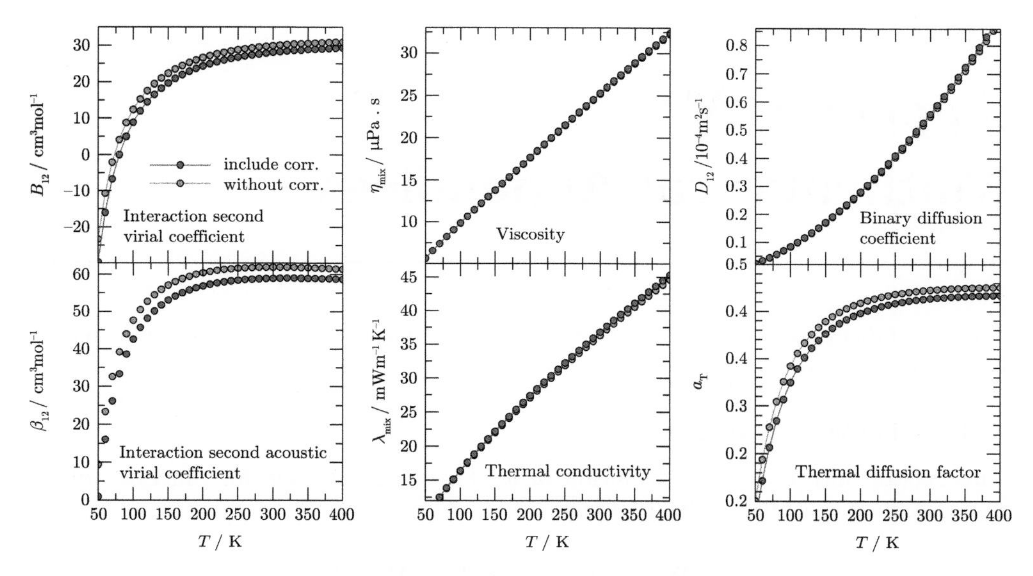

Fig. 5.11 The calculated cross second pressure virial coefficients, interaction second acoustic virial coefficients, and the interaction transport properties of the relevant dilute equimolar mixtures of xenon and helium including or without the correction (the V_{core}, V_{rel}, $V_{\text{T}-(\text{T})}$, $V_{(\text{Q})-\text{T}}$, and $V_{\text{T}-(\text{T})/\text{core}}$) from 50 K to 400 K. Image from (Liu *et al.*, 2022), with permission of the publisher.

coefficient, $D_{12}(T)$, and the thermal diffusion factor, α_{T}, for equimolar mixtures of Xe and He, both with and without the correlation correction described earlier. The blue lines correspond to the thermophysical properties computed from PEFs at the CCSD(T) level with frozen core approximation. As may be seen from this figure, these thermophysical properties are affected by the nature of the repulsive wall of the PEF, especially the cross second virial coefficient, the interaction second acoustic virial coefficient and the thermal diffusion factor. Hence, the corrections mentioned previously are necessary in order to achieve the level of accuracy required for the energies at short internuclear separations.

In summary, the PEF plays a crucial role in determining the spectroscopic and thermophysical properties of matter. As the *ab initio* approaches have matured sufficiently to enable highly accurate interaction energies to be calculated for multi-electron systems, shows its superiority for obtaining accurate spectroscopic and thermophysical properties of the rare gas dimers. Additional details of MLR PEFs and computational programs for rare gas dimers may be found at https://huiligroup.org.

Appendix A
Mathematical Appendices

The appendices contain material of a more formal nature that has been used mainly in Chapters 1 and 2.

A.1 Maxwellian Averages

This section introduces the definition of a Maxwellian average for functions of the dimensionless (peculiar) velocity $\mathbf{W}$ as

$$\langle \phi(\mathbf{W}) \rangle \equiv \pi^{-\frac{3}{2}} \int_0^\infty \mathrm{e}^{-W^2} \phi(\mathbf{W}) \, \mathrm{d}\mathbf{W} \,. \tag{A.1}$$

Some commonly occurring examples of Maxwellian averages given in Chapter 1 are those of the average (relative) speed $\bar{c}$ and average (relative) kinetic energies $\langle W_i^2 \rangle$, $i = x, y, z$, and $\langle W^2 \rangle$ of eqns (1.77). Additional examples of Maxwellian averages are provided by the various powers of W, such as

$$\langle 1 \rangle = 1, \quad \langle W \rangle = \frac{2}{\sqrt{\pi}}, \quad \langle W^2 \rangle = \tfrac{3}{2}, \quad \langle W^4 \rangle = \tfrac{15}{4} \,.$$

These averages can be employed to obtain a related average, given by the variance $\langle (E - \langle E \rangle)^2 \rangle$ of the reduced kinetic energy E, given by

$$\langle (E - \langle E \rangle)^2 \rangle = \langle (W^2 - \tfrac{3}{2})^2 \rangle = \tfrac{3}{2} \,. \tag{A.2}$$

More generally, Maxwellian averages of integer powers of W are given by

$$\langle W^n \rangle = \frac{2}{\sqrt{\pi}} \Gamma(\tfrac{n+3}{2}) = \frac{(n+1)!!}{2^{n/2}} \,, \quad n \text{ even} \,, \tag{A.3a}$$

and

$$\langle W^n \rangle = \frac{2}{\sqrt{\pi}} \left(\frac{n+1}{2} \right)! \,, \quad n \text{ odd} \,. \tag{A.3b}$$

Some Maxwellian averages that arise in the determination of transport properties (Chapman and Cowling, 1970; Hirschfelder *et al.*, 1964; Maitland *et al.*, 1981; McCourt *et al.*, 1990) of noble gases are listed below. Maxwellian averages relevant to the viscosity typically employ Maxwellian averages, such as

$$\langle \mathbf{WW} \rangle = \tfrac{1}{2} \boldsymbol{\delta} \tag{A.4a}$$

and

$$\langle \overline{\mathbf{WW}}\,\overline{\mathbf{WW}} \rangle = \tfrac{1}{2}\boldsymbol{\Delta} \tag{A.4b}$$

involving the second-rank tensors $\mathbf{WW}$ and $\overline{\mathbf{WW}}$ and the isotropic second-rank unit tensor $\boldsymbol{\delta}$ and the isotropic fourth-rank tensor $\boldsymbol{\Delta} \equiv \boldsymbol{\Delta}^{(2)}$ that is symmetric and traceless in both the front and back pairs of indices. More generally, Maxwellian averages of higher-order tensors constructed from $\mathbf{W}$ take forms similar to

$$\langle W_i W_j W_k W_l \rangle = \tfrac{1}{4}(\delta_{ij}\delta_{kl} + \delta_{ik}\delta_{jl} + \delta_{il}\delta_{jk})\,, \tag{A.5a}$$

with $i, j, k, l \in \{x, y, z\}$, and

$$\langle \overline{\mathbf{WW}^{l}}\,\overline{\mathbf{WW}^{l'}} \rangle = l!\, 2^{-l} \delta_{ll'} \boldsymbol{\Delta}^{(l)}\,, \tag{A.5b}$$

with $l, l' \in \{x, y, z\}$ and $\boldsymbol{\Delta}^{(l)}$ an isotropic tensor of rank $2l$ that is symmetric and traceless in the first l indices and in the final l indices.

A.2 Special Functions

(McCourt *et al.*, 1990).

A.2.1 Associated Laguerre polynomials

The special functions most commonly encountered in descriptions of the transport properties of noble gases are the associated Laguerre polynomials $L_n^{(\alpha)}(x)$. These polynomials are defined on the interval $[0, \infty)$ and are given by

$$\begin{aligned} L_n^{(\alpha)}(x) &\equiv \frac{1}{n!}\, x^{-\alpha} e^{x}\, \frac{d^n}{dx^n}(x^{n+\alpha} e^{-x}) \\ &= \sum_{k=0}^{n} \frac{(-1)^k\, \Gamma(\alpha + n + 1)}{k!(n-k)!\, \Gamma\,(\alpha + k + 1)}\, x^k\,. \end{aligned} \tag{A.6}$$

The associated Laguerre functions have been defined here in terms of gamma functions (see Section A.2.3) as α may assume both integer and half-odd-integer values.[1] The ordinary Laguerre polynomials $L_n^{(0)}(x) \equiv L_n(x)$ of degree n correspond to associated Laguerre polynomials with $\alpha = 0$. The associated Laguerre polynomials are orthogonalized according to

$$\int_0^{\infty} x^{\alpha} e^{-x} L_n^{(\alpha)}(x) L_{n'}^{(\alpha)}(x)\, dx = \frac{\Gamma(n + 1 + \alpha)}{n!}\, \delta_{nn'}\,, \tag{A.7a}$$

and obey the recurrence relation

$$(n+1)L_{n+1}^{(\alpha)}(x) = (2n + 1 + \alpha - x)L_n^{(\alpha)}(x) - (n + \alpha)L_{n-1}^{(\alpha)}(x)\,. \tag{A.7b}$$

[1] Note that the associated Laguerre polynomials are often referred to as Sonine polynomials, designated $S_{\alpha}^{(n)}(x)$, in much of the earlier traditional kinetic theory literature (Chapman and Cowling, 1970; Hirschfelder *et al.*, 1964; Maitland *et al.*, 1981)

A.2.2 Legendre and associated Legendre polynomials

Legendre polynomials are defined on the closed interval $[-1, 1]$ and are traditionally obtained directly from the Rodrigues formula

$$P_l(x) = \frac{1}{2^l l!} \frac{\mathrm{d}^l}{\mathrm{d}x^l}(x^2 - 1)^l \tag{A.8a}$$

or from the more explicit expression

$$P_l(x) = \frac{1}{2^l} \sum_{m=0}^{[l/2]} (-1)^m \frac{(2l - 2m)!}{m!(l - m)!(l - 2m)!} x^{l-2m}, \tag{A.8b}$$

in which the upper limit $[l/2]$ for the summation equals $l/2$ should l be an even integer and $(l-1)/2$ should l be an odd integer. Legendre polynomials are normalized so that $P_l(1) = 1$, they are orthogonal, that is,

$$\int_{-1}^{1} P_l(x) P_n(x)\, \mathrm{d}x = \delta_{nl} \frac{2}{2l + 1}, \tag{A.9a}$$

and they satisfy the recurrence relation

$$(l + 1)P_{l+1}(x) = (2l + 1)xP_l(x) - lP_{l-1}(x). \tag{A.9b}$$

Associated Legendre polynomials, $P_{lm}(x)$, are obtained from Legendre polynomials via

$$P_{lm}(x) = (1 - x^2)^{m/2} \frac{\mathrm{d}^m}{\mathrm{d}x^m} P_l(x), \qquad m = 1, 2, \cdots, l, \tag{A.10}$$

and satisfy the orthogonality condition

$$\int_{-1}^{1} P_{lm}(x) P_{nm}(x)\, \mathrm{d}x = \delta_{nl} \frac{2}{2l + 1} \frac{(l + m)!}{(l - m)!}. \tag{A.11}$$

A.2.3 Gamma functions

The gamma function that appears in Eqs.(A.6–7) is itself defined formally by the integral

$$\Gamma(z + 1) = \int_0^\infty t^z \mathrm{e}^{-t}\, \mathrm{d}t, \quad \mathrm{Re}\, z > 1, \tag{A.12}$$

and has the general property

$$\Gamma(z + 1) = z\Gamma(z). \tag{A.13}$$

If z is a real non-negative integer n, then $\Gamma(n + 1)$ and $\Gamma(n + \frac{1}{2})$ are given by

$$\Gamma(n + 1) = n(n - 1) \cdots 3 \cdot 2 \cdot 1 \equiv n! \tag{A.14a}$$

and

$$\Gamma(n + \tfrac{1}{2}) = \frac{(2n - 1)!!}{2^n} \sqrt{\pi} \tag{A.14b}$$

in terms of the factorial $n!$ of n for even-integer values of n and in terms of the double-factorial $n!!$ of n for odd-integer values of n. Although $n!$ is fairly common for both even and odd integer values of n, the double factorial $(2n+1)!!$ for an odd integer, given by

$$(2n+1)!! \equiv (2n+1)(2n-1)\ \cdots\ 5\cdot 3\cdot 1 = \frac{(2n+1)!}{2^n n!}\,, \tag{A.15}$$

is less common. Note also that $1!! \equiv 1$ and $(-1)!! \equiv 1$.

A.2.4 Spherical harmonics

The spherical harmonic functions $Y_{lm}(\mathbf{u})$ employed here correspond to those defined, for example, by Edmonds (1960). They are given in terms of associated Legendre polynomials by

$$Y_{lm}(\mathbf{u}) \equiv Y_{lm}(\theta,\phi) = (-1)^m \left[\frac{(2l+1)(l-m)!}{4\pi(l+m)!}\right]^{\frac{1}{2}} P_{lm}(\cos\theta)\mathrm{e}^{im\phi}\,, \tag{A.16}$$

for $m = 0, 1, 2, \cdots l$, and $l = 0, 1, 2, \cdots$: the unit vector $\mathbf{u}$ is defined in terms of the (constant) Cartesian unit vectors $\mathbf{e}_x$, $\mathbf{e}_y$, $\mathbf{e}_z$ lying along the three Cartesian axes via $\mathbf{u} = \sin\theta\cos\phi\,\mathbf{e}_x + \sin\theta\sin\phi\,\mathbf{e}_y + \cos\theta\,\mathbf{e}_z$. This definition may be extended to $m < 0$ via

$$Y_{l-m}(\mathbf{u}) = (-1)^m Y_{lm}^*(\mathbf{u})\,. \tag{A.17}$$

For $m = 0$, the spherical harmonics $Y_{lm}(\theta,\phi)$ reduce to multiples of the Legendre polynomials, specifically,

$$Y_{l0}(\theta) = \left(\frac{2l+1}{4\pi}\right)^{\frac{1}{2}} P_l(\cos\theta)\,. \tag{A.18}$$

As the $Y_{lm}(\mathbf{u})$ are normalized spherical components of the tensor $\overline{\mathbf{u}^l}$ (see Section A.4.1), they behave under parity as

$$\widehat{\Pi}\,Y_{lm}(\mathbf{u}) \equiv Y_{lm}(-\mathbf{u}) = Y_{lm}(\pi-\theta,\phi+\pi) = (-1)^l Y_{lm}(\mathbf{u})\,. \tag{A.19}$$

Moreover, as spherical harmonics are also orthogonal, they satisfy the orthonormality relation

$$\int Y_{lm}^*(\mathbf{u})Y_{l'm'}(\mathbf{u})\,\mathrm{d}\mathbf{u} = \int_{-1}^{1}\int_0^{2\pi} Y_{lm}^*(\theta,\phi)Y_{l'm'}(\theta,\phi)\,\mathrm{d}\phi\mathrm{d}(\cos\theta) = \delta_{ll'}\delta_{mm'}\,. \tag{A.20}$$

A.3 Vectors and Tensors

A.3.1 Definitions and notation

(Hess, 2015; McCourt *et al.*, 1990) Consider two sets of noncoincident right-handed Cartesian axis systems located at a common origin in a three-dimensional coordinate space. Denote the components of a vector $\mathbf{a}$ in terms of one of these Cartesian axis systems as a_i $(i = 1, 2, 3)$ and the components of the same vector, labeled $\mathbf{a}'$, in the other

Cartesian axis system as a'_j ($j = 1, 2, 3$). Upon performing a set of rotations to bring the two coordinate axis systems into coincidence, the components of **a** transform as

$$a'_j = \sum_{i=1}^{3} D_{ji} a_i \,. \tag{A.21}$$

This relation can be expressed in a coordinate-invariant form as

$$\mathbf{a}' = \mathsf{D} \cdot \mathbf{a} \,, \tag{A.22}$$

with D an orthogonal matrix (with determinant +1) that represents the transformation between the two Cartesian coordinate axis systems.

For a second-rank tensor A, the transformation analogous to the component-form transformation of a vector given by eqn (A.21) is

$$A'_{kl} = \sum_{i,j} D_{ki} D_{lj} A_{ij} \tag{A.23a}$$

or, because the components of an orthogonal matrix D and its inverse D^{-1} are related by $D_{ij} = (\mathsf{D}^{-1})_{ji}$, the transformation may be expressed in a coordinate-invariant form as

$$\mathsf{A}' = \mathsf{D} \cdot \mathsf{A} \cdot \mathsf{D}^{-1} \,. \tag{A.23b}$$

This concept can readily be extended to tensors of rank l, should it be needed.

The concept of a tensor contraction, or scalar product, plays an important role in the kinetic theory. A single contraction between two vectors (tensors of rank 1) **a** and **b** is commonly referred to as a “dot product” between the two vectors, and is defined as

$$\mathbf{a} \cdot \mathbf{b} = \sum_i a_i b_i \,. \tag{A.24a}$$

Two tensors A and B of rank 2 can form both a single tensor contraction (or dot product), which may be expressed as

$$(\mathsf{A} \cdot \mathsf{B})_{ik} = \sum_j A_{ij} B_{jk} \,, \tag{A.24b}$$

and a double tensor contraction (referred to as a double dot product), defined via

$$\mathsf{A} : \mathsf{B} = \sum_{i,j} A_{ij} B_{ji} \,. \tag{A.24c}$$

Higher tensor contractions may generally be defined for higher-rank tensors. It will suffice to provide an example of a triple contraction between two tensors of rank 4:

$$(\mathsf{A} \odot^3 \mathsf{B})_{im} \equiv \sum_{j,k,l} A_{ijkl} B_{lkjm} \,.$$

Note that the procedure employed in this definition of the triple dot product between the two rank 4 tensors involves contractions between the nearest pairs of adjacent uncontracted indices, sometimes referred to as an "inside-out" contraction, as opposed to a "parallel" contraction, which involves contractions between respective left-to-right indices on the two tensors. By convention, a full contraction between two tensors A and B of the same rank is indicated by a $\odot$ without an exponent: for example, for two rank 4 tensors A and B,

$$\mathsf{A} \odot \mathsf{B} \equiv \sum_{ijkl} A_{ijkl} B_{lkji} \,.$$

Contraction between a pair of indices of the same tensor is referred to as a trace (denoted by *tr*) over that pair of indices. An example is provided by the contraction of a tensor A of rank 2, designated $tr\,\mathsf{A}$, and given by

$$tr\,\mathsf{A} \equiv \sum_{i} A_{ii} \,. \tag{A.25a}$$

Single and multiple traces of tensors of higher rank may be defined in the same fashion. For example, for a tensor A of rank 4, single and double traces are defined as

$$(tr\,\mathsf{A})_{jk} = \sum_{i} A_{ijki} \,, \tag{A.25b}$$

and

$$tr^2\,\mathsf{A} = \sum_{ij} A_{ijji} \,, \tag{A.25c}$$

in which the agreed format for traces over tensors is to form the appropriate internal contractions (traces) starting from the outermost Cartesian indices.

A tensor is termed (totally) symmetric if it does not change sign under an arbitrary permutation of its Cartesian indices, and is termed totally antisymmetric if it changes sign under permutation of any pair of indices. Similarly, if a tensor vanishes upon contraction between any pair of indices, it is said to be completely traceless. Finally, if its components remain invariant under rotations of the coordinate system, it is termed an isotropic tensor.

Because tensors are typically classified according to their behaviors under rotation, additional information regarding their representation may be obtained upon considering the irreducible representations of the rotation group O(3). As O(3) has irreducible representations having dimensions 1, 3, 5, $\cdots$, $2\ell+1$, a tensor of rank ℓ, which belongs to a (reducible) representation of dimension 3^ℓ, may therefore be decomposed into a linear combination of its irreducible components, each of which spans one of the irreducible representations of O(3). Thus, for example, an arbitrary second-rank tensor, with $3^2 = 9$ linearly independent components, may be decomposed into a sum of three irreducible tensors, one with 5, one with 3, and one with 1, linear independent components. These three irreducible tensors are said to have weights 2, 1, 0, respectively. Because a fully symmetric and traceless tensor has a weight that is equal to its rank, it is said to be a tensor in its natural form.[2]

[2]The specifications "totally symmetric traceless," "irreducible of maximal weight," and "natural form" as applied to tensors are all equivalent.

A.3.2 Isotropic, symmetric-traceless, and antisymmetric tensors

All scalars may be treated as isotropic tensors of rank 0, as they are simply multiples of 1, which is the fundamental isotropic tensor of rank 0. As all tensors of rank 1 (vectors) change under rotation, there are no isotropic rank 1 tensors. There is a single linearly independent isotropic rank 2 tensor, $\boldsymbol{\delta}$, that may be represented by the unit matrix, namely,

$$\boldsymbol{\delta} = \begin{pmatrix} 1 & 0 & 0 \\ 0 & 1 & 0 \\ 0 & 0 & 1 \end{pmatrix}, \tag{A.26}$$

with components represented by Kronecker deltas δ_{ij} $(i, j \in \{x, y, z\})$.

The tensor product of two isotropic second-rank tensors is a fourth-rank isotropic tensor $\boldsymbol{\delta\delta}$. A single contraction between the two delta tensors gives

$$\boldsymbol{\delta} \cdot \boldsymbol{\delta} = \boldsymbol{\delta}, \tag{A.27a}$$

while a double contraction between them gives

$$\boldsymbol{\delta} : \boldsymbol{\delta} = tr\, \boldsymbol{\delta} = 3. \tag{A.27b}$$

There is also a single linearly independent third-rank isotropic tensor $\boldsymbol{\varepsilon}$ that is totally antisymmetric: its components are given in terms of the Levi–Cività symbol by

$$\varepsilon_{ijk} = \begin{cases} 1, \; ijk &= 123;\; 312;\; 231; \\ -1, \; ijk &= 213;\; 132;\; 321; \\ 0, & \text{otherwise.} \end{cases} \tag{A.28a}$$

As the tensor product of two $\boldsymbol{\varepsilon}$ tensors is an isotropic tensor of rank six, a single contraction between them gives an isotropic tensor of rank four that is antisymmetric in both the front and back pairs of indices, namely,

$$\{\boldsymbol{\varepsilon} \cdot \boldsymbol{\varepsilon}\}_{ijkl} = \delta_{ik}\delta_{jl} - \delta_{il}\delta_{jk}, \tag{A.28b}$$

while a double contraction between two $\boldsymbol{\varepsilon}$-tensors results in a multiple of the second-rank isotropic tensor $\boldsymbol{\delta}$, namely,

$$\boldsymbol{\varepsilon} : \boldsymbol{\varepsilon} = -2\boldsymbol{\delta}. \tag{A.28c}$$

Finally, a full (or triple) contraction between them yields a scalar:

$$\boldsymbol{\varepsilon} \odot \boldsymbol{\varepsilon} = -6. \tag{A.28d}$$

Perhaps the most important use of the $\boldsymbol{\varepsilon}$-tensor is to represent components of the vector product $\mathbf{a} \times \mathbf{b}$ as $-\varepsilon_{ijk} a_k b_j$, which in component-free form is becomes

$$-\boldsymbol{\varepsilon} : \mathbf{ab} = \mathbf{b} \cdot \boldsymbol{\varepsilon} \cdot \mathbf{a} = \mathbf{a} \times \mathbf{b}. \tag{A.29}$$

Note that the minus sign in these equations is due to the "inside-out" convention for the contraction index order.

A.3.3 Reduction of arbitrary second-rank tensors

The quadratic tensor product $\mathbf{rr}$ of the familiar position vector $\mathbf{r}$ (which has Cartesian components $\{x, y, z\}$) may be used to illustrate the concept of tensor products of vectors and a number of tensor properties. Typical tensor products[3] of $\mathbf{r}$ may be written as $\mathbf{r}^2 \equiv \mathbf{rr}$, $\mathbf{r}^3 \equiv \mathbf{rrr}$, and so on. As the tensor product $\mathbf{rr}$ is inherently symmetric, it has only six independent elements, as opposed to the more general tensor product $\mathbf{ab}$ of two arbitrary vectors $\mathbf{a}$ and $\mathbf{b}$, which has nine independent elements.

Subtraction of the isotropic second rank tensor $\boldsymbol{\delta}$ multiplied by one-third the trace r^2 of $\mathbf{rr}$ from $\mathbf{rr}$ results in a tensor

$$\overline{\mathbf{rr}} \equiv \mathbf{rr} - \tfrac{1}{3} r^2 \,\boldsymbol{\delta} \tag{A.30a}$$

that is both symmetric and traceless. This new tensor has only five linearly independent components, and serves also as an example of an irreducible tensor of rank 2. The tensor $\mathbf{rr}$ can be reduced in this manner into a combination of its rank 2, $\overline{\mathbf{rr}}$, and rank 0, $tr\,\mathbf{rr} = r^2$, irreducible components as

$$\mathbf{rr} = \overline{\mathbf{rr}} + \tfrac{1}{3}(tr\,\mathbf{rr})\,\boldsymbol{\delta}\,. \tag{A.30b}$$

More generally, an irreducible tensor of rank ℓ has $2\ell + 1$ independent components. An arbitrary tensor of rank ℓ can similarly be converted into an irreducible tensor of rank ℓ with $2\ell + 1$ linearly independent components, by symmetrizing over all pairs of indices and removing all traces.

The irreducible parts of a second-rank tensor A can be expressed as the sum of a symmetric traceless (weight 2) tensor, to be denoted by $\overline{\mathsf{A}}$; an antisymmetric (weight 1) tensor $\widehat{\mathsf{A}}$, and an isotropic (weight 0) tensor $\overset{\circ}{\mathsf{A}}$. Because $\overset{\circ}{\mathsf{A}}$ is given by $\frac{1}{3}(\mathrm{tr}\,\mathsf{A})\boldsymbol{\delta}$, A can be written as

$$\mathsf{A} = \overline{\mathsf{A}} + \widehat{\mathsf{A}} + \tfrac{1}{3}(\mathrm{tr}\,\mathsf{A})\boldsymbol{\delta}\,, \tag{A.31}$$

with the antisymmetric part being equivalent to the 3-component vector

$$\mathbf{A} = -\tfrac{1}{2}\boldsymbol{\varepsilon}:\widehat{\mathsf{A}} = -\tfrac{1}{2}\boldsymbol{\varepsilon}:\mathsf{A} = \tfrac{1}{2}(A_{yz} - A_{zy}, A_{zx} - A_{xz}, A_{xy} - A_{yx})\,.$$

Likewise, an isotropic tensor is equivalent to a scalar. It is customary to refer to $\widehat{\mathsf{A}} = \boldsymbol{\varepsilon}\cdot\mathbf{A}$ and $\overset{\circ}{\mathsf{A}} = \frac{1}{3}(\mathrm{tr}\,\mathsf{A})\boldsymbol{\delta}$ as embeddings of the rank 1 tensor $\mathbf{A}$ and the rank 0 tensor $\frac{1}{3}(\mathrm{tr}\mathsf{A})$ into rank two. Of course, any irreducible tensor A of rank l and weight $w < l$ can be regarded as the embedding of a symmetric traceless tensor of rank w and, vice versa, any symmetric traceless tensor of rank w can be embedded into a tensor of the same weight and having any rank $l > w$.

[3]The vector $\mathbf{r}$ is itself a tensor of rank 1. Recall that $\mathbf{r}$ is classified as a true vector as it changes sign upon inversion of the coordinate system. Similarly $\mathbf{W}$, which is essentially the time derivative of the relative vector $\mathbf{c} - \mathbf{v}$, is also a true vector. However, as the cross-product vector $\mathbf{r} \times \mathbf{W}$ formed from $\mathbf{r}$ and $\mathbf{W}$ does not change sign under space inversion, it is referred to as a *pseudovector*. Tensor products of vectors are similarly referred to as "true" or "pseudo" tensors depending upon whether or not they behave in the same manner under space inversion as the tensor product $\mathbf{r}^\ell$. Under space inversion true tensors thus change sign depending upon their rank ℓ as $(-1)^\ell$, and are said to have parity $(-1)^\ell$.

Decomposition of a high-rank tensor may yield more than one irreducible tensor of a given weight. A simple example is provided by a reducible third-rank tensor with $3^3 = 27$ components: it can in general be decomposed into a sum of third-rank tensors, one having seven linearly independent components, two each having five linearly independent components, three each having three linearly independent components, and one having one linearly independent component. A decomposition of the tensor product of two irreducible tensors in this way can often prove useful.

It will be instructive to examine the reduction of the dyadic (tensor product of two vectors) **ab**, and then to apply the same procedure to a general rank 2 tensor A. The three irreducible components of a dyadic may be constructed in the following manner. The symmetric-traceless second-rank tensor

$$\overline{\mathbf{ab}} \equiv \tfrac{1}{2}(\mathbf{ab} + \mathbf{ba}) - \tfrac{1}{3}\mathbf{a}\cdot\mathbf{b}\,\boldsymbol{\delta}\,, \tag{A.32}$$

with its five linearly independent components represents the natural form for the weight 2 component of the dyadic **ab**. Similarly, the cross-product vector $\mathbf{a}\times\mathbf{b}$, with three linearly independent components, represents the natural form for the weight 1 component of the dyadic **ab**, and the dot product $\mathbf{a}\cdot\mathbf{b}$, with only a single component, represents the natural form for the weight 0 component of **ab**. These latter two expressions for the weight 1 and weight 0 components do not, however, have the forms of rank 2 tensors, as the cross-product vector is a rank 1 tensor and the dot product is a rank 0 tensor (i.e. a scalar). The weight 0 and weight 1 natural forms may be converted into rank 2 tensors of appropriate symmetries upon multiplying the dot product $\mathbf{a}\cdot\mathbf{b}$ by the rank 2 isotropic tensor $\boldsymbol{\delta}$ and forming a dot product between the cross-product vector and the rank 3 isotropic tensor $\boldsymbol{\varepsilon}$, thereby giving **ab** as

$$\mathbf{ab} = \overline{\mathbf{ab}} + \tfrac{1}{2}\boldsymbol{\varepsilon}\cdot(\mathbf{a}\times\mathbf{b}) + \tfrac{1}{3}\mathbf{a}\cdot\mathbf{b}\,\boldsymbol{\delta}\,. \tag{A.33}$$

This combining process is referred to as "embedding" the lower weight natural forms for the components of a tensor into tensors that have both the correct weights and rank.

The rank 4 isotropic tensor $\boldsymbol{\Delta}^{(2)}$, with components

$$\Delta^{(2)}_{ijkl} = \tfrac{1}{2}(\delta_{ik}\delta_{jl} + \delta_{il}\delta_{jk}) - \tfrac{1}{3}\delta_{ij}\delta_{kl}\,, \tag{A.34}$$

is symmetric and traceless in both the front and back pairs of indices. This isotropic tensor may be employed to obtain the weight 2 tensor $\overline{\mathbf{ab}}$ as

$$\overline{\mathbf{ab}} = \boldsymbol{\Delta}^{(2)}:\mathbf{ab}\,. \tag{A.35}$$

If, in addition, the defining relation $\mathbf{a}\times\mathbf{b} \equiv -\boldsymbol{\varepsilon}:\mathbf{ab}$ given in eqn (A.30) is used for the cross-product vector formed from vectors **a** and **b** it is possible to express **ab** as

$$\mathbf{ab} = (\boldsymbol{\Delta}^{(2)} - \tfrac{1}{2}\boldsymbol{\varepsilon}\cdot\boldsymbol{\varepsilon} + \tfrac{1}{3}\boldsymbol{\delta}\boldsymbol{\delta}):\mathbf{ab}\,. \tag{A.36}$$

Note that if eqn (A.28b) for the contraction between a pair of isotropic rank 3 tensors is employed in eqn (A.36), the manifestly antisymmetric nature of the weight 1 component of the dyadic **ab** becomes more prominently displayed as

$$(\widehat{\mathbf{ab}})_{ij} = a_i b_j - b_i a_j = [\mathbf{ab} - (\mathbf{ab})^t]_{ij} \,,$$

in terms of the dyadics **ab** and its transpose $(\mathbf{ab})^{\mathrm{t}}$ ($\equiv$ **ba**).[4]

The procedure for the corresponding reduction of an arbitrary tensor A of rank 2 into three irreducible components of weights $\ell = 2$, 1, and 0 parallels that employed for the reduction of the dyadic **ab**. The weight 2 component of A is obtained from A, its transpose A^{t}, with components $(\mathsf{A}^{\mathrm{t}})_{ij} \equiv \mathsf{A}_{ji}$, and the trace of A as

$$\overline{\mathsf{A}} = \tfrac{1}{2}(\mathsf{A} + \mathsf{A}^{\mathrm{t}}) - (\tfrac{1}{3} tr \mathsf{A})\boldsymbol{\delta} \,, \tag{A.37a}$$

while the weight 1 component of A, designated $\widehat{\mathsf{A}}$, is given by

$$\widehat{\mathsf{A}} = \tfrac{1}{2}(\mathsf{A} + \mathsf{A}^{\mathrm{t}}) \,, \tag{A.37b}$$

and the weight 0 component, denoted $\mathring{\mathsf{A}}$, is given by

$$\mathring{\mathsf{A}} = (\tfrac{1}{3} tr \mathsf{A})\boldsymbol{\delta} \,. \tag{A.37c}$$

A.3.4 Special tensor forms

Expression (A.36) for rank 2 tensors may be interpreted in either of two ways: as an embedding of the natural forms for the lower-weight components of the tensor using the fourth-rank isotropic tensors $\boldsymbol{\Delta}^{(2)}$, $\boldsymbol{\varepsilon}{\cdot}\boldsymbol{\varepsilon}$, and $\boldsymbol{\delta\delta}$, as represented by eqn (A.33), or as employing rank 4 isotropic tensors to project out the (fully) symmetric traceless weight 2 component, the antisymmetric weight 1 component, and the weight 0 (isotropic) component of a rank 2 tensor. Viewed in the latter way, $\boldsymbol{\Delta}^{(2)}$, $\boldsymbol{\varepsilon} \cdot \boldsymbol{\varepsilon}$, and $\boldsymbol{\delta\delta}$ thus serve as isotropic projection tensors that produce the weight 2, 1, and 0 components of an arbitrary rank 2 tensor. The three rank 4 projection tensors all consist of linear combinations of the three (primitive) binary products $\delta_{ij}\delta_{kl}$, $\delta_{ik}\delta_{jl}$, $\delta_{il}\delta_{jk}$ of Kronecker deltas that can be constructed from four indices i, j, k, l.

Although generalizations of the isotropic projection tensors for the lower weight components of an arbitrary tensor of rank ℓ are difficult to construct, the generalization of $\boldsymbol{\Delta}^{(2)}$ to ranks higher than 2 is relatively straightforward. The isotropic tensor $\boldsymbol{\Delta}^{(\ell)}$ of rank 2ℓ that is symmetric and traceless in the first and final ℓ indices gives the fully symmetric component of an arbitrary tensor A of rank ℓ as

$$\overline{\mathsf{A}} = \boldsymbol{\Delta}^{(\ell)} \odot \mathsf{A} \,. \tag{A.38}$$

In addition to being symmetric and traceless in the first and final ℓ indices, $\boldsymbol{\Delta}^{(\ell)}$ is symmetric to the interchange of the two sets of ℓ indices.

[4]Some care must be exercised should vectors be visualized as column or row vectors, as the tensor product **ab** is represented by the square matrix $\mathbf{ab}^{\mathrm{t}}$ for **a**, **b** represented by column vectors and by the square matrix $\mathbf{a}^{\mathrm{t}}\mathbf{b}$ for **a**, **b** represented by row vectors.

The projection tensors $\mathbf{\Delta}^{(\ell)}$ have a number of useful special properties, such as the projector property

$$\mathbf{\Delta}^{(\ell)} \odot^{\ell} \mathbf{\Delta}^{(\ell)} = \mathbf{\Delta}^{(\ell)}\,, \tag{A.39a}$$

the simple trace

$$tr\mathbf{\Delta}^{(\ell)} = \frac{2\ell+1}{2\ell-1}\,\mathbf{\Delta}^{(\ell-1)}\,, \tag{A.39b}$$

and the total trace

$$tr^{\ell}\mathbf{\Delta}^{(\ell)} = \sum_{ij\cdots m} \mathbf{\Delta}^{(\ell)}_{ij\cdots m,m\cdots ji} = (2\ell+1)\,. \tag{A.39c}$$

A.4 Spherical Harmonics and Spherical Tensors

A.4.1 Spherical tensors

A vector $\mathbf{a}$ with Cartesian components a_x, a_y, a_z can be represented in terms of spherical components a_μ, $\mu = 0, \pm 1$ via the introduction of basis vectors

$$\mathbf{e}_0 = \mathbf{e}_z\,, \qquad \mathbf{e}_{\pm 1} = \mp\tfrac{1}{\sqrt{2}}(\mathbf{e}_x \pm i\,\mathbf{e}_y)\,, \tag{A.40}$$

with properties

$$\mathbf{e}^*_\mu \cdot \mathbf{e}_\nu = \delta_{\mu\nu}\,, \qquad \sum_{\mu=-1}^{1} \mathbf{e}_\mu \mathbf{e}^*_\mu = \boldsymbol{\delta}\,, \qquad \mathbf{e}^*_\mu = (-1)^\mu \mathbf{e}_{-\mu}\,. \tag{A.41}$$

The spherical vector components of $\mathbf{a}$ are then given as

$$a_0 = \mathbf{a}\cdot\mathbf{e}_0 = a_z\,, \qquad a_{\pm 1} = \mathbf{a}\cdot\mathbf{e}_{\pm 1} = \mp\tfrac{1}{\sqrt{2}}(a_x \pm i\,a_y)\,, \tag{A.42a}$$

while the inverse translation, giving the Cartesian components of $\mathbf{a}$ in terms of spherical components, is

$$\mathbf{a} = \sum_{\mu=-1}^{1} a_\mu \mathbf{e}^*_\mu\,. \tag{A.42b}$$

The basis elements for rank l tensors are given by

$$\mathbf{e}^{(l)}_\mu = \left[\frac{l!(2\mu)!(2l-1)!!}{(l-\mu)!\mu!(l+\mu)!(2\mu-1)!!}\right]^{\frac{1}{2}} \overline{\mathbf{e}_0^{l-\mu}\mathbf{e}_1^{\mu}}\,, \qquad \mu \geq 0\,, \tag{A.43a}$$

with basis elements for $\mu < 0$ obtained via the extension

$$\mathbf{e}^{(l)}_{-\mu} = (-1)^\mu \mathbf{e}^{(l)*}_\mu\,. \tag{A.43b}$$

As an example, the basis element for $\mu = l$ is obtained from eqn (A.43a) as

$$\mathbf{e}^{(l)}_l = \mathbf{e}^l_1 = (-1)^l 2^{-\frac{l}{2}} (\mathbf{e}_x + i\,\mathbf{e}_y)^l\,.$$

Orthogonality and completeness of these basis elements give

$$\mathbf{e}_\mu^{(l)*} \odot \mathbf{e}_\nu^{(l)} = \delta_{\mu\nu}\,, \quad \text{and} \quad \sum_\mu \mathbf{e}_\mu^{(l)} \mathbf{e}_\mu^{(l)*} = \mathbf{\Delta}^{(l)}\,. \tag{A.44}$$

The spherical components of a real symmetric traceless tensor A of rank l may be expressed in the form

$$A_\mu = \mathsf{A} \odot \mathbf{e}_\mu^{(l)}\,, \qquad A_\mu^* = \mathsf{A} \odot \mathbf{e}_\mu^{(l)*} = (-1)^\mu A_{-\mu}^*\,, \tag{A.45a}$$

while the inverse relation, giving the tensor A in terms of its spherical components, is

$$\mathsf{A} = \sum_\mu A_\mu \mathbf{e}_\mu^{(l)*}\,. \tag{A.45b}$$

Consider doubly spherical components of rank $(l' + l)$ tensors that are symmetric traceless with respect to both the initial l' and the final l indices, that is,

$$B_{\mu\nu} = \mathbf{e}_\mu^{(l')*} \odot \mathsf{B} \odot \mathbf{e}_\nu^{(l)}\,, \qquad \mathsf{B} = \sum_{\mu\nu} B_{\mu\nu} \mathbf{e}_\mu^{(l')} \mathbf{e}_\nu^{(l)*}\,, \tag{A.46a}$$

an important example of which is provided by the projection tensors $\mathbf{\Delta}^{(l)}$. These projection tensors behave as unit tensors in doubly spherical representation, namely,

$$\Delta_{\mu\nu}^{(l)} = \delta_{\mu\nu}\,. \tag{A.46b}$$

The symmetric traceless part of an l-fold tensor product of a unit vector $\mathbf{e}_\mu^l$ has the properties[5]

$$\{\overline{\mathbf{u}^l}\}_\mu = \left[\frac{4\pi\, l!}{(2l+1)!!}\right]^{\frac{1}{2}} Y_{l\mu}(\mathbf{u})\,. \tag{A.47a}$$

$$\overline{\mathbf{u}^l} \odot \overline{\mathbf{u}^l} = \frac{l!}{(2l-1)!!}\,, \tag{A.47b}$$

$$\int \mathrm{d}\mathbf{u}\, \overline{\mathbf{u}^l}\, \overline{\mathbf{u}'^l} = \frac{4\pi\, l!}{(2l+1)!!}\, \delta_{ll'} \mathbf{\Delta}^{(l)}\,. \tag{A.47c}$$

A.4.2 Second-rank tensors

$$\mathbf{e}_0^{(2)} = \left(\tfrac{3}{2}\right)^{\frac{1}{2}} \overline{\mathbf{e}_z \mathbf{e}_z} = \frac{1}{\sqrt{6}} \begin{pmatrix} -1 & 0 & 0 \\ 0 & -1 & 0 \\ 0 & 0 & 2 \end{pmatrix}\,, \tag{A.48a}$$

[5]Different conventions from those given here can be found. For example, the basis tensors adopted by Chen *et al.* (1972) involve a phase factor $\mathbf{e}^{l\mu} = i^l \mathbf{e}_\mu^{(l)}$, while Hess and Köhler (1980) employ complex conjugate basis vectors $\mathbf{e}_{\pm1} = \mp(\mathbf{e}_x \mp i\,\mathbf{e}_y)/\sqrt{2}$, so that $a_{\pm1} = \mathbf{a} \cdot \mathbf{e}_{\pm1}^*$ and $\mathbf{a} = \sum a_\mu \mathbf{e}_\mu$.

$$\mathbf{e}^{(2)}_{\pm 1} = \mp\overline{\mathbf{e}_x\mathbf{e}_z} \pm i\,\overline{\mathbf{e}_y\mathbf{e}_z}) = \frac{1}{2}\begin{pmatrix} 0 & 0 & \mp 1 \\ 0 & 0 & -i \\ \mp 1 & -i & 0 \end{pmatrix}, \tag{A.48b}$$

$$\mathbf{e}^{(2)}_{\pm 2} = \tfrac{1}{2}[(\mathbf{e}_x\mathbf{e}_x - \mathbf{e}_y\mathbf{e}_y) \pm i(\mathbf{e}_x\mathbf{e}_y + \mathbf{e}_y\mathbf{e}_x)] = \frac{1}{2}\begin{pmatrix} 1 & \pm i & 0 \\ \pm i & -1 & 0 \\ 0 & 0 & 0 \end{pmatrix}. \tag{A.48c}$$

$$A_0 = \left(\tfrac{3}{2}\right)^{\frac{1}{2}} A_{zz} = -\left(\tfrac{3}{2}\right)^{\frac{1}{2}} (A_{xx} + A_{yy}), \tag{A.49a}$$

$$A_{\pm 1} = \mp A_{xz} - i\,A_{yz}, \tag{A.49b}$$

$$A_{\pm 2} = \tfrac{1}{2}(A_{xx} - A_{yy}) \pm i\,A_{xy}, \tag{A.49c}$$

$$A_{xx} = \mathrm{Re}\,(A_2) - \tfrac{1}{\sqrt{6}}A_0, \quad A_{yy} = -\mathrm{Re}\,(A_2) - \tfrac{1}{\sqrt{6}}A_0, \quad A_{zz} = \left(\tfrac{2}{3}\right)^{\frac{1}{2}} A_0, \tag{A.50a}$$

$$A_{xy} = A_{yx} = \mathrm{Im}\,(A_2), \quad A_{yz} = A_{zy} = -\mathrm{Im}\,(A_1), \quad A_{zx} = A_{xz} = -\mathrm{Re}\,(A_1). \tag{A.50b}$$

A.4.3 Scalar products

Scalar products of vectors:

$$\mathbf{a}\cdot\mathbf{b} = \sum_{i=1}^{3} a_i b_i = \sum_{\mu=-1}^{1} a^*_\mu b_\mu. \tag{A.51}$$

Full scalar product of two symmetric traceless tensors:

$$\mathsf{A}\odot\mathsf{B} = \sum_{ij\cdots m=1}^{3} A_{ij\cdots m} B_{m\cdots ji} = \sum_{\mu=-l}^{l} A^*_\mu B_\mu. \tag{A.52}$$

Further examples of scalar products for unit vectors **u**, **v**, **w**:

$$\overline{\mathbf{u}^l}\odot\overline{\mathbf{v}^l} = 4\pi\frac{l!}{(2l+1)!!}\sum_m Y^*_{lm}(\mathbf{u})Y_{lm}(\mathbf{v}) = \frac{l!}{(2l-1)!!}P_l(\mathbf{u}\cdot\mathbf{v}), \tag{A.53}$$

$$\mathbf{u}^{l_1}\mathbf{v}^{l_2}\odot\overline{\mathbf{w}^{l_1+l_2}}\begin{pmatrix} l_1 & l_2 & l_3 \\ 0 & 0 & 0 \end{pmatrix}^2 = \frac{(l_1+l_2)!}{(2l_1+2l_2-1)!!}P_{l_1,l_2,l_1+l_2}(\mathbf{u},\mathbf{v},\mathbf{w}). \tag{A.54}$$

Examples:

$$\overline{\mathbf{u}\mathbf{v}} : \overline{\mathbf{w}\mathbf{w}} = 5P_{112}(\mathbf{u},\mathbf{v},\mathbf{w}), \tag{A.55a}$$

$$\overline{\mathbf{uu}}\,\overline{\mathbf{vv}} \odot \overline{\mathbf{w}^4} = 4P_{224}(\mathbf{u},\mathbf{v},\mathbf{w})\,. \tag{A.55b}$$

Similarly,

$$\overline{\mathbf{u}^l\mathbf{v}} \odot \overline{\mathbf{u}^l\mathbf{v}} = \frac{l!}{(2l+1)!!}\left(1 + \frac{l}{2}(1 + (\mathbf{u}\cdot\mathbf{v})^2)\right)\,, \tag{A.56a}$$

of which the simplest example is

$$\overline{\mathbf{uv}} : \overline{\mathbf{uv}} = \tfrac{1}{6}[3 + (\mathbf{u}\cdot\mathbf{v}]^2)\,. \tag{A.56b}$$

References

Abramowitz, M. and Stegun, I. A. (1972). *Handbook of Mathematical Functions.* Dover Publications, New York.

Achtermann, H. J., Hong, J. G., Magnus, G., Aziz, R. A., and Slaman, M. J. (1993). Experimental determination of the refractivity virial coefficients of atomic gases. *The Journal of Chemical Physics*, **98**, 2308–2318.

Acton, A. and Kellner, K. (1977). The low temperature thermal conductivity of ^{4}He: I. Measurements between 3.3 and 20 K on the dilute gas, the dense gas and the liquid. *Physica B*, **90**(2), 192–204.

Ahlrichs, R., Penco, R., and Scoles, G. (1977). Intermolecular forces in simple systems. *Chemical Physics*, **19**, 119–130.

Anderson, J. B. (2001). An exact quantum Monte Carlo calculation of the helium–helium intermolecular potential. II. *The Journal of Chemical Physics*, **115**, 4546–4548.

Aquilanti, V. and Grossi, G. (1980). Angular momentum coupling schemes in the quantum mechanical treatment of P-state atom collisions. *The Journal of Chemical Physics*, **73**(3), 1165–1172.

Arora, P. S., Carson, P. J., and Dunlop, P. J. (1978). Determination of potential parameters for the systems Ne–Ar and Ar–Kr from the temperature dependence of their binary diffusion coefficients. *Chemical Physics Letters*, **54**(1), 117–119.

Arora, P. S., Robjohns, H. L., and Dunlop, P. J. (1979). Use of accurate diffusion and second virial coefficients to determine (m, 6, 8) potential parameters for nine binary noble gas systems. *Physica A: Statistical Mechanics and its Applications*, **95**(3), 561–571.

Assael, M. J., Dix, M., Lucas, A., and Wakeham, W. A. (1981). Absolute determination of the thermal conductivity of the noble gases and two of their binary mixtures as a function of density. *Journal of the Chemical Society, Faraday Transactions*, **77**, 439–464.

Assael, M. J., Nieto de Castro, C. A., Roder, H. M., and Wakeham, W. A. (1991). Transient methods for thermal conductivity. In *Experimental Thermodynamics. III. Measurement of the Transport Properties of Fluids* (eds. W. A. Wakeham, A. Nagashima, and J. V. Sengers), Chapter 7. Blackwell, Oxford.

Axilrod, B. M. and Teller, E. (1943). Interaction of the van der Waals type between three atoms. *The Journal of Chemical Physics*, **11**(6), 299–300.

Aziz, R. A. (1993). A highly accurate interatomic potential for argon. *The Journal of Chemical Physics*, **99**, 4518–4525.

Aziz, R. A., Janzen, A. R., and Moldover, M. R. (1995). *Ab initio* calculations for helium: A standard for transport property measurements. *Physical Review Letters*, **74**, 1586–1589.

Aziz, R. A., Janzen, A. R., and Simmons, R. O. (2004). Rare gases. In *Encyclopedia of Applied Physics*, Volume 16, p. 71–96. Wiley–VCH, Weinheim.

Aziz, R. A., McCourt, F. R. W., and Wong, C. C. K. (1987). A new determination of the ground state interatomic potential for He_2. *Molecular Physics*, **61**(6), 1487–1511.

Aziz, R. A., Presley, J., Buck, U., and Schleusenerc, J. (1979). An accurate intermolecular potential for Ar–Kr. *The Journal of Chemical Physics*, **70**, 4737–4741.

Aziz, R. A. and Slaman, M. J. (1986*a*). On the Xe–Xe potential energy curve and related properties. *Molecular Physics*, **57**(4), 825–840.

Aziz, R. A. and Slaman, M. J. (1986*b*). The argon and krypton interatomic potentials revisited. *Molecular Physics*, **58**, 679–697.

Aziz, R. A. and Slaman, M. J. (1989*a*). Accurate transport properties and second virial coefficients for krypton based on a state-of-the-art interatomic potential. *Chemical Engineering Communications*, **78**(1), 153–165.

Aziz, R. A. and Slaman, M. J. (1989*b*). The Ne–Ne interatomic potential revisited. *Chemical Physics*, **130**, 187–194.

Aziz, R. A., Slaman, M. J., Koide, A., Allnattc, A. R., and Meath, W. J. (1992). Exchange-Coulomb potential energy curves for He–He, and related physical properties. *Molecular Physics*, **77**, 321–337.

Aziz, R. A. and van Dalen, A. (1983*a*). An improved potential for Ar–Kr. *The Journal of Chemical Physics*, **78**(5), 2413–2418.

Aziz, R. A. and van Dalen, A. (1983*b*). Comparison of the predictions of literature intermolecular potentials for Ar–Xe and Kr–Xe with experiment: Two new potentials. *The Journal of Chemical Physics*, **78**, 2402–2412.

Barker, J. A., Bobetic, M. V., and Pompe, A. (1971). An experimental test of the Boltzmann equation: Argon. *Molecular Physics*, **20**, 347–355.

Barker, J. A., Fock, W., and Smith, F. (1964). Calculation of gas transport properties and the interaction of argon atoms. *The Physics of Fluids*, **7**, 897–903.

Barrow, D. A. and Aziz, R. A. (1988). The neon–argon potential revisited. *The Journal of Chemical Physics*, **89**(10), 6189–6194.

Barrow, D. A., Slaman, M. J., and Aziz, R. A. (1989). Simple accurate potentials for Ne–Kr and Ne–Xe. *The Journal of Chemical Physics*, **91**(10), 6348–6358.

Barrow, D. A., Slaman, M. J., and Aziz, R. A. (1992). Erratum: Simple accurate potentials for Ne–Kr and Ne–Xe [J. Chem. Phys. 91, 6348 (1989)]. *The Journal of Chemical Physics*, **96**(7), 5555.

Bartlett, R. J. (1981). Many-body perturbation theory and coupled cluster theory for electron correlation in molecules. *Annual Review of Physical Chemistry*, **32**(1), 359–401.

Bartlett, R. J. (1989). Coupled-cluster approach to molecular structure and spectra: A step toward predictive quantum chemistry. *The Journal of Physical Chemistry*, **93**(5), 1697–1708.

Bearden, J. A. (1939). A precision determination of the viscosity of air. *Physical Review*, **56**, 1023–1040.

Beattie, J. A., Barriault, R. J., and Brierley, J. S. (1951). The compressibility of gaseous xenon. I. An equation of state for xenon and the weight of a liter of xenon. *The Journal of Chemical Physics*, **19**, 1219–1221.

Beattie, J. A., Brierley, J. S., and Barriault, R. J. (1952). Compressibility of Gaseous Krypton—II The Virial Coefficients and Potential Parameters of Krypton. *Journal of Chemical Physics*, **20**, 1615–1618.

Becker, E. W. and Misenta, R. (1955). Die Zähigkeit von HD und He3 zwischen 14° K und 20° K. *Zeitschrift für Physik*, **140**, 535–539.

Becker, E. W., Misenta, R., and Schmeissner, F. (1954). Die Zähigkeit von gasförmigem He3 und He4 zwischen 1,3° K und 4,2° K. *Zeitschrift für Physik*, **137**, 126–136.

Benedetto, G., Gavioso, R. M., Spagnolo, R., Marcarinoc, P., and Merlone, A. (2004). Acoustic measurements of the thermodynamic temperature between the triple point of mercury and 380 K. *Metrologia*, **41**, 74–98.

Beneventi, L., Casavecchia, P., and Volpi, G. G. (1986). Observation of high frequency quantum oscillations in elastic differential cross sections: A critical test of the Ne–Ar interaction potential. *The Journal of Chemical Physics*, **84**(9), 4828–4832.

Berg, R. F. (2005). Simple flow meter and viscometer of high accuracy for gases. *Metrologia*, **42**(1), 11.

Berg, R. F. (2006). Erratum: Simple flow meter and viscometer of high accuracy for gases. *Metrologia*, **43**(1), 183.

Berg, R. F. and Burton, W. C. (2013). Noble gas viscosities at 25 °C. *Molecular Physics*, **111**(2), 195–199.

Berg, R. F. and Moldover, M. R. (2012). *Journal of Physical and Chemical Reference Data*, **41**, 043104.

Bernath, P. F. (2017). *Spectra of Atoms and Molecules* (3rd edn). Oxford University Press, New York.

Beser, B., Sovkov, V. B., Bai, J., Ahmed, E. H., Tsai, C. C., Xie, F., Li, Li, Ivanov, V. S., and Lyyra, A. M. (2009). Experimental investigation of the $^{85}Rb_2$ $a^3\Sigma_u^+$ triplet ground state: Multiparameter Morse long range potential analysis. *The Journal of Chemical Physics*, **131**(9), 094505.

Bhatnagar, P. L., Gross, E. P., and Krook, M. (1954). A model for collision processes in gases. I. Small amplitude processes in charged and neutral one-component systems. *Physical Review*, **94**, 511–525.

Bich, E., Hellmann, R., and Vogel, E. (2007). *Ab initio* potential energy curve for the helium atom pair and thermophysical properties of the dilute helium gas. II. Thermophysical standard values for low-density helium. *Molecular Physics*, **105**, 3035–3049.

Bich, E., Hellmann, R., and Vogel, E. (2008). *Ab initio* potential energy curve for the neon atom pair and thermophysical properties for the dilute neon gas. II. Thermophysical properties for low-density neon. *Molecular Physics*, **106**, 813–825.

Bich, E., Millat, J., and Vogel, E. (1990). The viscosity and thermal conductivity of pure monatomic gases from their normal boiling point up to 5000 K in the limit of zero density and at 0.101325 MPa. *Journal of Physical and Chemical Reference Data*, **19**(6), 1289–1305.

Bich, E. and Vogel, E. (1991). The initial density dependence of transport properties: Noble gases. *International Journal of Thermophysics*, **12**, 27–42.

Dymond, J. H., Bich, E., Vogel, E., Wakeham, W. A., Vesovic, V., and Assael, M. J. (1996). Dense fluids. In *Transport Properties of Fluids: Their Correlation, Prediction, and Estimation* (eds. J. Millat, J. H. Dymond, and C. A. Nieto de Castro), pp. 66–112. Cambridge University Press.

Bird, G. A. (1994). *Molecular Gas Dynamics and the Driect Simulation of Gas Flows.* Oxford University Press, Oxford.

Bishop, D. M. and Pipin, J. (1993). Dipole, quadrupole, octupole, and dipole–octupole polarizabilities at real and imaginary frequencies for H, He, and H_2 and the dispersion-energy coefficients for interactions between them. *International Journal of Quantum Chemistry*, **45**(4), 349–361.

Boley, C. D. and Yip, S. (1972). Modeling theory of the linearized collision operator for a gas mixture. *The Physics of Fluids*, **15**, 1424–1433.

Boltzmann, L. (1872). Weitere studien über das wärmegleichgewicht unter gas-molekülen. *Wiener Berichte*, **66**, 275–370.

Born, M. and Mayer, J. E. (1932). Zur gittertheorie der ionenkristalle. *Zeitschrift für Physik*, **75**, 1–18.

Born, M. and Oppenheimer, J. R. (1927). *Annalen der Physik (Leipzig)*, **74**, 1.

Bowman Jr., J. D., Hirschfelder, J. O., and Wahl, A. C. (1970). Extended Hartree–Fock calculations for the ground state and Hartree–Fock calculations for the first excited state of H_2. *The Journal of Chemical Physicsc*, **53**, 2743–2749.

Boyd, R. J. (2019). Theoretical and computational chemistry. In *Reference Module in Chemistry, Molecular Sciences and Chemical Engineering.* https://www.sciencedirect.com/science/article/pii/B9780124095472116890

Boyes, S. J. (1994). The interatomic potential of argon. *Chemical Physics Letters*, **221**(5), 467–472.

Boys, S. F. and Bernardi, F. (1970). The calculation of small molecular interactions by the differences of separate total energies. Some procedures with reduced errors. *Molecular Physics*, **19**(4), 553–566.

Brenner, D. M. and Peters, D. W. (1982). State-to-state dynamics of IR multiphoton absorption in the a1a2 state of thiophosgene: Absorption cross sections. *The Journal of Chemical Physics*, **76**(1), 197–218.

Buck, U. (1971). Determination of Intermolecular potentials by the inversion of molecular beam scattering data. I. The inversion procedure. *The Journal of Chemical Physics*, **54**, 1923–1928.

Buck, U. (1974). Inversion of molecular scattering data. *Reviews of Modern Physics*, **46**, 369–389.

Buck, U. (1976). Elastic scattering. *Advances in Chemical Physics*, **30**, 313–388.

Buck, U. (1986). Inversion of molecular scattering data. *Computer Physics Reports*, **5**, 1–58.

Buck, U., Hoppe, H. G., Huisken, F., and Pauly, H. (1974). Intermolecular potentials by the inversion of molecular beam scattering data. IV. Differential cross sections and potential for LiHg. *The Journal of Chemical Physics*, **60**, 4925–4929.

Buck, U., Huisken, F., Pauly, H., and Schleusener, J. (1978). Intermolecular potentials by the inversion of differential cross sections. V. ArKr. *The Journal of Chemical Physics*, **68**, 3334–3338.

Buck, U., Kick, M., and Pauly, H. (1972). Determination of intermolecular potentials by the inversion of molecular beam scattering data. III. High resolution measurements and potentials for K–Hg and Cs–Hg. *The Journal of Chemical Physics*, **56**, 3391–3397.

Buck, U., Köhler, K. A., and Pauly, H. (1971). Messung der Glorienstreuung des totalen Streuquerschnitts von Na–Hg. *Zeitschrift für Physik A Hadrons and nuclei*, **244**, 180–189.

Buck, U. and Pauly, H. (1971). Determination of intermolecular potentials by the inversion of molecular beam scattering data. II. High resolution measurements of differential scattering cross sections and the inversion of the data for Na–Hg. *The Journal of Chemical Physics*, **54**, 1929–1936.

Buckingham, A. D., Cole, R. H., and Sutter, H. (1970). Direct determination of the imperfect gas contribution to dielectric polarization. *The Journal of Chemical Physics*, **52**, 5960–5961.

Buckingham, A. D. and Graham, C. (1974). The density dependence of the refractivity of gases. *Proceedings of the Royal Society of London. Series A, Mathematical and Physical Sciences*, **337**, 275–291.

Buckingham, R. A. (1938). The classical equation of state of gaseous helium, neon and argon. *Proceedings of the Royal Society of London. Series A, Mathematical and Physical Sciences*, **168**, 264–283.

Buckingham, R. A. and Corner, J. (1947). Tables of second virial and low-pressure Joule–Thomson coefficients for intermolecular potentials with exponential repulsion. *Proceedings of the Royal Society of London. Series A, Mathematical and Physical Sciences*, **189**, 118–129.

Burcl, R., Krems, R. V., Buchachenko, A. A., Szcześniak, M. M., Chałasiński, G., and Cybulski, S. M. (1998). RG+Cl(^{2}P) (RG=He, Ne, Ar) interactions: *Ab initio* potentials and collision properties. *The Journal of Chemical Physics*, **109**(6), 2144–2154.

Bytautas, L. and Ruedenberg, K. (2008). Correlation energy and dispersion interaction in the *ab initio* potential energy curve of the neon dimer. *The Journal of Chemical Physics*, **128**(21), 214308.

Callaway, J. and Bauer, E. (1965). *Inelastic collisions of slow atoms. Physical Review A*, **140**, 1072–1084.

Cambi, R., Cappelletti, D., Liuti, G., and Pirani, F. (1991). Generalized correlations in terms of polarizability for van der Waals interaction potential parameter calculations. *The Journal of Chemical Physics*, **95**, 1852–1861.

Case, K. M. and Zweifel, P. F. (1967). *Linear Transport Theory.* Addison–Wesley Publishing Company, Massachusetts.

Casimir, H. B. G. and Polder, D. (1948). The influence of retardation on the London-van der Waals forces. *Physical Review*, **73**, 360–372.

Cencek, W., Harvey, G. A. H., McLinden, M. O., and Szalewicz, K. (2013). Three-body nonadditive potential for argon with estimated uncertainties and third virial coefficient. *The Journal of Physical Chemistry A*, **117**(32), 7542–7552.

Cencek, W., Przybytek, M., Komasa, J., Mehl, J. B., Jeziorski, B., and Szalewicz, K. (2012). Effects of adiabatic, relativistic, and quantum electrodynamics interactions on the pair potential and thermophysical properties of helium. *The Journal of Chemical Physics*, **136**(22), 224303.

Cencek, W. and Szalewicz, K. (2008). Ultra-high accuracy calculations for hydrogen molecule and helium dimer. *International Journal of Quantum Chemistry*, **108**(12), 2191–2198.

Ceperley, D. M. and Partridge, H. (1986). The He_2 potential at small distances. *The Journal of Chemical Physics*, **84**(2), 820–821.

Cercignani, C. (1975). *Theory and Application of the Boltzmann Equation.* Scottish Academic Press, Edinburgh.

Chałasiński, G. and Gutowski, M. (1988). Weak interactions between small systems. Models for studying the nature of intermolecular forces and challenging problems for *ab initio* calculations. *Chemical Reviews*, **88**(6), 943–962.

Chałasiński, G. and Szcześniak, M. M. (1994). Origins of structure and energetics of van der Waals clusters from *ab initio* calculations. *Chemical Reviews*, **94**(7), 1723–1765.

Chałasiński, G. and Szcześniak, M. M. (2000). State of the art and challenges of the *ab initio* theory of intermolecular interactions. *Chemical Reviews*, **100**(11), 4227–4252.

Chandrasekhar, S. (1950). *Radiative Transfer.* Oxford University Press, Oxford.

Chapman, S. (1916). On the law of distribution of molecular velocities, and on the theory of viscosity and thermal conduction, in a non–uniform simple monatomic gas. *Philosophical Transactions of the Royal Society of London. Series A, Containing Papers of a Mathematical or Physical Character*, **216**, 279–348.

Chapman, S. and Cowling, T. G. (1970). *Mathematical Theory of Nonuniform Gases* (3rd edn). Cambridge University Press, Cambridge, UK.

Chen, C. H., Siska, P. E., and Lee, Y. T. (1973). Intermolecular potentials from crossed beam differential elastic scattering measurements VIII. He+Ne, He+Ar, He+Kr, and He+Xe. *The Journal of Chemical Physics*, **59**, 601–610.

Chen, F. M., Moraal, H., and Snider, R. F. (1972). On the evaluation of kinetic theory collision integrals: Diamagnetic diatomic molecules. *The Journal of Chemical Physics*, **57**, 542–561.

Child, M. S. (1974). *Molecular Collision Theory.* Academic Press, London.

Child, M. S. (1991). *Semiclassical Mechanics with Molecular Applications.* Oxford University Press, Oxford.

Clancy, P., Gough, D. W., Matthews, G. P., Smith, E. B., and Maitland, G. C. (1975). Simplified methods for the inversion of thermophysical data. *Molecular Physics*, **30**(5), 1397–1407.

Clarke, A. G. and Smith, E. B. (1968). Low-temperature viscosities of argon, krypton, and xenon. *The Journal of Chemical Physics*, **48**(9), 3988–3991.

Clarke, A. G. and Smith, E. B. (1969). Low-temperature viscosities and intermolecular forces of simple gases. *The Journal of Chemical Physics*, **51**(9), 4156–4161.

Clifford, A. A., Fleeter, R., Kestin, J., and Wakeham, W. A. (1979). Thermal conductivity of some mixtures of monatomic gases at room temperature and at pressures up to 15 mpa. *Physica A: Statistical Mechanics and its Applications*, **98**(3), 467–490.

Clifford, A. A., Gray, P., and Scott, A. C. (1975). Viscosities of gaseous argon, oxygen and carbon monoxide between 273 and 1300 K. *Journal of the Chemical Society, Faraday Transactions 1*, **71**, 875–882.

Cohen, E. G. D. (ed). (1971). *Statistical Mechanics at the Turn of the Decade.* Marcel Dekker, New York.

Cohen, E. G. D. (1972). The generalization of the Boltzmann equation to higher densities. In *The Boltzmann Equation* (eds. E. G. D. Cohen and W. Thirring pp. 157–176. Springer, New York.

Colbourn, E. A. and Douglas, A. E. (1976). The spectrum and ground state potential curve of Ar_2. *The Journal of Chemical Physics*, **65**, 1741–1745.

Coremans, J. M. J., van Itterbeek, A, Beenakker, J. J. M., Knaap, H. F. P., and Zandbergen, P. (1958). The viscosity of gaseous He, Ne, H_2 and D_2 below 80°K. *Physica*, **24**(6), 557–576.

Cowan, R. D. and Griffin, D. C. (1976). Approximate relativistic corrections to atomic radial wave functions*. *Journal of the Optical Society of America*, **66**(10), 1010–1014.

Cox, H. E., Crawford, F. W., Smith, E. B., and Tindell, A. R. (1980). A complete iterative inversion procedure for second virial coefficient data. *Molecular Physics*, **40**(3), 705–712.

Curtiss, C. F. (1981). Classical, diatomic molecule, kinetic theory cross sections. *The Journal of Chemical Physics*, **75**, 1341–1346.

Cybulski, S. M. and Toczyłowski, R. R. (1999). Ground state potential energy curves for He_2, Ne_2, Ar_2, He–Ne, He–Ar, and Ne–Ar: A coupled-cluster study. *The Journal of Chemical Physics*, **111**(23), 10520–10528.

Czachorowski, P., Przybytek, M., Lesiuk, M., Puchalski, M., and Jeziorski, B. (2020). Second virial coefficients for ^{4}He and ^{3}He from an accurate relativistic interaction potential. *Physical Review A*, **102**, 042810.

Dalgarno, A. and Lewis, J. T. (1955). The exact calculation of long-range forces between atoms by perturbation theory. *Proceedings of the Royal Society of London. Series A, Mathematical and Physical Sciences*, **233**, 70–74.

Danielson, L. J. and Keil, M. (1988). Interatomic potentials for HeAr, HeKr, and HeXe from multiproperty fits. *The Journal of Chemical Physics*, **88**(2), 851–870.

Dattani, N. S. and Le Roy, R. J. (2011). A DPF data analysis yields accurate analytic potentials for $Li_2(a^3\Sigma_u^+)$ and $Li_2(1^3\Sigma_g^+)$ that incorporate 3-state mixing near the $1^3\Sigma_g^+$ state asymptote. *Journal of Molecular Spectroscopy*, **268**, 199–210.

Dawe, R. A. and Smith, E. B. (1970). Viscosities of the inert gases at high temperatures. *The Journal of Chemical Physics*, **52**(2), 693–703.

De Paz, M., Turi, B., and Klein, M. L. (1967). New self-diffusion measurements in argon gas. *Physica*, **36**(1), 127–135.

Derevianko, A., Johnson, W. R., Safronova, M. S., and Babb, J. F. (1999). High-precision calculations of dispersion coefficients, static dipole polarizabilities, and atom–wall interaction constants for alkali-metal atoms. *Physical Review Letters*, **82**, 3589–3592.

Desclaux, J. P. (1973). Relativistic Dirac–Fock expectation values for atoms with z = 1 to z = 120. *Atomic Data and Nuclear Data Tables*, **12**(4), 311–406.

Dham, A. K., Allnatt, A. R., Meath, W. J., and Aziz, R. A. (1989). The Kr–Kr potential energy curve and related physical properties; the XC and HFD-B potential models. *Molecular Physics*, **67**, 1291–1307.

Dham, A. K., Meath, W. J., Allnatt, A. R., Aziz, R. A., and Slaman, M. J. (1990). XC and HFD–B potential energy curves for Xe–Xe and related physical properties. *Chemical Physics*, **142**, 173–189.

Dillard, D. D., Waxman, M., and Robinson, Jr., R. L. (1978). Volumetric data and virial coefficients for helium, krypton, and helium–krypton mixtures. *Journal of Chemical & Engineering Data*, **23**(4), 269–274.

Dorfman, J. R. and van Beijeren, H. (1977). In *Statistical Mechanics, Part B: Time-Dependent Processes* (ed. B. J. Berne), Chapter 3. Plenum Press, New York.

Dorfman, R. (1963). Note on the linearized Boltzmann integral equation for rigid sphere molecules. *Proceedings of the National Academy of Sciences*, **50**, 804–806.

Douketis, C. and Scoles, G. (1982). Intermolecular forces via hybrid Hartree–Fock–SCF plus damped dispersion (HFD) energy calculations. An improved spherical model. *The Journal of Chemical Physics*, **76**, 3057–3063.

Duderstadt, J. J. and Martin, W. R. (1979). *Transport Theory.* Wiley, New York.

Dunning, T. H. (1989). Gaussian basis sets for use in correlated molecular calculations. I. The atoms boron through neon and hydrogen. *The Journal of Chemical Physics*, **90**(2), 1007–1023.

Dunning, T. H., Peterson, K. A., and Wilson, A. K. (2001). Gaussian basis sets for use in correlated molecular calculations. X. The atoms aluminum through argon revisited. *The Journal of Chemical Physics*, **114**(21), 9244–9253.

Dymond, J. H., Marsh, K. N., Wilhoit, R. C., and Wong, K. C. (2002*a*). In *Landolt–Börnstein, New Series, Group IV: Physical Chemistry 21A, Virial Coefficients of Pure Gases.* Springer, Berlin.

Dymond, J. H., Marsh, K. N., Wilhoit, R. C., and Wong, K. C. (2002*b*). *Landolt–Börnstein, New Series, Group IV: Physical Chemistry 21B, Virial Coefficients of Mixtures.* Springer, Berlin.

Dymond, J. H., Marsh, K. N., Wilhoit, R. C., and Wong, K. C. (2003). *Landolt-Börnstein: Numerical Data and Functional Relationships in Science and Technology, New Series, Group IV: Physical Chemistry.* Volume 21. Springer, Berlin.

Edmonds, A. R. (1957). *Angular Momentum in Quantum Mechanics.* Princeton University Press, Princeton.

Edmonds, A. R. (1974). *Angular Momentum in Quantum Mechanics* (3rd edn). Princeton University Press, Princeton.

Eisenschitz, R. and London, F. (1930). Über das Verhältnis der van der Waalsschen Kräfte zu den homöopolaren Bindungskräften. *Zeitschrift für Physik*, **60**, 491–527.

Enskog, D. (1917). *Kinetische Theorie der Vorgänge in mässig verdünnten Gasen.* Ph.D. thesis, Fysiska institutionen.

Enskog, D. (1922). *Kungliga Svenska Vatenskapsakademiens Handlingar*, **63**, No. 4.

Ern, A. and Giovangigli, V. (1994). *Multicomponent Transport Algorithms.* Springer–Verlag, Berlin.

Estrada-Alexanders, A. F. and Trusler, J. P. M. (1995). The speed of sound in gaseous argon at temperatures between 110 K and 450 K and at pressures up to 19 MPa. *The Journal of Chemical Thermodynamics*, **27**(10), 1075–1089.

Evers, C., Lösch, H. W., and Wagner, W. (2002). An absolute viscometer-densimeter and measurements of the viscosity of nitrogen, methane, helium, neon, argon, and krypton over a wide range of density and temperature. *International Journal of Thermophysics*, **23**, 1411–1439.

Ewing, M. B. and Goodwin, A. R. H. (1992). An apparatus based on a spherical resonator for measuring the speed of sound in gases at high pressures. Results for argon at temperatures between 255 K and 300 K and at pressures up to 7 MPa. *The Journal of Chemical Thermodynamics*, **24**(5), 531–547.

Ewing, M. B., Owusu, A. A., and Trusler, J. P. M. (1989). Second acoustic virial coefficients of argon between 100 and 304 K. *Physica A: Statistical Mechanics and its Applications*, **156**(3), 899–908.

Ewing, M. B. and Trusler, J. P. M. (1992*a*). Interaction second acoustic virial coefficients of (N_2 + Ar) between 90 and 373 K. *Physica A.*, **184**, 437–450.

Ewing, M. B. and Trusler, J. P. M. (1992*b*). Second acoustic virial coefficients of nitrogen between 80 and 373 K. *Physica A: Statistical Mechanics and its Applications*, **184**(3), 415–436.

Faas, S., Lenthe, J. H. van, and Snijders, J. G. (2000). Regular approximated scalar relativistic correlated ab initio schemes: applications to rare gas dimers. *Molecular Physics*, **98**(18), 1467–1472.

Fano, U. (1957). Description of states in quantum mechanics by density matrix and operator techniques. *Reviews of Modern Physics*, **29**, 74–93.

Feltgen, R. (1981). Potential model for the interaction of two like S state atoms involving spin symmetry. *The Journal of Chemical Physics*, **74**, 1186–1199.

Fernández, B. and Koch, H. (1998). Accurate *ab initio* rovibronic spectrum of the $X^1\Sigma_g^+$ and $B^1\Sigma_u^+$ states in Ar_2. *The Journal of Chemical Physics*, **109**(23), 10255–10262.

Ferziger, J. H. and Kaper, H. G. (1972). *Mathematical Theory of Transport Processes in Gases.* North Holland, Amsterdam.

Firsov, O. B. (1953). *Zhurnal Eksperimental'noi i Teoreticheskoi Fiziki*, **24**, 279.

Flynn, G. P., Hanks, R. V., and Lemaire, N. A. (1963). Viscosity of Nitrogen, Helium, Neon, and Argon from —78.5° to 100°C below 200 Atmospheres. *The Journal of Chemical Physics*, **38**(1), 154–162.

Fokkens, K., Vermeer, W., Taconis, K.W., and Ouboter, R. De Bruyn (1964). The heat conductivity of ^{3}He and ^{4}He and their mixtures in the gaseous state between 0.5 and 3°K. *Physica*, **30**(12), 2153–2174.

Fox, R. L. (1968*a*). Calculation of low-energy electron–atom transport cross sections from total cross-section measurements. *Physical Review*, **168**, 1–3.

Fox, R. L. (1968*b*). Calculation of low energy electron–atom transport cross sections from total cross section measurements: Erratum. *Physical Review*, **173**, 325–325.

Freeman, D. E., Yoshino, K., and Tanaka, Y. (1974). Vacuum ultraviolet absorption spectrum of the van der Waals molecule Xe_2. I. Ground state vibrational structure, potential well depth, and shape. *The Journal of Chemical Physics*, **61**, 4880–4889.

Friend, D. G. and Rainwater, J. C. (1984). Transport properties of a moderately dense gas. *Chemical Physics Letters*, **107**(6), 590–594.

Frisch, H. L. and Helfand, E. (1960). Conditions imposed by gross properties on the intermolecular potential. *The Journal of Chemical Physics*, **32**(1), 269–270.

Fuchs, R. R., McCourt, F. R. W., Thakkar, A. J., and Grein, F. (1984). Two new anisotropic potential energy surfaces for nitrogen-helium: The use of Hartree–Fock SCF calculations and a combining rule for anisotropic long-range dispersion coefficients. *The Journal of Physical Chemistry*, **88**, 2036–2045.

Gaiser, C. and Fellmuth, B. (2009). Helium virial coefficients—a comparison between new highly accurate theoretical and experimental data. *Metrologia*, **46**(5), 525.

Gambhir, R. S. and Saxena, S. C. (1966). Thermal conductivity of binary and ternary mixtures of krypton, argon, and helium. *Molecular Physics*, **11**(3), 233–241.

Gavioso, R. M., Benedetto, G., Ripa, D. M., Albo, P. A. G., Guianvarc'h, C., Merlone, A., L. Pitre, D. Truong, Moro, F., and Cuccaro, R. (2011). Progress in INRiM experiment for the determination of the Boltzmann constant with a quasi-spherical resonator. *International Journal of Thermophysics*, **32**, 1339.

Gdanitz, R. J. (2001). An accurate interaction potential for neon dimer (Ne_2). *Chemical Physics Letters*, **348**(1), 67–74.

Gerber, R. B. and Karplus, M. (1970). Determination of the phase of the scattering amplitude from the differential cross section. *Physical Review D*, **1**, 998–1012.

Gerber, R. B. and Shapiro, M. (1976). A numerical method for the determination of atom–atom scattering amplitudes from the measured differential cross sections. *Chemical Physics*, **13**(3), 227–233.

Gerber, R. B., Shapiro, M., Buck, U., and Schleusener, J. (1978). Quantum-mechanical inversion of the differential cross section: Determination of the He–Ne potential. *Physical Review Letters*, **41**, 236–239.

Geum, N., Jeung, G.-H., Derevianko, A., Côté, R., and Dalgarno, A. (2001). Interaction potentials of LiH, NaH, KH, RbH, and CsH. *The Journal of Chemical Physics*, **115**(13), 5984–5988.

Gilgen, R., Kleinrahm, R., and Wagner, W. (1994). Measurement and correlation of the (pressure, density, temperature) relation of argon I. The homogeneous gas and liquid regions in the temperature range from 90 K to 340 K at pressures up to 12 MPa. *The Journal of Chemical Thermodynamics*, **26**(4), 383–398.

Goldman, E. and Sirovich, L. (1967). Equations for gas mixtures. *The Physics of Fluids*, **10**, 1928–1940.

Gough, D. W., Maitland, G. C., and Smith, E. B. (1972). The direct determination of intermolecular potential energy functions from gas viscosity measurements. *Molecular Physics*, **24**(1), 151–161.

Gough, D. W., Matthews, G. P., and Smith, E. B. (1976). Viscosity of nitrogen and certain gaseous mixtures at low temperatures. *Journal of the Chemical Society, Faraday Transactions 1: Physical Chemistry in Condensed Phases*, **72**, 645–653.

Gough, D. W., Matthews, G. P., Smith, E. B., and Maitland, G. C. (1975). The direct determination of pair potential energy functions for mixed interactions: Ar–Kr. *Molecular Physics*, **29**, 1759–1765.

Gough, D. W., Smith, E. B., and Maitland, G. C. (1974). The pair potential energy function for krypton. *Molecular Physics*, **27**(4), 867–872.

Grabow, J.-U., Pine, A. S., Fraser, G. T., Lovas, F. J., Suenram, R. D., Emilsson, T., Arunan, E., and Gutowsky, H. S. (1995). Rotational spectra and van der Waals potentials of Ne–Ar. *The Journal of Chemical Physics*, **102**(3), 1181–1187.

Gracki, J. A., Flynn, G. P., and Ross, J. (1969). Viscosity of nitrogen, helium, hydrogen, and argon from −100 to 25°C up to 150–250 atm. *The Journal of Chemical Physics*, **51**(9), 3856–3863.

Grad, H. (1949). On the kinetic theory of rarefied gases. *Communications on Pure and Applied Mathematics*, **2**, 331–407.

Grad, H. (1952). Statistical mechanics, thermodynamics, and fluid dynamics of systems with an arbitrary number of integrals. *Communications on Pure and Applied Mathematics*, **5**, 455–494.

Grad, H. (1958). Principles of the kinetic theory of gases. In *Handbuch der Physik* (ed. S. Flügge), Volume 12. Springer–Verlag, Berlin.

Grad, H. (1963). Asymptotic theory of the Boltzmann equation, II. Proceedings of the 3rd International Symposium on Rarefied Gas Dynamics, 1962, pp. 25–59. New York: Academic Press.

Grisenti, R. E., Schöllkopf, W., Toennies, J. P., Hegerfeldt, G. C., Köhler, T., and Stoll, M. (2000). Determination of the bond length and binding energy of the helium dimer by diffraction from a transmission grating. *Physical Review Letters*, **85**, 2284–2287.

Gross, E. P. and Jackson, E. A. (1959). Kinetic models and the linearized Boltzmann equation. *The Physics of Fluids*, **2**, 432–441.

Gross, E. P. and Krook, M. (1956). Model for collision processes in gases: Small-amplitude oscillations of charged two-component systems. *Physical Review*, **102**, 593–604.

Groot, S. R. de and Mazur, P. (1962). *Nonequilibrium Thermodynamics.* North Holland, Amsterdam.

Guevara, F. A., McInteer, B. B., and Wageman, W. E. (1969). High-temperature viscosity ratios for hydrogen, helium, argon, and nitrogen. *The Physics of Fluids*, **12**(12), 2493–2505.

Haarman, J. W. (1973). Thermal conductivity measurements of He, Ne, Ar, Kr, N_2 and CO_2 with a transient hot wire method. *American Institute of Physics Conference Proceedings*, **11**(1), 193–202.

Hahn, R., Schäfer, K., and Schramm, B. (1974). Messungen zweiter Virialkoeffizienten im Temperaturbereich von 200–300 K. *Berichte der Bunsengesellschaft für physikalische Chemie*, **78**(3), 287–289.

Haire, M. A. and Vargo, D. D. (2007). Review of helium and xenon pure component and mixture transport properties and recommendation of estimating approach for project Prometheus (viscosity and thermal conductivity). *AIP Conference Proceedings*, **880**, 559–570.

Haley, T. P. and Cybulski, S. M. (2003). Ground state potential energy curves for He–Kr, Ne–Kr, Ar–Kr, and Kr_2: Coupled-cluster calculations and comparison with experiment. *The Journal of Chemical Physics*, **119**(11), 5487–5496.

Halkier, A., Helgaker, T., Jørgensen, P., Klopper, W., Koch, H., Olsen, J., and Wilson, A. K. (1998). Basis-set convergence in correlated calculations on Ne, N_2, and H_2O. *Chemical Physics Letters*, **286**(3), 243–252.

Halls, M. D., Schlegel, H. B., DeWitt, M. J., and Drake, G. W. F (2001). *Ab initio* calculation of the $a^3\Sigma_u^+$ interaction potential and vibrational levels of $^7Li_2$. *Chemical Physics Letters*, **339**(5), 427–432.

Hamel, B. B. (1965). Kinetic model for binary gas mixtures. *The Physics of Fluids*, **8**, 418–425.

Hanni, M., Lantto, P., Runeberg, N., Jokisaari, J., and Vaara, J. (2004). Calculation of binary magnetic properties and potential energy curve in xenon dimer: Second virial coefficient of ^{129}Xe nuclear shielding. *The Journal of Chemical Physics*, **121**(12), 5908–5919.

Hättig, C., Cacheiro, J. L., Fernández, B., and Rizzo, A. (2003). *Ab initio* calculation of the refractivity and hyperpolarizability second virial coefficients of neon gas. *Molecular Physics*, **101**(13), 1983–1995.

Hättig, C. and Hess, B. A. (1996). TDMP2 calculation of dynamic multipole polarizabilities and dispersion coefficients of the noble gases Ar, Kr, Xe, and Rn. *The Journal of Physical Chemistry*, **100**(15), 6243–6248.

Hellemans, J. M., Kestin, J., and Ro, S. T. (1974). On the properties of multicomponent mixtures of monatomic gases. *Physica*, **71**(1), 1–16.

Hellmann, R., Bich, E., and Vogel, E. (2007). *Ab initio* potential energy curve for the helium atom pair and thermophysical properties of dilute helium gas. I. Helium–helium interatomic potential. *Molecular Physics*, **105**(23–24), 3013–3023.

Hellmann, R., Bich, E., and Vogel, E. (2008*a*). *Ab initio* potential energy curve for the neon atom pair and thermophysical properties of the dilute neon gas. I. Neon–neon interatomic potential and rovibrational spectra. *Molecular Physics*, **106**(1), 133–140.

Hellmann, R., Bich, E., and Vogel, E. (2008*b*). Erratum: *Ab initio* potential energy curve for the neon atom pair and thermophysical properties of the dilute neon gas. *Molecular Physics*, **106**(8), 1107–1122.

Hellmann, R., Gaiser, C., Fellmuth, B., Vasyltsova, T., and Bich, E. (2021). Thermophysical properties of low-density neon gas from highly accurate first-principles calculations and dielectric-constant gas thermometry measurements. *The Journal of Chemical Physics*, **154**(16), 164304.

Hellmann, R., Jäger, B., and Bich, E. (2017). State-of-the-art *ab initio* potential energy curve for the xenon atom pair and related spectroscopic and thermophysical properties. *The Journal of Chemical Physics*, **147**(3), 034304.

Hemminger, W. (1987). The thermal conductivity of gases: Incorrect results due to desorbed air. *Pure and Applied Chemistry*, **8**(3), 317–333.

Hepburn, J., Scoles, G., and Penco, R. (1975). A simple but reliable method for the prediction of intermolecular potentials. *Chemical Physics Letters*, **36**, 451–456.

Herman, P. R., LaRocque, P. E., and Stoicheff, B. P. (1988). Vacuum ultraviolet laser spectroscopy. V. Rovibronic spectra of Ar_2 and constants of the ground and excited states. *The Journal of Chemical Physics*, **89**, 4535–4549.

Herring, C. and Flicker, M. (1964). Asymptotic exchange coupling of two hydrogen atoms. *Physical Review*, **134**, A362–A366.

Hess, S. (2015). *Tensors for Physics.* Springer International Publishing, Basel.

Hess, S. and Köhler, W. E. (1980). *Formeln zur Tensorrechnung.* Palm and Enke, Erlangen.

Hirschfelder, J. O. (1967). Perturbation theory for exchange forces, I. *Chemical Physics Letters*, **1**, 325–329.

Hirschfelder, J. O., Curtiss, C. F., and Bird, R. B. (1964). *Molecular Theory of Gases and Liquids.* John Wiley, New York.

Hirschfelder, J. O. and Eliason, M. A. (1957). The estimation of the transport properties for electronically excited atoms and molecules. *Annals of the New York Academy of Sciences*, **67**(9), 451–461.

Hirschfelder, J. O. and Meath, W. J. (1967). The nature of intermolecular forces. *Advances in Chemical Physics: Intermolecular Forces*, **12**, 3–106.

Hoffmann–Ostenhof, M. and Hoffmann–Ostenhof, T. (1977). "Schrödinger inequalities" and asymptotic behavior of the electron density of atoms and molecules. *Physical Review A*, **16**, 1782–1785.

Hogervorst, W. (1971). Diffusion coefficients of noble-gas mixtures between 300 K and 1400 K. *Physica*, **51**(1), 59–76.

Holway Jr., L. H. (1966). New statistical models for kinetic theory: Methods of construction. *The Physics of Fluids*, **9**, 1658–1673.

Hu, Y. R., Zhai, Y., Li, H., and McCourt, F. R. W. (2022). *Ab initio* potential energy functions, spectroscopy and thermal physics for krypton-contained rare gas dimers. *Journal of Quantitative Spectroscopy and Radiative Transfer*, 108244.

Huber, K. P. and Herzberg, G. (1979). *Molecular spectra and molecular structure. IV. Constants of diatomic molecules.* Van Nostrand Reinhold, New York.

Hurly, J. J. and Mehl, J. B. (2007). ^{4}He thermophysical properties: New *ab initio* calculations. *Journal of Research of the National Institute of Standards and Technology*, **112**, 75–94.

Hurly, J. J. and Moldover, M. R. (2000). *Ab initio* values of the thermophysical properties of helium as standards. *Journal of Research of the National Institute of Standards and Technology*, **105**, 667.

Hurly, J. J., Schmidt, J. W., Boyes, S. J., and Moldover, M. R. (1997). Virial equation of state of helium, xenon, and helium–xenon mixtures from speed-of-sound and burnett p ρ t measurements. *International Journal of Thermophysics*, **18**(3), 579–634.

Hurly, J. J., Taylor, W. L., and Menke, D. A. (1991). Thermal diffusion factors for equimolar He–Ar from 80 to 640 K and equimolar He–Xe from 140 to 270 K. *The Journal of Chemical Physics*, **94**, 8282–8288.

Hutchinson, F. (1949). Self-diffusion in argon. *The Journal of Chemical Physics*, **17**(11), 1081–1086.

Hutson, J. M. and Soldán, P. (2007). Molecular collisions in ultracold atomic gases. *International Reviews in Physical Chemistry*, **26**(1), 1–28.

Ivanov, V. S., Sovkov, V. B., and Li, Li (2003). Joint analysis of the attractive and repulsive regions of the Na_2 $a^3\Sigma_u^+$ state potential: A new empirical potential energy curve. *The Journal of Chemical Physics*, **118**(18), 8242–8247.

Jacobi, N. and Csanak, Gy. (1975). Dispersion forces at arbitrary distances. *Chemical Physics Letters*, **30**, 367–372.

Jäger, B. and Bich, E. (2017). Thermophysical properties of krypton–helium gas mixtures from *ab initio* pair potentials. *The Journal of Chemical Physics*, **146**(21), 214302.

Jäger, B., Hellmann, R., Bich, E., and Vogel, E. (2009). *Ab initio* pair potential energy curve for the argon atom pair and thermophysical properties of the dilute argon gas. I. Argon–argon interatomic potential and rovibrational spectra. *Molecular Physics*, **107**(20), 2181–2188.

Jäger, B., Hellmann, R., Bich, E., and Vogel, E. (2010). Erratum: *Ab initio* pair potential energy curve for the argon atom pair and thermophysical properties of the dilute argon gas. I. Argon–argon interatomic potential and rovibrational spectra. *Molecular Physics*, **108**(1), 105–105.

Jäger, B., Hellmann, R., Bich, E., and Vogel, E. (2016). State-of-the-art *ab initio* potential energy curve for the krypton atom pair and thermophysical properties of dilute krypton gas. *The Journal of Chemical Physics*, **144**(11), 114304.

Jäger, W. and Gerry, M. C. L. (1993). Pure rotational spectra of the mixed rare gas van der Waals complexes Ne–Xe, Ar–Xe, and Kr–Xe. *The Journal of Chemical Physics*, **99**, 919–927.

Jäger, W., Xu, Y., and Gerry, M. C. L. (1993). Pure rotational spectra of the mixed rare gas van der Waals complexes Ne–Xe, Ar–Xe, and Kr–Xe. *The Journal of Chemical Physics*, **99**(2), 919–927.

Jamieson, M. J., Drake, G. W. F., and Dalgarno, A. (1995). Retarded dipole–dipole dispersion interaction potential for helium. *Physical Review A*, **51**, 3358–3361.

Janzen, A. R. and Aziz, R. A. (1997). An accurate potential energy curve for helium based on *ab initio* calculations. *The Journal of Chemical Physics*, **107**(3), 914–919.

Jeziorska, M., Cencek, W., Patkowski, K., Jeziorski, B., and Szalewicz, K. (2007). Pair potential for helium from symmetry-adapted perturbation theory calculations and from supermolecular data. *The Journal of Chemical Physics*, **127**(12), 124303.

Jeziorski, B., Moszynski, R., and Szalewicz, K. (1994). Perturbation theory approach to intermolecular potential energy surfaces of van der Waals complexes. *Chemical Reviews*, **94**(7), 1887–1930.

Jeziorski, B. and Szalewicz, K. (2003). Molecules in the physico-chemical environment: Spectroscopy, dynamics and bulk properties. In *Handbook of Molecular Physics and Quantum Chemistry*, Volume 3, Chapter 9. Wiley, Hoboken.

Johns, A. I., Rashid, S., Watson, J. T. R., and Clifford, A. A. (1986). Thermal conductivity of argon, nitrogen and carbon dioxide at elevated temperatures and pressures. *Journal of the Chemical Society, Faraday Transactions 1*, **82**, 2235–2246.

Johnston, H. L. and Grilly, E. R. (1942). Viscosities of carbon monoxide, helium, neon, and argon between 80° and 300°K. Coefficients of viscosity. *The Journal of Physical Chemistry*, **46**, 948–963.

Jones, R. C. (1940). On the theory of the thermal diffusion coefficient for isotopes. *Physical Review*, **58**, 111–122.

Jones, R. C. and Furry, W. H. (1940). On the calculation of the thermal diffusion constant from viscosity data. *Physical Review*, **57**, 547–547.

Kalelkar, A. S. and Kestin, J. (1970). Viscosity of He–Ar and He–Kr binary gaseous mixtures in the temperature range 25–720°C. *The Journal of Chemical Physics*, **52**(8), 4248–4261.

Kamerlingh Onnes, H. (1901). *Communications from the Physical Laboratory of the University of Leiden*, **No. 71**.

Kantorovich, L. V. and Krylov, V. I. (1958). *Approximate Methods of Higher Analysis.* Translated by C. D. Benster. Interscience, New York.

Kao, J. F. and Kobayashi, R. (1967). Viscosity of helium and nitrogen and their mixtures at low temperatures and elevated pressures. *The Journal of Chemical Physics*, **47**(8), 2836–2849.

Kaplan, I. G. (2003). Molecules in the physico-chemical environment: Spectroscopy, dynamics, and bulk properties. In *Handbook of Molecular Physics and Quantum Chemistry* (ed. S. Wilson), Volume 3, Chapter 8. Wiley, Hoboken.

Karton, A. and Martin, J. M. L. (2005). Comment on: "Estimating the Hartree–Fock limit from finite basis set calculations" [Jensen, F. (2005) *Theor Chem Acc* 113, 267]. *Theoretical Chemistry Accounts*, **115**, 330–333.

Kate, F. H. and Robinson, R. L. (1973). Vapor + solid equilibrium for helium + krypton. *The Journal of Chemical Thermodynamics*, **5**(2), 259–271.

Keil, M., Danielson, L. J., Buck, U., Schleusener, J., Huisken, F., and Dingle, T. W. (1988). The HeNe interatomic potential from multiproperty fits and Hartree–Fock calculations. *The Journal of Chemical Physics*, **89**(5), 2866–2880.

Keil, M., Danielson, L. J., and Dunlop, P. J. (1991). On obtaining interatomic potentials from multiproperty fits to experimental data. *The Journal of Chemical Physics*, **94**(1), 296–309.

Kendall, R. A., Dunning, T. H., and Harrison, R. J. (1992). Electron affinities of the first-row atoms revisited. Systematic basis sets and wave functions. *The Journal of Chemical Physics*, **96**(9), 6796–6806.

Kennard, E. H. (1938). *Kinetic Theory of Gases.* McGraw Hill, New York.

Kerrisk, J. F. and Keller., W. E. (1969). Thermal Conductivity of Fluid He^3 and He^4 at Temperatures between 1.5 and 4.0° K and for pressures up to 34 atm. *Physical Review*, **177**, 341–351.

Kestin, J. and Dorfman, J. R. (1971). *Statistical Thermodynamics.* Adademic Press, New York.

Kestin, J., Khalifa, H. E., and Wakeham, W. A. (1978). The viscosity and diffusion coefficients of the binary mixtures of xenon with the other noble gases. *Physica A*, **90**, 215–228.

Kestin, J., Knierim, K. D., Mason, E. A., Najafi, B., Ro, S. T., and Waldman, M. (1984). Equilibrium and transport properties of the noble gases and their mixtures at low density. *Journal of Physical and Chemical Reference Data*, **13**(1), 229–303.

Kestin, J., Kobayashi, Y., and Wood, R. T. (1966). The viscosity of four binary, gaseous mixtures at 20° and 30°C. *Physica*, **32**(6), 1065–1089.

Kestin, J. and Leidenfrost, W. (1959). An absolute determination of the viscosity of eleven gases over a range of pressures. *Physica*, **25**(7), 1033–1062.

Kestin, J. and Nagashima, A. (1964). Viscosity of neon–helium and neon–argon mixtures at 20° and 30°C. *The Journal of Chemical Physics*, **40**(12), 3648–3654.

Kestin, J., Paul, R., Clifford, A. A., and Wakeham, W. A. (1980). Absolute determination of the thermal conductivity of the noble gases at room temperature up to 35 mpa. *Physica A: Statistical Mechanics and its Applications*, **100**(2), 349–369.

Kestin, J., Ro, R. T., and Wakeham, W. A. (1972*a*). Viscosity of the binary gaseous mixture helium–nitrogen. *The Journal of Chemical Physics*, **56**(8), 4036–4042.

Kestin, J., Ro, S. T., and Wakeham, W. (1972*b*). An extended law of corresponding states for the equilibrium and transport properties of the noble gases. *Physica*, **58**(2), 165–211.

Kestin, J., Ro, S. T., and Wakeham, W. A. (1972*c*). Viscosity of the binary gaseous mixtures He–Ne and Ne–N_2 in the temperature range 25–700°C. *The Journal of Chemical Physics*, **56**(12), 5837–5842.

Kestin, J., Ro, S. T., and Wakeham, W. A. (1972*d*). Viscosity of the isotopes of hydrogen and their intermolecular force potentials. *Journal of the Chemical Society, Faraday Transactions 1*, **68**, 2316–2323.

Kestin, J., Ro, S. T., and Wakeham, W. A. (1972*e*). Viscosity of the noble gases in the temperature range 25–700°C. *The Journal of Chemical Physics*, **56**(8), 4119–4124.

Kestin, J. and Wakeham, W. A. (1983). The viscosity and diffusion coefficient of binary mixtures of nitrous oxide with He, Ne and CO. *Berichte der Bunsengesellschaft für physikalische Chemie*, **87**(4), 309–311.

Kestin, J., Wakeham, W. A., and Watanabe, K. (1970). Viscosity, thermal conductivity, and diffusion coefficient of Ar–Ne and Ar–Kr gaseous mixtures in the temperature range 25–700°C. *The Journal of Chemical Physics*, **53**, 3773–3780.

Kislyakov, I. M. (1999). Dynamic polarizabilities of rare-gas atoms at imaginary frequencies. *Optics and Spectroscopy*, **87**, 357–361.

Klein, M. and Smith, F. J. (1968). Tables of collision integrals for the (m,6) potential function for 10 values of m. *Journal of Research of the National Bureau of Standards. Section A, Physics and Chemistry*, **72A**, 359–423.

Kleinekathöfer, U., Tang, K. T., Toennies, J. P., and Yiu, C. L. (1997). Van der Waals potentials of He_2, Ne_2, and Ar_2 with the exchange energy calculated by the surface integral method. *The Journal of Chemical Physics*, **107**, 9502–9513.

Klimeck, J., Kleinrahm, R., and Wagner, W. (1998). An accurate single-sinker densimeter and measurements of the (p, ρ,T) relation of argon and nitrogen in the temperature range from 235 to 520 K at pressures up to 30 MPa. *The Journal of Chemical Thermodynamics*, **30**, 1571–1588.

Klopper, W. (1997). Simple recipe for implementing computation of first-order relativistic corrections to electron correlation energies in framework of direct perturbation theory. *Journal of Computational Chemistry*, **18**(1), 20–27.

Klopper, W. (2001). Highly accurate coupled-cluster singlet and triplet pair energies from explicitly correlated calculations in comparison with extrapolation techniques. *Molecular Physics*, **99**(6), 481–507.

Knoop, S., Schuster, T., Scelle, R., Trautmann, A., Appmeier, J., Oberthaler, M. K., Tiesinga, E., and Tiemann, E. (2011). Feshbach spectroscopy and analysis of the interaction potentials of ultracold sodium. *Physical Review A*, **83**, 042704.

Knowles, P. J., Hampel, C., and Werner, H.-J. (1993). Coupled cluster theory for high spin, open shell reference wave functions. *The Journal of Chemical Physics*, **99**(7), 5219–5227.

Knowles, P. J., Hampel, C., and Werner, H.-J. (2000). Erratum: "Coupled cluster theory for high spin, open shell reference wave functions" [*J. Chem. Phys.* 99, 5219 (1993)]. *The Journal of Chemical Physics*, **112**(6), 3106–3107.

Knowles, P. J. and Meath, W. J. (1986). Non-expanded dispersion energies and damping functions for Ar_2 and Li_2. *Chemical Physics Letters*, **124**, 164–171.

Köhler, W. E. and 't Hooft, G. W. (1979). Waldmann–Snider collision integrals for mixtures of polyatomic gases. exact and approximate relations. *Zeitschrift für Naturforschung*, **34a**, 1255–1268.

Koide, A., Meath, W. J., and Allnatt, A. R. (1981). Second order charge overlap effects and damping functions for isotropic atomic and molecular interactions. *Chemical Physics*, **58**, 105–119.

Kolos, W. and Wolniewicz, L. (1974). Variational calculation of the long–range interaction between two ground-state hydrogen atoms. *Chemical Physics Letters*, **24**, 457–460.

Komasa, J. (1999). Exponentially correlated Gaussian functions in variational calculations: Energy expectation values in the ground state helium dimer. *The Journal of Chemical Physics*, **110**(16), 7909–7916.

Komasa, J., Cencek, W., and Rychlewski, J. (1999). Adiabatic corrections of the helium dimer from exponentially correlated Gaussian functions. *Chemical Physics Letters*, **304**(3), 293–298.

Komasa, J. and Thakkar, A. J. (1995). Accurate Heitler–London interaction energy for He_2. *Journal of Molecular Structure: THEOCHEM*, **3443**, 43–48.

Korek, M., Bleik, S., and Allouche, A. R. (2007). Theoretical calculation of the low laying electronic states of the molecule NaCs with spin-orbit effect. *The Journal of Chemical Physics*, **126**(12), 124313.

Korona, T., Williams, H. L., Bukowski, R., Jeziorski, B., and Szalewicz, K. (1997). Helium dimer potential from symmetry-adapted perturbation theory calculations using large Gaussian geminal and orbital basis sets. *The Journal of Chemical Physics*, **106**(12), 5109–5122.

Kramer, H. L. and Herschbach, D. R. (1970). Combination rules for van der Waals force constants. *The Journal of Chemical Physics*, **53**, 2792–2800.

Krauss, M. and Neumann, D. B. (1979). Charge overlap effects in dispersion energies. *The Journal of Chemical Physics*, **71**, 107–112.

Krauss, M. and Stevens, W. J. (1982). *Ab initio* determination of the ground-state potential energy curve for Ar_2. *Chemical Physics Letters*, **85**, 423–427.

Krauss, M., Stevens, W. J., and Neumann, D. B. (1980). The dispersion damping functions and interaction energy curves for Xe–Xe. *Chemical Physics Letters*, **71**, 500–502.

Kreek, H. and Meath, W. J. (1969). Charge-overlap effects. Dispersion and induction forces. *The Journal of Chemical Physics*, **50**, 2289–2302.

Kumar, A. and Thakkar, A. J. (2010). Dipole oscillator strength distributions with improved high-energy behavior: Dipole sum rules and dispersion coefficients for Ne, Ar, Kr, and Xe revisited. *The Journal of Chemical Physics*, **132**(7), 074301.

Kutzelnigg, W. (1980). The "primitive" wave function in the theory of intermolecular interactions. *The Journal of Chemical Physics*, **73**(1), 343–359.

Kutzelnigg, W. (1989). Perturbation theory of relativistic corrections. *Zeitschrift für Physik D: Atoms, Molecules and Clusters*, **11**, 15–28.

Kutzelnigg, W., Ottschofski, E., and Franke, R. (1995). Relativistic Hartree–Fock by means of stationary direct perturbation theory. I. General theory. *The Journal of Chemical Physics*, **102**(4), 1740–1751.

Kuščer, I. and Williams, M. M. R. (1967). Relaxation constants of a uniform hard-sphere gas. *The Physics of Fluids*, **10**, 1922–1927.

López Cacheiro, J., Fernández, B., Marchesan, D., Coriani, S., Hättig, C., and Rizzo, A. (2004). Coupled cluster calculations of the ground state potential and interaction induced electric properties of the mixed dimers of helium, neon and argon. *Molecular Physics*, **102**(1), 101–110.

Landau, L. D. and Lifshitz, E. M. (1977). *Quantum Mechanics.* Pergamon Press, London.

LaRocque, P. E., Lipson, R. H., Herman, P. R., and Stoicheff, B. P. (1986). Vacuum ultraviolet laser spectroscopy. IV. Spectra of Kr_2 and constants of the ground and excited states. *The Journal of Chemical Physics*, **84**, 6627–6641.

Laschuk, E. F., Martins, M. M., and Evangelisti, S. (2003). *Ab initio* potentials for weakly interacting systems: Homonuclear rare gas dimers. *International Journal of Quantum Chemistry*, **95**(3), 303–312.

Lau, J. A., Toennies, J. P., and Tang, K. T. (2016). An accurate potential model for the $a^3\Sigma_u^+$ state of the alkali dimers Na_2, K_2, Rb_2, and Cs_2. *The Journal of Chemical Physics*, **145**, 194308.

Le Neindre, B., Tufeu, R., and Sirota, A. M. (1991). Transient methods for thermal conductivity. In *Experimental Thermodynamics. III. Measurement of the Transport Properties of Fluids* (eds. W. A. Wakeham, A. Nagashima, and J. V. Sengers), Chapter 6. Blackwell, Oxford.

Le Roy, R. J. (2010). Determining equilibrium structures and potential energy functions for diatomic molecules. In *Equilibrium Structures of Molecules* (eds. J. Demaison and A. G. Csaszar), Chapter 6. Taylor & Francis, London.

Le Roy, R. J. (2017a). dPotFit: A computer program to fit diatomic molecule spectral data to potential energy functions. *Journal of Quantitative Spectroscopy and Radiative Transfer*, **186**, 179–196.

Le Roy, R. J. (2017b). LEVEL: A computer program for solving the radial Schrödinger equation for bound and quasibound levels. *Journal of Quantitative Spectroscopy and Radiative Transfer*, **186**, 167–178.

Le Roy, R. J., Dattani, N. S., Coxon, J. A., Crozet, P., and Linton, C. (2009). Accurate analytic potentials for $Li_2(X^1\Sigma_g^+)$ and $Li_2(A^1\Sigma_u^+)$ from 2 to 90 Å, and the radiative lifetime of Li(2p). *The Journal of Chemical Physics*, **131**, 204309.

Le Roy, R. J., Haugen, C. C., Tao, J., and Li, H. (2011). Long-range damping functions improve the short–range behaviour of "MLR" potential energy functions. *Molecular Physics*, **109**, 435–446.

Le Roy, R. J. and Henderson, R. D. E. (2007). A new potential function form incorporating extended long-range behaviour: Application to ground-state Ca_2. *Molecular Physics*, **105**, 663–677.

Le Roy, R. J. and Pashkov, A. (2017). betaFIT: A computer program to fit pointwise potentials to selected analytic functions. *Journal of Quantitative Spectroscopy and Radiative Transfer*, **186**, 210–220.

Lee, J. K., Henderson, D., and Barker, J. A. (1975). Intermolecular interactions and excess thermodynamic properties of argon–krypton and krypton–xenon mixtures. *Molecular Physics*, **29**, 429–435.

Lee, J. S. (2005). Accurate *ab intio* determination of binding energies for rare-gas dimers by basis set extrapolation. *Theoretical Chemistry Accounts*, **113**, 87–94.

Lennard–Jones, J. E. (1924*a*). On the determination of molecular fields. I. From the variation of the viscosity of a gas with temperature. *Proceedings of the Royal Society of London. Series A, Containing Papers of a Mathematical and Physical Character*, **106**, 441–462.

Lennard–Jones, J. E. (1924*b*). On the determination of molecular fields. II. From the equation of state of a gas. *Proceedings of the Royal Society of London. Series A, Containing Papers of a Mathematical and Physical Character*, **106**, 463–477.

Levin, E., Partridge, H., and Stallcop, J. R. (1990). Collision integrals and high temperature transport properties for N–N, O–O, and N–O. *Journal of Thermophysics and Heat Transfer*, **4**(4), 469–477.

Li, S. F. Y., Papadaki, M., and Wakeham, W. A. (1994). Thermal conductivity of low-density polyatomic gases. In *Proceedings of the 22nd International Thermal Conductivity Conference*, (ed. T. W. Tong), 531–542. Technomic Publications, Lancaster, PA.

Lide, D. R. (2001). *CRC Handbook of Chemistry and Physics.* CRC Press, Boca Raton.

Lin, H., Che, J., Zhang, J. T., and Feng, X. J. (2016). Measurements of the viscosities of Kr and Xe by the two-capillary viscometry. *Fluid Phase Equilibria*, **418**, 198–203.

Lin, H., Feng, X. J., Zhang, J. T., and Liu, C. (2014). Using a two-capillary viscometer with preheating to measure the viscosity of dilute argon from 298.15 K to 653.15 K. *The Journal of Chemical Physics*, **141**(23), 234311.

Linse, C. A., van den Biesen, J. J. H., van Veen, E. H., and van den Meijdenberg, C. J. M. (1979). Measurements on the glory structure in the total cross section of noble gas systems II. Ne and Ar scattered by Ar, Kr and Xe. *Physica A: Statistical Mechanics and its Applications*, **99**, 166–183.

Linton, C., Martin, F., Ros, A. J., Russier, I., Crozet, P., Yiannopoulou, A., Li, L., and Lyyra, A. M. (1999). The high–lying vibrational levels and dissociation energy of the $a^3\Sigma_u^+$ State of $^7Li_2$. *Journal of Molecular Spectroscopy*, **196**, 20–28.

Liu, B. and McLean, A. D. (1973). Accurate calculation of the attractive interaction of two ground state helium atoms. *The Journal of Chemical Physics*, **59**(8), 4557–4558.

Liu, B. and McLean, A. D. (1989). The interacting correlated fragments model for weak interactions, basis set superposition error, and the helium dimer potential. *The Journal of Chemical Physics*, **91**(4), 2348–2359.

Liu, J., Zhai, Y., Li, H., and McCourt, F. R. W. (2022). *Ab initio* Morse/long-range potential energy functions plus spectroscopic and thermophysical properties of heteronuclear diatomic complexes of xenon with the rare gases. *Journal of Quantitative Spectroscopy and Radiative Transfer*, **285**, 108169.

London, F. (1930). Zur theorie und systematik der molekularkräfte. *Zeitschrift für Physik*, **63**, 245–279.

London, F. (1937). The general theory of molecular forces. *Transactions of the Faraday Society*, **33**, 8b–26.

Longuet-Higgins, H. C. (1956). The electronic states of composite systems. *Proceedings of the Royal Society of London. Series A, Mathematical and Physical Sciences*, **235**, 537–543.

Luo, F., Giese, C. F., and Gentry, W. R. (1996). Direct measurement of the size of the helium dimer. *The Journal of Chemical Physics*, **104**(3), 1151–1154.

Luo, F., McBane, G. C., Kim, G., Giese, C. F., and Gentry, W. R. (1993). The weakest bond: Experimental observation of helium dimer. *The Journal of Chemical Physics*, **98**(4), 3564–3567.

López Cacheiro, J., Fernández, B., Marchesan, D., Coriani, S., Hättig, C., and Rizzo, A. (2004). Coupled cluster calculations of the ground state potential and interaction induced electric properties of the mixed dimers of helium, neon and argon. *Molecular Physics*, **102**(1), 101–110.

Maitland, G. C., Mason, E. A., Viehland, L. A., and Wakeham, W. A. (1978). A justification of methods for the inversion of gas transport coefficients. *Molecular Physics*, **36**(3), 797–816.

Maitland, G. C., Rigby, M., Smith, E. B., and Wakeham, W. A. (1980). *Intermolecular Forces: Their Origin and Determination.* Oxford University Press, Oxford.

Maitland, G. C., Rigby, M., Smith, E. B., and Wakeham, W. A. (1981). *Intermolecular Forces: Their Origin and Determination.* Clarendon Press, Oxford.

Maitland, G. C. and Smith, E. B. (1971). The intermolecular pair potential of argon. *Molecular Physics*, **22**, 861–868.

Maitland, G. C. and Smith, E. B. (1972). The direct determination of potential energy functions from second virial coefficients. *Molecular Physics*, **24**(6), 1185–1201.

Maitland, G. C. and Smith, E. B. (1973). A simplified representation of intermolecular potential energy. *Chemical Physics Letters*, **22**, 443–446.

Maitland, G. C. and Smith, E. B. (1974). Viscosities of binary gas mixtures at high temperatures. *Journal of the Chemical Society, Faraday Transactions 1: Physical Chemistry in Condensed Phases*, **70**, 1191–1211.

Maitland, G. C., Vesovic, V., and Wakeham, W. A. (1985). The inversion of thermophysical properties. *Molecular Physics*, **54**(2), 287–300.

Maitland, G. C. and Wakeham, W. A. (1978*a*). Direct determination of intermolecular potentials from gaseous transport coefficients alone. *Molecular Physics*, **35**, 1443–1469.

Maitland, G. C. and Wakeham, W. A. (1978*b*). Direct determination of intermolecular potentials from gaseous transport coefficients alone. *Molecular Physics*, **35**(5), 1429–1442.

Malinauskas, A. P. (1965). Gaseous diffusion. The systems He–Ar, Ar–Xe, and He–Xe. *The Journal of Chemical Physics*, **42**(1), 156–159.

Martin, A. (1969). Construction of the scattering amplitude from the differential cross-sections. *Nuovo Cimento A*, **59**, 131–152.

Mason, E. A. (1957). Higher approximations for the transport properties of binary gas mixtures. I. General formulas. *The Journal of Chemical Physics*, **27**, 75–84.

Mason, E. A. and McDaniel, E. W. (1988). *Transport Properties of Ions in Gases.* John Wiley, New York.

Mason, E. A., Munn, R. J., and Smith, F. J. (1966). Thermal diffusion in gases. *Advances in Atomic and Molecular Physics*, **2**, 33.

Mason, E. A. and Spurling, T. H. (1969). The virial equation of state. In *The International Encyclopedia of Physical Chemistry and Chemical Physics*, Volume 2, Topic 10. Pergamon Press, Oxford.

Mason, E. A. and Von Ubisch, H. (1960). Thermal conductivities of rare gas mixtures. *The Physics of Fluids*, **3**(3), 355–361.

Massey, H. S. W. and Mohr, C. B. O. (1933). Free paths and transport phenomena in gases and the quantum theory of collisions. I. The rigid sphere model. *Proceedings of the Royal Society of London. Series A, Containing Papers of a Mathematical and Physical Character*, **141**, 434–453.

Maxwell, J. C. (1860). Illustrations of the dynamical theory of gases. In *Philosophical Transactions of the Royal Society of London (1860); reprinted in The Scientific Papers of James Clerk Maxwell (1965)* (ed. W. D. Niven), Volume 1, pp. 377–391. Dover Publications, New York.

Maxwell, J. C. (1867). On the dynamical theory of gases. *Philosophical Transactions of the Royal Society of London*, **157**, 49–88.

May, E. F., Berg, R. F., and Moldover, M. R. (2007). Reference viscosities of H_2, CH_4, Ar, and Xe at low densities. *International Journal of Thermophysics*, **28**, 1085–1110.

May, E. F., Moldover, M. R., Berg, R. F., and Hurly, J. J. (2006). Transport properties of argon at zero density from viscosity-ratio measurements. *Metrologia*, **43**, 247–258.

McCormack, F. J. (1973). Construction of linearized kinetic models for gaseous mixtures and molecular gases. *The Physics of Fluids*, **16**, 2095–2105.

McCormack, F. J. and Craven, D. E. (1974). Kinetic theory of sound propagation in gaseous mixtures. I. Two-fluid 5-moment, 13-moment, and Navier–Stokes theories. *The Journal of the Acoustical Society of America*, **55**, 775–782.

McCourt, F. R. W. (2003). Molecules in the physico-chemical environment: Spectroscopy, dynamics, and bulk properties. In *Handbook of Molecular Physics and Quantum Chemistry* (ed. S. Wilson), Volume 3, Chapter 27. Wiley, Chichester.

McCourt, F. R. W., Beenakker, J. J. M., Köhler, W. E., and Kuščer, I. (1990). *Nonequilibrium Phenomena in Polyatomic Gases.* Oxford University Press, Oxford.

McLinden, M. O. and Lösch-Will, C. (2007). Apparatus for wide-ranging, high-accuracy fluid (p, ρ, t) measurements based on a compact two-sinker densimeter. *The Journal of Chemical Thermodynamics*, **39**, 507–530.

McWeeney, R. (1960). Some recent advances in density matrix theory. *Reviews of Modern Physics*, **32**, 335–369.

Meath, W. J. and Hirschfelder, J. O. (1966). Long-range (retarded) intermolecular forces. *The Journal of Chemical Physics*, **44**(9), 3210–3215.

Meath, W. J. and Koulis, M. (1991). On the construction and use of reliable two- and many-body interatomic and intermolecular potentials. *Journal of Molecular Structure: THEOCHEM*, **226**, 1–37.

Meeks, F. R., Cleland, T. J., Hutchinson, K. E., and Taylor, W. L. (1994). On the quantum cross sections in dilute gases. *The Journal of Chemical Physics*, **100**, 3813–3820.

Mehl, J. B. (2009). *Ab initio* properties of gaseous helium. *Comptes Rendus Physique*, **10**(9), 859–865.

Messiah, A. (1966). *Quantum Mechanics.* North Holland, Amsterdam.

Michels, A., Levelt, J. M., and De Graaff, W. (1958). Compressibility isotherms of argon at temperatures between −25°C and −155°C, and at densities up to 640 Amagats (pressures up to 1050 atmospheres). *Physica*, **24**(6), 659–671.

Michels, A., Wassenaar, T., and Louwerse, P. (1954). Isotherms of xenon at temperatures between 0°C and 150°C and at densities up to 515 Amagats (pressures up to 2800 atmospheres). *Physica*, **20**(1), 99–106.

Michels, A., Wijker, H., and Wijker, H. (1949). Isotherms of argon between 0°C and 150°C and pressures up to 2900 atmospheres. *Physica*, **15**(7), 627–633.

Mie, G. (1903). Zur kinetischen Theorie der einatomigen Körper. *Annalen der Physik*, **316**, 657–697.

Millat, J., Mustafa, M., Ross, M., Wakeham, W. A., and Zalaf, M. (1987). The thermal conductivity of argon, carbon dioxide and nitrous oxide. *Physica A: Statistical Mechanics and its Applications*, **145**(3), 461–497.

Millat, J., Ross, M., Wakeham, William A., and Zalaf, M. (1988). The thermal conductivity of neon, methane and tetrafluoromethane. *Physica A: Statistical Mechanics and Its Applications*, **148**, 124–152.

Miller, W. H. (1969). WKB solution of inversion problems for potential scattering. *The Journal of Chemical Physics*, **51**, 3631–3638.

Moelwyn-Hughes, E. A. (1957). *Physical Chemistry.* Pergamon, New York.

Mohr, P. J., Taylor, B. N., and Newell, D. B. (2008). CODATA recommended values of the fundamental physical constants: 2006. *Journal of Physical and Chemical Reference Data*, **37**(3), 1187–1284, Table XLVIII, p. 1259.

Moldover, M. R., Boyes, S. J., Meyer, C. W., and Goodwin, A. R. H. (1999). Thermodynamic temperatures of the triple points of mercury and gallium and in the interval 217 K to 303 K. *Journal of Research of the National Institute of Standards and Technology*, **104**(1), 11–46.

Moldover, M. R., Gills, K. A., Hurly, J. J., Hehl, J. B., and Wilhelm, J. (2001). Measurements in gases. In *Modern Acoustical Techniques for the Measurement of Mechanical Properties* (eds. M. Levy, H. E. Bass, and R. Stern), Volume 39, Chapter 10. Academic Press, New York.

Moldover, M. R. and Trusler, J. P. M. (1988). Accurate acoustic thermometry I: The triple point of gallium. *Metrologia*, **25**(3), 165.

Monchick, L., Yun, K. S., and Mason, E. A. (1963). Formal kinetic theory of transport phenomena in polyatomic gas mixtures. *The Journal of Chemical Physics*, **39**, 654–669.

Moraal, H. and McCourt, F. R. W. (1972). Sound propagation in dilute polyatomic gases. *Zeitschrift für Naturforschung*, **27a**, 583–592.

Moran, T. I. and Watson, W. W. (1958). Thermal diffusion factors for the noble gases. *Physical Review*, **109**, 1184–1190.

Moszynski, R. (2007). Theory of intermolecular forces: An introductory account. In *Molecular materials with specific applications* (ed. W. A. Sokalski), Chapter 1. Springer, Dordrecht.

Munn, R. J., Smith, F. J., Mason, E. A., and Monchick, L. (1965). Transport collision integrals for quantum gases obeying a 12–6 potential. *The Journal of Chemical Physics*, **42**, 537–539.

Myatt, P. T., Dham, A. K., Chandrasekhar, P., McCourt, F. R. W., and Le Roy, R. J. (2018). A new empirical potential energy function for Ar_2. *Molecular Physics*, **116**(12), 1598–1623.

Najafi, B., Mason, E. A., and Kestin, J. (1983). Improved corresponding states principle for the noble gases. *Physica A: Statistical Mechanics and its Applications*, **119**(3), 387–440.

Nasrabad, A. E. and Deiters, U. K. (2003). Prediction of thermodynamic properties of krypton by Monte Carlo simulation using *ab initio* interaction potentials. *The Journal of Chemical Physics*, **119**(2), 947–952.

Nasrabad, A. E., Laghaei, R., and Deiters, U. K. (2004). Prediction of the thermophysical properties of pure neon, pure argon, and the binary mixtures neon–argon and argon–krypton by Monte Carlo simulation using *ab initio* potentials. *The Journal of Chemical Physics*, **121**(13), 6423–6434.

Naumkin, F. Y. and McCourt, F. R. W. (1998). Contributions of the two conformers to the microwave spectrum and scattering cross-section of the He–Cl_2 van der Waals system, evaluated from an *ab initio* potential energy surface. *The Journal of Chemical Physics*, **108**(22), 9301–9312.

Nelkin, M. and Yip, S. (1966). Brillouin scattering by gases as a test of the Boltzmann equation. *The Physics of Fluids*, **9**, 380–381.

Nesterov, N. A. and Sudnik, V. M. (1976). Thermal conductivity of gaseous neon and krypton at reduced temperatures and atmospheric pressure. *Inzhenerno-Fizicheskii Zhurnal*, **30**, 572–575.

Newton, R. G. (1966). *Scattering Theory of Waves and Particles.* McGraw-Hill, New York.

Newton, R. G. (1968). Determination of the amplitude from the differential cross section by unitarity. *Journal of Mathematical Physics*, **9**(12), 2050–2055.

Ng, K.-C., Meath, W. J., and Allnatt, A. R. (1978). A simple reliable approximation for isotropic intermolecular forces. A critical test using H–H($^3\Sigma_u^+$) as a model. *Chemical Physics*, **32**, 175–182.

Ng, K.-C., Meath, W. J., and Allnatt, A. R. (1979). A reliable semi-empirical approach for evaluating the isotropic intermolecular forces between closed-shell systems. *Molecular Physics*, **37**, 237–253.

Nieuwoudt, J. C. and Shankland, I. R. (1991). Oscillating body viscometers. In *Experimental Thermodynamics. III. Measurement of the Transport Properties of Fluids* (eds. W. A. Wakeham, A. Nagashima, and J. V. Sengers), Chapter 2. Blackwell, Oxford.

O'Hara, H. and Smith, F. J. (1971). Transport collision integrals for a dilute gas. *Computer Physics Communications*, **2**, 47–54.

Olson, R. E. and Smith, F. T. (1972). Collision spectroscopy. IV. Semiclassical theory of inelastic scattering with application to He^+ + Ne. *Physical Review A*, **6**, 526.

Oppenheim, A. (1965). Nonlinear model of Boltzmann's equations for a many-component gas. *The Physics of Fluids*, **8**, 992–993.

Orcutt, R. H. and Cole, R. H. (1967). Dielectric constants of imperfect gases. III. Atomic gases, hydrogen, and nitrogen. *The Journal of Chemical Physics*, **46**, 697–702.

Ottschofski, E. and Kutzelnigg, W. (1995). Relativistic Hartree–Fock by means of stationary direct perturbation theory. II. Ground states of rare gas atoms. *The Journal of Chemical Physics*, **102**(4), 1752–1757.

Pack, R. T. and Hirschfelder, J. O. (1968). Separation of rotational coordinates from the N-electron diatomic Schrödinger equation. *The Journal of Chemical Physics*, **49**(9), 4009–4020.

Pack, R. T. and Hirschfelder, J. O. (1970*a*). Adiabatic corrections to long-range Born–Oppenheimer interatomic potentials. *The Journal of Chemical Physics*, **52**(8), 4198–4211.

Pack, R. T. and Hirschfelder, J. O. (1970*b*). Energy corrections to the Born–Oppenheimer approximation. The best adiabatic approximation. *The Journal of Chemical Physics*, **52**(2), 521–534.

Pack, R. T., Valentini, J. J., Becker, C. H., Buss, R. J., and Lee, Y. T. (1982). Multiproperty empirical interatomic potentials for ArXe and KrXe. *The Journal of Chemical Physics*, **77**, 5475–5485.

Partridge, H., Langhoff, S. R., and Bauschlicher, C. W. (1986). Theoretical study of the $^7\Sigma_u^+$ state of N_2. *The Journal of Chemical Physics*, **84**(12), 6901–6906.

Partridge, H., Stallcop, J. R., and Levin, E. (2001). Potential energy curves and transport properties for the interaction of He with other ground-state atoms. *The Journal of Chemical Physics*, **115**(14), 6471–6488.

Pashov, A., Docenko, O., Tamanis, M., Ferber, R., Knöckel, H., and Tiemann, E. (2007). Coupling of the $X^1\Sigma^+$ and $a^3\Sigma^+$ states of KRb. *Physical Review A*, **76**, 022511.

Pathak, R. K. and Thakkar, A. J. (1987). Very short–range interatomic potentials. *The Journal of Chemical Physics*, **87**, 2186–2190.

Patkowski, K. (2012). On the accuracy of explicitly correlated coupled-cluster interaction energies—Have orbital results been beaten yet? *The Journal of Chemical Physics*, **137**(3), 034103.

Patkowski, K., Murdachaew, G., Fou, C.-M., and Szalewicz, K. (2005). Accurate *ab initio* potential for argon dimer including highly repulsive region. *Molecular Physics*, **103**(15-16), 2031–2045.

Patkowski, K. and Szalewicz, K. (2007). Frozen core and effective core potentials in symmetry-adapted perturbation theory. *The Journal of Chemical Physics*, **127**(16), 164103.

Patkowski, K. and Szalewicz, K. (2010). Argon pair potential at basis set and excitation limits. *The Journal of Chemical Physics*, **133**(9), 094304.

Paul, R., Howard, A. J., and Watson, W. W. (1963). Isotopic thermal-diffusion factor of argon. *The Journal of Chemical Physics*, **39**(11), 3053–3056.

Peterson, K. A. and Dunning, T. H. (2002). Accurate correlation consistent basis sets for molecular core–valence correlation effects: The second row atoms Al–Ar, and the first row atoms B–Ne revisited. *The Journal of Chemical Physics*, **117**(23), 10548–10560.

Phelps, A. V., Greene, C. H., and Burke, Jr., J. P. (2000). Collision cross sections for argon atoms with argon atoms for energies from 0.01 eV to 10 keV. *Journal of Physics B: Atomic, Molecular and Optical Physics*, **33**, 2965.

Piela, L. (2020). *Ideas of Quantum Chemistry.* Elsevier, Amsterdam.

Pirani, F., Albertí, M., Castro, A., Teixidor, M. M., and Cappelletti, D. (2004). Atom-bond pairwise additive representation for intermolecular potential energy surfaces. *Chemical Physics Letters*, **394**, 37–44.

Pirani, F., Brizi, S., Roncaratti, L. F., Casavecchia, P., Cappellitti, D., and Vecchiocattivi, F (2008). Beyond the Lennard-Jones model: a simple and accurate potential function probed by high resolution scattering data useful for molecular dynamics simulations. *Physical Chemistry Chemical Physics*, **10**, 5489–5503.

Piticco, L., Merkt, F., Cholewinski, A. A., McCourt, F. R. W., and Le Roy, R. J. (2010). Rovibrational structure and potential energy function of the $X0^+$ ground electronic state of ArXe. *Journal of Molecular Spectroscopy*, **264**, 83–93.

Pitre, L., Moldover, M. L., and Tew, W. L. (2006). Acoustic thermometry: New results from 273K to 77K and progress towards 4K. *Metrologia*, **43**(1), 142.

Pollard, C. A. (1971). Thermodynamic properties of rare gases at low temperatures. PhD thesis, University of London.

Pople, J. A., Head-Gordon, M., and Raghavachari, K. (1987). Quadratic configuration interaction. A general technique for determining electron correlation energies. *The Journal of Chemical Physics*, **87**(10), 5968–5975.

Porsev, S. G. and Derevianko, A. (2003). Accurate relativistic many-body calculations of van der Waals coefficients C_8 and C_{10} for alkali-metal dimers. *The Journal of Chemical Physics*, **119**(2), 844–850.

Przybytek, M., Cencek, W., Komasa, J., Łach, G., Jeziorski, B., and Szalewicz, K. (2010). Relativistic and quantum electrodynamics effects in the helium pair potential. *Physical Review Letters*, **104**, 183003.

Przybytek, M., Cencek, W., Komasa, J., Łach, G., Jeziorski, B., and Szalewicz, K. (2012). Erratum: Relativistic and quantum electrodynamics effects in the helium pair potential [*Phys. Rev. Lett.* 104, 183003 (2010)]. *Physical Review Letters*, **108**, 129902.

Pérez, S., Schmiedel, H., and Schramm, B. (1980). Second interaction virial coefficients of the noble gas–hydrogen mixtures Ar–H_2, Kr–H_2, and Xe–H_2. *Zeitschrift für Physikalische Chemie*, **123**(1), 35–38.

Quéméner, G., Honvault, P., Launay, J.-M., Soldán, P., Potter, D. E., and Hutson, J. M. (2005). Ultracold quantum dynamics: Spin-polarized $K + K_2$ collisions with three identical bosons or fermions. *Physical Review A*, **71**, 032722.

Rae, A. I. M. (1975). A calculation of the interaction between pairs of rare-gas atoms. *Molecular Physics*, **29**, 467–483.

Rainwater, J. C. (1984). On the phase space subdivision of the second virial coefficient and its consequences for kinetic theory. *The Journal of Chemical Physics*, **81**(1), 495–510.

Rainwater, J. C. and Friend, D. G. (1987). Second viscosity and thermal-conductivity virial coefficients of gases: Extension to low reduced temperature. *Physical Review A*, **36**, 4062–4066.

Ranganathan, S. and Yip, S. (1966). Time-dependent correlations in a Maxwell gas. *The Physics of Fluids*, **9**, 372–379.

Reid, R. G. H. (1973). Transitions among the $3p^{2P}$ states of sodium induced by collisions with helium. *Journal of Physics B: Atomic, Molecular and Optical Physics*, **6**, 2018–2039.

Reid, R. G. H. and A. Dalgarno (1969). Fine-structure transitions and shape resonances. *Physical Review Letters*, **22**, 1029–1030.

Rentschler, H.-P. and Schram, B. (1977). Eine Apparatur zur Messung von zweiten Virialkoeffizienten bei hohen Temperaturen. *Berichte der Bunsengesellschaft für physikalische Chemie*, **81**, 319–321.

Resibois, P. M. V. and de Leener, M. (1977). *Classical Kinetic Theory of Fluids*. John Wiley, New York.

Riesz, F. and Nagy, B. (1955). *Functional Analysis*. Frederick Ungar Publishing Company, New York.

Rietveld, A. O., Itterbeek, A. Van, and Velds, C. A. (1959). Viscosity of binary mixtures of hydrogen isotopes and mixtures of He and Ne. *Physica*, **25**(1), 205–216.

Roman, P. (1965). *Advanced Quantum Theory*. Addison-Wesley Publishing Company, Boston, MA.

Runeberg, N. and Pyykkö, P. (1998). Relativistic pseudopotential calculations on Xe_2, RnXe, and Rn_2: The van der Waals properties of radon. *International Journal of Quantum Chemistry*, **66**, 131–140.

Rutherford, W. M. (1973). Isotopic thermal diffusion factors of Ne, Ar, Kr, and Xe from column measurements. *The Journal of Chemical Physics*, **58**(4), 1613–1618.

Rutkowski, A. (1986). Relativistic perturbation theory. I. A new perturbation approach to the Dirac equation. *Journal of Physics B: Atomic and Molecular Physics*, **19**(2), 149.

Rybak, S., Jeziorski, B., and Szalewicz, K. (1991). Many-body symmetry-adapted perturbation theory of intermolecular interactions. H_2O and HF dimers. *The Journal of Chemical Physics*, **95**(9), 6576–6601.

Sabatier, P. C. (1965). On the asymptotic approximation for the elastic scattering by a potential. *Il Nuovo Cimento*, **37**, 1180–1227.

Santafe, J., Urieta, J. S., and Losa, C. G. (1976). Second virial coefficients of F_6S + Ar and F_6S + Kr mixtures. *Chemical Physics*, **18**(3), 341–344.

Schmidt, G., Köhler, W. E., and Hess, S. (1981). On the kinetic theory of the Enskog fluid. Viscosity and viscoelasticity, heat conduction and thermal pressure. *Zeitschrift für Naturforschung*, **36a**, 545–553.

Schmiedel, H., Gehrmann, R., and Schramm, B. (1980). Die zweiten Virialkoeffizienten verschiedener Gasmischungen im Temperaturbereich von 213 bis 475 K. *Berichte der Bunsengesellschaft für physikalische Chemie*, **84**(8), 721–724.

Schöllkopf, W. and Toennies, J. P. (1994). Nondestructive mass selection of small van der Waals clusters. *Science*, **266**(5189), 1345–1348.

Schramm, B., Schmiedel, H., Gehrmannc, R., and Bartl, R. (1977). Die Virialkoeffizienten der schweren Edelgase und ihrer binären Mischungen. *Berichte der Bunsengesellschaft für physikalische Chemie*, **81**, 316–318.

Schöllkopf, W. and Toennies, J. P. (1996). The nondestructive detection of the helium dimer and trimer. *The Journal of Chemical Physics*, **104**(3), 1155–1158.

Scoles, G. (1988). In *Atomic and Molecular Beam Methods* (ed. G. Scoles), Volume 1, Chapters 19, 20. Oxford University Press, Oxford.

Sengers, J. V., Bolk, W. T., and Stigter, C. J. (1964). The thermal conductivity of neon between 25°C and 75°C at pressures up to 2600 atmospheres. *Physica*, **30**(5), 1018–1026.

Shapiro, M. and Gerber, R. B. (1976). Extraction of interaction potentials from the elastic scattering amplitudes: An accurate quantum mechanical procedure. *Chemical Physics*, **13**(3), 235–242.

Shapiro, M., Gerber, R. B., Buck, U., and Schleusener, J. (1977). Deconvolution of differential cross sections obtained from molecular beam experiments. *The Journal of Chemical Physics*, **67**(8), 3570–3576.

Sharipov, F. and Benites, V. J. (2015). Transport coefficients of helium–argon mixture based on *ab initio* potential. *The Journal of Chemical Physics*, **143**(15), 154104.

Sharipov, F. and Benites, V. J. (2017). Transport coefficients of helium-neon mixtures at low density computed from *ab initio* potentials. *The Journal of Chemical Physics*, **147**(22), 224302.

Sharipov, F. and Strapasson, J. L. (2012*a*). *Ab initio* simulation of transport phenomena in rarefied gases. *Physical Review E*, **86**(4), 049903.

Sharipov, F. and Strapasson, J. L. (2012*b*). Direct simulation Monte Carlo method for an arbitrary intermolecular potential. *Physics of Fluids*, **24**(1), 011703.

Shashkov, A. G., Abramenko, T. N., Nesterov, N. A., Joshi, R. K., Afshar, R., and Saxena, S. C. (1978). Thermal conductivity of argon, krypton and their mixtures at low temperatures (90–270 K). *Chemical Physics*, **29**(3), 373–381.

Shavitt, I. and Bartlett, R. J. (2009). *Many-Body Methods in Chemistry and Physics. MBPT and Coupled-Cluster Theory.* Cambridge University Press, Cambridge, UK.

Simon, S., Duran, M., and Dannenberg, J. J. (1996). How does basis set superposition error change the potential surfaces for hydrogen-bonded dimers? *The Journal of Chemical Physics*, **105**, 11024–11031.

Simoni, A., Launay, J.-M., and Soldán, P. (2009). Feshbach resonances in ultracold atom–molecule collisions. *Physical Review A*, **79**, 032701.

Sirovich, L. (1962). Kinetic modeling of gas mixtures. *The Physics of Fluids*, **5**, 908–918.

Sirovich, L. (1963). Dispersion relations in rarefied gas dynamics. *The Physics of Fluids*, **6**, 10–20.

Sirovich, L. and Thurber, J. (1961). *Sound Propagation According to Kinetic Models.* AEC Development and Research Reoprt (sic). New York University, Institute of Mathematical Sciences, Magneto-Fluid Dynamics Division.

Slater, J. C. and Kirkwood, J. G. (1931). The van der Waals forces in gases. *Physical Review*, **37**, 682–697.

Slavíček, P., Kalus, R., Paška, P., Odvárková, I., Hobza, P., and Malijevský, A. (2003). State-of-the-art correlated *ab initio* potential energy curves for heavy rare gas dimers: Ar_2, Kr_2, and Xe_2. *The Journal of Chemical Physics*, **119**(4), 2102–2119.

Smirnov, B. M. and Chibisov, M. I. (1965). The breaking up of atomic particles by an electric field and by electron collisions. *Soviet Physics, JETP*, **21**, 624.

Smith, E. B., Tindell, A. R., Wells, B. H., and Crawford, F. W. (1981). A complete iterative inversion procedure for second virial coefficient data. *Molecular Physics*, **42**(4), 937–942.

Smith, E. B., Tindell, A. R., Wells, B. H., and Tildesley, D. J. (1980). On the inversion of second virial coefficient data derived from an undisclosed potential energy function. *Molecular Physics*, **40**(4), 997–998.

Smith, K. M., Rulis, A. M., Scoles, G., Aziz, R. A., and Nain, V. (1977). Intermolecular forces in mixtures of helium with the heavier noble gases. *The Journal of Chemical Physics*, **67**, 152–163.

Snel, J. A. A., Trappeniers, N. J., and Botzen, A. (1979). *Proceedings of the Koninklijke Nederlandse Akademie van Wetenschappen. Series B*, **82**, 316.

Snider, R. F. (1964). Variational methods for solving the Boltzmann equation. *The Journal of Chemical Physics*, **41**, 591–595.

Soldán, P. (2010). Potential energy surface for spin-polarized rubidium trimer. *The Journal of Chemical Physics*, **132**(23), 234308.

Soldán, P., Cvitaš, M. T., and Hutson, J. M. (2003). Three-body nonadditive forces between spin-polarized alkali-metal atoms. *Physical Review A*, **67**, 054702.

Soldán, P. and Špirko, V. (2007). Tuning *ab initio* data to scattering length: The $a^3\Sigma^+$ state of KRb. *The Journal of Chemical Physics*, **127**(12), 121101.

Song, B., Wang, X., and Liu, Z. (2016). Recommended gas transport properties of argon at low density using *ab initio* potential. *Molecular Simulation*, **42**(1), 9–13.

Song, B., Wang, X., Wu, J., and Liu, Z. (2010). Prediction of transport properties of pure noble gases and some of their binary mixtures by *ab initio* calculations. *Fluid Phase Equilibria*, **290**(1), 55–62.

Song, B., Wang, X., Wu, J., and Liu, Z. (2011). Calculations of the thermophysical properties of binary mixtures of noble gases at low density from *ab initio* potentials. *Molecular Physics*, **109**(12), 1607–1615.

Springer, G. and Wingeier, E. W. (1973). Thermal conductivity of neon, argon, and xenon at high temperatures. *The Journal of Chemical Physics*, **59**(5), 2747–2750.

Springer, G. and Wingeier, E. W. (2000). Absolute steady-state thermal conductivity measurements by use of a transient hot-wire system. *Journal of Research of the National Institute of Standards and Technology*, **105**(2), 221.

Staemmler, V. and Jaquet, R. (1985). CEPA calculations of potential energy surfaces for open-shell systems: III. Van der Waals interaction between $O(^3P)$ and $He(^1S)$. *Chemical Physics*, **92**(1), 141–153.

Staker, G. R. and Dunlop, P. J. (1976). The pressure dependence of the mutual diffusion coefficients of binary mixtures of helium and six other gases at 300 K: Tests of Thorne's equation. *Chemical Physics Letters*, **42**(3), 419–422.

Stallcop, J. R., Bauschlicher Jr., C. W., Partridge, H., Langhoff, S. R., and Levin, E. (1992). Theoretical study of hydrogen and nitrogen interactions: N–H transport cross sections and collision integrals. *The Journal of Chemical Physics*, **97**(8), 5578–5585.

Stallcop, J. R., Partridge, H., and Levin, E. (2000*a*). Effective potential energies and transport cross sections for interactions of hydrogen and nitrogen. *Physical Review A*, **62**, 062709.

Stallcop, J. R., Partridge, H., Pradhan, A., and Levin, E. (2000*b*). Potential energies and collision integrals for interactions of carbon and nitrogen atoms. *Journal of Thermophysics and Heat Transfer*, **14**(4), 480–488.

Starkschall, G. and Gordon, R. G. (1972). Calculation of coefficients in the power series expansion of the long-range dispersion force between atoms. *The Journal of Chemical Physics*, **56**(6), 2801–2806.

Stevens, G. A. and De Vries, A. E. (1968). The influence of the distribution of atomic masses within the molecule on thermal diffusion: II. Isotopic methane and methane/argon mixtures. *Physica*, **39**(3), 346–360.

Stier, L. G. (1942). The coefficients of thermal diffusion of neon and argon and their variation with temperature. *Phys. Rev.*, **62**, 548–551.

Stokvis, J. J. H., van den Biesenand, F. A., van Veen, E. H., and van den Meijdenberg, C. J. N. (1980). Measurements on the glory structure in the total cross section of noble gas systems: III. Ar–Kr, Kr–Kr, Kr–Xe and Xe–Xe; application of the n(x)-6 potential. *Physica A: Statistical Mechanics and its Applications*, **100**, 375–396.

Stone, A. J. (1996). *The Theory of Intermolecular Forces.* Clarendon, Oxford.

Strehlow, T. (1987). Ph.D. thesis, University of Rostock. Re-analyzed results courtesy of E. Vogel (Private Communication, 2018).

Sugawara, A., Yip, S., and Sirovich, L. (1968). Spectrum of density fluctuations in gases. *The Physics of Fluids*, **11**, 925–932.

Sutherland, W. (1893). The viscosity of gases and molecular force. *The Philosophical Magazine*, **36**, 507–531.

Sutherland, W. (1909). Molecular diameters. *The Philosophical Magazine*, **17**, 320–321.

Szalewicz, K. and Jeziorski, B. (1979). Symmetry-adapted double-perturbation analysis of intramolecular correlation effects in weak intermolecular interactions. *Molecular Physics*, **38**(1), 191–208.

Szegö, G. (1959). *Orthogonal Polynomials.* American Mathematical Society, New York.

Tal, Y. (1978). Asymptotic behavior of the ground-state charge density in atoms. *Physical Review A*, **18**, 1781–1783.

Tanaka, Y. and Yoshino, K. (1970). Absorption spectrum of the argon molecule in the vacuum-uv region. *The Journal of Chemical Physics*, **53**, 2012–2030.

Tanaka, Y. and Yoshino, K. (1972). Absorption spectra of Ne_2 and HeNe molecules in the vacuum–uv region. *The Journal of Chemical Physics*, **57**, 2964–2976.

Tanaka, Y., Yoshino, K., and Freeman, D. E. (1973). Vacuum ultraviolet absorption spectra of the van der Waals molecules Kr_2 and ArKr. *The Journal of Chemical Physics*, **59**, 5160–5183.

Tang, K. T., Toennies, J. P, and Meyer, W. (1991). A simple predictive model of chemical potentials: $H_2(^1\Sigma_g)$ and $Li_2(^1\Sigma_g)$. *The Journal of Chemical Physics*, **95**, 1144–1150.

Tang, K. T. and Toennies, J. P. (1977). A simple theoretical model for the van der Waals potential at intermediate distances. I. Spherically symmetric potentials. *The Journal of Chemical Physics*, **66**, 1496–1506.

Tang, K. T. and Toennies, J. P. (1984). An improved simple model for the van der Waals potential based on universal damping functions for the dispersion coefficients. *The Journal of Chemical Physics*, **80**, 3726–3741.

Tang, K. T. and Toennies, J. P. (1986). New combining rules for well parameters and shapes of the van der Waals potential of mixed rare gas systems. *Zeitschrift für Physik D: Atoms, Molecules and Clusters*, **1**, 91–101.

Tang, K. T. and Toennies, J. P. (1991). A combining rule calculation of the van der Waals potentials of the rare-gas hydrides. *Chemical Physics*, **156**(3), 413–425.

Tang, K. T. and Toennies, J. P. (2003). The van der Waals potentials between all the rare gas atoms from He to Rn. *The Journal of Chemical Physics*, **118**, 4976–4983.

Tang, K. T., Toennies, J. P., and Yiu, C. L. (1993). Exchange energy of H_2 calculated by the surface integral method in zeroth order approximation. *The Journal of Chemical Physics*, **99**, 377–388.

Tang, K. T., Toennies, J. P., and Yiu, C. L. (1994). The perturbation calculation of van der Waals potentials. *Theoretica chimica acta*, **88**, 169–181.

Tang, K. T., Toennies, J. P., and Yiu, C. L. (1998). The generalized Heitler–London theory for interatomic interaction and surface integral method for exchange energy. *International Reviews in Physical Chemistry*, **17**, 363–406.

Tao, F.-M. (1999). *Ab initio* calculation of the interaction potential for the krypton dimer: The use of bond function basis sets. *The Journal of Chemical Physics*, **111**(6), 2407–2413.

Taylor, W. L. (1975). Erratum: Experimental thermal diffusion factors for ^{20}Ne–^{22}Ne and their application as a test of the neon interatomic potential: Isotopic thermal diffusion factors for argon and krypton. *The Journal of Chemical Physics*, **62**(9), 3837–3838.

Taylor, W. L. (1979). Algorithms and FORTRAN programs to calculate classical collision integrals for realistic intermolecular potentials. Technical report.

Taylor, W. L. and Cain, D. (1983). Temperature dependence of the mutual diffusion coefficients of He–Ar, Ne–Ar, and Xe–Ar from 350 to 1300 K. *The Journal of Chemical Physics*, **78**, 6220–6227.

Taylor, W. L. and Weissman, S. (1973). Isotopic thermal diffusion factors for argon and krypton. *The Journal of Chemical Physics*, **59**(3), 1190–1195.

Tenti, G., Boley, C. D., and Desai, R. C. (1974). On the kinetic model description of Rayleigh–Brillouin scattering from molecular gases. *Canadian Journal of Physics*, **52(4)**, 285–290.

ter Haar, D. (1960). *Elements of Statistical Mechanics*. Holt, New York.

Thakkar, A. J. (1988). Higher dispersion coefficients: Accurate values for hydrogen atoms and simple estimates for other systems. *The Journal of Chemical Physics*, **89**, 2092–2098.

Thakkar, A. J. (2001). Intermolecular Interactions. In *Encyclopedia of Chemical Physics and Physical Chemistry, Vol. 1, Fundamental,* Chapter A1.5. IOP Publishing, Bristol.

Thompson, W. J. (1994). *Angular Momentum.* Wiley–Interscience, New York.

Toennies, J. P. (1973). On the validity of a modified Buckingham potential for the rare gas dimers at intermediate distances. *Chemical Physics Letters*, **20**, 238–241.

Tolman, R. C. (1979). *The Principles of Statistical Mechanics.* Dover, New York.

Trappeniers, N. J., Botzen, A., van Oosten, J., and van den Berg, H. R. (1965). The viscosity of krypton between 25°C and 75°C and at pressures up to 2000 atm. *Physica*, **31**(6), 945–952.

Trappeniers, N. J., Wassenaar, T., and Wolkers, G. J. (1966). Isotherms and thermodynamic properties of krypton at temperatures between 0° and 150°C and at densities up to 620 Amagat. *Physica*, **32**(9), 1503–1520.

Trengove, R. D., Robjohns, H. L., Bell, T. N., Martin, M. L., and Dunlop, P. J. (1981). Thermal diffusion factors at 300 K for seven binary noble gas systems containing helium or neon. *Physica A: Statistical Mechanics and its Applications*, **108**(2), 488–501.

Trusler, J. P. M. (1991). *Physical Acoustics and Metrology of Fluids.* Hilger, Bristol.

Ubbink, J. B. and de Haas, W. J. (1943). Thermal conductivity of gaseous helium. *Physica*, **10**(7), 465–470.

Uehling, E. A. and Uhlenbeck, G. E. (1933). Transport phenomena in Einstein–Bose and Fermi–Dirac gases. I. *Physical Review*, **43**, 552–561.

Van Dael, W. and Cauwenbergh, H. (1968). Measurements of the thermal conductivity of gases: II. Data for binary mixtures of He, Ne and Ar. *Physica*, **40**(2), 173–181.

van de Bovenkamp, J. and van Duijneveldt, F. B. (1999). CCSD(T) investigation of the interaction in neon dimer. *Chemical Physics Letters*, **309**(3), 287–294.

van den Berg, H. J. (1979). Ph.D. thesis, University of Amsterdam.

van den Biesen, J. J. H., Hermans, R. M., and van den Meijdenberg, C. J. N. (1982). Experimental total collision cross sections in the glory region for noble gas systems. *Physica A: Statistical Mechanics and its Applications*, **115**(3), 396–439.

van Heijningen, R. J. J., Harpe, J. P., and Beenakker, J. J. M. (1968). Determination of the diffusion coefficients of binary mixtures of the noble gases as a function of temperature and concentration. *Physica*, **38**(1), 1–34.

van Mourik, T. and Dunning, T. H. (1999). A new *ab initio* potential energy curve for the helium dimer. *The Journal of Chemical Physics*, **111**(20), 9248–9258.

Vargaftik, N. B. and Vasilevskaya, Yu. D. (1984). Viscosity of monatomic gases at temperatures up to 5000–6000°K. *Journal of Engineering Physics*, **46**(1), 30–34.

Velarde, M. P. (1974). On the Enskog hard-sphere kinetic equation and the transport phenomena of Dense simple gases. In *Transport Phenomena* (eds. G. Kirczenow and J. Marro), Volume 31 p. 288–336. Springer–Verlag, Berlin.

Viehland, L. A. (1983). Interaction potentials for Li^+—rare-gas systems. *Chemical Physics*, **78**(2), 279–294.

Viehland, L. A. (1984). Interaction potentials for the alkali ion—rare-gas systems. *Chemical Physics*, **85**(2), 291–305.

Viehland, L. A., Harrington, M. M., and Mason, E. A. (1976). Direct determination of ion-neutral molecule interaction potentials from gaseous ion mobility measurements. *Chemical Physics*, **17**(4), 433–441.

Viehland, L. A., Janzen, A. R., and Aziz, R. A. (1995). High approximations to the transport properties of pure atomic gases. *The Journal of Chemical Physics*, **102**, 5444–5450.

Vogel, E. (1972). *Z. University of Rostock*, **21**(M2), 169.

Vogel, E. (1984). Präzisionsmessungen des Viskositätskoeffizienten von Stickstoff und den Edelgasen zwischen Raumtemperatur und 650 K. *Berichte der Bunsengesellschaft für physikalische Chemie*, **88**(10), 997–1002.

Vogel, E. (2010). Reference viscosity of argon at low density in the temperature range from 290 K to 680 K. *International Journal of Thermophysics*, **31**(3), 447–461.

Vogel, E. (2016). The viscosities of dilute Kr, Xe, and CO_2 revisited: New experimental reference data at temperatures from 295 K to 690 K. *International Journal of Thermophysics*, **37**(6), 63.

Vogel, E., Jäger, B., Hellmann, R., and Bich, E. (2008). *Ab initio* pair potential energy curve for the argon atom pair and thermophysical properties for the dilute argon gas. II. Thermophysical properties for low-density argon. *Molecular Physics*, **108**, 3335 – 3352.

Vogel, E., Jäger, B., Hellmann, R., and Bich, E. (2010). *Molecular Physics* Longuet–Higgins Young Author's Prize 2009 winner's profile. *Molecular Physics*, **108**(24), 3333–3334.

Vollmer, G. (1969). *Zeitschrift für Physik B*, **22**, 423.

Vugts, H. F., Boerboom, A. J. H., and Los, J. (1969). Measurements of relative diffusion coefficients of argon. *Physica*, **44**(2), 219–226.

Waldmann, L. (1957). Die Boltzmann–Gleichung für Gase mit rotierenden Molekülen. *Zeitschrift für Naturforschung*, **13a**, 660–662.

Waldmann, L. (1958*a*). Transporterscheinungen in Gasen von mittlerem Druck. In *Handbuch der Physik* (ed. S. Flügge), Volume 12, p. 452–469. Springer–Verlag, Berlin.

Waldmann, L. (1958*b*). Transporterscheinungen in Gasen von mittlerem Druck. In *Handbuch der Physik* (ed. S. Flügge), Volume 12, p. 370–377. Springer–Verlag, Berlin.

Waldmann, L. (1958*c*). Transporterscheinungen in Gasen von mittlerem Druck. In *Handbuch der Physik* (ed. S. Flügge), Volume 12, p. 364–384. Springer–Verlag, Berlin.

Waldmann, L. (1958*d*). Transporterscheinungen in Gasen von mittlerem Druck. In *Handbuch der Physik* (ed. S. Flügge), Volume 12, p. 394. Springer–Verlag, Berlin.

Waldmann, L. (1958*e*). Transporterscheinungen in Gasen von mittlerem Druck. In *Handbuch der Physik* (ed. S. Flügge), Volume 12, p. 378. Springer–Verlag, Berlin.

Waldmann, L. and Kupatt, H. D. (1963). Diffusion von Spin-Teilchen im Magnetfeld. *Zeitschrift für Naturforschung*, **18a**, 86–87.

Waldrop, J. M., Song, B., Patkowski, K., and Wang, X. (2015). Accurate *ab initio* potential for the krypton dimer and transport properties of the low-density krypton gas. *The Journal of Chemical Physics*, **142**(20), 204307.

Wang Chang, C. S. and Uhlenbeck, G. E. (1970*a*). In *Studies in Statistical Mechanics* (eds. J. de Boer and G. E. Uhlenbeck), Volume 5, p. 83–85. North–Holland Publishing Company, Amsterdam.

Wang Chang, C. S. and Uhlenbeck, G. E. (1970*b*). In *Studies in Statistical Mechanics* (ed. J. de Boer and G. E. Uhlenbeck), Volume 5, p. 71–73. North–Holland Publishing Company, Amsterdam.

Wei, H. and Le Roy, R. J. (2006). Calculation of absolute scattering phase shifts. *Molecular Physics*, **104**, 147–150.

Whalley, E., Lupien, Y., and Schneider, W. G. (1955). The compressibility of gases at high temperatures: X. Xenon in the temperature range 0° to 700 °C and the pressure range 8 to 50 atmospheres. *Canadian Journal of Chemistry*, **33**(4), 633–636.

Wheatley, R. J. and Meath, W. J. (1993*a*). Dispersion energy damping functions, and their relative scale with interatomic separation, for (H, He, Li)–(H, He, Li) interactions. *Molecular Physics*, **80**, 25–54.

Wheatley, R. J. and Meath, W. J. (1993*b*). On the relationship between first-order exchange and Coulomb interaction energies for closed shell atoms and molecules. *Molecular Physics*, **79**, 253–275.

Wigner, E. (1932). On the quantum correction for thermodynamic equilibrium. *Physical Review*, **40**, 749–759.

Wilhelm, J. and Vogel, E. (2000). Viscosity measurements on gaseous argon, krypton, and propane[1]. *International Journal of Thermophysics*, **21**(2), 301–318.

Williams, H. L., Szalewicz, K., Jeziorski, B., Moszynski, R., and Rybak, S. (1993). Symmetry-adapted perturbation theory calculation of the Ar–H_2 intermolecular potential energy surface. *The Journal of Chemical Physics*, **98**(2), 1279–1292.

Williams, M. M. R. (1972). *Mathematical Methods in Particle Transport Theory.* Butterworths, London.

Wilson, A. K., Woon, D. E., Peterson, K. A., and Dunning, T. H. (1999). Gaussian basis sets for use in correlated molecular calculations. IX. The atoms gallium through krypton. *The Journal of Chemical Physics*, **110**(16), 7667–7676.

Winn, E. B. (1950). The temperature dependence of the self-diffusion coefficients of argon, neon, nitrogen, oxygen, carbon dioxide, and methane. *Physical Review*, **80**, 1024–1027.

Woon, D. E. (1994). Benchmark calculations with correlated molecular wave functions. V. The determination of accurate *ab initio* intermolecular potentials for He_2, Ne_2, and Ar_2. *The Journal of Chemical Physics*, **100**(4), 2838–2850.

Woon, D. E. and Dunning, T. H. (1993). Gaussian basis sets for use in correlated molecular calculations. III. The atoms aluminum through argon. *The Journal of Chemical Physics*, **98**(2), 1358–1371.

Wüest, A., Hollenstein, U., de Bruin, K. G., and Merkt, F. (2004). High-resolution vacuum ultraviolet laser spectroscopy of the C $0_u^+ \leftarrow$ X 0_g^+ transition of Xe_2. *Canadian Journal of Chemistry*, **82**(6), 750–761.

Wüest, A. and Merkt, F. (2003). Determination of the interaction potential of the ground electronic state of Ne_2 by high-resolution vacuum ultraviolet laser spectroscopy. *The Journal of Chemical Physics*, **118**(19), 8807–8812.

Xie, F., Sovkov, V. B., Lyyra, A. M., Li, D., Ingram, S., Bai, J., Ivanov, V. S., Magnier, S., and Li, L. (2009). Experimental investigation of the Cs_2 $a^3\Sigma_u^+$ triplet ground state: Multiparameter Morse long-range potential analysis and molecular constants. *The Journal of Chemical Physics*, **130**(5), 051102.

Xu, Y., Jäger, W., Djauhari, J., and Gerry, M. C. L. (1995). Rotational spectra of the mixed rare gas dimers Ne–Kr and Ar–Kr. *The Journal of Chemical Physics*, **103**(8), 2827–2833.

Yang, D. D., Li, P., and Tang, K. T. (2009). The ground state van der Waals potentials of the calcium dimer and calcium rare-gas complexes. *The Journal of Chemical Physics*, **131**(15), 154301.

Yip, S. and Nelkin, M. (1964). Application of a kinetic model to time-dependent density correlations in fluids. *Physical Review*, **135**, A1241–A1247.

Zare, R. N. (1988). *Angular Momentum.* Wiley–Interscience, New York.

Zhai, Y. and Li, H. (2022). Basis sets dependency in constructing spectroscopy-accuracy *ab initio* global electric dipole moment functions. *Chinese Journal of Chemical Physics*, **35**, 52.

Zhai, Y., Li, Y., Li, H., and McCourt, F. R. W. (2023). Peng: A program for transport properties of low–density binary gas mixtures. *Computer Physics Communications*, **287**, 108712.

Zhao, G., Zemke, W. T., Kim, J. T., Ji, B., Wang, H., Bahns, J. T., Stwalley, W. C., Li, L., Lyyra, A. M., and Amiot, C. (1996). New measurements of the $a^3\Sigma_u^+$ state of K_2 and improved analysis of long-range dispersion and exchange interactions between two K atoms. *The Journal of Chemical Physics*, **105**(18), 7976–7985.

Ziebland, H. (1981). Recommended reference materials for realization of physicochemical properties. Section: Thermal conductivity of fluid substances. *Pure and Applied Chemistry*, **53**(10), 1863–1877.

Index